Ternary Alloys Based on II-VI Semiconductor Compounds

Ternary Alloys Based on II-VI Semiconductor Compounds

Vasyl Tomashyk
Petro Feychuk
Larysa Shcherbak

CRC Press
Taylor & Francis Group
Boca Raton London New York

CRC Press is an imprint of the
Taylor & Francis Group, an **informa** business

CRC Press
Taylor & Francis Group
6000 Broken Sound Parkway NW, Suite 300
Boca Raton, FL 33487-2742

First issued in paperback 2019

© 2014 by Taylor & Francis Group, LLC
CRC Press is an imprint of the Taylor & Francis Group, an Informa business

No claim to original U.S. Government works

ISBN-13: 978-1-4398-9566-5 (hbk)
ISBN-13: 978-0-367-86693-8 (pbk)

Library of Congress Cataloging-in-Publication Data

Tomashyk, Vasyl.
    Ternary alloys based on II-IV semiconductor compounds / Vasyl Tomashyk, Petro Feychuk, Larysa Shcherbak.
        pages cm
    Includes bibliographical references and index.
    ISBN 978-1-4398-9566-5 (hardback)
    1. Semiconductors--Materials. 2. Phase diagrams, Ternary. 3. Ternary alloys. I. Feychuk, Petro. II. Shcherbak, Larysa. III. Title.

QC611.8.A44T67 2013
541'.377--dc23   2013025170

Visit the Taylor & Francis Web site at
http://www.taylorandfrancis.com

and the CRC Press Web site at
http://www.crcpress.com

Contents

Preface

A number of device applications based on II–VI semiconductors are currently in use. In most cases, their properties were modified by isovalent or heterovalent foreign impurities (F) doping that resulted in the formation of solid solution or in the creation of a separate electronic level in the semiconductor lattice. Such ternary II–VI–F materials provided control of optical properties, from blue-green up to infrared spectral range, and electronic properties ranging from n-type to p-type. Doped by various impurities, II–VI semiconductor compounds are used as γ-ray and x-ray detectors, as refractive materials, as active elements of solar cells, and so on.

The problem of reproducible manufacturing of the doped materials with predicted and desired properties cannot be successfully resolved without knowledge of appropriate ternary system phase diagram, which is "a map" for technologists. Reference books that were published earlier contained restricted data, and the main information can be found scattered in various papers. This book aims to collect and systematize all available data on the II–VI–F ternary systems.

This book consists of critically compiled ternary systems based on II–VI semiconductor compounds. It also includes data about phase equilibriums on the cross sections, which have been derived from several papers, and contains tables and figures. The book is divided into nine chapters according to the number of possible combinations of Zn, Cd, and Hg with chalcogens S, Se, and Te. Hence, the chapters are structured in the order at first of d-element number in the periodic system increasing, that is, from Zn to Hg compounds, and then in the order of the chalcogen number increasing, that is, from sulfides to tellurides. The same principle is used for a further description of the systems in every chapter, that is, in the order of the third component number in the periodic system increasing.

Each ternary system database description contains brief information in the following order: the diagram type, possible phase transformation and physical–chemical interaction of the components, equilibrium investigation methods, thermodynamic characteristics, and the sample preparation method. The solid- and liquid-phase equilibriums with vapor are illustrated in some cases because of their importance to crystal growth from the vapor, the melt, and the vapor–liquid–solid technique.

The homogeneity range is of much importance to govern the crystal defect structure. Therefore, the book collects all such available data. In addition, it presents data on baric and temperature dependences of solubility impurities both in the semiconductor lattice and the liquid phase, as well as the pressure–composition relationship. As the semiconductors and metals' mutual solubilities are usually small, the figures present a restricted concentration range (in mol. %).

Most of the figures are presented in their original form, although some are a little corrected. If the existing data varied essentially, several versions were presented in comparison. The content of system components is presented in mol. %. If the original phase diagram is given in mass %, the axis with the content in mol. % is provided below the figure.

This book is meant for researchers at industrial and national laboratories and for university and graduate students majoring in materials science and engineering. It is also suitable for phase diagram researchers, inorganic chemists, and solid state physicists.

1

Phase Equilibria in the Systems Based on ZnS

1.1 Zinc–Sodium–Sulfur

ZnS–Na$_2$S: The more reliable phase diagram belongs to the eutectic type with peritectic transformations (Figure 1.1) (Kopylov et al. 1978). The compounds Na$_6$ZnS$_4$ and Na$_2$ZnS$_2$ are melted incongruently at 820°C ± 5°C and 615°C ± 5°C correspondingly (Kopylov et al. 1978, Polyvyanny and Lata 1985). The eutectic composition and temperature are 54.5 mol. % (49 mass. %) Na$_2$S and 605°C, respectively. Other compounds were not found in this system.

Na$_6$ZnS$_4$ crystallizes in an orthorhombic structure with lattice parameters $a = 642.1$, $b = 1133.3$, and $c = 594.2$ pm, and Na$_2$ZnS$_2$ crystallizes in a hexagonal structure with lattice parameters $a = 892.4$ and $c = 688.9$ pm (Klepp and Bronger 1983).

The isotherm at 910°C in the field of ZnS primary crystallization corresponds to the sphalerite–wurtzite transformation (Kopylov et al. 1978). An additional phase in a small quantity was found in the field of Na$_2$ZnS$_2$ existence. Apparently this phase was formed at the oxidizing of Na$_2$ZnS$_2$, and it can be referred to as an oxysulfide type. Its composition can be determined by a figurative point in the system Zn–Na–S–O.

The ZnS–Na$_2$S system was investigated by other authors also. The compounds Na$_8$ZnS$_5$, Na$_6$ZnS$_4$, Na$_2$ZnS$_2$, and Na$_2$Zn$_2$S$_3$, which are melted incongruently at 810°C, 680°C, 610°C, and 870°C correspondingly, were found through DTA and metallography (Shishkin et al. 1968). These compounds, with the exception of Na$_6$ZnS$_4$, were not found using DTA in Kopylov and Kodzoeva (1969). Na$_6$ZnS$_4$ melts with decomposition at 805°C. According to Shishkin et al. (1968), the eutectic contains 42.5 mol. % (48 mass. %) ZnS and crystallizes at 590°C. The thermal effects at 900°C were fixed in the field of ZnS primary crystallization that corresponds to the sphalerite–wurtzite transformation. There were small thermal effects found at 625°C in the region of 23.8–50.5 mol. % (20–45 mass. %), the nature of which was not revealed.

The compounds Na$_2$Zn$_2$S$_3$ and Na$_8$ZnS$_5$, which were determined through differential thermal analysis (DTA), metallography, x-ray diffraction (XRD),

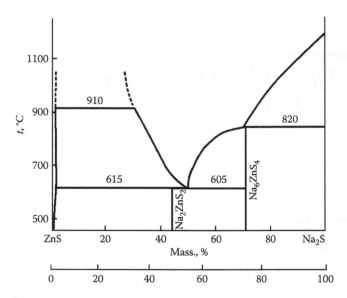

FIGURE 1.1
The ZnS–Na₂S phase diagram. (From Kopylov, N.I. et al., *Zhurn. neorgan. khimii*, 23(11), 3095, 1978.)

and chemical analysis, melt incongruently at 870°C and 810°C, respectively (Maslov et al. 1972). The eutectic contains 50.5 mol. % (45 mass. %) ZnS and crystallizes at 615°C. The compounds Na_6ZnS_4 and Na_8ZnS_5 are not stable and decompose below 600°C with Na_2ZnS_2 forming.

1.2 Zinc–Copper–Sulfur

The melt based on sulfur (L_S) and phases based on Cu_2S and metallic alloys are in equilibrium with ZnS at 900°C in the Zn–Cu–S ternary system (Figure 1.2) (Lott et al. 1981). The greatest solubility of Cu in ternary system is along the section ZnS–Cu_2S. The temperature dependence of copper solubility in this region can be express by the equation (700°C–1070°C) x (cm⁻³) = 4.17 · 10²³ exp (−9168 ± 407)/T.

ZnS–Cu: The phase diagram is not constructed. The isotherms of Cu solubility in ZnS depending on the zinc pressure are given in Figure 1.3 (Lott et al. 1981). The solubility in isothermal condition is maximum at equilibrium with a phase based on Cu_2S and does not depend on zinc vapor pressure. The increase of zinc vapor pressure leads to significant decrease of the solubility.

The diffusion saturation of ZnS by copper was realized in a two-zone evacuated quartz ampoule at fixed temperature and at a given partial pressure of zinc or copper.

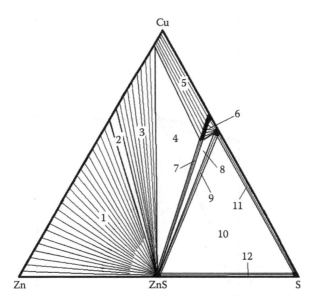

FIGURE 1.2
The isothermal section of Zn–Cu–S system at 900°C: 1, ZnS+L(Cu–Sn); 2, ZnS+α-brass+ L(Cu–Sn); 3, ZnS+α-brass; 4, ZnS+α-brass+Cu$_{2-x}$S; 5, Cu$_{2-x}$S+α-brass; 6, Cu$_{2-x}$S–L(Cu$_{2-x}$S); 7, ZnS+Cu$_{2-x}$S; 8, ZnS+Cu$_{2-x}$S+L(Cu$_{2-x}$S); 9, ZnS+L(Cu$_{2-x}$S); 10, ZnS+L(Cu$_{2-x}$S)+L(S); 11, L(Cu$_{2-x}$S)+L(S); 12, ZnS+L(S). (From Lott, K.P. et al., *Zhurn. Neorgan. Khimii*, 26(7), 1894, 1981.)

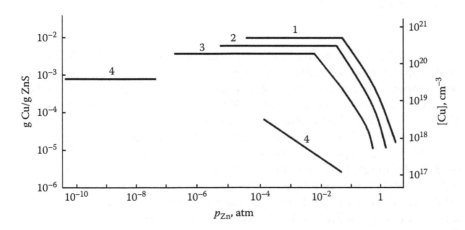

FIGURE 1.3
The isotherms of Cu solubility in ZnS depending on the Zn pressure: 1, 1070°C; 2, 980°C; 3, 900°C; 4, 700°C. (From Lott, K.P. et al., *Zhurn. Neorgan. Khimii*, 26(7), 1894, 1981.)

ZnS–CuS: The phase diagram is not constructed. The attempts to synthesize ternary sulfides in the ZnS–CuS system at different CuS contents had not given any results (Charbonnier 1973). According to the data of XRD, the mixtures of zinc and copper sulfides existed in the ingots, prepared by annealing at 500°C–1350°C for 30-6 days.

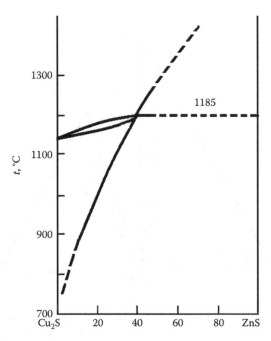

FIGURE 1.4
The ZnS–Cu₂S phase diagram. (From Mizetskaya, I.B. et al., *Izv. AN SSSR. Neorgan. materialy*, 18(11), 1792, 1982.)

ZnS–Cu₂S: The phase diagram is a peritectic type (Figure 1.4) (Novoselov 1955, Kopylov et al. 1976, Mizetskaya et al. 1982). The peritectic composition and temperature are 36 mol. % ZnS and 1185°C, respectively (Mizetskaya et al. 1982) [34 mol. % or 24 mass. % ZnS and 1210°C (Kopylov et al. 1976); 41 mol. % ZnS and 1200°C (Novoselov 1955)].

The solubility of ZnS decreases rapidly with temperature decreasing from 42.3 mol. % (31 mass. %) at 1210°C [42 mol. % at 1185°C (Mizetskaya et al. 1982, Trishchuk et al. 1985)] to 1.6 mol. % (1.0 mass. %) at 400°C (Kopylov et al. 1976). At 800°C, the solubility of ZnS in Cu₂S is equal to 8 mol. % and the solubility of Cu₂S in ZnS is insignificant (Gashurov and Banks 1967). The solubility of Cu₂S in ZnS can be described by the equations given in Table 1.1. The sphalerite–wurtzite transformation in microcrystalline ZnS containing Cu₂S is inhibited with the amount of blended Cu₂S in atmospheres of He, HCl, H₂S, or their mixtures (Sakaguchi et al. 1977).

This system was investigated through DTA, metallography, and XRD (Novoselov 1955, Gashurov and Banks 1967, Kopylov et al. 1976, Mizetskaya et al. 1982, Trishchuk et al. 1985). The solubility of Cu₂S in ZnS was determined by diffusion saturation of the ingots (Bundel' et al. 1970, Mihalev et al. 1970).

TABLE 1.1

Solubility of Cu_2S in ZnS

Equation	Comments
$\log x$ (mol. %) $= -4033/T + 2.968$	800°C–1100°C (Bundel' et al. 1970)
$\log x$ (mol. %) $= -(4200 \pm 820)/T - (2.03 \pm 0.69)$	600°C–1100°C (Sinel'nikov et al. 1984)
$\log x$ (mol. %) $= (1279 \pm 517)/T - (1.971 \pm 0.471)$	600°C–1000°C (Shevtsov and Sinel'nikov 1989)
$\log x$ (g-at Cu/g-mol. ZnS) $= -5058/T + 2.06$	700°C–1100°C (Mihalev et al. 1970)

1.3 Zinc–Silver–Sulfur

ZnS–Ag₂S: The ZnS–Ag₂S phase diagram is a eutectic type (Figure 1.5) (Murach 1947). The solubility of Ag_2S in ZnS determined by saturation of thin films and at 600°C, 700°C, and 750°C is equal to 0.011, 0.027, and 0.052 mol. % Ag_2S, respectively (Vishniakov and Iofis 1974). The ZnS single crystals containing less than $2 \cdot 10^{-5}$ mol. % Ag_2S are transparent as undoped crystals (Atroshchenko and Kolodiazhnyi 1977). They become at first yellowish and then brown at the increase of Ag_2S content. According to the microhardness measurement, the solubility of Ag_2S in ZnS at room temperature consists $2 \cdot 10^{-5}$ mol. %.

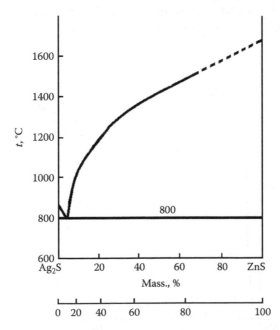

FIGURE 1.5
The ZnS–Ag₂S phase diagram. (From Murach, N.N., *Metallurgizdat*, 2, 784, 1947.)

The attempts to synthesize of ternary sulfides in the $ZnS–Ag_2S$ system at the Zn contents lower than 50 at. % had not given any results (Charbonnier 1973). According to the data of XRD, the mixture of zinc and copper sulfides existed in the ingots, prepared by annealing at 500°C–1350°C for 30-6 days. The structure investigations of the ingots, obtained by the recrystallization of the mixture $Ag_2S + 10$ at. % Zn in H_2 flow at 700°C for 15 days, showed that they are two phased.

$ZnS–Ag_2S$ composite particles have been synthesized from the mixture of $AgNO_3$ and $Zn(NO_3)_2$ solutions by a homogeneous precipitation method using thioacetamide at 70°C (Kim et al. 1997). The precipitates were spherical composite particles with narrow size distribution when Ag^+ concentrations $\leq 4 \cdot 10^{-3}$ mol L^{-1}. At Ag^+ concentrations $\geq 4 \cdot 10^{-3}$ mol L^{-1}, $ZnS–Ag_2S$ composite particles became aggregated ones. The formation of spherical $ZnS–Ag_2S$ composite particles occurred by the deposition of ZnS at the initial stage on the surface of Ag_2S nuclei that were formed by reaction between Ag^+ ions and thioacetamide.

1.4 Zinc–Magnesium–Sulfur

ZnS–MgS: The phase diagram is a eutectic type (Figure 1.6) (Brightwell et al. 1984). According to the XRD, the solubility of MgS in ZnS composes 32, 37, and 42 mol. % at 850°C, 1000°C, and 1050°C, respectively (Brightwell et al. 1984)

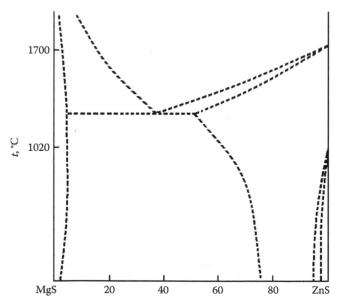

FIGURE 1.6
The ZnS–MgS phase diagram. (From Brightwell, J.W. et al., *J. Mater. Sci. Lett.*, 3(11), 951, 1984.)

[30–32 and 36–38 mol. % at 800°C and 900°C (Dmitrenko et al. 1990); 22 mol. % at 1100°C (Kondrashev and Omel'chenko 1964); 25 mol. % at 1100°C (Smith 1952)]. The increase of MgS solubility leads to the increase of lattice parameters of forming solid solutions (Smith 1952, Kondrashev and Omel'chenko 1964, Atroshchenko et al. 1979, Charifi et al. 2005), and the lattice parameters closely follow Vegard's law (Charifi et al. 2005). The solid solutions have the wurtzite structure. The solubility of ZnS in MgS composes 5 mol. % (Kondrashev and Omel'chenko 1964).

The bulk modulus of the $Zn_{1-x}Mg_xS$ solid solutions decreases on the increase of the Mg concentration, which is due to the increase in the ionicity of bonds (Charifi et al. 2005). A significant deviation of the bulk modulus from linear concentration dependence was found.

According to the metallography data, unstrained crystals are formed at 8–16 mol. % (5–10 mass. %) MgS (Sysoev and Obuhova 1973). The internal strains in the crystals increase at the increasing of MgS content, and two phases appear at the MgS content more than 30–43 mol. % (20–30 mass. %). The single crystals based on ZnS and growing by the oriented crystallization have the sphalerite structure (Obuhova et al. 1977, Atroshchenko et al. 1979). The defect concentration increases strongly at the increasing of MgS content and the crystals with wurtzite structure are formed at 11.2 mol. % MgS. The wurtzite structure is retained up to 48.9 mol. % MgS. The ingots are two phased at the higher MgS concentrations. The energy gap of the solid solutions with wurtzite structure changes from 3.53 to 3.95 eV (Sysoev and Obuhova 1974, Atroshchenko et al. 1979). The distribution coefficient of MgS in ZnS is $k < 1$ (Obuhova et al. 1977).

The gradual transition from sphalerite structure with 10% defect package to wurtzite structure takes place in the ZnS–MgS grown single crystals (Atroshchenko et al. 1977).

Solid solutions of crystalline $Zn_xMg_{1-x}S$ thin films have been prepared in the composition range x between 0.13 and 1.0 (Inoue et al. 1998). Lattice constant and energy gap of these solid solutions decrease linearly with increasing x.

The ZnS–MgS ingots were obtained by a caking of ZnO and MgO in CS_2 flow at 1050°C–1100°C for 2 h with cooling at air (Kondrashev and Omel'chenko 1964). The starting mixtures were annealed at 800°C and 900°C for 2 h (Dmitrenko et al. 1990). Brightwell et al. (1984) obtained the ZnS–MgS ingots from $ZnCl_2$ and $MgCl_2$ $6H_2O$ by heating in H_2 and then in H_2S flow.

1.5 Zinc–Calcium–Sulfur

ZnS–CaS: The phase diagram is not constructed. The mutual solubility of ZnS and CaS at 1100°C was not found out (Kondrashev and Omel'chenko 1964).

1.6 Zinc–Barium–Sulfur

ZnS–BaS: The phase diagram is not constructed. The compound Ba_2ZnS_3 was obtained at 800°C according to the next exchange reaction (Schnering and Hoppe 1961): $2BaZnO_2 + 4H_2S = Ba_2ZnS_3 + ZnS + 4H_2O$ or by heating of $2BaO + ZnO$ mixtures with $BaCO_3$ and ZnO or ZnS impurities in the H_2S flow (Malur 1964).

This compound crystallizes in an orthorhombic structure with lattice parameters $a = 1205$, $b = 1265$, and $c = 421$ pm (Schnering and Hoppe 1961) [$a = 1205.4$, $b = 1264.7$, and $c = 421$ pm (Scott et al. 1977)]. The calculation and experimental density of Ba_2ZnS_3 are 4.52 and 4.50 g cm^{-3}, respectively (Schnering and Hoppe 1961, Scott et al. 1977). Its melting temperature, determined using DTA, is 997°C (Scott et al. 1977) [977°C (Nicholas et al. 1977), 1050°C–1100°C (Malur 1964)], and energy gap lies within the interval of 3.3–3.5 eV (Scott et al. 1977). The Ba_2ZnS_3 crystals decompose in the air at 800°C (Nicholas et al. 1977).

The $BaZnS_2$ compound (orthorhombic structure, $a = 1207$, $b = 1265$, $c = 422$ pm), which was found in this system (Hoppe 1959) and obtained by the same method as the Ba_2ZnS_3 according to the reaction (Schnering and Hoppe 1961) $BaZnO_2 + 2H_2S = BaZnS_2 + 2H_2O$, is the mixture of ZnS and Ba_2ZnS_3 (Scott et al. 1977). According to the data of Hoppe (1959), single crystals of $BaZnS_2$ can be obtained by the annealing of this compound in the H_2S flow at 1040°C.

1.7 Zinc–Strontium–Sulfur

ZnS–SrS: The phase diagram is not constructed. The compound $SrZnS_2$ was obtained according to the next exchange reaction (Hoppe 1959): $SrZnO_2 + 2H_2S = SrZnS_2 + 2H_2O$

Single crystals of $BaZnS_2$ can be obtained by the annealing of this compound in the H_2S flow at 1200°C.

1.8 Zinc–Cadmium–Sulfur

The isothermal section of Zn–Cd–S ternary system at 80°C–100°C was constructed using metallography, XRD, and microhardness measurement (Figure 1.7) (Palatnik et al. 1965).

ZnS–CdS: The phase diagram is shown in Figure 1.8a and b. The melting diagram (Figure 1.8a) was estimated in Tomashik (1981), and phase relations in the solid state were investigated by many authors. According to

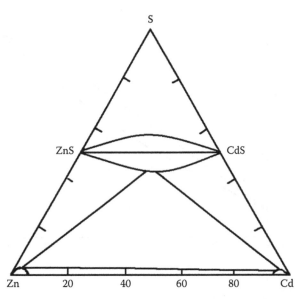

FIGURE 1.7
The isothermal section of Zn–Cd–S ternary system at 80°C–100°C (the fields of solid solutions are enlarged). (From Palatnik, L.S. et al., *Fizika tverdogo tela*, 7(9), 2850, 1965.)

the data of XRD, solid solutions with sphalerite structure form at the ZnS contents up to 30 mol. % and with wurtzite structure within the interval of 0–45 mol. % CdS (Kozhina et al. 1965). The mixture of both solid solutions forms at the contents of CdS within the interval 50–60 mol. %. Existence of the second phase in the ingots containing more than 3.4 mol. % (5 mass. %) CdS (Andreeva et al. 1973) can be explained by insufficient annealing.

The solid solutions obtained by other methods have both cubic (Dmitrienko and Kniazev 1975, Bol'shakov et al. 1979) and hexagonal (Charbonnier 1973, Bol'shakov et al. 1979) structure. The scheme of the crystallization fields was constructed in Dmitrienko et al. (1977). The cubic phase was found in the $Cd_{0.1}Zn_{0.9}S$ crystals, and it presents in small quantities up to $Cd_{0.01}Zn_{0.99}S$. The investigation of $Zn_xCd_{1-x}S$ $(0 \leq x \leq 1)$ single crystals has shown that this solid solution crystallizes in the wurtzite structure (Cherin et al. 1970). Solid solutions with sphalerite structure are formed only at higher ZnS concentrations $(x \approx 1)$. The lattice parameters of wurtzite phase change linearly with composition changing. The existence of 2H, 3C, 4H, 6H, and 10H polytype was determined using XRD in the solid-solution single crystals (Paszkowicz et al. 1996, Palosz and Przedmojski 1976). According to the data of Sakaguchi et al. (1977), the sphalerite–wurtzite transformation in ZnS crystal containing CdS is promoted with a certain amount of blended CdS at 850°C in the atmospheres of He, HCl, H_2S, or their mixtures. Using the data of XRD and electron probe microanalyzer (EPMA), Yonemura and Kotera (1986) assumed that at the formation of the $Zn_xCd_{1-x}S$ solid solutions, CdS sublimates on the first stage of solid-solution formation especially when

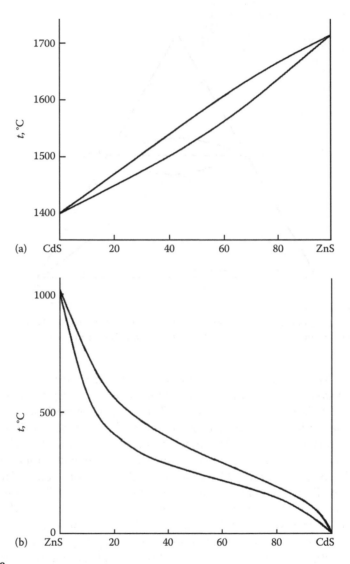

FIGURE 1.8

The phase diagram (a) and subsolidus equilibria (b) in the ZnS–CdS system. (a: From Tomashik, V.N., *Izv. AN SSSR. Neorgan. materialy*, 17(6), 1116, 1981; b: From Gan'shin, V.A. et al., *Zhurn. neorgan. khimii*, 38(12), 2026, 1993.)

CdS is a form of fine powder. This assumption was supported by the observation of CdS on the outer surface of ZnS (Dmitrienko et al. 1977, Yonemura and Kotera 1986). The next mutual diffusion proceeds to form solid solution. The thermodynamic calculations indicate that $Zn_xCd_{1-x}S$ solid solutions could decompose at temperature <290°C forming two solid solutions (Sanitarov et al. 1979).

At high pressures, $Zn_xCd_{1-x}S$ solid solutions undergo phase transitions to a cubic structure of NaCl type (Skums et al. 1992). The pressure that corresponds to the beginning of the phase transition is a function of composition according to the linear function p_i (GPa) = 2.51 + 13.68x.

Investigating the kinetics of equimolar solid-solution formation through XRD, it was shown that such solid solution crystallizes in hexagonal structure and at 750°C solid solution forms for 70 min (Dmitrienko et al. 1979). $Zn_xCd_{1-x}S$ solid solutions were detected in the mixture of ZnS and CdO heated up to 1050°C (Vasiliev et al. 1964). The vapor phase above $Zn_xCd_{1-x}S$ solid solutions consists of ZnS, CdS, and S_2 molecules and Zn and Cd atoms (Bundel' et al. 1967, Bundel' and Zhukova 1967). The concentration dependence of vapor pressure at 900°C exhibits low deviations from ideality, and this permits to think that $Zn_xCd_{1-x}S$ solid solutions are regular (Vishniakov and Bundel' 1966).

Susa et al. (1980) and Kaneko et al. (1984) investigated ZnS–CdS system at high pressures and temperatures and showed that at such conditions, a mixed phase of solid solutions with wurtzite and sphalerite structures was obtained.

Gan'shin et al. (1993) studied the phase equilibria in the subsolidus region of ZnS–CdS system by the investigation of ion-exchange processes $Cd^{2+} \rightarrow Zn^{2+}$ in the mixture "ZnS powder + melt (solution) of the salts." These authors gave the full thermodynamic description of the ZnS–CdS phase diagram in the subsolidus region and obtained a good coincidence of theoretical phase diagram with experimental data (Figure 1.8b). The ZnS–CdS phase diagram in the solid state (350°C–1050°C) belongs to the diagrams with solid-solution degradation in consequence of polymorphous transformation (Tauson and Chernyshev 1981, Chen et al. 1988, Gan'shin et al. 1993). The ZnS-rich region of this phase diagram up to 18 mol. % was studied by Chen et al. (1988) using XRD over temperature range 700°C–950°C.

The concentration dependence of optical energy gap of $Zn_xCd_{1-x}S$ thin films has a small deviation from linearity (Semenov et al. 1990). The values of the bulk modulus for $Zn_xCd_{1-x}S$ alloys as a function of x slightly deviate from Vegard's law (Noor et al. 2010).

$Zn_xCd_{1-x}S$ solid solutions were obtained by precipitation from water solutions (Kozhina et al. 1965, Dmitrienko and Kniazev 1975), from chemical elements by annealing at 800°C for 8 h (Charbonnier 1973) and from ZnS and CdS powder previously heated at 1100°C and 700°C (Dmitrienko et al. 1979). The $Zn_xCd_{1-x}S$ single crystals were obtained by the vapor chemical reaction and using hydrothermal method (Cherin et al. 1970). High-pressure investigations of the ZnS–CdS system were carried out at 2.0 GPa and 800°C (Susa et al. 1980). In the hydrothermal study of this system, the pressure was varied from 1.6 MPa (16 atm) at 200°C to 20 MPa (200 atm) at 400°C (Kaneko et al. 1984). Phase transition in the solid solutions was determined by investigation of exchange process between cations from II–VI semiconductor crystals and cations from the melts or solutions of salts. Such process takes place under equilibrium conditions since the composition of the produced solid solutions is independent of the process duration (Gan'shin et al. 1993).

1.9 Zinc–Mercury–Sulfur

ZnS–HgS: The phase diagram is not constructed. Continuous solid solutions with sphalerite structure are formed in this system (Kremheller et al. 1960, Wachtel 1960, Charbonnier and Murat 1974). The concentration dependence of lattice parameters changes according to Vegard's law (Charbonnier and Murat 1974, Tauson and Abramovich 1980). Hexagonal phase appears in the $Zn_{0.05}Hg_{0.95}S$ solid solution, and its contents increase with the increase of HgS concentration (Charbonnier 1973).

Tauson and Abramovich (1980) and Tauson and Chernyshev (1981) investigated the phase relations in this system at temperature drop 200°C–280°C and 100 MPa (1000 atm) using a hydrothermal method. There are two immiscibility gaps in the ZnS–HgS system (Figure 1.9). The first is determined by the decomposition of cubic solid solutions sphalerite–metacinnabarite, and the second is caused by the metacinnabarite–cinnabar transformation. The temperature and composition of the decomposition dome are 330°C and 46 mol. % HgS. The temperature of three-phase equilibria was estimated as 150°C at 100 MPa (1000 atm) and 100°C–120°C at 0.1 MPa (1 atm).

At 250°C–280°C and 100 MPa (1000 atm), sphalerite contains approximately 28 mol. % HgS, and metacinnabarite contains 23–26 mol. % ZnS (Tauson and Abramovich 1980). At 200°C–280°C and 100 MPa (1000 atm), metacinnabarite in equilibrium with cinnabar contains 9–13 mol. % ZnS, and cinnabar contains not more than 0.2 mol. % ZnS.

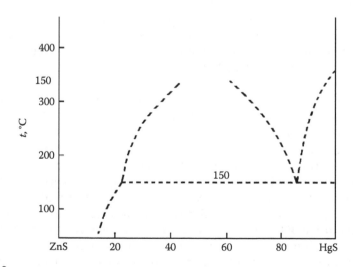

FIGURE 1.9

Phase relation in the system ZnS–HgS at 1000 atm (100 MPa). (From Tauson, V.L. and Abramovich, M.G., *Geokhimia*, (6), 808, 1980; Tauson, V.L. and Chernyshev, L.V., *Novosibirsk, Nauka Publish.*, 190, 1981.)

The concentration dependence of energy gap in the ZnS–HgS system is linear (Kremheller et al. 1960).

The $Zn_xHg_{1-x}S$ solid solutions were obtained by hydrothermal method at 350°C and 10 MPa for 2–3 days (Kremheller et al. 1960). Mineral guadalcazarite ($Zn_xHg_{1-x}S$, cubic structure, $a = 583.0 \pm 0.8$ pm) is a variety of metacinnabarite containing 2.69 mass. % Zn (Vasil'ev 1963).

1.10 Zinc–Aluminum–Sulfur

ZnS–Al: The phase diagram is not constructed. Atroshchenko et al. (1972) showed using XRD that the Al solubility in ZnS is 7.2 10^{-2} at. %.

The ZnS single crystals were obtained by the crystallization from the melt at 1750°C and 10 MPa (100 atm) in the Ar atmosphere (Atroshchenko et al. 1972).

According to the data of Gamidov and Kulieva (1969) $ZnAlS_2$ exists in the Zn–Al–S ternary system. This compound crystallizes in a tetragonal structure of the chalcopyrite type with lattice parameters $a = 539.8$ and $c = 1057.0$ pm.

ZnS–Al$_2$S$_3$: The phase diagram is shown in Figure 1.10 (Bonsall and Hummel 1978). The $ZnAl_2S_4$ ternary compound is formed in this system. The eutectic on Al_2S_3-rich side crystallizes at 950°C–1000°C. The peritectic interaction at 1050°C includes the equilibrium between solid solutions with spinel and wurtzite structure based on $ZnAl_2S_4$ and liquid. The eutectoid interaction in the solid state exists at 740°C and 15 mol. % Al_2S_3. $ZnAl_2S_4$ melts at 1160°C ± 10°C.

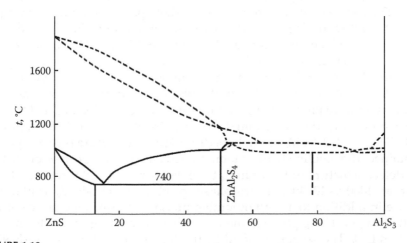

FIGURE 1.10
The ZnS–Al$_2$S$_3$ phase diagram. (From Bonsall, S.B. and Hummel, F.A., *J. Solid State Chem.*, 25(4), 379, 1978.)

The solubility of Al_2S_3 in ZnS (sphalerite) reaches 6.5 mol. % at 720°C–750°C, and the solubility of Al_2S_3 in α-$ZnAl_2S_4$ constitutes 56 mol. % at 800°C–950°C. The wurtzite modification of ZnS forms continuous solid solutions with β-$ZnAl_2S_4$ (Bonsall and Hummel 1978).

$ZnAl_2S_4$ ingots, obtained at 800°C, crystallize in a cubic structure of spinel type with lattice parameter $a = 996.8 \pm 8$ pm (Hahn and Frank 1952, Hahn et al. 1955) [$a = 999$ pm (Flahaut 1952); 1000.5 pm (Steigmann 1967), 1000.9 pm (Berthold et al. 1983); 1000.1 pm (Moldovyan 1991), 1000.93 pm (Berthold and Köhler 1981)]. The calculation and experimental density of $ZnAl_2S_4$ are 3.306 [3.29 (Steigmann 1967)] and 3.30 g cm^{-3}, respectively (Hahn and Frank 1952, Hahn et al. 1955) [3.28 and 3.25 g cm^{-3} (Flahaut 1952)]. $ZnAl_2S_4$ ingots, obtained at 1100°C, crystallize in the wurtzite structure with lattice parameters $a = 375.6$ and $c = 613$ pm. According to the data of Steigmann (1967), β-$ZnAl_2S_4$ crystallizes in an orthorhombic structure with lattice parameters $a = 1282.5$, $b = 750.0$, and $c = 609.9$ pm. Their calculation and experimental density are 2.727 [2.63 (Steigmann 1967)] and 2.65 g cm^{-3}, respectively (Hahn and Frank 1952, Hahn et al. 1955).

According to the data of XRD, the single-phase spinel forms in the $(8-3x)$ ZnS-$(8+x)Al_2S_3$ ingots at $x = 0, 0.3, 0.6, 0.9, 1.2, 1.8$ and 2.1 (Kovaliv and Lisniak 1975, Tretiakov et al. 1976). The spinel lattice constant decreases linearly at the transition from stoichiometric compound $ZnAl_2S_4$ to spinel with maximum vacancy concentration.

Single crystals of α-$ZnAl_2S_4$ with spinel structure ($a = 1000.93$ pm) have been obtained by chemical transport reactions at 740°C (Berthold and Köhler 1981). Heating of the compound to 800°C–900°C leads to decomposition and formation of a ZnS-poor defect spinel phase and a ZnS-rich phase with a defect wurtzite structure. The boundaries of the two-phase region at 830°C–860°C are approximately $Zn_{0.98}Al_{2.01}S_4$ (cubic α-phase, $a = 1000.72$ pm at 25°C) and $Zn_{1.8}Al_{1.47}S_4$ (hexagonal wurtzite phase, $a = 376$ and $c = 615$ pm at 25°C). Mixture of ZnS, Al, and S with the composition $Zn_xAl_{8/3-2x/3}S_4$ and $0.33 \leq x \leq 0.98$, which are heat treated at 830°C–860°C (70–140 h), yields after cooling to room temperature homogenous products with a defect spinel structure. The miscibility gap at the composition ZnS Al_2S_3 continues at higher temperatures with a shift of the phase boundaries and formation of high-temperature phases. A high-temperature modification of $ZnAl_2S_4$ does not exist up to 1080°C. When mixture with $0.44 \leq x \leq 0.85$ is heat treated at 1060°C–1080°C (72–96 h), a rhombohedral high-temperature phase (γ-phase) is obtained after cooling to room temperature. Its structure can be described as a defect structure of the $ZnIn_2S_4$ type. With $x = 1.00$, after thermal treatment at 1060°C–1080°C, a two-phase product is obtained, containing γ-phase in addition to an orthorhombic phase (β-phase, superlattice of the wurtzite type). The β-phase is the only phase occurring in products with $1.40 \leq x \leq 1.70$. The solubility of Al_2S_3 in ZnS (wurtzite) at 1060°C–1080°C with formation of a defect wurtzite structure reaches $Zn_{1.70-1.80}Al_{1.53-1.47}S_4$ (Berthold and Köhler 1981).

Moldovyan et al. (1989, 1993) and Moldovyan (1991) have obtained a layered hexagonal modification of $ZnAl_2S_4$ with lattice parameters $a = 362$ and $c = 1207$ pm and energy gap $E_g = 4.36$ eV. Chemical transport of $ZnAl_{2-x}Ga_xS_4$ at $x \geq 0.2$ leads to receiving single crystals with tetragonal structure and lattice parameters $a = 530$ and $c = 1040$ pm (Moldovyan 1991).

The ingots of the ZnS-Al_2S_3 system, annealed at different temperatures for 24–240 h followed by cooling in water, were investigated using XRD. $ZnAl_2S_4$ was obtained by the annealing of ZnS and Al_2S_3 mixtures at 800°C–1100°C for 24–72 h (Hahn and Frank 1952, Hahn et al. 1955) or by the chemical transport reactions (Berthold and Köhler 1981, Berthold et al. 1983, Moldovyan 1991). α-$ZnAl_2S_4$ was obtained by heating of appropriate amounts of ZnS, aluminum powder, and sulfur at 600°C for 15 h and then at 900°C for further 15 h, and β-$ZnAl_2S_4$ was obtained by heating of α-$ZnAl_2S_4$ at 950°C for 43 h and then at 1050°C for 24 h (Steigmann 1967).

1.11 Zinc–Gallium–Sulfur

Scheme of the Zn–Ga–S isothermal section at 1000°C was constructed by Lott et al. (1984a) (Figure 1.11; the proportions are not observed in this figure).

ZnS–Ga: The solubility of Ga into ZnS at 1000°C as a function of zinc vapor pressure was determined by full saturation and is presented in Figure 1.12 (Lott et al. 1984a).

ZnS–Ga$_2$S$_3$: The tentative phase diagram is shown in Figure 1.13 (Krämer et al. 1987). At high temperatures, a complete solid solution with defect wurtzite structure exists in this system. The $ZnGa_2S_4$ ternary compound is formed. Besides it, a second phase with a phase width from 80 to 92 mol. % Ga_2S_3 was found. Various attempts to prepare single crystals of this second phase failed.

According to the data of Malevski (1967) at 1300°C, the solid solutions with wurtzite structure contain up to 50 mol. % ½Ga_2S_3. The addition of Ga_2S_3 to ZnS reduces the temperature of phase transition sphalerite–wurtzite from 1020°C for pure ZnS to 950°C for solid solution with 5 mol. % ½Ga_2S_3 and to 850°C at 10 mol. % ½Ga_2S_3. At 650°C, the solid solutions with sphalerite structure contain up to 20 mol. % ½Ga_2S_3. The unreacted mixture of ZnS and Ga_2S_3 was found in the solid solutions at the annealing at 600°C even for 500 h. Therefore, the solubility at temperature lower than 600°C is not determined.

According to the data of XRD, at 900°C the solubility of Ga_2S_3 in ZnS attains 30 mol. % and the solubility of ZnS in Ga_2S_3 is equal to 20 mol. % (Hahn et al. 1955). The $ZnGa_2S_4$ compound was found in this system (Hahn et al. 1955, Beun et al. 1961, Chess et al. 1984, Solans et al. 1988, Wu et al. 1988, Carpenter et al. 1989, Zhang et al. 1990, Rigan and Stasiuk 1991). It melts at 1248°C±30°C (Chess et al. 1984) and crystallizes in a tetragonal structure with lattice parameters $a = 529.7$ and $c = 1036.3$ pm (Carpenter et al. 1989) [$a = 529.0 \pm 0.1$ and

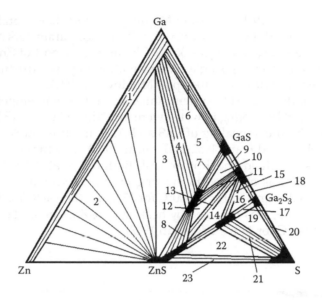

FIGURE 1.11

Isothermal section of the ternary system Zn–Ga–S at 1000°C: 1, L(Ga–Zn); 2, ZnS+L(Ga–Zn); 3, ZnS+ L(Ga–Zn)+L(GaS–ZnS); 4, L(Ga–Zn)+L(GaS–ZnS); 5, GaS+L(Ga–Zn)+L(GaS–ZnS); 6, GaS+L(Ga–Zn); 7, GaS+L(GaS–ZnS); 8, ZnS+L(GaS–ZnS); 9, GaS+L(GaS+Ga$_2$S$_3$); 10, GaS+L(GaS+Ga$_2$S$_3$)+L(GaS– ZnS); 11, L(GaS–ZnS)+L(GaS+Ga$_2$S$_3$); 12, ZnS+L(GaS–ZnS)+L(GaS+Ga$_2$S$_3$); 13, ZnS+L(GaS+Ga$_2$S$_3$); 14, ZnS+L(GaS+Ga$_2$S$_3$)+ZnGa$_2$S$_4$; 15, ZnGa$_2$S$_4$+L(GaS+Ga$_2$S$_3$); 16, ZnGa$_2$S$_4$+L(GaS+Ga$_2$S$_3$)+Ga$_2$S$_3$; 17, ZnGa$_2$S$_4$+Ga$_2$S$_3$; 18, Ga$_2$S$_3$+L(GaS+Ga$_2$S$_3$); 19, ZnGa$_2$S$_4$+Ga$_2$S$_3$+L(S); 20, Ga$_2$S$_3$+L(S); 21, ZnGa$_2$S$_4$+L(S); 22, ZnS+ZnGa$_2$S$_4$ +L(S); 23, ZnS+L(S). (From Lott, K.P. et al., *Tr. Tallinsk. Politekhn. In-ta*, 587, 17, 1984a.)

$c=1035.4\pm0.5$ pm (Krämer et al. 1987); $a=527.8$ and $c=1037.7$ pm (Solans et al. 1988); $a=525$ and $c=1038$ pm (Rigan and Stasiuk 1991); $a=526.3$ and $c=1042$ pm (Hahn et al. 1955)] and calculation and experimental density 3.76 and 3.70 g cm^{-3}, respectively (Krämer et al. 1987) [3.808 and 3.75 g cm^{-3} (Hahn et al. 1955); 3.825 (Solans et al. 1988) and 3.70–3.73 g cm^{-3} (Rigan and Stasiuk 1991)].

ZnGa$_2$S$_4$ starts to sublimate at 760°C and the more intensive evaporation takes place at $t > 980$°C (Dovgoshey et al. 2001). The enthalpy of evaporation for this compound is equal to 578.6 kJ mol^{-1}. The main vapor components are S$_2^+$, Ga$_2$S$^+$, GaS$^+$, ZnS$_2^+$, and ZnS$^+$ ions.

Edwards et al. (1989) have synthesized ZnGa$_8$S$_{13}$ and shown that this compound evaporates incongruently, forming vapor from Ga$_2$S, Zn, and S$_2$.

Solid solutions and binary compounds of probable composition 3ZnS·Ga$_2$S$_3$, ZnS Ga$_2$S$_3$, 2ZnS 3Ga$_2$S$_3$, and ZnS 3Ga$_2$S$_3$ were obtained in the ZnS–Ga$_2$S$_3$ system at coprecipitation of zinc and gallium sulfides from water solutions (Chaus et al. 1973). However, some of these compounds could not exist.

ZnGa$_2$S$_4$ was synthesized by the annealing of pressed ZnS+Ga$_2$S$_3$ mixtures at 900°C for 12–24 h followed by annealing at 600°C for 12 h (Hahn et al. 1955) or by the annealing of necessary quantities of Zn, Ga, and S at 900°C

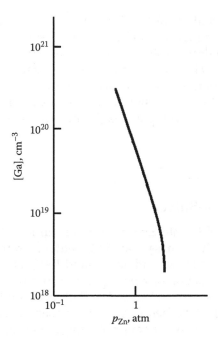

FIGURE 1.12
Solubility of Ga into ZnS at 1000°C as a function of zinc vapor pressure. (From Lott, K.P. et al., *Tr. Tallinsk. Politekhn. In-ta*, 587, 17, 1984a.)

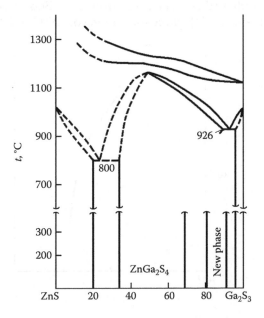

FIGURE 1.13
Tentative phase diagram of the system ZnS–Ga$_2$S$_3$. (From Krämer, V. et al., *Thermochim. Acta*, 112(1), 88, 1987.)

for 10 days (Solans et al. 1988). Solid solutions in this system were obtained by the annealing of ingots at 660°C–1300°C for 7–400 h (Malevski 1967). The $ZnGa_2S_4$ single crystals were obtained by chemical transport reactions (Beun et al. 1961, Wu et al. 1988, Carpenter et al. 1989, Rigan and Stasiuk 1991) and by vacuum sublimation (Rigan and Stasiuk 1991).

Samples of the $ZnS–Ga_2S_3$ system were prepared via solid-state diffusion from powders of the pure components (Zhang et al. 1990).

1.12 Zinc–Indium–Sulfur

ZnS–In: Solubility of In into ZnS increases from less than 0.15 mass. % at 600°C to approximately 2.3 mass. % at 862°C, the temperature of the sphalerite–wurtzite inversion (Boorman and Sutherland 1969). The solubility of In into ZnS at 1000°C as a function of zinc vapor pressure was determined by full saturation and is presented in Figure 1.14 (Lott et al. 1984b).

ZnS–In₂S₃: In this system, there exist several ternary phases of the general formula $Zn_mIn_2S_{3+m}$ that present various polytypic modifications. Although these compounds differ in composition, they are similar in some structural pecularities, that is, a slightly distorted packing layer structure of the sulfur atoms and vacant cationic sites (Anagnostopoulos et al. 1986).

The most reliable phase diagram was constructed by Boorman and Sutherland (1969) and Barnett et al. (1971) within the interval of 600°C–1080°C (Figure 1.15).

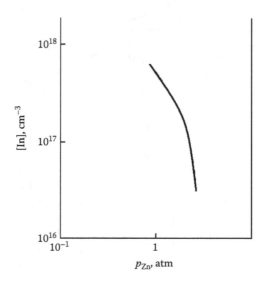

FIGURE 1.14
Solubility of In into ZnS at 1000°C as a function of zinc vapor pressure. (From Lott, K.P. et al., *Tr. Tallinsk. Politekhn. In-ta*, 587, 29, 1984b.)

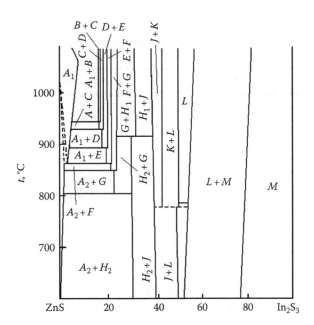

FIGURE 1.15
Phase equilibria in the ZnS–In₂S₃ system in the solid state at 600°C–1080°C. (From Boorman, R.S. and Sutherland, J.K., *J. Mater. Sci.*, 4(8), 658, 1969; Barnett, D.E. et al., *Phys. Status Solidi A*, 4(1), K49, 1971. With permission.)

There were synthesized 11 ternary phases. The seven most ZnS-rich compounds form a series (Boorman and Sutherland 1969, Barnett et al. 1971, Nitsche 1971). All investigated single crystals have hexagonal or rhombohedral layer structure with lattice parameter $a = 385 \pm 1$ pm. The c dimension varies from 3747 to 1863 pm with a periodicity of 314 pm, which can be related to a decrease in the ZnS content. For $Zn_3In_2S_6$, the lattice parameter is equal to $c = 2179$ pm (Boorman and Sutherland 1969) [$c = 1870$ pm (Nitsche 1971)]. The lattice parameters for other ternary phases are given in Table 1.2 (Boorman and Sutherland 1969, Barnett et al. 1971).

The c parameters, determined using XRD, atomic absorption, and electron microprobe analysis, are twice as many (Barnett et al. 1971). The H_1, J_1, and J_2 phases crystallize in hexagonal (Barnett et al. 1971) but not triclinic structure (Boorman and Sutherland 1969). The existence of 2H, 4H, and 6H polytypes was confirmed for J_1 and J_2 phases.

It was determined using XRD that cubic structure in the $(ZnS)_{3x}-(In_2S_3)_{1-x}$ system remains up to 10 mol. % In_2S_3 (Radautsan et al. 1969b). Solid-solution region on β-In_2S_3-rich side do not exceed 6.25 mol. % (Donika 1972).

The phase relations in the ZnS–In₂S₃ system were investigated also by many other authors Beun et al. (1960), Lappe et al. (1962), Zhitar' et al. (1965), Zhitar' et al. (1968), Cnehm et al. (1969), Imamov et al. (1970), Radautsan et al. (1970, 1987), Mustia et al. (1972), Chaus et al. (1973), Sobolev (1976),

TABLE 1.2

Phase Compositions and c Parameter of Ternary Phases
in the $ZnS-In_2S_3$ System

Phase	x_{ZnS}, mol. % (Boorman and Sutherland 1969)	c, pm	x_{ZnS}, mol. % (Barnett et al. 1971)	c, pm
B	82.9±0.2	3747	83.40±0.8	3739±8
	82.6±0.1	3435		6856±10
C	82.9±0.2			
	81.2±0.2	3118	81.64±0.7	3118±8
D	82.1±0.4			
	79.4±0.2	2802	79.30±0.6	5604±3
E	80.1±0.1			
	77.6±0.1	2492	77.64±0.5	2486±4
F	78.1±0.3			
	75.0±0.1	2179	75.54±0.7	4345±5
G	75.3±0.1			
H_1	67.9±0.2	Triclinic	66.48±1.1	5584±8
H_2	69.3±0.1	1863	70.08±1.2	1862±4
J_1			60.03±0.6	3100±3
J_2	60.7±0.1		59.78±0.9	4643±10
K	56.0±0.2		55.11±1.3	5559±2
L	49.6±0.2	3695	49.36±1.0	3690±3

Donika et al. (1980), Kalomiros et al. (1987), Biyushkina et al. (1989b), Biyushkina et al. (1989a), but their results in most cases are contradictory.

The $ZnIn_2S_4$, $Zn_2In_2S_5$, $Zn_3In_2S_6$, and $Zn_5In_2S_8$ compounds have been obtained in this system. $ZnIn_2S_4$ crystallize in the layered hexagonal structure with experimental density 4.38 g cm^{-3} (Lappe et al. 1962). The elementary cell contains three formula units (Sobolev 1976). Sixteen polytypes are known for $ZnIn_2S_4$ with $a = 385 \pm 2$ pm and c parameter correspondingly: I, 1234 pm; II_a and II_b, 2468 pm; III_a [$a = 387.3 \pm 0.2$ and $c = 3706.7 \pm 0.4$ pm (Tinoco et al. 1999)], III_b and III_c, 3702 pm [$c = 3706$ pm (Lappe et al. 1962), $c = 3695$ pm (Boorman and Sutherland 1969)]; IV, 4936 pm; V, 6170 pm; VI_a, VI_b, and VI_c, 7404 pm; XII_a and XII_b, 14808 pm; XIV, 17276 pm; and XXIV_a and XXIV_b, 29616 pm (Donika et al. 1970b, 1971, 1972b, 1982, Radautsan et al. 1970, Donika 1972, Biyushkina et al. 1989b). Indium occupied both tetrahedral and octahedral hollows (Lappe et al. 1962, Radautsan et al. 1969b, Imamov et al. 1970, Sobolev 1976). Any phase transitions were observed for this compound up to 18 GPa (Tinoco et al. 1999).

The melt temperature and optical energy gap of $ZnIn_2S_4$ are 1120°C and 2.6 eV (Beun et al. 1960, Zhitar' et al. 1965), respectively. Approximately 2 mol. % ZnS is soluble in $ZnIn_2S_4$ at 600°C and its solubility increases up to 8 mol. % at 1080°C (Boorman and Sutherland 1969). The single crystals of this compound have high transparence in the visible spectrum, high photoconductivity, and low electroconductivity and sluggishness (Zhitar' et al. 1968). This compound can dissolve up to 10 mol. % In_2S_3 at 750°C whereby the color turns from yellow to deep red, the lattice remaining unchanged (Nitsche 1971).

$ZnIn_2S_4$ is stable at heating in air up to 550°C, and oxidizing and sublimation begin at this temperature and end at 1040°C (Lyalikova and Mirovich 1970, Donika et al. 1980).

$Zn_2In_2S_5$ crystallizes in a hexagonal structure and has six polytypes (Radautsan et al. 1970, Donika 1972, Donika et al. 1980). The c parameter for these polytypes is equal correspondingly: II_a, 3085 pm; III_a and III_b, 4627 pm [$c = 4620 \pm 5$ pm (Radautsan et al. 1969a)]; IV, 6170 pm; and VI_a and VI_b, 9255 pm (Donika et al. 1970a, 1972a, 1980). The energy gap of this compound estimated from fundamental absorption edge is equal to 2.65 eV at 77 K and 2.50 eV at 300 K (Radautsan et al. 1969a).

The structure of $Zn_3In_2S_6$ is similar to the $ZnIn_2S_4$ structure (Imamov et al. 1970). It crystallizes in hexagonal structure with lattice parameters $a = 385$ and $c = 1850$ pm (Donika et al. 1967, Donika 1972) [$a = 387 \pm 2$ and $c = 1890 \pm 5$ pm (Biyushkina et al. 1989a)] and calculation and experimental density 4.31 and 4.23 g cm^{-3}, respectively (Donika et al. 1967, Donika 1972). This compound has two polytypes, I_a and I_b, with the same c parameter (Radautsan et al. 1970, Donika et al. 1980) and order–disorder transition (Radautsan et al. 1987). Their melting temperature and optical energy gap are 1200°C and 2.8 eV, respectively (Zhitar' et al. 1965). According to the data of Anagnostopoulos et al. (1986), the optical energy gap of this compound at 77 and 300 K is equal to 2.88 and 2.77 eV, respectively. It is stable at heating in air up to 650°C and in vacuum up to 1000°C (Lyalikova and Mirovich 1970, Donika et al. 1980).

$Zn_5In_2S_8$ crystallizes in hexagonal structure with lattice parameters $a = 385$ and $c = 2480$ pm [$a = 386$ and $c = 2478$ pm (Machuga et al. 2000)] and its energy gap is equal to 2.84 eV (Kalomiros et al. 1987).

According to the data of Cnehm et al. (1969), single crystals in the ZnS–In_2S_3 system have hexagonal structure with constant value of $a = 383 \pm 3$ pm, and c parameter changes stepwise with the interval of 310 ± 3 pm.

Binary compounds of probable composition $4ZnS \cdot In_2S_3$, $2ZnS \cdot In_2S_3$, $ZnS \cdot In_2S_3$, $ZnS \cdot 2In_2S_3$, and $ZnS \cdot 4In_2S_3$ were obtained in the ZnS–In_2S_3 system at coprecipitation of zinc and indium sulfides from water solutions (Chaus et al. 1973). However, some of these compounds could not exist.

The ingots for investigations were obtained by annealing (Lappe et al. 1962, Zhitar' et al. 1965) and melting (Donika 1972) of mixtures of binary compounds and by chemical transport reactions (Beun et al. 1960, Zhitar' et al. 1968, Cnehm et al. 1969, Imamov et al. 1970, Donika 1972, Mustia et al. 1972). Single crystals of $ZnIn_2S_4$, $Zn_2In_2S_5$, $Zn_3In_2S_6$, and $Zn_5In_2S_8$

were obtained by chemical transport reactions (Beun et al. 1960, Lappe et al. 1962, Donika et al. 1970b, 1980, Mustia et al. 1972, Kalomiros et al. 1987, Machuga et al. 2000). From ZnS side, single crystals of the sphalerite, wurtzite, and 4H and 6H polytypes containing up to 2 mol. % In_2S_3 have been grown (Nitsche 1971).

1.13 Zinc–Thallium–Sulfur

ZnS–Tl: The phase diagram is not constructed. Addition of $5 \cdot 10^{-4}$ at. % Tl leads to formation of prismatic inclusions in ZnS crystals (Atroshchenko et al. 1972).

Zinc sulfide was thallium doped at single crystal growing at 1750°C and 10 MPa (100 atm) (Atroshchenko et al. 1972).

ZnS–TlS: The phase diagram is not constructed. According to the DTA, the thermograms for ZnS–TlS system have only two effects at 265°C and 235°C within the 30°C–1280°C (Guseinov et al. 1982). A field of Tl_4S_3 secondary crystallization and peritectic equilibria $L + Tl_4S_3 = ZnS + TlS$ exist in this system. The mutual solubility of ZnS and TlS is less than 1 mol. %.

This system was investigated through DTA, XRD, and measuring of microhardness. The ingots were annealed at temperatures 20°C–30°C lower than solidus temperature for 400–500 h (Guseinov et al. 1982).

ZnS–Tl₂S: The phase diagram is a eutectic type (Figure 1.16) (Chaus et al. 1985). The eutectic composition and temperature are 30 mol. % ZnS

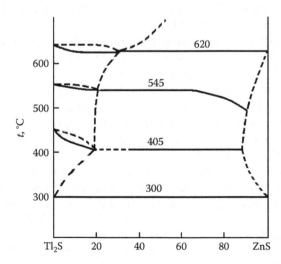

FIGURE 1.16
The ZnS–Tl₂S phase diagram. (From Chaus, I.S. et al., *Ukr. khim. zhurn.*, 51(4), 355, 1985.)

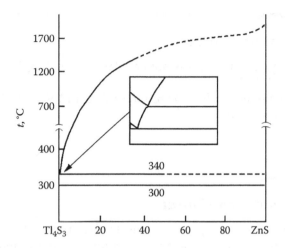

FIGURE 1.17

Phase relations in the ZnS–Tl$_4$S$_3$ system. (From Guseinov, F.H. et al., *Azerb. khim. zhurn.*, (4), 108, 1985.)

and $620 \pm 5°C$, respectively. Thermal effects at $545°C \pm 5°C$, $405°C$, and $300°C$ correspond to polymorph transitions in the solid solutions based on Tl$_2$S.

This system was investigated through DTA, metallography, XRD, and measuring of microhardness. The ingots were annealed at $550°C–600°C$ for 15 h (Chaus et al. 1985).

ZnS–Tl$_4$S$_3$: This section is a nonquasibinary section of the Zn–Tl–S ternary system (Figure 1.17) (Guseinov et al. 1985). Thermal effects at $300°C$ and $340°C$ were observed in this system. Peritectic transformation L + Tl$_2$S ⇔ Tl$_4$S$_3$ + ZnS occurs at $300°C$ and the Tl$_2$S secondary crystallization takes place at $340°C$. Two phases ZnS and Tl$_4$S$_3$ exist in the ingots in the solid state. The mutual solubility of ZnS and Tl$_4$S$_3$ is negligible.

This system was investigated using DTA, XRD, and measuring of microhardness and emf of concentration chains. The ingots were annealed at $260°C$ for 300 h (Guseinov et al. 1985).

1.14 Zinc–Scandium–Sulfur

ZnS–Sc$_2$S$_3$: The phase diagram is not constructed. ZnSc$_2$S$_4$ is formed in this system, which melts at $1750°C \pm 30°C$ (Chess et al. 1984) and crystallizes in cubic structure of spinel type with lattice parameter $a = 1048.3$ pm and energy gap $E_g = 2.1$ eV (Yim et al. 1973).

ZnSc$_2$S$_4$ was obtained by the interaction of ZnS and Sc$_2$S$_3$ in CS$_2$ + He flow at $1000°C–1200°C$ for 3–6 h. Its single crystals were grown by chemical transport reactions (Yim et al. 1973).

1.15 Zinc–Yttrium–Sulfur

ZnS–Y₂S₃: The phase diagram is not constructed. ZnY_2S_4 is formed in this system (Yim et al. 1973).

ZnY_2S_4 was obtained by the interaction of ZnS and Y_2S_3 in $CS_2 + He$ flow at 1000°C–1200°C for 3–6 h.

1.16 Zinc–Lanthanum–Sulfur

ZnS–La₂S₃: The phase diagram belongs to the eutectic type (Figure 1.18) (Aliev and Mamedov 2000). The eutectic composition and temperature are 20 mol. % La_2S_3 and 1527°C, respectively (the data are taken from the figure). $ZnLa_2S_4$ is formed in this system (Yim et al. 1973, Aliev and Mamedov 2000) that melts incongruently at 1635°C and crystallizes in the orthorhombic structure with the lattice parameters $a = 869$, $b = 1328$, and $c = 652$ pm and calculation density 3.892 g cm⁻³ (Aliev and Mamedov 2000). It is necessary to note that the polymorph transformation of the solid solutions based on ZnS is not indicated in Figure 1.18 and that La_2S_3 melting temperature was not correct in the author version and was revised.

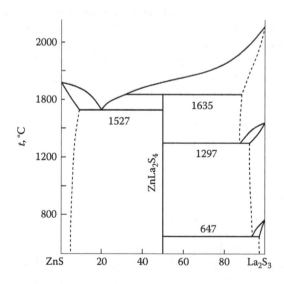

FIGURE 1.18
The ZnS–La₂S₃ phase diagram. (From Aliev, K.A. and Mamedov, G.G., Investigation of the phase equilibria in the ZnS–La₂S₃ system [in Azerbaijani], in: 6th Resp. konf., *Fiz.-khim. analiz i neorgan. materialoved.*, Baku, Azerbaijan, pp. 75–78, 2000.)

This system was investigated through DTA, XRD, metallography, and measuring of microhardness (Aliev and Mamedov 2000). $ZnLa_2S_4$ was obtained by the interaction of ZnS and La_2S_3 in CS_2 + He flow at 1000°C–1200°C for 3–6 h (Yim et al. 1973).

1.17 Zinc–Cerium–Sulfur

$ZnS–Ce_2S_3$: The phase diagram is not constructed. $ZnCe_2S_4$ is formed in this system (Yim et al. 1973).

$ZnCe_2S_4$ was obtained by the interaction of ZnS and Ce_2S_3 in CS_2 + He flow at 1000°C–1200°C for 3–6 h.

1.18 Zinc–Praseodymium–Sulfur

$ZnS–Pr_2S_3$: The phase diagram is not constructed. $ZnPr_2S_4$ is formed in this system (Yim et al. 1973).

$ZnPr_2S_4$ was obtained by the interaction of ZnS and Pr_2S_3 in CS_2 + He flow at 1000°C–1200°C for 3–6 h.

1.19 Zinc–Neodymium–Sulfur

$ZnS–Nd_2S_3$: The phase diagram is shown in Figure 1.19 (Aliev et al. 1992). The eutectic composition and temperature are 28.5 mol. % Nd_2S_3 and 1530°C, respectively. $ZnNd_2S_4$ is formed in this system (Yim et al. 1973, Aliev et al. 1992). It melts incongruently at 1700°C and crystallizes in a cubic structure of Th_3P_4 type with lattice parameter $a = 860$ pm (Aliev et al. 1992). The solubility of Nd_2S_3 in ZnS at 1530°C is equal to 17 mol. % and it decreases to 6 mol. % at room temperature, and the solubility of ZnS in α-Nd_2S_3 is equal to 2 mol. %. The solid solutions based on $ZnNd_2S_4$ are negligible. Thermal effects at 1060°C and 1360°C correspond to the phase transitions of the solid solutions based on Nd_2S_3.

This system was investigated through DTA, XRD, and measuring of microhardness. The ingots were annealed at 830°C for 1 month (Aliev et al. 1992). $ZnNd_2S_4$ was obtained by the interaction of ZnS and Nd_2S_3 in CS_2 + He flow at 1000°C–1200°C for 3–6 h (Yim et al. 1973).

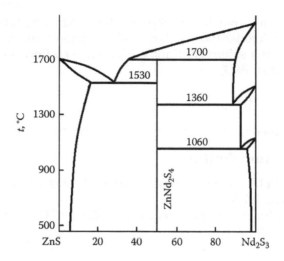

FIGURE 1.19
The ZnS–Nd$_2$S$_3$ phase diagram. (From Aliev, O.M. et al., *Zhurn. neorgan. khimii*, 37(12), 2792, 1992.)

1.20 Zinc–Samarium–Sulfur

ZnS–Sm$_2$S$_3$: The phase diagram is shown in Figure 1.20 (Aliev et al. 1991a). The eutectic compositions and temperatures are 37 and 56 mol. % Sm$_2$S$_3$ and 1250°C and 1320°C, respectively. ZnSm$_2$S$_4$ and ZnSm$_4$S$_7$ are formed in this system (Yim et al. 1973, Aliev et al. 1991a). ZnSm$_2$S$_4$ melts congruently at 1395°C and crystallizes in a cubic structure of Th$_3$P$_4$ type with lattice

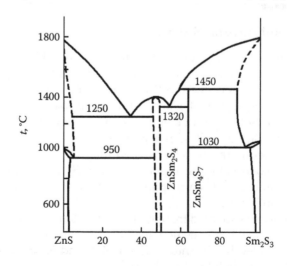

FIGURE 1.20
The ZnS–Sm$_2$S$_3$ phase diagram. (From Aliev, O.M. et al., *Zhurn. neorgan. khimii*, 36(12), 2628, 1991a.)

parameter $a = 856$ pm (Aliev et al. 1991a). Its homogeneity range is within the interval of 49–51.5 mol. % Sm_2S_3. $ZnSm_4S_7$ melts incongruently at 1450°C.

The solubility based on α- and β-ZnS is equal to 2 and 6 mol. % Sm_2S_3, respectively, and solid solutions based on α'- and γ-Sm_2S_3 contain correspondingly 4 and 8 mol. % ZnS (Aliev et al. 1991a).

This system was investigated through DTA, metallography, and XRD. The ingots were annealed at 1100°C for 250–1000 h (Aliev et al. 1991a). $ZnSm_2S_4$ was obtained by the interaction of ZnS and Sm_2S_3 in $CS_2 + He$ flow at 1000°C–1200°C for 3–6 h (Yim et al. 1973). Its single crystals were grown by the chemical transport reactions (Aliev et al. 1991a).

1.21 Zinc–Gadolinium–Sulfur

$ZnS–Gd_2S_3$: The phase diagram is shown in Figure 1.21 (Aliev 2003). It is necessary to note that ZnS melting temperature was not correct in the author version and was revised. The eutectic compositions and temperatures are 22 and 57 mol. % Gd_2S_3 and 1340°C and 1480°C, respectively. $ZnGd_2S_4$ and $ZnGd_4S_7$ are formed in this system (Yim et al. 1973, Aliev 2003). $ZnGd_2S_4$ melts congruently at 1580°C and crystallizes in a cubic structure of Th_3P_4 type with a lattice parameter $a = 840$ pm (Aliev 2003). $ZnGd_4S_7$ melts incongruently at 1680°C.

Zinchenko et al. (2010) noted that $ZnGd_2S_4$ was not observed at the investigation of the $ZnS–Gd_2S_3$ alloys.

The solubility based on α-ZnS is equal to approximately 2 mol. % Gd_2S_3, and solid solution based on α-Gd_2S_3 contains 3 mol. % ZnS (Aliev 2003).

$ZnGd_2S_4$ was obtained by the interaction of ZnS and Gd_2S_3 in $CS_2 + He$ flow at 1000°C–1200°C for 3–6 h (Yim et al. 1973).

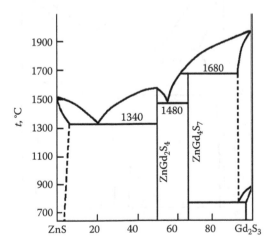

FIGURE 1.21
The $ZnS–Gd_2S_3$ phase diagram. (From Aliev, K.A., *Zhurn. neorgan. khimii*, 48(11), 1902, 2003.)

1.22 Zinc–Terbium–Sulfur

ZnS–Tb$_2$S$_3$: The phase diagram is not constructed. ZnTb$_2$S$_4$ ternary compound is formed in this system (Yim et al. 1973).

ZnTb$_2$S$_4$ was obtained by the interaction of ZnS and Tb$_2$S$_3$ in CS$_2$ + He flow at 1000°C–1200°C for 3–6 h.

1.23 Zinc–Dysprosium–Sulfur

ZnS–Dy$_2$S$_3$: The phase diagram is shown in Figure 1.22 (Aliev et al. 1991b). The eutectic compositions and temperatures are 31 and 66 mol. % Dy$_2$S$_3$ and 1325°C and 1420°C, respectively. ZnDy$_2$S$_4$ is formed in this system (Yim et al. 1973, Aliev et al. 1991b). It melts congruently at 1530°C and crystallizes in an orthorhombic structure with lattice parameter $a = 752$, $b = 1340$, and $c = 634$ pm (Aliev et al. 1991b) and energy gap $E_g = 3.1$ eV (Yim et al. 1973). Its homogeneity range at room temperature is within the interval of 47–53.5 mol. % Dy$_2$S$_3$ and has maximum at the eutectic temperature (14 mol. %).

Zinchenko et al. (2010) noted that ZnDy$_2$S$_4$ was not observed at the investigation of the ZnS–Dy$_2$S$_3$ alloys.

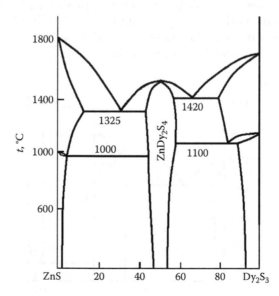

FIGURE 1.22
The ZnS–Dy$_2$S$_3$ phase diagram. (From Aliev, O.M. et al., *Zhurn. neorgan. khimii*, 36(10), 3194, 1991b.)

The solubility based on α-ZnS is equal to 1.5 mol. % Sm_2S_3. Solid solutions based on β-ZnS contain at the eutectic temperature 13 mol. % Sm_2S_3. Maximum solubility range on Dy_2S_3-rich side reaches 20 mol. % ZnS at 1420°C. The solubility of ZnS in β-Dy_2S_3 at room temperature consists 8 mol. % ZnS. Crystals of solid solutions with β-Dy_2S_3 structure were not obtained (Aliev et al. 1991b).

This system was investigated through DTA, metallography, XRD, and measuring of microhardness. The ingots were baked at 1000°C–1050°C, melted, and annealed (Aliev et al. 1991b). $ZnDy_2S_4$ was obtained by the interaction of ZnS and Dy_2S_3 in CS_2 + He flow at 1000°C–1200°C for 3–6 h (Yim et al. 1973).

1.24 Zinc–Holmium–Sulfur

ZnS–Ho_2S_3: The phase diagram is not constructed. $ZnHo_2S_4$ ternary compound is formed in this system (Yim et al. 1973).

$ZnHo_2S_4$ was obtained by the interaction of ZnS and Ho_2S_3 in CS_2 + He flow at 1000°C–1200°C for 3–6 h.

1.25 Zinc–Erbium–Sulfur

ZnS–Er_2S_3: The phase diagram is not constructed. $ZnEr_2S_4$ ternary compound is formed in this system (Yim et al. 1973), which crystallizes in an orthorhombic structure of olivine type with lattice parameters $a = 1334.4 \pm 0.4$, $b = 778.6 \pm 0.2$, and $c = 629.9 \pm 0.1$ pm (Vollebregt and Ijdo 1982).

$ZnEr_2S_4$ was obtained by the interaction of ZnS and Er_2S_3 in CS_2 + He flow at 1000°C–1200°C for 3–6 h (Yim et al. 1973) or by sintering an equimolar mixture of the binary sulfides in a stream of hydrogen sulfide at 1200°C (Vollebregt and Ijdo 1982).

1.26 Zinc–Thulium–Sulfur

ZnS–Tm_2S_3: The phase diagram is not constructed. $ZnTm_2S_4$ ternary compound is formed in this system, which crystallizes in an orthorhombic structure of olivine type with lattice parameters $a = 1330.8$, $b = 776.9$, and $c = 628.5$ pm (Lemoine et al. 1990) [$a = 1322.7$, $b = 773.4$, and $c = 626.3$ pm (Yim et al. 1973); $a = 1330.1 \pm 0.4$, $b = 774.7 \pm 0.2$, and $c = 628.1 \pm 0.1$ pm (Vollebregt and

Ijdo 1982)]; calculation and experimental density 5.42 and 5.43 g cm^{-3}, respectively (Lemoine et al. 1990), and energy gap $E_g = 3.6$ eV (Yim et al. 1973).

ZnTm$_2$S$_4$ was obtained by the interaction of ZnS and Tm$_2$S$_3$ in CS$_2$ + He flow at 1000°C–1200°C for 3–6 h (Yim et al. 1973), by the interaction of binary compounds at 950°C (Lemoine et al. 1990), or by sintering an equimolar mixture of the binary sulfides in a stream of hydrogen sulfide at 1050°C (Vollebregt and Ijdo 1982). Its single crystals were grown by the chemical transport reactions (Yim et al. 1973).

1.27 Zinc–Ytterbium–Sulfur

ZnS–Yb$_2$S$_3$: The phase diagram is not constructed. ZnYb$_2$S$_4$ ternary compound is formed in this system (Yim et al. 1973), which crystallizes in an orthorhombic structure of olivine type with lattice parameters $a = 1326.1 \pm 0.4$, $b = 771.8 \pm 0.2$, and $c = 627.8 \pm 0.2$ pm (Vollebregt and Ijdo 1982) and energy gap $E_g = 2.5$ eV (Yim et al. 1973).

ZnYb$_2$S$_4$ was obtained by the interaction of ZnS and Yb$_2$S$_3$ in CS$_2$ + He flow at 1000°C–1200°C for 3–6 h (Yim et al. 1973) or by sintering an equimolar mixture of the binary sulfides in a stream of hydrogen sulfide at 1050°C (Vollebregt and Ijdo 1982).

1.28 Zinc–Lutetium–Sulfur

ZnS–Lu$_2$S$_3$: The phase diagram is not constructed. ZnLu$_2$S$_4$ ternary compound is formed in this system, which crystallizes in an orthorhombic structure of olivine type with lattice parameters $a = 1318.3$, $b = 765.8$, and $c = 624.6$ pm [$a = 1321.80 \pm 0.09$, $b = 768.48 \pm 0.05$, and $c = 626.06 \pm 0.04$ pm (Vollebregt and Ijdo 1982)] and energy gap $E_g = 3.7$ eV (Yim et al. 1973).

ZnLu$_2$S$_4$ was obtained by the interaction of ZnS and Lu$_2$S$_3$ in CS$_2$ + He flow at 1000°C–1200°C for 3–6 h (Yim et al. 1973) or by sintering an equimolar mixture of the binary sulfides in a stream of hydrogen sulfide at 1050°C (Vollebregt and Ijdo 1982). Its single crystals were grown by chemical transport reactions (Yim et al. 1973).

1.29 Zinc–Carbon–Sulfur

ZnS–C: The phase diagram is not constructed. At 1200°C–1500°C, ZnS interacts with carbon forming CS$_2$ and zinc vapor (Tsygoda et al. 1967).

1.30 Zinc–Silicon–Sulfur

Ternary compounds, including Zn_4SiS_6, in the Zn–Si–S system were not found (Kaldis et al. 1967, Dubrovin et al. 1989a).

1.31 Zinc–Germanium–Sulfur

Ternary compounds in the Zn–Ge–S system were not found (Kaldis et al. 1967).

ZnS–GeS₂: The phase diagram is a eutectic type (Figure 1.23) (Dubrovin et al. 1989b). The eutectic composition and temperature are 18 mol. % ZnS and 810°C ± 5°C. The mutual solubility of ZnS and GeS_2 is not higher than 1 mol. %.

According to the data of Hahn and Lorent (1958), Zn_2GeS_4 was obtained in this system. It crystallizes in a cubic structure of the sphalerite type with lattice parameter $a = 543.6$ pm and calculation and experimental density 3.427 and 3.26 g cm⁻³, respectively. This compound was obtained by heating of $2ZnS + GeS_2$ mixture at 500°C–700°C, but later its existence was not confirmed (Kaldis et al. 1967, Dubrovin et al. 1989b).

This system was investigated through DTA and XRD (Dubrovin et al. 1989b). The ingots were annealed at 750°C and 790°C for 100 and 80 h, respectively.

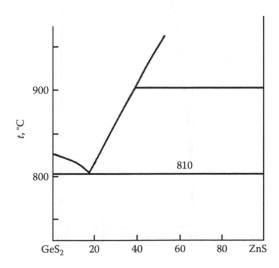

FIGURE 1.23
The ZnS–GeS₂ phase diagram. (From Dubrovin, I.V. et al., *Izv. AN SSSR. Neorgan. materialy,* 25(8), 1386, 1989b.)

1.32 Zinc–Tin–Sulfur

At 300°C–500°C and 100 MPa (1000 atm), four phases in hydrothermal conditions exist in this system: α- and β-ZnS, SnS, and SnS$_2$ (Nekrasov et al. 1977). At 500°C, α-ZnS dissolves not higher than 0.086 mol. % SnS in equilibrium with SnS and Sn and not higher than 0.018 mol. % SnS in equilibrium with SnS and SnS$_2$. Ternary compounds were not obtained in the Zn–Sn–S system.

ZnS–Sn: The phase diagram is not constructed. The ZnS solubility in liquid tin (Figure 1.24) (Rubenstein 1968, Kimura 1971, Babanski et al. 1980) was studied and a part of Zn–Sn–S liquidus surface on the Sn-rich side (Figure 1.25) (Kimura 1971).

Liquidus temperatures were determined by high-temperature filtration (Rubenstein 1968) and visual observation of solid phase appearance and disappearance for heating and cooling of ingots (accuracy of measurement is ± 5°C) (Kimura 1971).

ZnS–SnS: The phase diagram is not constructed. The ZnS solubility in liquid tin sulfide was studied (Figure 1.26) (Chernyshev and Babanski 1977,

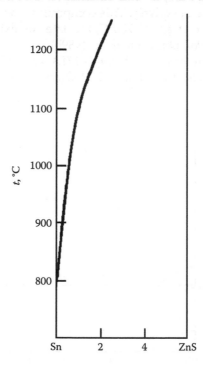

FIGURE 1.24
Temperature dependence of ZnS solubility in liquid Sn. (From Kimura, S., *J. Chem. Thermodyn.*, 3(1), 7, 1971. With permission.)

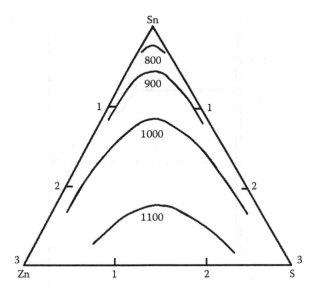

FIGURE 1.25
Part of the Zn–Sn–S liquidus surface on the Sn-rich side. (From Kimura, S., *J. Chem. Thermodyn.*, 3(1), 7, 1971. With permission.)

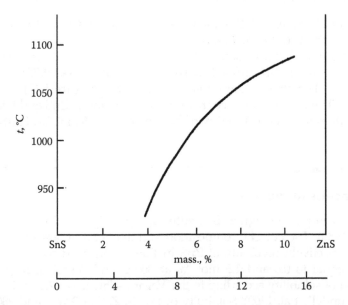

FIGURE 1.26
Temperature dependence of ZnS solubility in liquid SnS. (From Chernyshev, A.I. and Babanski, M.D., Dissolution and crystallization of sphalerite from the solution in monosulfide melt [in Russian], in *Reaktsionnaya sposobnost' veshchestv*, Tomsk, Russia, Un-t, pp. 34–37, 1977.)

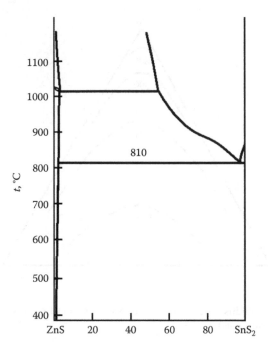

FIGURE 1.27
The ZnS–SnS$_2$ phase diagram. (From Olekseyuk, I.D. et al., *J. Alloys Compd.*, 368(1–2), 135, 2004.)

Babanski et al. 1980). α-ZnS, which was recrystallized from the solution in SnS melt, contains $6 \cdot 10^{-3}$ at. % Sn.

The solubility of ZnS in SnS was determined by loss in ZnS weight at its dissolution in liquid tin sulfide (Chernyshev and Babanski 1977).

ZnS–SnS$_2$: The phase diagram is a eutectic type (Figure 1.27) (Olekseyuk et al. 2004). The eutectic point coordinates are 2.5 mol. % ZnS and 800°C. The solubility does not exceed 1 mol. % of ZnS in SnS$_2$ and 5 mol. % of SnS$_2$ in ZnS.

1.33 Zinc–Lead–Sulfur

ZnS–PbS: The phase diagram is a eutectic type (Figure 1.28) (Murach 1947, Dutrizac 1980). The eutectic composition and temperature are 13 mol. % and 1050°C, respectively. Solubility of ZnS in PbS, determined by diffusion saturation, is equal to 0.6 and 1.5 mol. % at 800°C and 900°C correspondingly. Estimation of solubility according to the Vegard's law gives 2.1 mol. % ZnS at 1000°C (Bundel' et al. 1969). Solubility of PbS in ZnS at 800°C and 980°C constitutes correspondingly $(6–8) \cdot 10^{-6}$ and $1.5 \cdot 10^{-5}$ g Pb/g ZnS (Zubkovskaya and Vishniakov 1981).

This system was investigated through DTA, metallography, and XRD (Murach 1947, Dutrizac 1980).

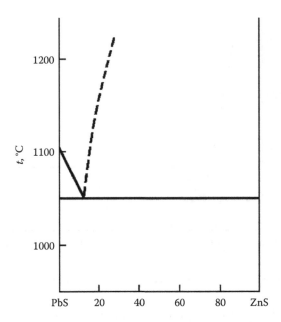

FIGURE 1.28
The ZnS–PbS phase diagram. (From Dutrizac, J.E., *Can. J. Chem.*, 58(7), 739, 1980.)

1.34 Zinc–Titanium–Sulfur

ZnS–TiS₂: The phase diagram is not constructed. $Zn_2Ti_{18}S_{32}$ was obtained in this system by heating of TiS_2 (4H polytype), ZnS, and Ti at 1200°C followed by quenching (Saeki and Onoda 1982, Kawada et al. 1985). It crystallizes in cubic face-centered structure with lattice parameter $a = 984.3$ pm and calculation and experimental density 3.49 and 3.51 g cm⁻³, respectively. Saeki and Onoda (1983) described synthesis of $(Zn_xTi_{1-x})_yS_{32}$ compound with $x = 0.08$–0.11 (when $x = 0.11$, $y = 19.4$–21.3). This compound was obtained by mixing Ti and/or titanium sulfide with Zn and/or ZnS at the atomic ratio $Zn/Ti = 0.087$–0.124 and $S/(Ti + Zn) = 1.5$–1.7 and heating this mixture at temperatures higher than 1100°C in vacuum.

1.35 Zinc–Phosphorus–Sulfur

The $Zn_4(P_2S_6)_3$ compound was synthesized by the interaction of the chemical elements at 600°C in the Zn–P–S ternary system (Bouchetiere et al. 1978). It crystallizes in a monoclinic structure with lattice parameters $a = 2141$, $b = 655.2$, $c = 1068$ pm, and $\beta = 122.28°$.

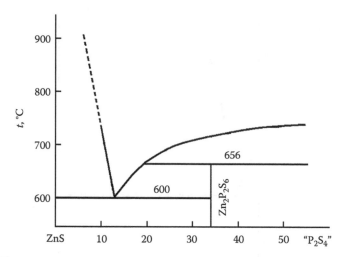

FIGURE 1.29
The ZnS–"P$_2$S$_4$" phase diagram. (From Kovach, A.P. et al., *Ukr. khim. zhurn.*, 63(2), 92, 1997.)

ZnS–"P$_2$S$_4$": The phase diagram is shown in Figure 1.29 (Kovach et al. 1997). The eutectic contains 12.5 mol. % "P$_2$S$_4$" and crystallizes at 600°C±5°C. The Zn$_2$P$_2$S$_6$ (ZnPS$_3$) compound was synthesized in this section (Klinger et al. 1970, 1973, Brec et al. 1980, Prouzet et al. 1986, Kovach et al. 1997). It melts incongruently at 656°C±5°C and crystallizes in a monoclinic structure with lattice parameters $a = 598 \pm 3$, $b = 1023 \pm 5$, $c = 672.1 \pm 0.5$ pm, and $\beta = 107.1°$ (Kovach et al. 1997) [$a = 597.17$, $b = 1034.24$, $c = 675.65$ pm, and $\beta = 107.139°$ (Brec et al. 1980, Prouzet et al. 1986); $a = 596$, $b = 1028$, $c = 673$ pm, and $\beta = 107.1°$ (Klinger et al. 1970, 1973)] and calculation and experimental density 3.24 [3.207 (Prouzet et al. 1986)] and 3.18 g cm^{-3} (Klinger et al. 1973), respectively.

ZnS–P$_2$S$_5$: The phase diagram is not constructed. Zn$_3$(PS$_4$)$_2$ ternary compound was synthesized in this system (Soklakov and Nechaeva 1970, Hozhainov et al. 1972). It crystallizes in a cubic structure with lattice parameter $a = 1295 \pm 7$ pm (Soklakov and Nechaeva 1970) and energy gap $E_g = 3.3$ eV (Hozhainov et al. 1972). This compound was obtained by alloying ZnS and P$_2$S$_5$ (Soklakov and Nechaeva 1970).

1.36 Zinc–Arsenic–Sulfur

There are nine fields of primary phase crystallization on the Zn–As–S liquidus surface: ZnS and α- and β-solid solutions based on Zn$_3$As$_2$, ZnAs$_2$, As, As$_2$S$_2$, As$_2$S$_3$, S, and Zn (Figure 1.30) (Olekseyuk and Tovtin 1987). The greatest of them is the field of ZnS primary crystallization. These fields are

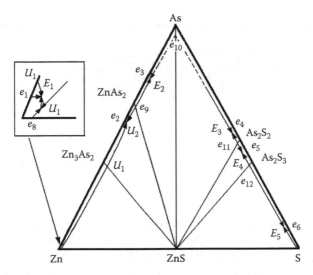

FIGURE 1.30
The scheme of liquidus surface of Zn–As–S ternary system. (From Olekseyuk, I.D. and Tovtin, N.A., *Zhurn. neorgan. khimii*, 32(10), 2562, 1987.)

TABLE 1.3

Invariant Equilibria in the Zn–As–S Ternary System

Symbol	Reaction	t, °C
E_1	$L \Leftrightarrow ZnS + \beta\text{-}Zn_3As_2 + Zn$	417
E_2	$L \Leftrightarrow ZnS + ZnAs_2 + As$	710
E_3	$L \Leftrightarrow ZnS + As_2S_2 + As$	290
E_4	$L \Leftrightarrow ZnS + As_2S_3 + As_2S_2$	275
E_5	$L \Leftrightarrow ZnS + As_2S_3 + S$	100
U_1	$L + \beta\text{-}Zn_3As_2 \Leftrightarrow \alpha\text{-}Zn_3As_2 + Zn$	418
U_2	$L + ZnS \Leftrightarrow ZnAs_2 + \beta\text{-}Zn_3As_2$	760

Source: Olekseyuk, I.D. and Tovtin, N.A., *Zhurn. neorgan. khimii*, 32(10), 2562, 1987.

divided by 15 lines of monovariant equilibria. There are also seven invariant equilibria in this system (Table 1.3). The quasibinary sections in this system are ZnS–Zn_3As_2, ZnS–$ZnAs_2$, ZnS–As, ZnS–As_2S_2, and ZnS–As_2S_3 (Olekseyuk and Tovtin 1987).

Isothermal section at 230°C is characterized by existence in general of three-phase regions (Figure 1.31) (Olekseyuk and Tovtin 1987). One-phase region is only on the Zn_3As_2-rich side. The small glassy field forms on the As_2S_3 side and contains nearly 0.2 at. % Zn.

ZnS–ZnAs₂: The phase diagram is a eutectic type (Olekseyuk and Tovtin 1987). The eutectic composition and temperature are 0.5 mol. % ZnS and 762°C, respectively.

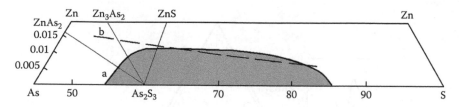

FIGURE 1.31
The region of glass formation at the ingots quenching from 400°C to 450°C (a) and the liquidus isotherm at 400°C (b) in the Zn–As–S ternary system. (From Olekseyuk, I.D. et al., *Kvantovaya elektronika*, (13), 93, 1977; Olekseyuk, I.D. and Tovtin, N.A., *Zhurn. neorgan. khimii*, 32(10), 2562, 1987.)

This system was investigated through DTA, metallography, XRD, and measuring of microhardness. The ingots were annealed at 300°C for 650 h and then at 240°C for 250 h (Olekseyuk and Tovtin 1987).

ZnS–Zn₃As₂: The phase diagram is a peritectic type (Figure 1.32) (Olekseyuk and Tovtin 1987). The peritectic composition and temperature are 10.5 mol. % ZnS and 1035°C, respectively. There is a eutectoid transformation at 610°C.

The solubility of ZnS in β-Zn₃As₂ constitutes 16 mol. % at 1035°C, 9 mol. % at 800°C and 7 mol. % at 610°C. The solution region based on Zn₃As₂ diminishes from 4 mol. % ZnS at 610°C to 1 mol. % ZnS at 240°C. Golovey et al. (1972) did not determine the solid solutions based on ZnS and Zn₃As₂ in this system.

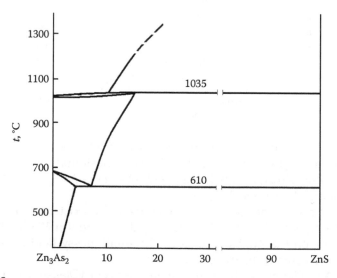

FIGURE 1.32
The ZnS–Zn₃As₂ phase diagram. (From Olekseyuk, I.D. and Tovtin, N.A., *Zhurn. neorgan. khimii*, 32(10), 2562, 1987.)

This system was investigated through DTA, metallography, XRD, and measuring of microhardness. The ingots were annealed at 300°C for 650 h and then at 240°C for 250 h (Olekseyuk and Tovtin 1987).

ZnS–As: The phase diagram is a eutectic type (Olekseyuk and Tovtin 1987). The eutectic composition and temperature are 0.7 mol. % ZnS and 786°C, respectively.

This system was investigated through DTA, metallography, XRD, and measuring of microhardness. The ingots were annealed at 300°C for 650 h and then at 240°C for 250 h (Olekseyuk and Tovtin 1987).

ZnS–As$_2$S$_2$: The phase diagram is a eutectic type (Olekseyuk and Tovtin 1987). The eutectic composition and temperature are 0.9 mol. % ZnS and 302°C, respectively.

This system was investigated through DTA, metallography, XRD, and measuring of microhardness. The ingots were annealed at 300°C for 650 h and then at 240°C for 250 h (Olekseyuk and Tovtin 1987).

ZnS–As$_2$S$_3$: The phase diagram is a eutectic type (Olekseyuk and Tovtin 1987). The eutectic composition and temperature are 3 mol. % ZnS and 300°C, respectively. The ingots of this system were not in equilibrium. Small thermal effects were observed at 207°C, which belonged to vitreous As$_2$S$_3$.

This system was investigated through DTA, metallography, XRD, and measuring of microhardness. The ingots were annealed at 300°C for 650 h and then at 240°C for 250 h (Olekseyuk and Tovtin 1987).

ZnS–As$_2$S$_5$: The phase diagram is not constructed. Effects at 130°C–140°C belonged to vitreous As$_2$S$_5$ and, at 420°C corresponding to Zn melting, were found on thermograms of ingots with ZnAs$_2$S$_6$ composition (Luzhnaya et al. 1966). It is evident that at the alloying of Zn, S, and As$_2$S$_5$ in the proportion, corresponding to the ZnAs$_2$S$_6$, the starting materials do not interact.

1.37 Zinc–Antimony–Sulfur

ZnS–Sb$_2$S$_3$: The phase diagram is not constructed. The ZnS solubility in liquid Sb$_2$S$_3$ was studied (Figure 1.33) (Babanski et al. 1980).

1.38 Zinc–Bismuth–Sulfur

ZnS–Bi: The phase diagram is not constructed. The ZnS solubility in liquid bismuth was studied (Figure 1.34) and a part of Zn–Bi–S liquidus surface on the Bi-rich side (Figure 1.35) (Rubenstein 1966, Kimura and Panish 1970).

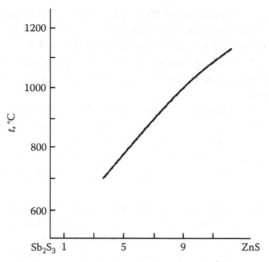

FIGURE 1.33

Temperature dependence of ZnS solubility in liquid Sb_2S_3. (From Babanski, M.D. et al., Sphalerite crystallization in solution-melts [in Russian], *6th Mezhdunarodn. konf. po rostu kristallov, Moscow, Russia, September 10–16, 1980, Rassh. tez., V. 3, Rost iz rasplavov i vysokotemperatur. rastvorov (Metody, materialy)*, Moscow, Russia, PIK VINITI, pp. 236–237, 1980.)

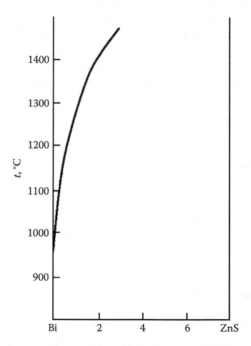

FIGURE 1.34

Temperature dependence of ZnS solubility in liquid Bi. (From Rubenstein, M., *J. Electrochem. Soc.*, 113(6), 623, 1966; Kimura, S. and Panish, M.B., *J. Chem. Thermodyn.*, 2(1), 77, 1970. With permission.)

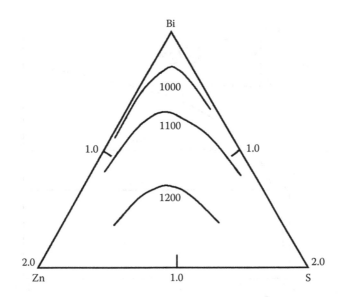

FIGURE 1.35
Part of the Zn-Bi-S liquidus surface on the Bi-rich side. (From Kimura, S. and Panish, M.B., *J. Chem. Thermodyn.*, 2(1), 77, 1970. With permission.)

The solubility of ZnS in liquid Bi at 1000°C, 1100°C, 1200°C, and 1300°C is equal to 0.17, 0.37, 0.80, and 1.22 mol. % correspondingly (Rubenstein 1966). The ternary eutectic in the Zn–Bi–S system is practically degenerated, and the field of ZnS primary crystallization occupies almost all liquidus surface (Kimura and Panish 1970).

The temperature dependence of ZnS solubility in liquid Bi was determined by high-temperature filtration. The mixtures of ZnS and Bi were annealed at low temperature for 16 h and at 1300°C for 6 h. The quantity of ZnS dissolved in Bi was determined by weighting of the ingots before and after its treatment (Rubenstein 1966). Liquidus temperatures in the Zn–Bi–S ternary system were determined by visual observation of solid phase appearance and disappearance for heating and cooling of ingots (accuracy of measurement is ±5°C) (Kimura and Panish 1970).

ZnS–Bi$_2$S$_3$: The phase diagram is shown in Figure 1.36 (Odin and Marugin 1991). Liquidus and solidus have maximum at 3.5±0.5 mol. % ZnS and 792°C. The eutectic composition and temperature are 6±1 mol. % ZnS and 774°C, respectively. Thermal effects at 1050°C are determined by polymorphous transition α-ZnS → β-ZnS. The solubility of ZnS in Bi$_2$S$_3$ constitutes 3.7±0.3 mol. % at 670°C and 2.8±0.3 mol. % at 500°C, and the solubility of Bi$_2$S$_3$ in α-ZnS is equal to less than 0.5 mol. % at 670°C.

This system was investigated through DTA, metallography, and XRD. The ingots were annealed 670°C for 500 h and next at 500°C for 720 h (Odin and Marugin 1991).

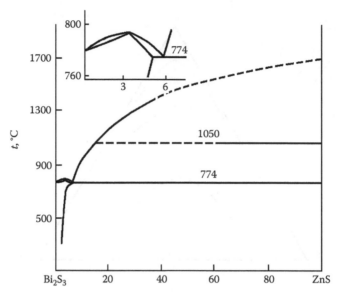

FIGURE 1.36
The ZnS–Bi$_2$S$_3$ phase diagram. (From Odin, I.N. and Marugin, V.V., *Zhurn. neorgan. khimii*, 36(7), 1842, 1991.)

1.39 Zinc–Oxygen–Sulfur

ZnS–ZnO: The phase diagram is not constructed. The solubility of ZnO in ZnS, determined through XRD and chemical analysis, constitutes correspondingly 2.38, 1.96, and 1.58 mol. % at 1200°C, 1100°C, and 1000°C (accuracy of measurement is ±0.05 mol. %) (Chechiotkina et al. 1978) [according to the data of Kröger and Dikhoff (1952), the mutual solubility of ZnS and ZnO at 1200°C is equal to 1 mol. % ZnO and 0.3 mol. % ZnS]. The addition of 0.8 mol. % (0.7 mass. %) ZnO to sphalerite reduces the unit cell edge from 540.93 ± 0.02 to 540.65 ± 3 pm (Skinner and Barton 1960).

There is (1–2) 10^{-2} mol. % ZnO in the ZnS single crystals, obtained using Bridgman method (Zuyev et al. 1981). The primary product of ZnS oxidation is ZnO, and the processes of ZnS thermal dissociation and sulfate formation are insignificant (Kozhahmetov et al. 1982, Kozhahmetov et al. 1983).

ZnS–ZnSO$_4$: ZnS and ZnSO$_4$ interact at heating in two independent stages (Malinowski 1992):

I: $11ZnSO_4 + ZnS = 4(ZnO \cdot 2ZnSO_4) + 4SO_2$

II: $3(ZnO \cdot 2ZnSO_4) + ZnS = 11ZnO + 8SO_2$

The studies carried out on the preparation of different mole ratio of $ZnSO_4$ and ZnS have shown that with an increase in this ratio to the value 3:1, a decrease in the ZnS oxidation rate is observed. An increase in the ZnS oxidation rate can again be observed above this value.

This system was investigated through DTA, TGA, XRD, and chemical analysis (Malinowski 1992).

1.40 Zinc–Selenium–Sulfur

ZnS–ZnSe: The phase diagram was estimated in Tomashik (1981) (Figure 1.37). It was determined that deviations from Raul's law in this system are small and this system is almost ideal (Koreniev et al. 1972). Zn atoms and S_2, Se_2, SSe molecules were determined in vapor phase over ZnS_xSe_{1-x} solid solutions. The dissociative saturation pressure over such solid solutions at 722°C and 828°C is shown in Figure 1.38 (Timoshin et al. 1975).

Using shock compression of the ZnS + Se mixtures, continuous series of ZnS_xSe_{1-x} solid solutions have been obtained (Batsanov et al. 1982). Single crystals of the solid solutions have been grown in all range of concentrations (Andreev et al. 1975). Concentration of stacking defects in ZnS_xSe_{1-x} single crystals is higher than in ZnS and ZnSe, and its maximum corresponds to composition with 37.5 mol. % ZnSe (Atroshchenko et al. 1977). Concentration dependence of lattice parameters, energy gap,

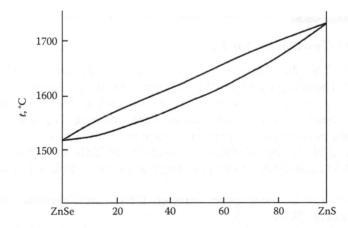

FIGURE 1.37
The ZnS–ZnSe phase diagram. (From Tomashik, V.N., *Izv. AN SSSR. Neorgan. materialy*, 17(6), 1116, 1981.)

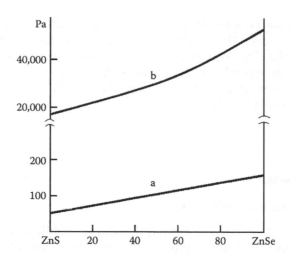

FIGURE 1.38
Dissociative saturation pressure over ZnS_xSe_{1-x} solid solutions at 722°C (a) and 828°C (b). (From Timoshin, I.A. et al., *Zhurn. phys. khimii*, 49(4), 1039, 1975.)

and microhardness are typical for the system with continuous solid solutions (Berger and Petrov 1969, Atroshchenko et al. 1977). The existence of 2H, 3C, 4H, 6H, and 10H polytypes was determined through XRD in the ZnS_xSe_{1-x} solid-solution single crystals (Palosz and Przedmojski 1976, Michalski et al. 1980, Paszkowicz et al. 1996).

1.41 Zinc–Tellurium–Sulfur

ZnS–ZnTe: The phase diagram is a eutectic type (Figure 1.39) (Tomashik et al. 1978). The eutectic composition and temperature are 26 mol. % ZnS and 1260°C, respectively. Solid solutions based on ZnS at 1000°C crystallize in sphalerite structure and at 1100°C in wurtzite structure. Transition sphalerite–wurtzite in two-phase region takes place at 1070°C ± 10°C.

According to the data of XRD, the solubility of ZnTe in ZnS and ZnS in ZnTe at 900°C constitutes correspondingly 8–10 and 5 mol. % (Larach et al. 1956, 1957).

This system was investigated through DTA, metallography, and XRD (Tomashik et al. 1978).

ZnS–Te: The phase diagram is not constructed. The solubility of ZnS in liquid tellurium at 950°C–1200°C was determined by the saturation of liquid Te with zinc sulfide (Figure 1.40) (Masaoko et al. 1979, Masaharu et al. 1980).

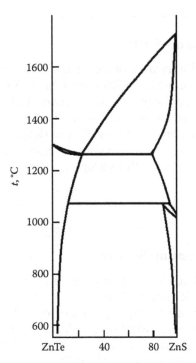

FIGURE 1.39
The ZnS–ZnTe phase diagram. (From Tomashik, V.N. et al., *Izv. AN SSSR. Neorgan. materialy*, 14(8), 1434, 1978.)

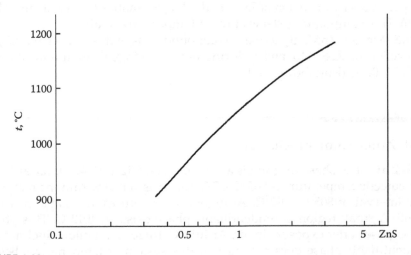

FIGURE 1.40
Temperature dependence of ZnS solubility in liquid Te. (From Masaoko, W. et al., *Jap. J. Appl. Phys.*, 18(5), 869, 1979; Masaharu, A., *J. Fac. Eng. Univ. Tokyo*, A(18), 38, 1980.)

1.42 Zinc–Chromium–Sulfur

ZnS–CrS: The phase diagram is not constructed. $ZnCr_2S_4$ ternary compound does not form at the interaction of ZnS and CrS at 950°C (Lutz et al. 1971).

ZnS–Cr$_2$S$_3$: The phase diagram is not constructed. $ZnCr_2S_4$ ternary compound is formed in this system, which crystallizes in a cubic structure of spinel type with lattice parameter $a = 998.6$ pm (Raccah et al. 1966).

$ZnCr_2S_4$ can be prepared by direct combination of the elements or by the reaction of zinc oxide spinel with hydrogen sulfide at elevated temperature (Raccah et al. 1966).

1.43 Zinc–Molybdenum–Sulfur

Ternary compound $Zn_xMo_{15}S_{19}$ (Tarascon and Hull 1986) and Chevrel phases $ZnMo_6S_8$ and $Zn_2Mo_6S_8$ (Gocke et al. 1987) exist in this ternary system. $ZnMo_6S_8$ and $Zn_2Mo_6S_8$ crystallize in a rhombohedral structure with the lattice parameters $a = 648.1$ pm, $\alpha = 94.56°$ and $a = 648.6$ pm, $\alpha = 94.58°$, respectively (in a hexagonal structure, $a = 952.4$, $c = 1029.4$ pm, and $a = 952.4$, $c = 1029.4$ pm, respectively) (Gocke et al. 1987).

AQ9　　$ZnMo_6S_8$ and $Zn_2Mo_6S_8$ were obtained via cathodic reduction in galvanic cells Zn/Zn^{2+} and CH_3CN (acetonitrile)/Mo_6S_8 (Gocke et al. 1987). In anodic oxidation of $Zn_2Mo_6S_8$, the compounds $ZnMo_6S_8$ and Mo_6S_8 appear again. $Zn_2Mo_6S_8$ undergoes decomposition upon heating at higher temperatures to products that have not been identified. $Zn_xMo_{15}S_{19}$ was prepared at low temperatures by diffusion of Zn into the $Mo_{15}S_{19}$ matrix (Tarascon and Hull 1986). This compound is stable at room temperature in air.

ZnS–Mo$_2$S$_3$: $ZnMo_2S_4$ ternary compound was not synthesized at the interaction of ZnS, Mo, and S during 6-day homogenization annealing at 600°C–1000°C (Kuleshov et al. 1981).

1.44 Zinc–Fluorine–Sulfur

ZnS–ZnF$_2$: The phase diagram is a eutectic type (Figure 1.41) (Linares 1963). The eutectic temperature is 805°C ± 5°C. α-ZnS is stable within the temperature interval of 805°C–1012°C. At higher temperatures, the stable phase is β-ZnS. A small region of unidentifiable phase exists at 1012°C. This phase decomposes after exposure in air. At 1012°C, there is a liquidus inclination. Unidentifiable phase does not exist in this system at temperatures below 1012°C. Apparently this phase melts incongruently.

This system was investigated through DTA and XRD (Linares 1963).

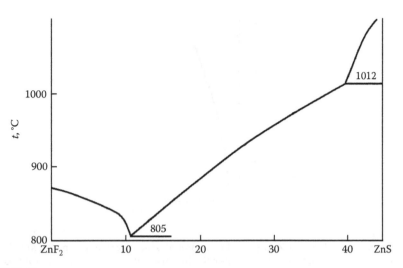

FIGURE 1.41
The $ZnS-ZnF_2$ phase diagram. (From Linares, R.C., Phase equilibrium and crystal growth in the system $ZnS-ZnF_2$, in *Metallurgy Advanced Electron. Material*, Interscience, London, U.K., pp. 329–334, 1963.)

1.45 Zinc–Chlorine–Sulfur

$ZnS-ZnCl_2$: The phase diagram is not constructed. The solubility of ZnS in liquid zinc chloride is shown in Figure 1.42 (Babanski et al. 1980).

$Zn_6S_5Cl_2$ ternary compound is formed in this system (Chen et al. 2010). It crystallizes in an orthorhombic structure with the lattice parameters $a = 1829.1 \pm 0.1$, $b = 918.7 \pm 0.4$, and $c = 927.2 \pm 0.7$ pm; calculation density 2.657 g cm^{-3}; and optical bandgap 2.71 eV. This compound is thermally stable up to 220°C and undergoes a three-step decomposition process.

$Zn_6S_5Cl_2$ could be prepared from the reaction of $ZnCl_2$, Zn, and S in vacuum at 500°C during 12 h (Chen et al. 2010).

1.46 Zinc–Iodine–Sulfur

$ZnS-I_2$: The phase diagram is not constructed. Gordeev and Karelin (1969) determined thermodynamic parameters of ZnS interaction with I_2 investigating the temperature dependence of a total pressure in this system. Probably the interaction in the $ZnS-I_2$ system can be described as follows:

$$ZnS + I_2 = ZnI_2 + \frac{1}{2}S_2$$

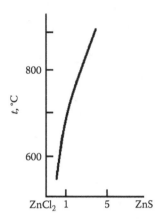

FIGURE 1.42
Temperature dependence of ZnS solubility in liquid zinc chloride. (From Babanski, M.D. et al., Sphalerite crystallization in solution-melts [in Russian], *6th Mezhdunarodn. konf. po rostu kristallov. Moscow, Russia, September 10–16, 1980. Rassh. tez. V. 3, Rost iz rasplavov i vysokotemperatur. rastvorov (Metody, materialy)*, Moscow, Russia, PIK VINITI, pp. 236–237, 1980.)

1.47 Zinc–Manganese–Sulfur

ZnS–Mn: The phase diagram is not constructed. Diffusion introduction of Mn in ZnS at 1170°C leads to transformation of ZnS cubic structure in hexagonal structure (Arhangel'ski et al. 1973). The annealing of the ZnS crystals with sphalerite structure at the same temperature does not lead to formation of hexagonal phase.

ZnS–MnS: The phase equilibria in this system have been discussed in Pajaczkowska (1978) based on some papers that have appeared up to 1977. The phase diagram is shown in Figure 1.43 (Neuhaus and Steffen 1970, Sombuthawee et al. 1978, Tauson and Chernyshev 1981). Phase relations in this system were investigated through XRD using ingots annealed at 310°C–1330°C for 15 min to 1 month (Sombuthawee et al. 1978). It was determined that ZnS dissolves 5 mol. % MnS at 630°C and solid solutions based on MnS contain 8 mol. % ZnS at 1050°C. Solid solutions with wurtzite structure at 1050°C contains 60 mol. % MnS (52 mol. % MnS at 1180°C Kröger 1938). These solid solutions are stable in a wide range of temperatures. The eutectiode composition and temperature are 40 mol. % MnS and 310°C (Sombuthawee et al. 1978) [26 mol. % MnS and 350°C at 1000 atm (100 MPa) (Tauson and Chernyshev 1981)].

According to the data of Juza et al. (1956), the solubility of MnS in ZnS at 600°C is equal to 43 mol. %, and such solid solutions crystallize in rock-salt structure. Kaneko et al. (1983) indicate that solid solubility of MnS in ZnS is equal to 50 mol. % and is almost independent of temperature, whereas ZnS is not dissolved into MnS at all. The mixtures of ZnS and MnS of various molar ratios were hydrothermally treated at 500°C or 600°C and 50 MPa (500 atm) for 24 h.

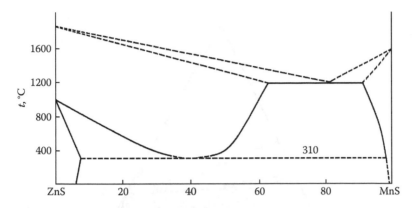

FIGURE 1.43
The ZnS–MnS phase diagram. (From Sombuthawee, C. et al., *J. Solid State Chem.*, 25(4), 391, 1978; Tauson, V.L. and Chernyshev, L.V., *Novosibirsk, Nauka Publish.*, 190, 1981.)

Solid solutions, obtained by coprecipitation from water solutions, crystallize in sphalerite structure. The limit solubility of MnS in ZnS is 10–11 mol. % (Juza et al. 1956, Blaton and Glibert 1975). Single crystals of the $Zn_xMn_{1-x}S$ solid solutions with $0 < x < 0.5$ were grown by chemical transport (Nitsche 1971, Giriat and Stankiewicz 1985). The crystal growth of the $Zn_{1-x}Mn_xS$ solid solutions is also given in Pajaczkowska (1978).

1.48 Zinc–Iron–Sulfur

Scheme of the Zn–Fe–S isothermal section at 580°C was constructed by Moh (1975) using existing experimental results (Figure 1.44). Phase equilibria in this ternary system at 580°C–850°C were also investigated by Barton and Toulmin (1966). At 850°C, a single-phase field of sphalerite does not deviate from ZnS–FeS section. Changes in the ratio (Fe + Zn):S = 1:1 by more than 0.1% of one of the components leads to a new phase appearance.

ZnS–FeS: The phase diagram is a eutectic type (Figure 1.45) (Murach 1947, Novoselov 1955, Avetisian and Gnatyshenko 1956, Dutrizac 1980). The eutectic composition and temperature are 10 mol. % ZnS and 1178°C, respectively (Dutrizac 1980) [3.6 mol. % ZnS and 1165°C (Novoselov 1955), 6.4 mol. % ZnS and 1180°C (Murach 1947), 6 mol. % ZnS and 1180°C (Avetisian and Gnatyshenko 1956)]. The solubility of FeS in ZnS at 1165°C is equal to 43 mol. %, and the solubility of ZnS in FeS at the same temperature is insignificant (Dutrizac 1980).

Phase relations involving sphalerite in the Zn–Fe–S system have been estimated below 300°C by combining the experimental data with EPMA of sphalerites from naturally occurring low-temperature assemblages (Scott and Rissin 1973). These results showed that sphalerite composition varies

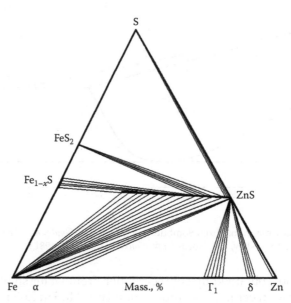

FIGURE 1.44
Isothermal section of the Zn–Fe–S system at 580°C. (From Moh, G.H., *Chem. Erde*, 34(1), 1, 1975.)

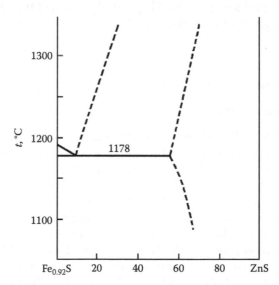

FIGURE 1.45
ZnS–Fe$_{0.92}$S phase diagram. (From Dutrizac, J.E., *Can. J. Chem.*, 58(7), 739, 1980.)

in response to a_{FeS} in similar manner to its behavior at higher temperatures producing lower FeS contents in sphalerites in the more S-rich iron sulfide assemblages.

Nonstoichiometric ternary chalcogenides, $Zn_{1-x}Fe_xS$, were prepared by coprecipitation of ZnS and FeS by Na_2S from aqueous solutions containing

$FeSO_4$ and $ZnSO_4$ and sintering of pellets of the coprecipitate repeatedly between 800°C and 1000°C (Deulkar et al. 2003). The chemical composition of the resulting solids as analyzed using x-ray fluorescence (XRF), energy dispersive spectrometry (EDAX), and EPMA reflected the composition of the solutions from which precipitation was carried out, with Fe contents up to $x = 0.4$. XRD indicated the presence of solid solutions, $Zn_{1-x}Fe_xS$ (sphalerite). Electrical resistivity studies indicated that the compounds were semiconducting. From the temperature dependence of the resistivity, thermal bandgaps were found to be decreasing with increasing Fe content, in agreement with values on optical bandgaps as estimated from diffuse optical reflectance measurements as a function of wavelength.

It was determined that at $pH = 5$ and 2.5, solid solutions based on ZnS and obtained by coprecipitation from water solutions contain 2 and 15 mol. % FeS, respectively (Blaton and Glibert 1975).

This system was investigated through DTA, metallography, and XRD (Dutrizac 1980).

1.49 Zinc–Cobalt–Sulfur

$ZnS–Co_{1-x}S$: The phase diagram is not constructed. It was determined that at $pH = 5$, solid solutions based on ZnS contain 3 mol. % $Co_{1-x}S$ and based on $Co_{1-x}S$ contain 0.4 mol. % ZnS (Blaton and Glibert 1975, 1976). At $pH = 4$, the solubility of $Co_{1-x}S$ in ZnS and ZnS in $Co_{1-x}S$ is equal correspondingly 8 and 0.5 mol. %. According to the data of XRD, the limit solubility of CoS in ZnS at 1000°C reached 41 mol. % [36.4 mol. % at 950°C (Niu et al. 1990)], and the solubility of ZnS in CoS is insignificant (Becker and Lutz 1978). The concentration dependence of the lattice parameters corresponds to Vegard's law.

The ingots of this system were obtained by coprecipitation from the water solutions at 25°C (Blaton and Glibert 1975, 1976). $Zn_xCo_{1-x}S$ single crystals with x up to about 0.285 (Paszkowicz et al. 1998) were grown by chemical vapor transport method using iodine as the transporting agent (Niu et al. 1990, Paszkowicz et al. 1998).

1.50 Zinc–Nickel–Sulfur

$ZnS–Ni_{1-x}S$: The phase diagram is not constructed. It was determined that at $pH = 4$, solid solutions based on ZnS contain 1 mol. % $Ni_{1-x}S$ and based on $Ni_{1-x}S$ contain 0.4 mol. % ZnS (Blaton and Glibert 1975, 1976).

The ingots of this system were obtained by coprecipitation from aqueous solutions at 25°C (Blaton and Glibert 1975, 1976).

References

Aliev K.A. ZnS–Gd$_2$S$_3$ system [in Russian], *Zhurn. neorgan. khimii*, **48**(11), 1902–1903 (2003).

Aliev O.M., Aliev K.A., Gadzhiev S.V. Interaction in ZnS–Sm$_2$S$_3$ and Zn–Sm$_2$S$_3$ systems [in Russian], *Zhurn. neorgan. khimii*, **36**(10), 2628–2630 (1991a).

Aliev O.M., Gadzhiev S.M., Aliev K.A. The ZnS–Dy$_2$S$_3$ system [in Russian], *Zhurn. neorgan. khimii*, **36**(12), 3194–3197 (1991b).

Aliev O.M., Gadzhiev S.M., Aliev K.A. The ZnS–Nd$_2$S$_3$ system [in Russian], *Zhurn. neorgan. khimii*, **37**(12), 2792–2794 (1992).

Aliev K.A., Mamedov G.G. Investigation of the phase equilibria in the ZnS–La$_2$S$_3$ system [in Azerbaijani], in: *6th Resp. konf. Fiz.-khim. analiz i neorgan. materialoved.* Baku, Azerbaijan: Baky Univ. Publishing, pp. 75–78 (2000).

Anagnostopoulos A., Kambas K., Spyridelis J. On the optical and electrical properties of the Zn$_3$In$_2$S$_6$ layered compound, *Mater. Res. Bull.*, **21**(4), 407–413 (1986).

Andreeva G.A., Bochkareva L.G., Fed'kovich L.N. X-ray investigations of ZnS–CdS solid solutions [in Russian], in: *Fizika tverdogo tela, Irkutsk.* Irkutsk, Russia: Polytehn. in-t, pp. 76–79 (1973).

Andreev A.A., Vukelich S.B., Kodzhespirov F.F. Obtaining and physico-chemical properties of ZnS$_x$Se$_{1-x}$ single crystals [in Russian], *Struktura zhidkosti i fazovyie perehody, vyp.* **3**, 70–73 (1975).

Arhangel'ski G.E., Voznesenskaya T.I., Fok M.V. Influence of manganese on transformation of cubic zinc sulfide in hexagonal structure [in Russian], *Kristallografiya*, **18**(3), 544–547 (1973).

Atroshchenko L.V., Brintsev F.I., Sarkisov L.A., Sysoev L.A. Stabilization of zinc sulfide hexagonal modification by the doping of donor impurities [in Russian], *Izv. AN SSSR. Neorgan. materialy*, **8**(4), 639–640 (1972).

Atroshchenko L.V., Kolodiazhnyi A.I. Structure and microhardness anisotropy of ZnS single crystals doped by the silver [in Russian], *Izv. AN SSSR. Neorgan. materialy*, **13**(3), 426–429 (1977).

Atroshchenko L.V., Kukol V.V., Lakin E.E., Nosachev B.G., Sysoev L.A. Growth from the melt and investigation of the crystal structure, defect and phase transformation of pure, doped and mixed A^{II}B^{VI} single crystals [in Russian], *Rost i legirovanie poluprovodn. kristallov i plenok*, Pt. 2, 298–301 (1977).

Atroshchenko L.V., Obuhova N.F., Pisarevski Y.V., Sil'vestrova I.M., Sysoev L.A. The regularity of properties change for ZnS–MgS, CdS–MgS and CdSe–MgSe solid solution single crystals [in Russian], in: *Fiz. slozhn. poluprovodn. soedin.* Kishinev, Moldova: Shtiinza, pp. 189–198 (1979).

Avetisian H.K., Gnatyshenko G.I. Thermal and metallographic investigation of PbS–ZnS–FeS system [in Russian], *Izv. AN KazSSR. Ser. gorn. dela, stroimaterialov i metallurgii*, (6), 11–25 (1956).

Babanski M.D., Stroitelev A.D., Hahlov A.V. Sphalerite crystallization in solution-melts [in Russian], *6th Mezhdunarodn. konf. po rostu kristallov. Moscow, Russia, 10–16 September, 1980. Rassh. tez. V. 3. Rost iz rasplavov i vysokotemperatur. rastvorov (Metody, materialy)*, Moscow, PIK VINITI, 236–237 (1980).

Barnett D.E., Bookman R.S., Sutherland J.K. New data on ternary phases in the system Zn-In- S, *Phys. status solidi (a)*, **4**(1), K49–K51 (1971).

Barton P.B., Tulmin P. Phase relations involving sphalerite in the Fe–Zn–S system, *Econ. Geol.*, **61**(5), 815–849 (1966).

Batsanov S.S., Kopaneva L.I., Lazareva E.V. Shock synthesis of the ZnS–ZnSe solid solutions [in Russian], *Zhurn. neorgan. khimii*, **27**(4), 1068–1070 (1982).

Becker W., Lutz H.D. Phase studies in the systems CoS–MnS, CoS–ZnS and CoS–CdS, *Mater. Res. Bull.*, **13**(9), 907–911 (1978).

Berger L.I., Petrov V.M. Analysis of the solid solutions of the sulfide zinc–selenide zinc system by the method of the diffuse reflection spectra [in Russian], *Tr. Vses. n.-i. in-ta. khim. reaktivov i osobo chist. khim. veshchestv*, (31), 424–426 (1969).

Berthold H.J., Köhler K. Präparative und röntgenographische Unterszchungen über das system Al_2S_3–ZnS (Temperaturbereich 800–1080°C), *Z. anorg. und allg. Chem.*, **475**(4), 45–55 (1981).

Berthold H.J., Köhler K., Wartchow R. Kristallstrukturverfeinerung des Zinkaluminium-sulfides $ZnAl_2S_4$ (normale Spinellstruktur) mit Röntgen-Einkristalldaten, *Z. anorg. und allg. Chem.*, **496**(1), 7–20 (1983).

Beun J.A., Nitsche R., Lichtensteiger M.L. Photoconductivity in ternary sulfides, *Physica*, **26**(8), 647–649 (1960).

Beun J.A., Nitsche R., Lichtensteiger M. Optical and electrical properties of ternary chalkogenides, *Physica*, **27**(5), 448–452 (1961).

Biyushkina A.V., Donika F.G., Mustia I.G., Radautsan S.I. Crystal structure of $Zn_3In_2S_6(I)_b$ one-packet polytype [in Russian], *Izv. AN MSSR. Ser. fiz.-tehn. i mat. nauk*, (3), 68–70 (1989a).

Biyushkina A.V., Donika F.G., Radautsan S.I. Crystal structure of $ZnIn_2S_4(VI)_a$ six-packet polytype [in Russian], *Dokl. AN SSSR*, **306**(3), 617–619 (1989b).

Blaton N., Glibert J. Contribution à l'étude des sulfures mixtes $ZnS–Co_{1-x}S$, $ZnS–Fe_{1-x}S$ et ZnS–MnS, *Bull. Soc. Chim. Belg.*, **84**(11), 1131–1138 (1975).

Blaton N., Glibert J. Contribution à l'étude des sulfures mixtes $ZnS–Co_{1-x}S$ et ZnS–$Ni_{1-x}S$, *Bull. Soc. Chim. France*, (1/2), pt 1, 23–27 (1976).

Bol'shakov A.F., Dmitrienko A.O., Abalduyev B.V. Polymorphic transformation in the polycrystalline ingots of the ZnS–CdS system [in Russian], *Izv. AN SSSR. Neorgan. materialy*, **15**(9), 1528–1531 (1979).

Bonsall S.B., Hummel F.A. Phase equilibria in the systems $ZnS–Al_2S_3$ and $ZnAl_2S_4$–$ZnIn_2S_4$, *J. Solid State Chem.*, **25**(4), 379–386 (1978).

Boorman R.S., Sutherland J.K. Subsolidus phase relations in the $ZnS–In_2S_3$ system: 600 to 1080°C, *J. Mater. Sci.*, **4**(8), 658–671 (1969).

Bouchetiere M., Toffoli P., Khodadad P. Structure cristalline de $Zn_4(P_2S_6)_3$, *Acta Crystallogr.*, **B34**(2), 384–387 (1978).

Brec R., Ouvrard G., Louisy A., Rouxel J. Propriétés structurales de phases $M^{2+}PX_3$ (X = S, Se), *Ann. Chim. (France)*, **5**(6), 499–512 (1980).

Brightwell J.W., Ray B., White S. Phase structure of $Zn_{1-x}Mg_xS$ prepared by reaction of mixed chlorides with H_2S, *J. Mater. Sci. Lett.*, **3**(11), 951–954 (1984).

Bundel' A.A., Vishniakov A.V., Zubkovskaya V.N. The solubility of Cu_2S in ZnS and CdS [in Russian], *Izv. AN SSSR. Neorgan. materialy*, **6**(7), 1248–1251 (1970).

Bundel' A.A., Vishniakov A.V., Zubkovskaya V.N., Klykova A.I. Solubility of zinc sulfide in lead sulfide in solid state [in Russian], *Tr. Mosk. khim.-tehnol. in-ta im. D.I.Mendeleeva*, vyp. **62**, 87–89 (1969).

Bundel' A.A., Zhukova V.N. Investigation of the equilibrium vapor phase composition above ZnS–CdS solid solutions at a calcination in hydrogen [in Russian], *Tr. Mosk. khim.-tehnol. in-ta im. D.I.Mendeleeva*, (54), 19–22 (1967).

Bundel' A.A., Zhukova V.N., Zhukov G.V. Investigation of the vapor phase composition above ZnS–CdS solid solutions [in Russian], *Zhurn. phys. khimii*, **41**(7), 1770–1774 (1967).

Carpenter G.B., Wu P., Gao Y.-M., Wold A. Redetermination of crystal structure of zinc thiogallate, *Mater. Res. Bull.*, **24**(9), 1077–1082 (1989).

Charbonnier M. Contribution à l'étude des sulfures simples et mixtes des métaux des groupes I B (Cu–Ag) et II B (Zn–Cd–Hg): These doct. sci. phys. Univ. Claude-Bernard. Lyon, 164pp (1973).

Charbonnier M., Murat M. Sur la détermination des diagrammes de phases à température ambiante des sulfures mixtes appartenant aux systèmes Zn–Cd–S, Zn–Hg–S, Cd–Hg–S, *C. r. Acad. sci.*, **C278**(4), 259–261 (1974).

Charifi Z., Hassan El Haj F., Baaziz H., Khosravizadeh Sh., Hashemifar S.J., Akbarzadeh H. Structural and electronic properties of the wide-gap $Zn_{1-x}Mg_xS$, $Zn_{1-x}Mg_xSe$ and $Zn_{1-x}Mg_xTe$ ternary alloys, *J. Phys.: Condens. Matter*, **17**(44), 7077–7088 (2005).

Chaus I.S., Kompanichenko N.M., Andreichenko V.G., Sheka I.A., Antishko A.N. The $Tl_2S–ZnS$ system [in Russian], *Ukr. khim. zhurn.*, **51**(4), 355–357 (1985).

Chaus I.S., Koval' L.B., Sheka I.A. Investigation of zinc, gallium and indium sulfides interaction at their co-precipitation from water solutions [in Russian], *Izv. AN SSSR. Neorgan. materialy*, **9**(2), 201–205 (1973).

Chechiotkina E.A., Hozhainov Yu.M., Galaktionov S.S., Bundel' A.A. Investigation of ZnO solubility in ZnS(wurtzite) [in Russian], *Izv. AN SSSR. Neorgan. materialy*, **14**(8), 1393–1396 (1978).

Chen W.-T., Kuang H.-M., Chen H.-L. Solid-state synthesis, crystal structures and properties of two novel metal sulfur chlorides—$Zn_6S_5Cl_2$ and $Hg_3ZnS_2Cl_4$, *J. Solid State Chem.*, **183**(10), 2411–2415 (2010).

Chen W.W., Zhang J.M., Ardell A.J., Dunn B. Solid-state phase equilibria in the ZnS–CdS system, *Mater. Res. Bull.*, **23**(11), 1667–1673 (1988).

Cherin P., Lind E.L., Davis E.A. The preparation and crystallography of cadmium zinc sulfide solid solutions, *J. Electrochem. Soc.*, **117**(2), 233–236 (1970).

Chernyshev A.I., Babanski M.D. Dissolution and crystallization of sphalerite from the solution in monosulfide melt [in Russian], in: *Reaktsionnaya sposobnost' veshchestv, Tomsk*, Tomsk, Russia: Un-t, pp. 34–37 (1977).

Chess D.L., Chess C.A., White W.B. Physical properties of ternary sulfide ceramics, *Mater. Res. Bull.*, **19**(12), 1551–1558 (1984).

Cnehm Ch., Nitsche R., Wild P. New phases in the system Zn-In-S, *Naturwissenschaften*, **56**(2), 86 (1969).

Deulkar S.H., Bhosale C.H., Sharon M., Neumann-Spallart M. Preparation of non-stoichiometric (Zn,Fe)S chalcogenides and evaluation of their thermal, optical and electrical properties, *J. Phys. Chem. Solids*, **64**(4), 539–344 (2003).

Dmitrienko A.O., Abalduyev B.V., Bol'shakov A.F., Kniazev Yu.V. Mechanism of the solid solution formation in the ZnS–CdS system [in Russian], *Izv. AN SSSR. Neorgan. materialy*, **13**(11), 1969–1971 (1977).

Dmitrenko A.O., Abramova L.V., Bukesov S.A. Solid solutions in the ZnS–CdS–MgS system [in Russian], *Izv. AN SSSR. Neorgan. materialy*, **26**(12), 2483–2487 (1990).

Dmitrienko A.O., Bol'shakov A.F., Abalduev B.V. Kinetics of equimolar solid solution formation in the ZnS–CdS system [in Russian], *Izv. AN SSSR. Neorgan. materialy*, **15**(1), 45–48 (1979).

Dmitrienko A.O., Kniazev Yu.V. About mechanism of solid solution formation in the ZnS–CdS system [in Russian], in: *Issled. v oblasti neorgan. i analit. khimii, Saratov.* Saratov, Russia: un-t, pp. 7–9 (1975).

Donika F.G. Crystal structure of ternary semiconductor phases in the ZnS-In$_2$S$_3$ system [in Russian], *Avtoref. dis. ... kand. fiz.-mat. nauk, Gor'kii*, 22 p., (1972).

Donika F.G., Kiosse G.A., Radautsan S.I., Semiletov S.A., Zhitar' V.F. Crystal structure of Zn$_3$In$_2$S$_6$ [in Russian], *Kristallografiya*, **12**(5), 854–859 (1967).

Donika F.G., Radautsan S.I., Kiosse G.A., Semiletov S.A., Donika T.V., Mustya I.G. Crystal structure of the Zn$_2$In$_2$S$_5$(III)$_a$ polytypical form, *Kristallografiya*, **15**(4), 816–817 (1970a).

Donika F.G., Radautsan S.I., Kiosse G.A., Semiletov S.A., Donika T.V., Mustya I.G. Crystal structure of the two-packet ZnIn$_2$S$_4$(II)$_a$ polytype and structure refinement of the three-packet ZnIn$_2$S$_4$(III)$_a$ polytype [in Russian], *Kristallografiya*, **16**(1), 235–237 (1971).

Donika F.G., Radautsan S.I., Semiletov S.A., Donika T.V., Mustya I.G., Zhitar' V.F. Crystal structure of the ZnIn$_2$S$_4$(I) polytypical form, *Kristallografiya*, **15**(4), 813–815 (1970b).

Donika F.G., Radautsan S.I., Semiletov S.A., Donika T.V., Mustya I.G. Crystal structure of the two-packet Zn$_2$In$_2$S$_5$(II)$_a$ polytype [in Russian], *Kristallografiya*, **17**(3), 666–667 (1972a).

Donika F.G., Radautsan S.I., Semiletov S.A., Kiosse G.A., Mustya I.G. Crystal structure of the two-packet ZnIn$_2$S$_4$(II)$_b$ polytype [in Russian], *Kristallografiya*, **17**(3), 663–665 (1972b).

Donika F.G., Zhitar' V.F., Radautsan S.I. *Semiconductors of the ZnS–In$_2$S$_3$ system.* Kishinev, Moldova: Shtiintsa Publishing, 148pp (1980).

Dovgoshey M.I., Popovych N.I., Kacher I.E. Mass-spectrometric investigations of the thermal and laser evaporation of ZnGa$_2$S$_4$ crystals [in Ukrainian], *Fiz. i khim. tv. tila*, **2**(3), 449–453 (2001).

Dubrovin I.V., Budionnaya L.D., Oleynik N.D., Sharkina E.V. Interaction of Cd$_4$SiS$_6$ with ZnS and SiS$_2$ along the section Cd$_4$SiS$_6$–(ZnS)$_{0.8}$(SiS$_2$)$_{0.2}$ [in Russian], *Izv. AN SSSR. Neorgan. materialy*, **25**(5), 722–725 (1989a).

Dubrovin I.V., Budionnaya L.D., Sharkina E.V. The GeS$_2$–ZnS system [in Russian], *Izv. AN SSSR. Neorgan. materialy*, **25**(8), 1386–1387 (1989b).

Dutrizac J.E. The Fe$_{1-x}$S–PbS–ZnS phase system, *Can. J. Chem.*, **58**(7), 739–743 (1980).

Edwards J.G., Ferro D., Weber J.K.R. The incongruent effusion of ZnGa$_8$S$_{13}$(s), *J. Less-Common Metals*, **156**(1–2), 369–385 (1989).

Flahaut J. Contribution à l'étude du sulfure d'aluminium, *Ann. Chim.*, **12**(7), 632–696 (1952).

Gamidov R.S., Kulieva T.Z. X-ray investigation of the thallium and zinc chalcoaluminates [in Russian], in: *Issled. v obl. neorgan. i fiz. khim. i ih rol' v khim. promyshlennosti*, Baku, Azerbaijan: AzINTI, pp. 32–34 (1969).

Gan'shin V.A., Korkishko Yu.N., Fedorov V.A. Structure phase diagram of the ZnS–CdS system [in Russian], *Zhurn. neorgan. khimii*, **38**(12), 2026–2031 (1993).

Gashurov G., Banks E. The role of copper in the transformation of hexagonal zinc sulfide, *J. Electrochem. Soc.*, **114**(11), 1143–1148 (1967).

Giriat W., Stankiewicz J. Preparation and optical absorption of $Mn_xZn_{1-x}S$, Progr. Cryst. Growth and Charact., **10**(1–4): *Proc 6th Int. Conf. Ternary and Multinary Compounds, Caracas*, August 15–17, 87–89 (1985).

Gocke E., Schramm W., Dolscheid P., Schöllhorn R. Molybdenum cluster chalcogenides Mo_6X_8: Electrochemical intercalation of closed shell ions Zn^{2+}, Cd^{2+}, and Na^+, *J. Solid State Chem.*, **70**(1), 71–81 (1987).

Golovey M.I., Olekseyuk I.D., Voroshilov Yu.V. Chemical bond and solubility in the systems $A^{II}_3B^V_2–A^{II}C^{VI}$ [in Russian], in: *Khim. svyaz' v poluprovodn. i polumetallah*. Minsk, Belarus: Nauka i Tekhnika Publishing, pp. 233–239 (1972).

Gordeev I.V., Karelin V.V. Equilibrium in the zinc sulfide–iodine system [in Russian], *Izv. AN SSSR. Neorgan. materialy*, **5**(7), 1190–1193 (1969).

Guseinov F.H., Babanly M.B., Kuliev A.A. Phase equilibria in the TlX–Cd(Zn)X (X – S, Se, Te) systems [in Russian], *Izv. AN SSSR. Neorgan. materialy*, **18**(5), 759–763 (1982).

Guseinov F.H., Kuliev A.A., Babanly M.B., Kuliev R.A. Investigation of the Tl_4S_3–Zn(Cd)S systems [in Russian], *Azerb. khim. zhurn.*, (4), 108–110 (1985).

Hahn H., Frank G. Über die Structur des $ZnAl_2S_4$, *Z. anorg. und allg. Chem.*, **269**(2), 227–231 (1952).

Hahn H., Frank G., Klingler W., Störger A.D., Störger G. Über ternäre Chalkogenide des Aluminiums, Galliums und Indiums mit Zink, Cadmium und Quecksilber, *Z. anorg. und allg. Chem.*, **279**(5/6), 241–270 (1955).

Hahn H., Lorent C. Untersuchungen über ternäre Chalkogenide. Über ternäre Sulfide und Selenide des Germaniums mit Zink, Cadmium und Quecksilber, *Naturwissenschaften*, **45**(24), 621–622 (1958).

Hoppe R. Untersuchungen an ternären Sulfiden, *Angew. Chem.*, **71**(14), 457 (1959).

Hozhainov Yu.M., Kazantsev I.A., Razvazhnoi E.M., Potapov I.I. Investigation of $ZnS–P_2S_5$ system [in Russian], Deposited in VINITI, N 4424–72Dep (1972).

Imamov R.M., Semiletov S.A., Pinsker S.G. Some questions of crystal chemistry of semiconductors with octahedral and mixed coordination of atoms [in Russian], *Kristallografiya*, **15**(2), 287–293 (1970).

Inoue R., Kitagawa M., Nishigaki T., Ichino K., Kobayashi H., Ohishi M., Saito H. Optical band gap of $Zn_xMg_{1-x}S$ thin films with composition x between 0.14 and 1.0, *J. Cryst. Growth*, **184/185**, 1076–1080 (1998).

Juza R., Rabenau A., Pascher G. Über teste Lösungen in den Systemen ZnS/MnS, ZnSe/MnSe und ZnTe/MnTe, *Z. anorg. und allg. Chem.*, **285**(1/2), 61–69 (1956).

Kaldis E., Krausbauer L., Widmer R. Cd_4SiS_6 and Cd_4SiSe_6, new ternary compounds, *J. Electrochem. Soc.*, **114**(10), 1074–1076 (1967).

Kalomiros J.A., Anagnostopoulos A.N., Spyrielis J. Growth and some properties of $Zn_5In_2S_8$ single crystal, *Mater. Res. Bull.*, **22**(10), 1307–1314 (1987).

Kaneko S., Aoki H., Kawahara Q., Imoto F. Solid solutions and phase transformations in the system ZnS–CdS under hydrothermal conditions, *J. Electrochem. Soc.*, **131**(6), 1445–1446 (1984).

Kaneko S., Aoki H., Nonaka I., Imoto F., Matsumoto K. Solid solutions and phase transformation in the system ZnS–MnS under hydrotermal conditions, *J. Electrochem. Soc.*, **130**(12), 2487–2489 (1983).

Kawada I., Onoda M., Saeki M. $Zn_2Ti_{18}S_{32}$, a new ternary sulfide, *Acta Crystallogr.*, **C41**(11), 1573–1575 (1985).

Kim Y.-D., Nagashima S., Maeda H., Kato A. Synthesis of $Ag_2S–ZnS$ composite powders [in Japanese], *J. Chem. Soc. Jap., Chem. Ind. Chem.*, (2), 159–162 (1997).

Kimura S. Phase equilibria in the systems Zn+S+Sn, Zn+Se+Sn and Cd+S+Sn, *J. Chem. Thermodyn.*, **3**(1), 7–17 (1971).

Kimura S., Panish M.B. Phase equilibria in the systems Zn+S+Bi, Zn+Se+Bi and Cd+S+Bi, *J. Chem. Thermodyn.*, **2**(1), 77–86 (1970).

Klepp K.O., Bronger W. Na_2ZnS_2 und Na_6ZnS_6, zwei neue Thiozinkate, *Rev. Chim. Miner.*, **20**(4–5), 682–688 (1983).

Klinger W., Eulenberger G., Hahn H. Über Hexachalkogeno-hypodiphosphate vom Typ $M_2P_2X_6$, *Naturwissenschaften*, **57**(2), 88 (1970).

Klinger W., Ott R., Hahn H. Über die Darstellung und Eigenschaften von Hexathio- und Hexaselenohypodiphosphaten, *Z. anorg. und allg. Chem.*, **396**(3), 271–278 (1973).

Kondrashev Yu.D., Omel'chenko Yu.A. X-ray investigations of some oxide and sulfide system [in Russian], *Zhurn. neorgan. khimii*, **9**(4), 937–943 (1964).

Kopylov N.I., Kodzoeva C.M. The $ZnS–Na_2S$ system [in Russian], *Zhurn. neorgan. khimii*, **14**(9), 2595 (1969).

Kopylov N.I., Polyviannyi I.R., Ivakina L.P., Antoniuk V.I. The $ZnS–Na_2S$ system [in Russian], *Zhurn. neorgan. khimii*, **23**(11), 3095–3101 (1978).

Kopylov N.I., Toguzov M.Z., Yarygin V.I. Investigation of the $PbS–Cu_{1.8}S–ZnS$ system [in Russian], *Izv. AN SSSR. Ser. Metally*, (6), 80–83 (1976).

Koreniev Yu.M., Karasev N.M., Timoshin I.A., Volkova T.A., Sidorov L.N., Novoselova A.V. Mass-spectrum thermodynamic investigation of binary system, forming by chalkogenides of second group. I. ZnS-ZnSe system [in Russian], *Zhurn. phys. khimii*, **46**(7), 1714–1717 (1972).

Kovach A.P., Potoriy M.V., Voroshilov Yu.V., Tovt V.V. Physico-chemical investigation of the $Cd(Zn)S(Se)–P_2S_4(Se_4)$ systems [in Russian], *Ukr. khim. zhurn.*, **63**(2), 92–96 (1997).

Kovaliv V.I., Lisniak S.S. Synthesis of spinels in the $ZnS–Al_2S_3$ system [in Russian], *Izv. AN SSSR. Neorgan. materialy*, **11**(1), 33–36 (1975).

Kozhahmetov S.M., Spivak M.M., Chokaev M.T., Tumarbekov Z.T. Application of partial pressure diagrams of zinc–sulfur–oxygen system [in Russian], *Vestn. AN KazSSR*, (2), 26–29 (1983).

Kozhahmetov S.M., Tumarbekov Z.T., Chokaev M.T. Thermodynamics and kinetics of interaction in the zinc-sulfur-oxygen system at 1000–1600°C [in Russian], in: *Sul'fid. rasplavy tiazh. met.* Moscow, Russia, pp. 89–96 (1982).

Kozhina I.I., Riskin I.V., Rogova T.V., Tolkachev S.S. Crystal structure and color in the Cd–Zn–S system [in Russian], *Vestn. Leningr. un-ta*, (4), 128–136 (1965).

Krämer V., Hirth W., Hofherr W., Trash H.-P. Phase studies in the systems $Ag_2Te–Ga_2Te_3$, $ZnSe–In_2Se_3$ and $ZnS–Ga_2S_3$. *Thermochim. Acta*, **112**(1), 88–94 (1987).

Kremheller A., Levine A.K., Gashurov G. Hydrothermal preparation of two-component solid solutions from II–VI compounds, *J. Electrochem. Soc.*, **107**(1), 12–15 (1960).

Kröger F.A. Formation of solid solutions in the system zinc sulfude–manganese sulfide, *Z. Kristallogr.*, **A100**, 543–545 (1938).

Kröger F.A., Dikhoff J.A.M. The function of oxygen in zinc sulfide phosphors, *J. Electrochem. Soc.*, **99**(4), 144–154 (1952).

Kuleshov I.V., Kesler Ya.A., Gordeev I.V., Mozhaev A.P., Tret'yakov Yu.D. Synthesis and investigation of the crystallochemical properties of gallium thiomolibdate with spinel structure [in Russian], *Izv. AN SSSR. Neorgan. materialy*, **17**(1), 145–147 (1981).

Lappe F., Niggli A., Nitsche R., White J.G. The crystal structure of In_2ZnS_4, Z. *Krist.*, **117**(2–3), 146–152 (1962).

Larach S., McCarroll W.H., Shrader R.E. Luminescence properties of zinc-interchalcogenides. I. Zinc sulfo-telluride phosphors, *J. Phys. Chem.*, **60**(5), 604–608 (1956).

Larach S., Schrader R.E., Stocker C.F. Anomalous variation of band gap with composition in zinc sulfo- and seleno-tellurides, *Phys. Rev.*, **108**(3), 587–589 (1957).

Lemoine P., Tomas A., Vovan T., Guittard M. Structure du sulfure de thallium et de zinc $ZnTm_2S_4$, *Acta Crystallogr.*, **C46**(3), 365–368 (1990).

Linares R.C. Phase equilibrium and crystal growth in the system $ZnS-ZnF_2$, in: *Metallurgy Advanced Electron. Material*, New York; London, U.K.: Interscience, pp. 329–334 (1963).

Lott K.P., Vishniakov A.V., Raukas M.M. Structure of the Zn–Cu–S phase diagram and solubility of copper in zinc sulfide [in Russian], *Zhurn. neorgan. khimii*, **26**(7), 1894–1899 (1981).

Lott K.P., Vishnyakov A.V., Raukas M.M. Solubility of gallium into zinc sulfide [in Russian], Tr. Tallinsk. Politekhn. In-ta, (587) (Physical chemistry of II-VI compounds), 17–27 (1984a).

Lott K.P., Vishnyakov A.V., Raukas M.M. Solubility of indium into zinc sulfide [in Russian], Tr. Tallinsk. Politekhn. In-ta, (587) (Physical chemistry of II-VI compounds), 29–33 (1984b).

Lutz H.D., Bertram K.-H., Becker R.-A., Becker W. Zur Kenntnis ternärer Chromsulfide, *Angew. Chem.*, **83**(22), 919 (1971).

Luzhnaya N.P., Nikol'skaya G.F., Kovaleva I.S. Investigation of phase diagrams for detection of ternary semiconductor compounds [in Russian], *Izv. AN SSSR. Neorgan. materialy*, **2**(8), 1390–1396 (1966).

Lyalikova P.Yu., Mirovich L.V. Application of thermogravimetry for investigation of complex semiconductor compounds [in Russian], in: *Nekot. voprosy khimii i fiz. poluprovodn. slozhn. sostava.* Uzhgorod, Ukraine: Uzhgorod. un-t, pp. 169–171 (1970).

Machuga A.I., Zhitar' V.F., Muntyan S.P., Arama E.D. Crystal growth and cathodoluminescence of the $Zn_5In_2S_8$, *Neorgan. materialy*, **36**(12), 1418–1420 (2000).

Malevski A.Yu. Investigation of isomorphous substitution limits in the $ZnS-Ga_2S_3$ system [in Russian], in: *Experiment. issled. v obl. mineralogii i geokhimii redkih elementov*, Moscow, Russia: Nauka Publishing, pp. 12–20 (1967).

Malinowski C. Phenomenologic analysis of chemical process in the systems $ZnSO_4-ZnS$ and $PbSO_4-ZnS$ [in Polish], *Zesz. nauk. AGH im. Stanisława Staszika. Met. i odlew.: [Monogr.]*, (141), 1–120 (1992).

Malur J. Zur Kristallisation von Zinksulfide aus Salzschmelzen, *Z. Chem.*, **4**(6), 232–233 (1964).

Masaharu A., Masaoko W., Hiroshi N. Growth of Zn and Cd chalkogenides from solutions using Te as a solvent, *J. Fac. Eng. Univ. Tokyo*, **A**(18), 38–39 (1980).

Masaoko W., Ken-ichi S., Masaharu A. Solution growth of ZnS, ZnSe, CdS and their mixed compounds using tellurium as a solvent, *Jap. J. Appl. Phys.*, **18**(5), 869–872 (1979).

Maslov V.I., Polyviannyi I.R., Turisbekov M.T. Investigation of Na_2S-ZnS and Na_2S-ZnO phase diagrams [in Russian], Deposited in VINITI, № 4653–72Dep (1972).

Michalski E., Demianiuk M., Kaczmarek S., Zmija J. Some new aspects of polytypic structures in $ZnS_{1-x}Se_x$ mixed crystals, *Acta Phys. Pol.*, **A58**(6), 711–718 (1980).

Mihalev A.A., Kalashnikova L.D., Levantsova N.S., Glagoleva A.A., Boiev E.I. The temperature dependence of copper sulfide solubility in the zinc and cadmium sulfides [in Russian], *Sb. nauch. tr./VNII liuminoforov i osobo chistyh veshchestv*, (3), 85–89 (1970).

Mizetskaya I.B., Oleinik G.S., Trishchuk L.I. Phase diagrams of the $Cu_{2-x}S$–ZnS and $Cu_{2-x}Se$–ZnSe systems [in Russian], *Izv. AN SSSR. Neorgan. materialy*, **18**(11), 1792–1794 (1982).

Moh G.H. Tin-containing mineral systems. Part II: Phase relations and mineral assemblages in the Cu–Fe–Zn–Sn–S system, *Chem. Erde*, **34**(1), 1–61 (1975).

Moldovyan N.A. Growth and investigation of photoconductivity of $ZnAl_2S_4$ and $CdAl_2S_4$ single crystals [in Russian], *Izv. AN SSSR. Neorgan. materialy*, **27**(9), 1969–1971 (1991).

Moldovyan N.A., Markus M.M., Radautsan S.I. Single crystals of $ZnAl_2S_4$ layer modification [in Russian], *Dokl. AN SSSR*, **304**(1), 151–153 (1989).

Moldovyan N.A., Pyshnaya N.B., Radautsan S.I. New multinary layered chalcogenides with octahedral and tetrahedral cation coordination, *Jap. J. Appl. Phys.*, **32**(Pt. 1, Suppl.), 781–783 (1993).

Murach N.N. (Ed.) Reference-book of metallurgist on non-ferrous metal [in Russian]. Moscow, Russia: Metallurgizdat, Vol. 2, 784 p. (1947).

Mustia I.G., Mirovich L.V., Zhitar' V.F. Growth of ternary phase single crystals in the ZnS–In_2S_3 system [in Russian], in: *Troinyie poluprovodn. $A^{II}B^{IV}C_2^V$ i $A^{II}B_2^{III}C_4^{VI}$.* Kishinev, Moldova: Shtiinza, pp. 211–212 (1972).

Nekrasov I.Ya., Sorokin V.I., Osadchiy E.G., Tikhomorova V.I. Experimental investigation of the Zn–Sn–S system at 300–500°C in hydrothermal conditions [in Russian], *Ocherki fiz.-khim. petrologii*, **6**, 200–207 (1977).

Neuhaus A., Steffen R. Über das Zustands- und Mischbarkeitsverhalten des Systems ZnS-MnS im Druckbereich bis 140 kbar, *Z. Phys. Chem.* (BRD), **73**(4/6), 188–214 (1970).

Nicholas D.M., Scott R.A.M., Shropshall M.R. Crystal data for Ba_2ZnS_3, *J. Appl. Crystallogr.*, **10**(4), 356 (1977).

Nitsche R. Crystal growth and phase investigations in multi-component systems by vapour transport, *J. Cryst. Growth*, **9**, 238–243 (1971).

Niu C.M., Kershaw R., Dwight K., Wold A. The preparation and properties of cobalt-doped II–VI chalcogenides, *J. Solid State Chem.*, **85**(2), 262–269 (1990).

Noor N.A., Ikram N., Ali S., Nazir S., Alay-e-Abbas S.M., Shaukat A. First-principles calculations of structural, electronic and optical properties of $Cd_xZn_{1-x}S$ alloys, *J. Alloys Compd.*, **507**(2), 356–363 (2010).

Novoselov S.S. Influence of zinc sulfide on the properties of copper matte [in Russian], *Tsvet. metally*, (3), 15–20 (1955).   AQ12

Obuhova N.F., Atroshchenko L.V., Kolodiazhnyi L.V., Obtaining and investigation of ZnS–MgS solid solutions with wurtzite structure [in Russian], *Izv. AN SSSR. Neorgan. materialy*, **13**(8), 1390–1393 (1977).

Odin I.N., Marugin V.V. Phase diagrams and electrophysical properties of ingots in the Bi_2X_3–ZnX (X – S, Se, Te) systems [in Russian], *Zhurn. neorgan. khimii*, **36**(7), 1842–1846 (1991).

Olekseyuk I.D., Dudchak I.V., Piskach L.V. Phase equilibria in the $Cu_2S–ZnS–SnS_2$ system, *J. Alloys Compd.*, **368**(1–2), 135–143 (2004).

Olekseyuk I.D., Tovtin N.A. Phase equilibria and glass formation in the Zn–As–S system [in Russian], *Zhurn. neorgan. khimii*, **32**(10), 2562–2565 (1987).

Olekseyuk I.D., Tsitrovskiy V.V., Turyanitsa I.D., Stoyka I.M., Chukhno T.A. Obtaining and properties of modulation and nonlinear materials based on some chalcogenides [in Russian], *Kvantovaya elektronika*, (13), 93–96 (1977).

Pajaczkowska A. Physicochemical properties and crystal growth of $A^{II}B^{VI}–MnB^{VI}$ systems, *Progr. Crystal Growth Caract.*, **1**(3), 289–326 (1978).

Palatnik L.S., Gladkih N.T., Naboka M.N. Investigation of ZnS–CdS and Zn–Cd–S condensed films of variable composition [in Russian], *Fizika tverdogo tela*, **7**(9), 2850–2852 (1965).

Palosz B., Przedmojski J. Application of theoretical intensity distribution curves to the analysis of disordered ZnS–CdS and ZnS–ZnSe crystals, *Acta Crystallogr.*, **A32**(3), 409–411 (1976).

Paszkowicz W., Domagala J., Golacki Z. X-ray characterization of $Zn_{1-x}Co_xS$ single crystals grown by chemical vapour transport, *J. Alloys Compd.*, **274**(1–2), 128–135 (1998).

Paszkowicz W., Spolnik Z., Firzt F., Meczynska H. H-polytype formation in $Zn_{1-x}Mg_xSe$, *Acta Crystallogr. A*, **52**, Suppl., 319 (1996).

Polyvyanny I.R., Lata V.A. Thermal analysis of certain sulphide–sodium systems, *Thermochim. Acta*, **92**, 747–749 (1985).

Prouzet E., Ouvrard G., Brec R. Structure determination of $ZnPS_3$, *Mater. Res. Bull.*, **21**(2), 195–200 (1986).

Raccah P.M., Bouchard R.J., Wold A. Crystallographic study of chromium spinels, *J. Appl. Phys.*, **37**(3), 1436–1437 (1966).

Radautsan S.I., Donika F.G., Kiosse G.A., Mustya I.G., Zhitar V.F. The new semiconductor compound $Zn_2In_2S_5$ in the system Zn–In–S, *Phys. status solidi*, **34**(2), K129–K131 (1969a).

Radautsan S.I., Donika F.G., Kyosse G.A., Mustya I.G. Polytypism of ternary phases in the system Zn-In- S, *Phys. status solidi*, **37**(2), K123–K127 (1970).

Radautsan S.I., Tsiulianu I.I., Lialikova R.Yu., Moldovyan N.A., Zhitar' V.F., Markus M.M. Polymorphous transformations in $Zn_3In_2(Ga,Al)S_6$ [in Russian], *Izv. AN SSSR. Neorgan. materialy*, **23**(5), 852–854 (1987).

Radautsan S.I., Zhitar' V.F., Tezlevan V.E., Donika F.G. Investigation of structure and some properties of alloys based on In_2S_3 and zinc and cadmium sulfides [in Russian], in: *Khimicheskaya sviaz' v kristallah*, Minsk, Belarus: Nauka i Tehnika Publishing, pp. 423–427 (1969b).

Rigan M.Yu., Stasiuk N.P. Some questions of charge synthesis and crystal growth of $ZnGa_2S_4$ and $ZnGa_2Se_4$ [in Russian], in: *Poluch. i svoistva slozhn. poluprovodn.*, *Uzhgorod. gos. un-t*, Kyiv, Ukraine: Nauk. Dumka Publishing, pp. 62–69 (1991).

Rubenstein M. Solubilities of some II–VI compounds in bismuth, *J. Electrochem. Soc.*, **113**(6), 623–624 (1966).

Rubenstein M. Solution growth of some II-VI compounds using tin as a solvent, *J. Cryst. Growth*, **3–4**, 309–312 (1968).

Saeki M., Onoda M. A new ternary phase $Zn_2Ti_{18}S_{32}$, *Chem. Lett.*, (9), 1329–1330 (1982).

Saeki M., Onoda M. Novel Zn–Ti–S compound and its manufacture Pat. 59-156917 (Japan). Published data September 6, 1984; Filing data February 22, 1983.

Sakaguchi M., Ohta M., Satoh M., Hirabayashi T. The phase transformation during crystallization of ZnS, *J. Electrochem. Soc.*, **124**(4), 550–553 (1977).

Sanitarov V.A., Kalinkin I.P., Dementyev A.G. Calculation of the phase diagram of the $Zn_xCd_{1-x}S$ soli solution [in Russian], *Zhurn. phys. khimii*, **53**(9), 2238–2241 (1979).

Schnering H.G., Hoppe R. Zur Kenntnis des Ba_2ZnS_3, *Z. anorg. und allg. Chem.*, **312**(1/2), 99–109 (1961).

Scott R.A.M., Nicholas D.M., Shropshall M.R. The growth of single crystals of Ba_2ZnS_3, *J. Cryst. Growth.*, **38**(2), 269–271 (1977).

Scott S.D., Rissin S.A. Sphalerite composition in the Zn–Fe–S system below 300°C, *Econ. Geol.*, **68**(4), 475–479 (1973).

Semenov V.N., Averbah E.M., Ugai Ya.A. Properties of $Cd_xZn_{1-x}S$ solid solution thin films [in Russian], *Izv. AN SSSR. Neorgan. materialy*, **26**(10), 2030–2032 (1990).

Shevtsov V.M., Sinel'nikov B.M. Thermodynamic activity of copper sulfide in the solid solutions with zinc sulfide [in Russian], in: *Termodyn. svoistva i analiz sistem perehodnyh metallov*, Krasnodar, Russia: Krasnodar. Gos. Un-t, pp. 40–41 (1989).

Shishkin V.I., Polyviannyi I.R., Demchenko R.S. The dissolution of lead and copper and phase diagram in the Na_2S–ZnS system [in Russian], *Vestn. AN KazSSR*, (9), 20–30 (1968).

Sinel'nikov B.M., Shevtsov V.M., Burylev V.P. Thermodynamic properties of ZnS–Cu solid solutions [in Russian], *Izv. vuzov. Khim. i khim. technol.*, **27**(12), 1437–1439 (1984).

Skinner B.J., Barton P.B. The substitution of oxygen for sulfur in wurtzite and sphalerite, *Amer. Mineralogist*, **45**(3), 612–625 (1960).

Skums V.F., Pan'ko E.P., Vecher A.A. Influence of high pressures on the electrical resistivity of cadmium and zinc chalcogenide solid solutions, *Inorg. Mater. (Engl. Trans.)*, **28**(4), 572–577 (1992), transl. from *Neorgan. Mater.*, **28**(4), 745–751 (1992).

Smith A.L. Zinc-magnesium oxide and zinc-magnesium sulfide phosphores, *J. Electrochem. Soc.*, **99**(1), 155–158 (1952).

Sobolev V.V. Zone structure of $CdIn_2S_4$-type crystals [in Russian], *Izv. AN MSSR. Ser. fiz.-tehn. i mat. nauk*, (2), 60–63 (1976).

Soklakov A.I., Nechaeva V.V. New Cu, Fe, Ni, Zn and Cd thiophosphates [in Russian], *Izv. AN SSSR. Neorgan. materialy*, **6**(5), 998–999 (1970).

Solans X., Font-Altaba M., Moreiras D., Trobajo-Fernandez M.C., Otero-Arean C. Redetermination and refinement of the crystal structure of $ZnGa_2S_4$, *AFINIDAD*, **45**(415), 277–279 (1988).

Sombuthawee C., Bonsall S.B., Hummel F.A. Phase equilibria in the systems ZnS–MnS, $ZnS–CuInS_2$ and $MnS–CuInS_2$, *J. Solid State Chem.*, **25**(4), 391–399 (1978).

Steigmann G.A. The crystal structure of $ZnAl_2S_4$, *Acta crystallogr.*, **23**(1), 142–147 (1967).

Susa K., Kobayashi T., Tanigushi S. High-pressures synthesis of rock-salt type CdS using metal sulfide additives, *J. Solid State Chem.*, **33**(2), 197–202 (1980).

Sysoev L.A., Obuhova N.F. Obtaining and investigation of some properties of ZnS–MgS single crystals [in Russian], *Monokristaly i tekhnika, vyp.* **2**(9), 54–57 (1973).

Sysoev L.A., Obuhova N.F. Optical properties of the ZnS–MgS single crystals [in Russian], *Monokristally i tekhnika, vyp.* **1**(10), 20–24 (1974).

Tarascon J.M., Hull G.W. On several new ternary molibdenum sulfide phases $M_{3,4}Mo_{15}S_{19}$ (M = vacancy, Li, Na, K, Zn, Cd, Sn and Tl), *Mater. Res. Bull.*, **21**(7), 859–869 (1986).

Tauson V.L., Abramovich M.G. Investigation of the ZnS–HgS system by hydrothermal method [in Russian], *Geokhimia*, (6), 808–820 (1980).

Tauson V.L., Chernyshev L.V. *Experimental Investigations on Zinc Sulfide Crystal Chemistry and Geochemistry* [in Russian]. Novosibirsk, Russia: Nauka Publishing, 190pp (1981).

Timoshin I.A., Pavlov O.B., Mamedov K.N., Mikhenkov A.F. Isothermic dissociative pressure of saturated vapor above the phases of variable composition of the quasibinary system Zn–S–Se [in Russian], *Zhurn. phys. khimii*, **49**(4), 1039–1040 (1975).

Tinoco T., Polian A., Itié J.P., López S.A. Structure of III_a-$ZnIn_2S_4$ under high pressures, *Phys. status. solidi (b)*, **211**(1), 385–387 (1999).

Tomashik V.N. The physico-chemical interaction in the Cd, Zn ‖ S, Se ternary reciprocal system [in Russian], *Izv. AN SSSR. Neorgan. materialy*, **17**(6), 1116–1117 (1981).

Tomashik V.N., Oleinik G.S., Mizetskaya I.B. Investigation of the CdTe + ZnS ⇔ CdS + ZnTe ternary mutual system [in Russian], *Izv. AN SSSR. Neorgan. materialy*, **14**(8), 1434–1436 (1978).

Tretiakov Yu.D., Lisniak S.S., Alferov V.A., Kovaliv V.I., Gordeev I.V. Crystal chemistry and chemical bond in nonstoichiometric spinel phases in the ZnS–Al_2S_3 system [in Russian], in: *Khimicheskaya sviaz' v kristallah i ih fiz. svoistva*. Minsk, Belarus: Nauka i Tehnika Publishing, Vol. 2, pp. 165–168 (1976).

Trishchuk L.I., Oleinik G.S., Mizetskaya I.B. Thermal analysis determination of $A^{II}B^{VI}$ solid solubility in $A_2^IB^{VI}$, *Thermochim. Acta*, **92**, 611–613 (1985).

Tsygoda I.M., Ponomarev V.D., Shkuridin I.S. Influence of carbon on zinc sulfide volatility at high temperatures [in Russian], *Sb. nauchn. tr. Vses. n.-i. gornometallurg. in-t tsvetn. met.*, (10), 106–111 (1967).

Vasil'ev V.I. About the zinc containing form of metacinabarite—Gvadalkatsarite in mercury ore of the Mountain Altai [in Russian], *Dokl. AN SSSR*, **153**(3), 676–678 (1963).

Vasiliev V.K., Kochkin V.P., Patrova N.P. Formation of (Zn,Cd)S solid solutions at the interaction of zinc and cadmium oxides and sulfides [in Russian], *Rentgenografia mineral'nogo syria*, (4), 84–87 (1964).

Vishniakov A.V., Bundel' A.A. Investigation of equilibria between ZnS–CdS solid solutions and Cd–Zn melts [in Russian], *Zhurn. phys. khimii*, **40**(3), 726–729 (1966).

Vishniakov A.V., Iofis B.G. Solubility of $Ag_2S(Se)$ in ZnS, CdS and CdSe [in Russian], *Izv. AN SSSR. Neorgan. materialy*, **10**(7), 1184–1186 (1974).

Vollebregt F.H.A., Ijdo D.J.W. Zinc rare-earth sulphides with the olivine structure, *Acta Crystallogr.*, **B38**(9), 2442–2444 (1982).

Wachtel A. (Zn,Hg)S and (Zn,Cd,Hg)S electroluminescent phosphors, *J. Electrochem. Soc.*, **107**(8), 682–688 (1960).

Wu P., He X.-C., Dwight K., Wold A. Growth and characterization of zinc and cadmium thiogallate, *Mater. Res. Bull.*, **23**(11), 1605–1609 (1988).

Yim W.M., Fan A.K., Stofko E.J. Preparation and properties of II–Ln_2–S_4 ternary sulfides, *J. Electrochem. Soc.*, **120**(3), 441–446 (1973).

Yonemura M., Kotera Y. Kinetic analysis of the formation of solid solution in ZnS–CdS system [in Japanese], *J. Nat. Chem. Lab. Ind.*, **81**(10), 491–497 (1986).

Zhang J., Chen W.W., Ardell A.J., Dunn B. Solid-state phase equilibria in the ZnS–Ga_2S_3 system, *J. Amer. Ceram. Soc.*, **73**(6), 1544–1547 (1990).

Zhitar' V.F., Goriunova N.A., Radautsan S.I. Alloy investigation of some sections of zinc–indium–arsenic–sulfur quaternary system [in Russian], *Izv. AN MSSR. Ser. fiz.-tehn. i mat. nauk*, (2), 9–14 (1965).

Zhitar' V.F., Lange T.I., Radautsan S.I. Detection of dislocations in $ZnIn_2S_4$ lamellar crystals [in Russian], *Izv. AN SSSR. Neorgan. materialy*, **4**(10), 1810–1812 (1968).

Zinchenko V.F., Chivireva N.O., Kocherba G.I., Markiv V.Ya., Belyavina N.M. Influence of Ln_2S_3 (Ln – Gd, Dy) dopant on the crystal structure and optical properties of zinc sulphide, *Chem. Met. Alloys*, 3(3–4), 75–82 (2010).

Zubkovskaya V.N., Vishniakov A.V. Equilibrium between zinc sulfide crystals and lead sulfide vapor [in Russian], *Zhurn. phys. khimii*, **55**(6), 1457–1460 (1981).

Zuyev A.P., Kulakov M.P., Fadeev A.V., Kireiko V.V. Oxygen contents in the ZnS, ZnSe and CdS crystals, obtained from the melt [in Russian], *Izv. AN SSSR. Neorgan. materialy*, **17**(7), 1159–1161 (1981).

2

Systems Based on ZnSe

2.1 Zinc–Copper–Selenium

ZnSe–Cu: The phase diagram is not constructed. The composition of metal phase does not change within the interval of Cu/ZnSe ratio $\alpha = 0.25$–16 g/g, and at 800°C, 900°C, 950°C, and 1000°C, α is equal correspondingly 2.08 ± 0.13, 5.6 ± 0.3, 8.0 ± 0.6, and 10.7 ± 0.2 (Harif and Vishniakov 1975). ZnSe disappears at $\alpha > 2.6$ (1000°C) and $\alpha > 6.1$ (900°C).

The equilibria in this system were investigated using a static method (Harif and Vishniakov 1975).

ZnSe–Cu$_{1.9}$Se: The phase diagram belongs to the peritectic type (Figure 2.1) (Mizetskaya et al. 1982). The peritectic composition and temperature are 46 mol. % ZnSe and 1142°C (Mizetskaya et al. 1982) or 45 ± 5 mol. % ZnSe and $1180°C \pm 15°C$ according to Vishniakov et al. (1981). According to x-ray diffraction (XRD) and metallography, ternary compounds do not form in this system (Harif and Vishniakov 1975).

The solubility of ZnSe in Cu$_2$Se at 1142°C, 1000°C, 900°C, 875°C, 575°C, and 400°C is equal to 55, 28, 18, 15, 2.5, and less than 1 mol. %, respectively (Mizetskaya et al. 1982, Trishchuk et al. 1985), or 20.2 ± 2.7 and 36.3 ± 4.6 mol. % at 900°C and 1000°C according to Harif and Vishniakov (1975). In the Zn–Cu–Se ternary system, solid solutions based on Cu$_2$Se are in equilibrium with ZnSe, and the solubility of ZnSe in Cu$_2$Se at room temperature is not higher than 0.5 mol. % (0.17 at. % Zn) (Abrikosov et al. 1985).

The solubility of Cu$_2$Se in ZnSe investigated using diffusion saturation of thin films is equal to 0.27, 0.54, and 1.90 mol. % at 700°C, 800°C, and 1000°C correspondingly (Vishniakov and Harif 1972). There is approximately $1.0 \cdot 10^{-2}$ at. % Cu in the ZnSe single crystals, growing at 1530°C and 10 MPa (100 atm) and using Cu$_2$Se as a dopant (Fadeev et al. 1977).

According to the data of Asadov et al. (2000), partly isomorphous substitution of Cu by Zn in the Cu$_2$Se compound stabilizes high-temperature face-centered cubic lattice at room temperature, and a low-temperature orthorhombic phase is stable up to 600°C. The ingot Zn$_{0.20}$Cu$_{1.80}$Se obtained by annealing

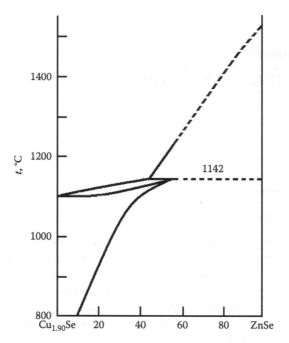

FIGURE 2.1
ZnSe–Cu$_{1.9}$Se phase diagram. (From Mizetskaya, I.B. et al., *Izv. AN SSSR. Neorgan. materialy*, 18(11), 1792, 1982.)

at 100°C for 350 h crystallizes in an orthorhombic structure with the lattice parameters $a = 411.7$, $b = 703.7$, and $c = 2039.4$ pm.

2.2 Zinc–Silver–Selenium

ZnSe–Ag$_2$Se: The phase diagram belongs to the eutectic type (Figure 2.2) (Trishchuk et al. 1982). The eutectic composition and temperature are 13.5 mol. % ZnSe and 850°C, respectively. The solubility of ZnSe in β-Ag$_2$Se at 850°C, 800°C, 700°C, 600°C, and 545°C is equal correspondingly to 12, 8, 4.5, 2, and 1 mol. % (Trishchuk et al. 1982, 1985). Thermal effects at 140°C correspond to a phase transition of Ag$_2$Se.

There is (6 ± 1) 10^{-3} at. % Ag in the ZnSe single crystals, growing at 1530°C and 10 MPa (100 atm) and using Ag$_2$Se as a dopant (Fadeev et al. 1977).

This system was investigated using differential thermal analysis (DTA) and metallography (Trishchuk et al. 1982, 1985).

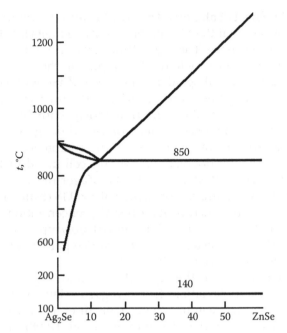

FIGURE 2.2

ZnSe–Ag$_2$Se phase diagram. (From Trishchuk, L.I. et al., *Izv. AN SSSR. Neorgan. materialy*, 18(11), 1795, 1982.)

2.3 Zinc–Beryllium–Selenium

ZnSe–BeSe: The phase diagram is not constructed. Zn$_{1-x}$Be$_x$Se solid solutions were grown from ZnSe and Be components by the high-pressure Bridgman method in the composition range $0 < x < 0.18$ (Firszt et al. 1998, Paszkowicz et al. 1998). They crystallize in the investigated range of composition with the lattice constant decreasing with x value increasing.

The calculated lattice constant of the Zn$_{1-x}$Be$_x$Se solid solutions scales linearly with composition, thus obeying Vegard's law (Paszkowicz et al. 1998, Kumar et al. 2009), and the calculated band gap could be fitted with a quadratic equation $E_g(x) = ax^2 + bx + c$ where a, b, and c are constants (Kumar et al. 2009). The single crystals were grown using the Bridgman method (Paszkowicz et al. 1998).

2.4 Zinc–Magnesium–Selenium

ZnSe–MgSe: The phase diagram is not constructed. X-ray investigations show that with increasing Mg content, the transition from sphalerite structure to wurtzite for the Zn$_{1-x}$Mg$_x$Se solid solutions occurs at

$x = 0.185 \pm 0.030$ (Firszt et al. 1995). The equilibrium distribution coefficient of Mg in ZnSe is near 1 at the Mg contents in ZnSe melts less than 0.6 at. % (Kulakov and Fadeev 1986). The crystalline structure of the solid-solution single crystals is very sensitive to the Mg doping in the range $0.03 < x < 0.06$ (Fedorov et al. 2001). Initial sphalerite lattice of ZnSe with comparatively large twinned blocks gains more ordered and thin twins and becomes closed to polytype modification simultaneously with Mg concentration increase. The most pronounced effect of Mg doping was observed at approximately $x = 0.06$. It was expressed in the emergence of well-ordered hexagonal structure with doubled elementary cell in c direction that corresponds to 4H polytype (Fedorov et al. 2001, Zagoruiko et al. 2005). Single crystals with cubic structure can be grown at $x < 0.18$ (Kulakov and Fadeev 1986) and with wurtzite structure at higher Mg concentrations (Paszkowicz et al. 1996, 1998). At $x = 0.15–0.21$, 4H and 8H polytypes were found [4H polytype at $x = 0.21$ is characterized by the lattice parameters $a = 404.1$ and $c = 1321.4$ pm (Paszkowicz et al. 1996)].

The lattice parameters of the $Zn_{1-x}Mg_xSe$ solid solutions change not linearly, and a small deviation from Vegard's law with lattice bowing equal to −3 pm has been observed (Charifi et al. 2005) (according to the data of Firszt et al. (1995), Paszkowicz et al. (1998), the sphalerite/ wurtzite lattice parameters change linearly with the magnesium content). The bulk modulus of these solid solutions decreases on increasing the Mg concentration, which is due to the increase in the ionicity of bonds. A significant deviation of the bulk modulus from linear concentration dependence was found.

According to the data of Falke et al. (1998), $Zn_{1-x}Mg_xSe$ solid solutions $(0.15 < x < 0.21)$ show a spatial transition from the wurtzite (2H) to the sphalerite (3C) structure related to Mg concentration gradient. In the transition range, the 4H and 8H structure was found, but with no other polytypes (Falke et al. 1998, Paszkowicz et al. 1999). The perfection of the structures decreases from 2H to 8H.

As the Mg concentration increases, the band-gap $Zn_{1-x}Mg_xSe$ single crystals increase from 2.7 eV at $x = 0.03$ to 3.0 eV at $x = 0.44$ (Zagoruiko et al. 2005).

High-pressure x-ray energy-dispersive measurements performed for $Zn_{0.53}Mg_{0.47}Se$ on uploading up to 5.17 GPa and downloading show that the wurtzite-type structure is conserved in the studied pressure range with possible first traces of a phase transition at the highest applied pressures (Paszkowicz et al. 2004). Bulk modulus value derived from the experimental unit-cell volume pressure dependence for this solid solution is found to be 48.0 GPa, and its first pressure derivative is 4.86.

The $Zn_{1-x}Mg_xSe$ single crystals were grown using the Bridgman method within the interval of $0 \le x \le 0.63$ [$0 \le x \le 0.21$ (Kulakov and Fadeev 1986)] (Paszkowicz et al. 1998) and using the modified Bridgman method under high argon pressure (2.03 MPa) within the interval of $0.03 \le x \le 0.44$ [$0.06 \le x \le 0.285$ (Firszt et al. 1995)] (Zagoruiko et al. 2005).

2.5 Zinc–Cadmium–Selenium

ZnSe–CdSe: The phase diagram is shown in Figure 2.3 (Mizetskaya et al. 1978, Oleinik et al. 1978, Kulakov et al. 1989, Gan'shin et al. 1993). At low temperature, solid solutions based on CdSe and ZnSe crystallize in the wurtzite and sphalerite structure correspondingly (Goriunova et al. 1955a, Savickas and Baršauskas 1960, Sirota and Yanovich 1972, Mizetskaya et al. 1978, Oleinik et al. 1978, Kulakov et al. 1989, Kulakov and Baliakina 1990, Gan'shin et al. 1993, Kirovskaya and Budanova 2001). Sphalerite–wurtzite transition for CdSe and ZnSe takes place at 95°C ± 5°C and 1425°C, respectively (Kulakov et al. 1976, Fedorov et al. 1991). At a decreasing temperature, the two-phase region moves to higher CdSe concentrations (Shalimova et al. 1969, Sirota and Yanovich 1972, Mizetskaya et al. 1978, Oleinik et al. 1978, Leute and Wulff 1992, Gan'shin et al. 1993). The width of the miscibility gap in the temperature range of 530°C–1020°C must be smaller than 5 mol. % (Leute and Wulff 1992).

Solid solutions of both structure types satisfy the model of regular solutions, and calculated phase diagram in the subsolidus region coincides with experimental data (Gan'shin et al. 1993).

At high pressures, $Zn_xCd_{1-x}Se$ solid solutions undergo phase transitions to a cubic structure of NaCl type (Skums et al. 1992). The pressure that corresponds to the beginning of the phase transition is a function of composition according to a linear function p_i (GPa) = 2.55 + 10.40x.

Lattice parameters of the solid solutions with both hexagonal and cubic structures change according to the Vegard's law (Zherdev and Ormont 1960,

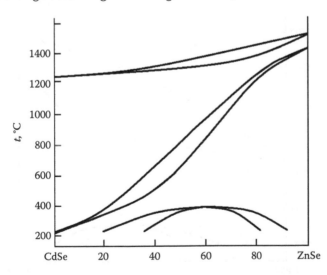

FIGURE 2.3
ZnSe–CdSe phase diagram. (From Mizetskaya, I.B. et al., *Ukr. khim. zhurn.*, 44(2), 163, 1978; Oleinik, G.S. et al., *Izv. AN SSSR. Neorgan. materialy*, 14(3), 441, 1978; Kulakov, M.P. et al., *Izv. AN SSSR. Neorgan. materialy*, 25(10), 1637, 1989; Gan'shin, V.A. et al., *Zhurn. neorgan. khimii*, 38(3), 533, 1993.)

Shalimova et al. 1969). Concentration dependence of energy gap for the solid solutions with wurtzite and sphalerite structure has small deviations from linearity (Zherdev and Ormont 1960a,b, Kot et al. 1962).

This system was investigated using DTA, metallography, and XRD (Mizetskaya et al. 1978, Oleinik et al. 1978, Kulakov et al. 1989). The ingots were annealed at 800°C, 900°C, 1000°C, 1100°C, and 1200°C (Shalimova et al. 1969). Phase transition in the solid solutions was determined using the investigation of exchange process between cations from II–VI semiconductor crystals and cations from the melts or solutions of salts. Such process takes place under equilibrium conditions, since the composition of the produced solid solutions is independent of the process duration (Fedorov et al. 1991, Gan'shin et al. 1993). Solid solutions $Zn_xCd_{1-x}Se$ could be obtained by a method of isothermal diffusion (Kirovskaya and Budanova 2001).

2.6 Zinc–Mercury–Selenium

ZnSe–HgSe: This system was investigated using the DTA. Phase diagram was calculated using the model of regular associated solutions. The solid/liquid equilibria were calculated on the assumption of the melt ideality (Gavaleshko et al. 1983, Leute and Plate 1989).

The phase diagram is shown in Figure 2.4 (Gavaleshko et al. 1983, Leute and Plate 1989). Solid solutions with sphalerite structure are formed

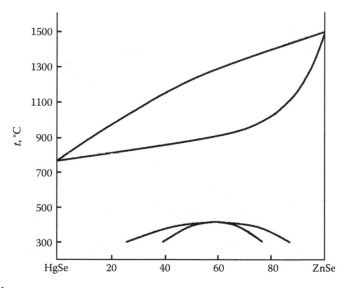

FIGURE 2.4
ZnSe–HgSe phase diagram. (From Gavaleshko, N.P. et al., *Izv. AN SSSR. Neorgan. materialy*, 10(2), 327, 1983; Leute, V. and Plate, H., *Ber. Bunsenges. Phys. Chem.*, 93(7), 757, 1989.)

over the entire concentrations range. Lattice parameter and energy gap change linearly with composition (Kot et al. 1962, Krucheanu et al. 1963, Kot and Simashkevich 1964, Leute and Plate 1989). The calculated critical values for spinodal demixing are 59 mol. % ZnSe and 443°C (Leute and Plate 1989).

Single crystals of $Zn_xHg_{1-x}Se$ solid solutions with $x=0.08–0.115$ were obtained using the Bridgman method (Cobb et al. 1991).

2.7 Zinc–Aluminum–Selenium

According to the data of Gamidov and Kulieva (1969), $ZnAlSe_2$ exists in the Zn–Al–Se ternary system. This compound crystallizes in a tetragonal structure of the chalcopyrite type with lattice parameters $a = 553.2$ and $c = 1091.4$ pm.

$ZnSe–Al_2Se_3$: The phase diagram is not constructed. $ZnAl_2Se_4$ is formed in this system. It crystallizes in a tetragonal structure of chalcopyrite type with lattice parameters $a = 549.2$ and $c = 1088$ pm (Hahn et al. 1955) [$a = 550$ and $c = 1090$ pm (Range et al. 1968), according to the calculations $a = 553$ and $c = 1095.5$ pm (Mishra and Ganguli 2011)] and calculation and experimental density 4.376 [4.37 (Range et al. 1968)] and 4.37 g cm^{-3} (Hahn et al. 1955). This compound is direct band-gap semiconductor with calculation energy band gap of 2.82 eV and bulk modulus of 52.06 GPa (Mishra and Ganguli 2011). At high temperature and pressure (4.5 GPa or 45 kbar and 400°C), this compound can exist in a cubic structure of a spinel type with lattice parameter $a = 1061$ pm and calculation density $d = 4.84$ g cm^{-3} (Range et al. 1968).

The Raman data analysis of $ZnAl_2Se_4$ reveals general trend for different defect chalcopyrite materials (Meenakshi et al. 2006). The line widths of the Raman peaks change at intermediate pressures between 4 and 6 GPa as indication of the pressure-induced two-stage order–disorder transitions observed in this compound.

$ZnAl_2Se_4$ was obtained from ZnSe and powderlike Al and Se in equimolar ratio. The mixtures were pressed and annealed at 800°C for 12–24 h (Hahn et al. 1955).

2.8 Zinc–Gallium–Selenium

Isothermal section of Zn–Ga–Se ternary system at 700°C–800°C is shown in Figure 2.5 (Dvoretskov et al. 1982).

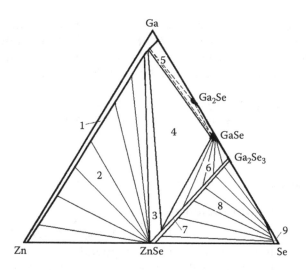

FIGURE 2.5

Isothermal section of Zn–Ga–Se ternary system at 700°C–800°C: 1, L(Ga + Zn); 2, ZnSe + L(Ga + Zn); 3, ZnSe + GaSe + α; 4, GaSe + α + L(Ga + Zn); 5, GaSe + L(Ga+Zn); 6, GaSe + α: 7, α; 8, α + L(Se); and 9, L(Se). (From Dvoretskov, G.A. et al., Solubility isotherm of gallium in cadmium and zinc selenides [in Russian], in: *Legir. poluprovodn.*, Moscow, Russia, Nauka, pp. 67–71, 1982.)

TABLE 2.1

Liquidus Temperatures in the ZnSe–Ga System

x_{ZnSe} mol. %	t, °C	x_{ZnSe} mol. %	t, °C
0.0506	607	1.20	912
0.120	658	2.36	1012
0.244	723	4.61	1106
0.485	795		

Source: Wagner, P. and Lorenz, M.R., *J. Phys. Chem. Solids*, 27(11/12), 1749, 1966.

ZnSe–Ga: The phase diagram is not constructed. The eutectic is degenerated from the Ga-rich side (Wagner and Lorenz 1966). Liquidus temperatures are given in Table 2.1. The liquidus line is straight in logarithmic coordinate; therefore, this system is almost ideal (Vasiliev and Novikova 1977).

Liquidus temperatures were determined by visual observation of solid-phase appearance and disappearance for heating and cooling of ingots (accuracy of measurement is ±10°C). Compositions of the ingots were investigated using XRD (Wagner and Lorenz 1966).

ZnSe–GaSe: The phase diagram is a eutectic type (Vishniakov et al. 1981). The eutectic composition and temperature are equal to 5 mol. % ZnSe and 907°C ± 10°C, respectively. The mutual solubility was not found.

ZnSe–Ga₂Se₃: The phase diagram is a peritectic type (Figure 2.6) (Olekseyuk et al. 1996). The peritectic composition and temperature are

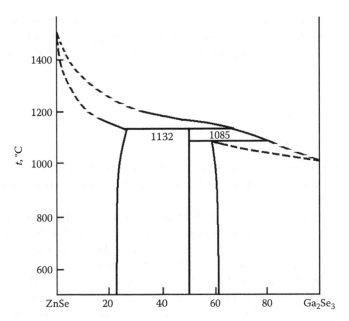

FIGURE 2.6
ZnSe–Ga$_2$Se$_3$ phase diagram. (From Olekseyuk, I.D. et al., *Pol. J. Chem.*, 70(9), 1111, 1996.)

equal to 67 mol. % Ga$_2$Se$_3$ and 1132°C, respectively. The peritectic reaction at 1085°C corresponds to the formation of the solid solutions based on Ga$_2$Se$_3$. The solubility of Ga$_2$Se$_3$ in ZnSe at 600°C is equal to 23 mol. % and Ga$_2$Se$_3$ dissolves 39 mol. % ZnSe at the same temperature.

According to Hahn et al. (1955), Beun et al. (1961), Veliev et al. (1976), Rigan and Stasiuk (1991), Olekseyuk et al. (1996), and Errandonea et al. (2008), ZnGa$_2$Se$_4$ is formed in the ZnSe–Ga$_2$Se$_3$ system. This compound melts incongruently at 1132°C (Olekseyuk et al. 1996) [at 1157°C (Veliev et al. 1976)] and crystallizes in a tetragonal structure with lattice parameters $a = 550.1 \pm 0.2$ and $c = 1099.9 \pm 0.6$ pm (Olekseyuk et al. 1996) [$a = 551$ and $c = 1103$ pm (Veliev et al. 1976), $a = 548.5$ and $c = 1097$ pm (Hahn et al. 1955), $a = 547$ and $c = 1088$ pm (Rigan and Stasiuk 1991)] and calculation and experimental density 5.215 and 5.13 [5.12–5.16 (Rigan and Stasiuk 1991)] g cm^{-3} (Hahn et al. 1955).

At a compression, ZnGa$_2$Se$_4$ exhibits a defect tetragonal stannite-type structure up to 15.5 GPa, and in the range from 15.5 to 18.5 GPa, the low-pressure phase coexists with a high pressure, which remains stable up to 23.0 GPa (Errandonea et al. 2008). High-pressure phase has been characterized as a defect-cubic NaCl-type structure.

According to the data of Goriunova et al. (1955b), Goriunova (1955), and Dvoretskov et al. (1982), the solid solutions over the entire range of concentrations are formed in the ZnSe–Ga$_2$Se$_3$ system.

Single crystals of $ZnGa_2Se_4$ were grown using the chemical transport reactions (Beun et al. 1961, Rigan and Stasiuk 1991). The homogenization of the alloys was carried out by their annealing at 600°C for 250 h (Olekseyuk et al. 1996).

2.9 Zinc–Indium–Selenium

ZnSe–In: The phase diagram is not constructed. The eutectic is degenerated from the In-rich side (Wagner and Lorenz 1966). Liquidus temperatures are given in Table 2.2.

Liquidus temperatures were determined by visual observation of solid-phase appearance and disappearance for heating and cooling of ingots (accuracy of measurement is ±10°C). Compositions of the ingots were investigated using XRD (Wagner and Lorenz 1966).

ZnSe–In$_2$Se$_3$: The phase diagram is shown in Figure 2.7 (Krämer et al. 1987). The $ZnIn_2Se_4$ ternary compound, which melts incongruently at 960°C [980°C (Veliev et al. 1976)], is formed in this system. It has a phase transition at 45°C±5°C (Krämer et al. 1987) and crystallizes in a tetragonal structure with the lattice parameters $a = 570.5$ and $c = 1144.8$ pm (Gastaldi et al. 1987, Marsh and Robinson, 1988) [$a = 570.95$ and $c = 1144.9$ pm (Trah and Krämer 1985), $a = 515.7$ and $c = 1143$ pm (Veliev et al. 1976), $a = 569.9$ and $c = 1140$ pm (Hahn et al. 1955), $a = 404.5$ and $c = 5229$ pm (Abdullayev et al. 1989)]. Its calculation and experimental density $d = 5.443$ and 5.36 g cm^{-3} (Hahn et al. 1955) and energy gap $E_g = 1.68$ eV (Abdullayev et al. 1989) [$E_g = 1.74$–1.82 eV (Beun et al. 1961); $E_g = 2.6$ eV (Busch et al. 1956)]. Superlattice with $a' = a \sqrt{3}$ was determined for $ZnIn_2Se_4$ (Abdullayev et al. 1989). Haeuseler and Himmrich (1986) report on a second intermediate compound of $2ZnSe \cdot 5In_2Se_3$ ($Zn_{0.4}In_2Se_{3.4}$) composition that cannot be confirmed unequivocally (Krämer et al. 1987).

Chemical changes during effusion of substances in the binary ZnSe–$ZnIn_2Se_4$ subsystem were observed using the simultaneous Knudsen and

TABLE 2.2

Liquidus Temperatures in the ZnSe–In System

x_{ZnSe}, mol. %	t, °C	x_{ZnSe}, mol. %	t, °C
0.0999	529	3.01	900
0.200	587	5.49	991
0.400	646	10.0	1086
0.779	722	20.0	1205
1.51	801		

Source: Wagner, P. and Lorenz, M.R., *J. Phys. Chem. Solids*, 27(11/12), 1749, 1966.

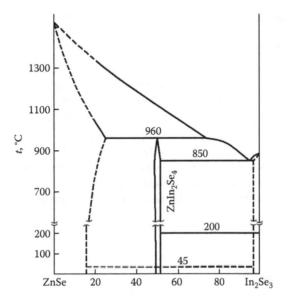

FIGURE 2.7
ZnSe–In$_2$Se$_3$ phase diagram. (From Krämer, V. et al., *Thermochim. Acta*, 112(1), 88, 1987.)

torsion-effusion method (Edwards et al. 1999). Principal vapor species were taken to be Zn(g), In$_2$Se(g), and Se$_2$(g). Two intermediate solid phases were found in the temperature range 830°C–930°C, ZnIn$_2$Se$_4$ and Zn$_5$In$_2$Se$_8$. The latter phase vaporized congruently throughout the temperature range, but it effused congruently only above 880°C; it was not observed at room temperature. ZnIn$_2$Se$_4$ effused incongruently to a vapor rich in ZnS and a condensed phase rich in indium selenide. Binary pressure–composition phase diagrams of this system have been proposed at approximately 850°C and 920°C, and equations for all effusion reactions in the system have been presented. Stability of the two intermediate condensed phases relative to the chemically binary constituents are small; $\Delta H^0_{298} = 12.5 \pm 3.5$ kJ mol^{-1} for ZnIn$_2$Se$_4$, and for the other phase, it was less than the experimental error.

A phase α with assumed formula (ZnSe)$_5$In$_2$Se$_3$ exists and effuses congruently above 880°C and incongruently below 880°C but becomes unstable below 830°C; a reasonable estimate of temperature at onset of this instability is 810°C (Edwards et al. 1999).

Solubility of In$_2$Se$_3$ in ZnSe, determined using the XRD, is equal to 20 mol. % (Hahn et al. 1955). There is approximately (16 ± 3) 10^{-2} at. % In in the ZnSe single crystals, growing at 1530°C and 10 MPa (100 atm) and using In$_2$Se$_3$ as a dopant (Fadeev et al. 1977).

ZnIn$_2$Se$_4$ was obtained by interaction of stoichiometric quantities of ZnSe and In$_2$Se$_3$ at 900°C for 4 days with next cooling to 600°C with the rate of 1°C h^{-1} (Gastaldi et al. 1987, Marsh and Robinson, 1988). Its single crystals were grown using the chemical transport reactions (Beun et al. 1961, Trah and Krämer 1985, Abdullayev et al. 1989).

2.10 Zinc–Thallium–Selenium

ZnSe–TlSe: The phase diagram is not constructed. According to the data of Guseinov et al. (1982), the phase diagram belongs to the eutectic type. The eutectic is degenerated from TlSe-rich side and crystallizes at 342°C. Solubility of TlSe in ZnSe and ZnSe in TlSe is not higher than 1 mol. %.

This system was investigated using DTA and XRD and measuring of microhardness. The ingots were annealed at the temperatures 20°C–30°C lower than solidus temperature for 400–500 h (Guseinov et al. 1982).

ZnSe–Tl$_2$Se$_3$: The phase diagram is not constructed. According to the data of Guseinov et al. (1981), ZnTl$_2$Se$_4$ ternary compound is not formed in this system.

2.11 Zinc–Lanthanum–Selenium

ZnSe–La$_2$Se$_3$: The phase diagram is not constructed. ZnLa$_2$Se$_4$ is formed in this system (Aliev 1998). It melts incongruently at 1427°C. This compound is semiconductor with energy gap 1.86 eV and calculation density 5.68 g cm^{-3}.

ZnLa$_2$Se$_4$ was obtained using the interaction of ZnSe and La$_2$Se$_3$ at 800°C–900°C with next annealing at 730°C for 150 h, cooling up to 530°C, and the annealing at this temperature for 250 h (Aliev 1998). Single crystals of this compound were grown using the chemical transport reactions using iodine as the transport agent.

2.12 Zinc–Carbon–Selenium

ZnSe–C: Carbon content in ZnSe crystals has been determined by direct chemical analysis as well as using the carbon concentrating by way of an acid treatment (Komar 1998). Carbon content on the level of 3×10^{-2} mass. % has been found.

2.13 Zinc–Silicon–Selenium

Ternary compounds in the Zn–Si–Se system were not found (Kaldis et al. 1967).

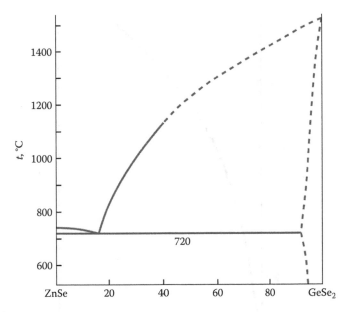

FIGURE 2.8
ZnSe–GeSe₂ phase diagram. (From Olekseyuk, I. et al., *Quasiternary chalcogenide systems* [in Ukrainian], Vol. 1, Luts'k, Ukraine, Vezha Publishing, 168p., 1999.)

2.14 Zinc–Germanium–Selenium

Ternary compounds in the Zn–Ge–Se system were not found (Kaldis et al. 1967).

$ZnSe–GeSe_2$: The phase diagram is a eutectic type (Figure 2.8) (Olekseyuk et al. 1999). The eutectic composition and temperature are 16 mol. % ZnSe and 720°C. $ZnGeSe_3$ and Zn_2GeSe_4 ternary compounds (Hahn and Lorent 1958, Radautsan and Ivanova 1961) were not found in this system (Kaldis et al. 1967, Koren' et al. 1984, Olekseyuk et al. 1999). To obtain the compounds in this system, four methods were used: (1) synthesis at 7 MPa and heating to 2000°C, (2) alloying of the components with vibrational mixing, (3) chemical transport reactions, and (4) sublimation. All ingots were two-phase and contained the mixtures of ZnSe and $GeSe_2$ (Koren' et al. 1984).

2.15 Zinc–Tin–Selenium

ZnSe–Sn: The phase diagram is not constructed. The ZnSe solubility in liquid tin (Figure 2.9) (Rubenstein 1968, Kimura 1971) and a part of Zn–Sn–Se liquidus surface on the Sn-rich side (Figure 2.10) (Kimura 1971) were studied.

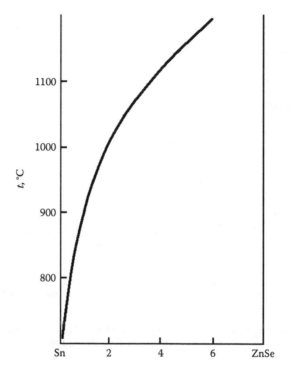

FIGURE 2.9
Temperature dependence of ZnSe solubility in liquid Sn. (From Kimura, S., *J. Chem. Thermodyn.*, 3(1), 7, 1971. With permission.)

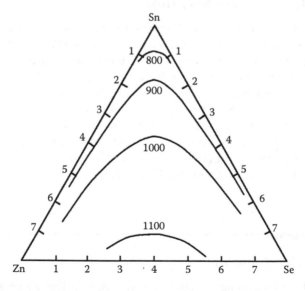

FIGURE 2.10
Part of the Zn–Sn–Se liquidus surface on the Sn-rich side. (From Kimura, S., *J. Chem. Thermodyn.*, 3(1), 7, 1971. With permission.)

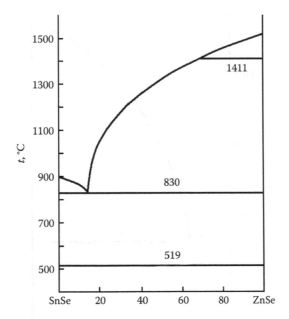

FIGURE 2.11
ZnSe–SnSe phase diagram. (From Dohnke, I. et al., *J. Cryst. Growth*, 198–199, 287, 1999.)

The eutectic is degenerated from the Sn-rich side. Liquidus is described using the equation of regular solutions with constant interaction parameter (Vasiliev and Novikova 1977).

Liquidus temperatures were determined using high-temperature filtration (Rubenstein 1968) and visual observation of solid-phase appearance and disappearance for heating and cooling of ingots (accuracy of measurement is ±5°C) (Kimura 1971).

ZnSe–SnSe: The phase diagram is a eutectic type (Figure 2.11) (Dohnke et al. 1999). The eutectic composition and temperature are 14 mol. % ZnSe and 830°C [850°C (Galiulin et al. 1982)]. At 519°C [517°C (Galiulin et al. 1982)], solid solutions based on SnSe undergo polymorphous transformation. Mutual solubility of ZnSe and SnSe is not higher than 0.5 mol. % (according to the data of Dohnke et al. (1999), there is practically no mutual solubility of SnSe and ZnSe).

This system was investigated using DTA, metallography, and XRD. The ingots were annealed at 550°C for 250 h (Galiulin et al. 1982, Dohnke et al. 1999).

ZnSe–SnSe₂: The phase diagram is a eutectic type (Figure 2.12) (Parasyuk et al. 2004). The eutectic composition and temperature are 13 mol. % ZnSe and 622°C [16 mol. % ZnSe and 642°C (Galiulin et al. 1982)], respectively. The solubility of SnSe₂ in ZnSe reaches 1 mol. % and the solubility of ZnSe in SnSe₂ is not higher than 0.5 mol. % (Galiulin et al. 1982, Parasyuk et al. 2004).

This system was investigated using DTA, metallography, and XRD (Galiulin et al. 1982, Parasyuk et al. 2004). The ingots were annealed at 550°C for 250 h (Galiulin et al. 1982).

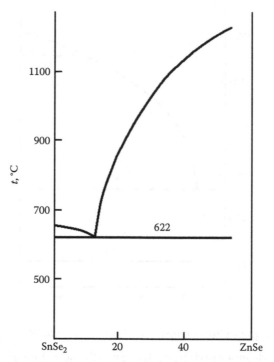

FIGURE 2.12
ZnSe–SnSe$_2$ phase diagram. (From Parasyuk, O.V. et al., *J. Alloys Compd.*, 379(1–2), 143, 2004.)

2.16 Zinc–Lead–Selenium

ZnSe–PbSe: The phase diagram is a eutectic type (Figure 2.13) (Oleinik et al. 1982). The eutectic composition and temperature are 22 mol. % ZnSe and 1010°C, respectively. Mutual solubility of ZnSe and PbSe is not higher than 1 mol. %.

This system was investigated using DTA, metallography, and XRD (Oleinik et al. 1982).

2.17 Zinc–Phosphorus–Selenium

ZnSe–"P$_2$Se$_4$": Ternary compounds were not found in this system (Kovach et al. 1997). The solubility of "P$_2$Se$_4$" in ZnSe reaches 28 mol. %. Only one exothermic effect at 606°C has been observed at the heating of the ingots of the ZnSe–"P$_2$Se$_4$" system that apparently is connected with the beginning of the ZnSe dissociation.

This system was investigated using DTA and XRD. The ingots were annealing at 530°C for 2 weeks (Kovach et al. 1997).

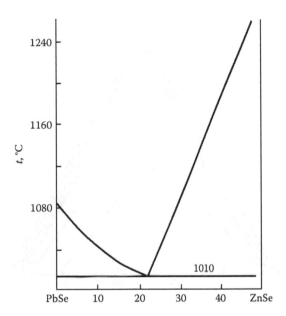

FIGURE 2.13

ZnSe–PbSe phase diagram. (From Oleinik, G.S. et al., *Izv. AN SSSR. Neorgan. materialy*, 18(5), 873, 1982.)

2.18 Zinc–Arsenic–Selenium

Liquidus surface of Zn–As–Se ternary system includes eight fields of primary crystallization of α- and β-solid solutions, Zn, $ZnAs_2$, As, AsSe, As_2Se_3, and Se (Figure 2.14) (Olekseyuk and Stoika 1978). There are also six invariant equilibria in this system (Table 2.3). The field of primary crystallization of solid solutions based on ZnSe occupies a greater part of liquidus surface. Ternary compounds were not obtained in the Zn–As–Se system.

Isothermal section at 340°C is characterized by the existence of two regions of solid solutions based on Zn_3As_2, which are divided by the two-phase region and elongated along the $ZnSe–Zn_3As_2$ system (Figure 2.15) (Olekseyuk and Stoika 1978).

ZnSe–ZnAs₂: The phase diagram is not constructed. According to Olekseyuk and Stoika (1978), the phase diagram belongs to the eutectic type, and the eutectic composition and temperature are 1.5 mol. % ZnSe and 753°C, respectively.

This system was investigated using DTA, metallography, and XRD. The ingots were annealed at the temperatures 20°C–40°C lower than solidus temperature for 200–1500 h (Olekseyuk and Stoika 1978).

ZnSe–Zn₃As₂: The phase diagram belongs to the peritectic type (Figure 2.16) (Olekseyuk and Stoika 1978, Olekseyuk et al. 1978). The peritectic temperature

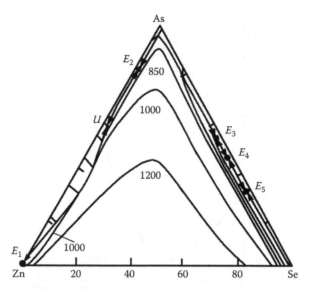

FIGURE 2.14

Liquidus surface of the Zn–As–Se ternary system. (From Olekseyuk, I.D. and Stoika, I.M., *Zhurn. neorgan. khimii*, 23(9), 2496, 1978.)

TABLE 2.3

Invariant Equilibria in the Zn–As–Se Ternary System

Symbol	Reaction	t, °C
E_1	$L \Leftrightarrow ZnSe + Zn_3As_2 + Zn$	418
E_2	$L \Leftrightarrow ZnSe + ZnAs_2 + As$	716
E_3	$L \Leftrightarrow ZnSe + AsSe + As$	230
E_4	$L \Leftrightarrow ZnSe + As_2Se_3 + AsSe$	155
E_5	$L \Leftrightarrow ZnSe + As_2Se_3 + S$	140
U	$L + ZnSe \Leftrightarrow Zn_3As_2 + ZnAs_2$	751

Source: Olekseyuk, I.D. and Stoika, I.M., *Zhurn. neorgan. khimii*, 23(9), 2496, 1978.

is 1055°C ± 5°C. At 340°C, solid solutions based on α- and β-Zn_3As_2 exist within the range of 0–5 and 7.5–16 mol. % ZnSe, respectively (according to the data of Golovey et al. (1972), the solubility of ZnSe in Zn_3As_2 reaches up to 10 mol. %). At the same temperature, the solubility of Zn_3As_2 in ZnSe is not higher than 1 mol. %.

This system was investigated using DTA, metallography, and XRD. The ingots were annealed at the temperatures 20°C–40°C lower than solidus temperature for 200–1500 h (Olekseyuk and Stoika 1978, Olekseyuk et al. 1978).

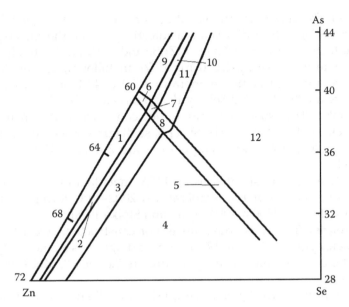

FIGURE 2.15
Isothermal section of Zn–As–Se ternary system at 340°C in the region of solid solutions based on Zn_3As_2: 1, α-Zn_3As_2+Zn; 2, α-Zn_3As_2+β-Zn_3As_2+Zn; 3, β-Zn_3As_2+Zn; 4, ZnSe+β-Zn_3As_2; 5, β-Zn_3As_2+ZnSe; 6, α-Zn_3As_2; 7, α-Zn_3As_2+β-Zn_3As_2; 8, β-Zn_3As_2; 9, α-Zn_3As_2 + $ZnAs_2$; 10, α-Zn_3As_2+β-Zn_3As_2+$ZnAs_2$; 11, β-Zn_3As_2+$ZnAs_2$; and 12, ZnSe + β-Zn_3As_2 + $ZnAs_2$. (From Olekseyuk, I.D. and Stoika, I.M., *Zhurn. neorgan. khimii*, 23(9), 2496, 1978.)

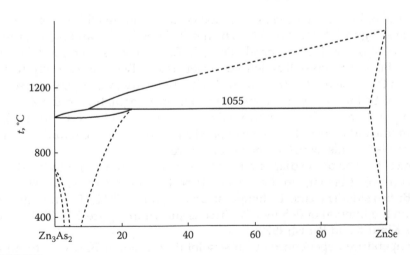

FIGURE 2.16
$ZnSe$–Zn_3As_2 phase diagram. (From Olekseyuk, I.D. and Stoika, I.M., *Zhurn. neorgan. khimii*, 23(9), 2496, 1978; Olekseyuk, I.D. et al., *Izv. AN SSSR. Neorgan. Materialy*, 14(8), 1383, 1978.)

ZnSe–As: The phase diagram is not constructed. According to Olekseyuk and Stoika (1978), a phase diagram is a eutectic type, and the eutectic composition and temperature are 0.7 mol. % ZnSe and 806°C, respectively.

This system was investigated using DTA, metallography, and XRD. The ingots were annealed at the temperatures 20°C–40°C lower than solidus temperature for 200–1500 h (Olekseyuk and Stoika 1978).

ZnSe–AsSe: The phase diagram is not constructed. According to the data of Olekseyuk and Stoika (1978), a phase diagram is a eutectic type, and the eutectic composition and temperature are 2 mol. % ZnSe and 270°C, respectively.

This system was investigated using DTA, metallography, and XRD. The ingots were annealed at the temperatures 20°C–40°C lower than solidus temperature for 200–1500 h (Olekseyuk and Stoika 1978).

ZnSe–As$_2$Se$_3$: The phase diagram is not constructed. According to the data of Olekseyuk and Stoika (1978), a phase diagram is a eutectic type, and the eutectic composition and temperature are 2.6 mol. % ZnSe and 360°C, respectively.

This system was investigated using DTA, metallography, and XRD. The ingots were annealed at the temperatures 20°C–40°C lower than solidus temperature for 200–1500 h (Olekseyuk and Stoika 1978).

2.19 Zinc–Bismuth–Selenium

The field of ZnSe primary crystallization occupies almost all Zn–Bi–Se ternary systems (Figure 2.17) (Odin 1996). The fields of BiSe, Bi$_2$Se$_3$, and Bi$_3$Se$_2$ primary crystallization form narrow bands along the Bi–Se binary system, and the fields of Zn, Bi, and Se crystallization are degenerated. Two immiscibility regions exist in this system: the first one at the Zn–Bi binary system and the second at the Bi–Se binary system. Three- and four-phase equilibria in the Zn–Bi–Se ternary system with liquid participation are given in Table 2.4 (Odin 1996).

Kimura and Panish (1970) investigated a part of Zn–Bi–Se liquidus surface on the Bi-rich side, which is shown in Figure 2.18.

ZnSe–Bi: The phase diagram is not constructed. According to the data of Odin (1996), it belongs to the eutectic type with degenerated eutectic from the Bi-rich side. The eutectic temperature is equal to 270°C. The liquidus line within the interval of 0–5 mol. % ZnSe is shown in Figure 2.19 (Rubenstein 1966, Kimura and Panish 1970).

Temperature dependence of ZnSe solubility in liquid Bi was determined using high-temperature filtration. The mixtures of ZnSe and Bi were annealed at low temperature for 16 h and at 1300°C for 6 h. The quantity of ZnSe dissolved in Bi was determined by weighting of the ingots before and after its treatment (Rubenstein 1966).

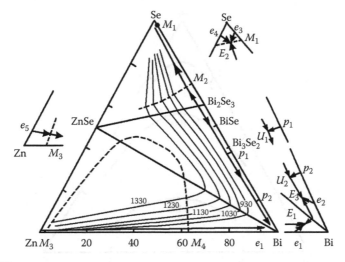

FIGURE 2.17

Liquidus surface of the Zn–Bi–Se ternary system. (From Odin, I.N., *Zhurn. neorgan. khimii*, 41(6), 941, 1996.)

TABLE 2.4

Three- and Four-Phase Equilibria with Liquid Participation in the Zn–Bi–Se Ternary System

Symbol	Reaction	t, °C
e_1	$L \Leftrightarrow Zn + Bi$	254
e_2	$L \Leftrightarrow Bi_3Se_2 + Bi$	270
e_3	$L \Leftrightarrow Bi_2Se_3 + Se$	217
e_4	$L \Leftrightarrow ZnSe + Se$	218
e_5	$L \Leftrightarrow ZnSe + Zn$	420
p_1	$L + Bi_2Se_3 \Leftrightarrow BiSe$	606
p_2	$L + BiSe \Leftrightarrow Bi_3Se_2$	470
m_2	$L + Bi_2Se_3 \Leftrightarrow L_1^*$	618
m_3	$L + Zn \Leftrightarrow L_2^*$	416
E_1	$L_2^* \Leftrightarrow Zn + ZnSe + Bi$	268
E_2	$L_1^* \Leftrightarrow ZnSe + Se + Bi_2Se_3$	215
E_3	$L_2^* \Leftrightarrow ZnSe + Bi_3Se_2 + Bi$	266
U_1	$L + Bi_2Se_3 \Leftrightarrow BiSe + ZnSe$	600
U_2	$L + BiSe \Leftrightarrow Bi_3Se_2 + ZnSe$	464

Source: Odin, I.N., *Zhurn. Neorgan. Khimii*, 41(6), 941, 1996.

L_1^*, Se-rich liquid.

L_2^*, Bi-rich liquid.

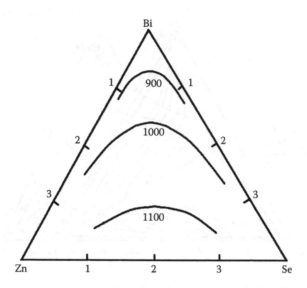

FIGURE 2.18
Part of the Zn–Bi–Se liquidus surface on the Bi-rich side. (From Kimura, S. and Panish, M.B., *J. Chem. Thermodyn.*, 2(1), 77, 1970. With permission.)

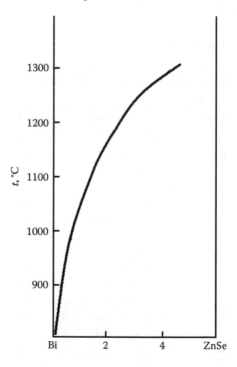

FIGURE 2.19
Temperature dependence of ZnSe solubility in liquid Bi. (From Rubenstein, M., *J. Electrochem. Soc.*, 113(6), 623, 1966; Kimura, S. and Panish, M.B., *J. Chem. Thermodyn.*, 2(1), 77, 1970. With permission.)

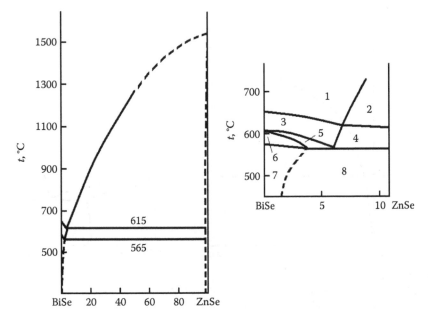

FIGURE 2.20
Phase relations in the ZnSe–BiSe system: 1, L; 2, L + ZnSe; 3, L + Bi$_2$Se$_3$; 4, L + Bi$_2$Se$_3$ + ZnSe; 5, L + BiSe + Bi$_2$Se$_3$; 6, L + BiSe; 7, BiSe; and 8, BiSe + ZnSe. (From Marugin, V.V. et al., *Zhurn. neorgan. khimii*, 28(8), 2104, 1983.)

ZnSe–BiSe: This section is a nonquasibinary section of the Zn–Bi–Se ternary system (Figure 2.20) (Marugin et al. 1983). There are lines of ZnSe and Bi$_2$Se$_3$ primary crystallization in this system. Secondary crystallization of ZnSe and Bi$_2$Se$_3$ takes place at 615°C. Peritectic equilibrium L + Bi$_2$Se$_3$ ⇔ ZnSe + BiSe exists in this system at 565°C. Solubility of ZnSe in BiSe at 500°C reaches 2 mol. %, and solubility of BiSe in ZnSe at the same temperature is less than 0.5 mol. %.

This system was investigated using DTA, metallography, and XRD. The ingots were annealed at 500°C for 500 h (Marugin et al. 1983).

ZnSe–Bi$_2$Se$_3$: The phase diagram is a eutectic type (Figure 2.21) (Marugin et al. 1983, Odin and Marugin 1991). The eutectic composition and temperature are 6 ± 1 mol. % ZnSe and 700°C ± 5°C. The solubility of ZnSe in Bi$_2$Se$_3$ at 500°C reaches 3.5 ± 0.3 mol. % and the solubility of Bi$_2$Se$_3$ in ZnSe at the same temperature is equal to less than 0.5 mol. %. Bi$_2$Se$_3$ has maximum melting temperature (706°C) at the content of 40.02 at. % Bi (Odin and Marugin 1991). Such small deviation from stoichiometry leads to too small deviation from quasibinarity.

This system was investigated using DTA, metallography, and XRD (Marugin et al. 1983, Odin and Marugin 1991). The ingots were annealed at 500°C for 720 h (Odin and Marugin 1991) [500 h (Marugin et al. 1983)].

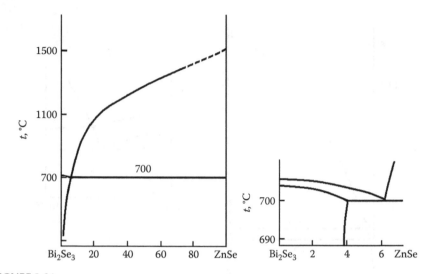

FIGURE 2.21
ZnSe–Bi$_2$Se$_3$ phase diagram. (From Marugin, V.V. et al., *Zhurn. neorgan. khimii*, 28(8), 2104, 1983; Odin, I.N. and Marugin, V.V., *Zhurn. neorgan. khimii*, 36(7), 1842, 1991.)

ZnSe–Bi$_3$Se$_2$: This section is a nonquasibinary section of the Zn–Bi–Se ternary system (Odin 1996). There are lines of ZnSe and BiSe primary crystallization in this system. Crystallization ends in the ternary peritectic L + BiSe ⇔ ZnSe + Bi$_3$Se$_2$. Mutual solubility of ZnSe and Bi$_3$Se$_2$ is insignificant.

This system was investigated using DTA, metallography, and XRD. The ingots were annealed at 20°C lower than temperature of nonvariant equilibrium with liquid participation for 1000 h (Odin 1996).

2.20 Zinc–Niobium–Selenium

Zn$_{0.50}$Nb$_6$Se$_8$ forms in the Zn–Nb–Se ternary system. This compound crystallizes in a hexagonal structure with lattice parameters $a = 999.5$ and $c = 346.49$ pm (Huan and Greenblatt 1987).

This compound was obtained by ion exchange in melts containing Tl$_x$Nb$_6$Se$_8$.

2.21 Zinc–Oxygen–Selenium

The phase diagram is not constructed. Isothermal section of the Zn–O–Se ternary system at room temperature, constructed according to the thermodynamic calculations, is shown in Figure 2.22 (Medvedev and Berchenko 1993).

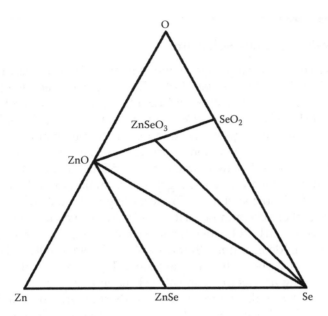

FIGURE 2.22
Isothermal section of the Zn–O–Se ternary system at room temperature. (From Medvedev, Yu.V. and Berchenko, N.N., *Zhurn. neorgan. khimii*, 38(12), 1940, 1993.)

ZnSe–ZnO: The phase diagram is not constructed. The solubility of ZnO in ZnSe at 1500°C is equal to 2 mol. % and decreases with decreasing temperature (Zuyev et al. 1981). At 300°C, the main product of ZnSe oxidation is $ZnSeO_3$. ZnO forms at the ZnSe oxidation at higher temperatures (Kulakov and Fadeev 1983).

ZnSe–SeO₂: The phase diagram is not constructed. $ZnSeO_3$ and $ZnSe_2O_5$ ternary compounds are formed in this system (Meunier and Bertaud 1974, Kohn et al. 1976, Hawthorne et al. 1986). $ZnSeO_3$ crystallizes in an orthorhombic structure with the lattice parameters $a = 623.3 \pm 0.1$, $b = 720.0 \pm 0.2$, and $c = 1198.7 \pm 0.3$ pm [$a = 592.31 \pm 0.04$, $b = 766.52 \pm 0.08$, and $c = 504.00 \pm 0.06$ pm (Kohn et al. 1976)] and calculation density 4.75 g cm^{-3} (Hawthorne et al. 1986). $ZnSe_2O_5$ also crystallizes in an orthorhombic structure with the lattice parameters $a = 679.7$, $b = 1041.2$, and $c = 606.8$ pm (Meunier and Bertaud 1974).

$ZnSeO_3$ and $ZnSe_2O_5$ were synthesized using direct solid-state reactions (Meunier and Bertaud 1974, Kohn et al. 1976). $ZnSeO_3$ could be obtained under high pressures and temperatures (Kohn et al. 1976). Equimolar amounts of ZnO and SeO₂ were intimately mixed and subjected to high pressures between 1.5 and 7.0 GPa (15 and 70 kbar). Reaction temperatures were between 50°C and 1000°C. Lengths of run time were varied from 20 to 60 min depending on the reaction temperature.

2.22 Zinc–Tellurium–Selenium

ZnSe–ZnTe: The phase diagram is shown in Figure 2.23 (Mizetskaya et al. 1978, Oleinik et al. 1978, Leute and Plate 1989). The composition and temperature of azeotropic point are 5 mol. % ZnSe and 1290°C [1275°C (Leute and Plate 1989)], respectively (Mizetskaya et al. 1978, Oleinik et al. 1978). Solid solutions are formed over the entire range of concentrations at the temperatures above 900°C (Larach et al. 1957). Concentration dependence of lattice parameters changes according to the Vegard's law (Larach et al. 1957, Leute and Plate 1989), and concentration dependence of energy gap has a minimum at 33 mol. % ZnSe (Larach et al. 1957, Rovnyi et al. 1977). According to the data of Kalinkin et al. (1984), solid solutions decompose on the solid solutions based on ZnSe and solid solutions based on ZnTe at temperatures less than 430°C. The calculated critical values for spinodal demixing are 50 mol. % ZnSe and 449°C (Leute and Plate 1989).

The activity of ZnTe in Te-saturated $ZnSe_xTe_{1-x}$ solid solutions has been determined using an electrochemical technique using molten salt electrolyte, and various partial, integral, and excess thermodynamic quantities have been calculated in Nasar (1995). It was shown that activities of ZnTe and ZnSe in the temperature range 437°C–572°C exhibit positive deviation from Raoult's law and decrease with the increase temperature. The continuous variation in activity and activity coefficient of both binary components with composition indicates that ZnSe and ZnTe is completely miscible in the solid state at the temperatures from 437°C to 572°C and consists of a single phase throughout the entire range of composition.

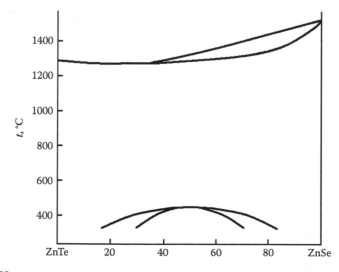

FIGURE 2.23

ZnSe–ZnTe phase diagram. (From Mizetskaya, I.B. et al., *Ukr. khim. zhurn.*, 44(2), 163, 1978; Oleinik, G.S. et al., *Izv. AN SSSR. Neorgan. materialy*, 14(3), 441, 1978; Leute, V. and Plate, H., *Ber. Bunsenges. Phys. Chem.*, 93(7), 757, 1989.)

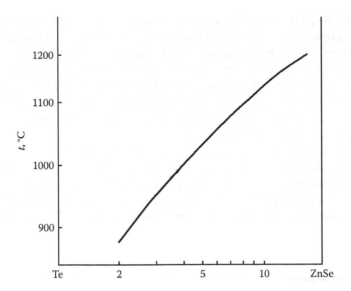

FIGURE 2.24
Temperature dependence of ZnSe solubility in liquid Te. (From Masako, W. et al., *Jap. J. Appl. Phys.*, 18(5), 869, 1979; Masaharu, A. *J. Fac. Eng. Univ. Tokyo*, A(18), 38, 1980.)

It was shown that Te dopant influences positively on the structural perfection of the ZnSe single crystals associated with a completeness of the wurtzite–sphalerite phase transition (Atroshchenko et al. 2004). The optimum Te concentration range is within the interval 0.3–0.6 mass. %.

This system was investigated using DTA, metallography, and XRD (Mizetskaya et al. 1978, Oleinik et al. 1978, Atroshchenko et al. 2004). Phase diagram was calculated using the model of regular associated solutions. The solid/liquid equilibria were calculated with the assumption of an ideal behaving melt (Leute and Plate 1989).

ZnSe–Te: The phase diagram is not constructed. Temperature dependence of ZnSe solubility in liquid Te was determined using the saturation method within the interval of 900°C–1200°C (Figure 2.24) (Masako et al. 1979, Masaharu et al. 1980).

2.23 Zinc–Chromium–Selenium

ZnSe–Cr: The phase diagram is not constructed. Cr-doped single crystals were grown using self-seeded physical vapor transport techniques in both vertical (stabilized) and horizontal configurations (Su et al. 1999). The chromium doping levels were determined to be in the range of (1.8–8.3) 10^{19} cm^{-3}.

ZnSe–Cr$_2$Se$_3$: The phase diagram is not constructed. ZnCr$_2$Se$_4$ ternary compound is formed in this system (Hahn and Schröder 1952, Zhukov et al. 1997). It crystallizes in a cubic structure of the NaCl type with lattice parameter $a = 1044.3 \pm 0.8$ pm (Hahn and Schröder 1952) [$a = 1044.0$ pm (Raccah et al. 1966)] and decomposes at heating in the air (Zhukov et al. 1997). At temperatures less than 650°C, ZnCr$_2$Se$_{4-y}$ compound loses selenium according to the next equation $y = (0.18 \pm 0.01) \, T - (0.060 \pm 0.005)$, and at higher temperatures, this compound oxidizes forming Cr$_2$O$_3$.

ZnCr$_2$Se$_4$ was obtained using the caking of mixtures containing ZnSe and Cr$_2$Se$_3$ in stoichiometric ratio (Hahn and Schröder 1952), or the direct combination of the elements, or the reaction of the zinc oxide spinels with hydrogen sulfide at elevated temperature (Raccah et al. 1966). Thermal stability of this compound was investigated using thermogravimetry and DTA (Zhukov et al. 1997).

2.24 Zinc–Molybdenum–Selenium

Chevrel phases ZnMo$_6$Se$_8$ and Zn$_2$Mo$_6$Se$_8$, which crystallize in a rhombohedral structure with the lattice parameters $a = 637.7$ pm, $\alpha = 94.46°$ and $a = 683.6$ pm, $\alpha = 94.69°$, respectively (in a hexagonal structure with the lattice parameters $a = 990.3$, $c = 1071.2$ pm and $a = 1005.3$, $c = 1082.4$ pm, respectively), exist in this ternary system (Gocke et al. 1987).

ZnMo$_6$Se$_8$ and Zn$_2$Mo$_6$Se$_8$ were obtained via cathodic reduction in a galvanic cell Zn/Zn^{2+}, CH$_3$CN (acetonitrile)/Mo$_6$Se$_8$ (Gocke et al. 1987).

2.25 Zinc–Manganese–Selenium

ZnSe–MnSe: The phase equilibria in this system have been discussed in Pajaczkowska (1978) based on papers that have appeared up to 1977. The phase diagram is not constructed. Cemič and Neuhaus (1974) investigated phase relations in this system at 500°C–1000°C and 10^2–10^7 kPa (Figure 2.25). The solubility of MnSe in ZnSe, determined using XRD, is equal to 35 mol. % at 600°C, and the solubility of ZnSe in MnSe is insignificant and increases with increasing temperature (Juza et al. 1956).

Zn$_{1-x}$Mn$_x$Se single crystals were grown using the Bridgman method within the interval of $0 \leq x \leq 0.57$ (Debska et al. 1984, Yoder-Short et al. 1985). According to XRD, the crystal structure belongs to sphalerite type for $0 \leq x \leq 0.30$ and to wurtzite type for $0.33 \leq x \leq 0.57$. According to the data of Demianiuk (1990), single crystals could be obtained up to 53 mol. % MnSe using the modified Bridgman method under argon pressure. MnSe pure sphalerite structure was observed in

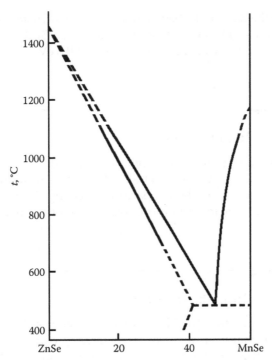

FIGURE 2.25
Phase relations in the ZnSe–MnSe system in the solid state. (From Juza, R. et al., *Z. anorg. und allg. Chem.*, 285(1/2), 61, 1956; Cemič, L. and Neuhaus, A., *High Temp.-High Pressures*, 6(2), 203, 1974.)

the concentration range below 14.4 mol. %. At 14.4–26.4 mol. % MnSe in the solid solutions, multi-polytypic structures with built-in hexagonal layers were observed and the hexagonal layer concentration increases with the increase in Mn content. Above 26.4 mol. %, only the wurtzite structure was obtained.

Using the results of electron probe microanalyzer (EPMA) and XRD, the lattice constants were found to increase linearly with manganese concentration obeying Vegard's law. At $x > 0.57$, the ingots contain $Zn_{0.43}Mn_{0.57}Se$, MnSe, and $MnSe_2$ (Yoder-Short et al. 1985). The crystal growth of the $Zn_{1-x}Mn_xSe$ solid solutions is given in Pajaczkowska (1978).

2.26 Zinc–Iron–Selenium

ZnSe–FeSe: The phase diagram is not constructed. Single crystals of the $Zn_{1-x}Fe_xSe$ solid solutions up to 22 mol. % FeSe were grown using the modified Bridgman method under argon pressure (Demianiuk 1990). The crystalline structure of the obtained single crystals was found to be cubic over the whole concentration range.

According to the data of Il'ichev et al. (2010), the diffusion coefficient of Fe in single crystalline ZnSe at 995°C is equal to (4.7 ± 0.5) 10^{-10} cm^2 s^{-1}, and the average activation energy of such process is 2.9 ± 0.3 eV.

2.27 Zinc–Cobalt–Selenium

ZnSe–CoSe: The phase diagram is not constructed. $Zn_{1-x}Co_xSe$ thin films were grown using epitaxy (Jonker et al. 1988). These thin films were single crystals at $x < 0.1$; moreover, lattice parameter changes linearly at $x \leq 0.036$. They exhibit a tendency to decompose with CoSe formation.

2.28 Zinc–Palladium–Selenium

Isothermal section of the Zn–Pd–Se ternary system at 340°C is shown in Figure 2.26 (Goesmann et al. 1998). Two ternary compounds are formed in this system: $ZnPd_5Se$ and ternary phase of approximate composition $Zn_{32}Pd_{62}Se_6$. The second compound is definitely not just a ternary solubility of Se in Pd_2Zn because some samples contained three distinctly different phases: $ZnPd_5Se$, Pd_2Zn, and $Zn_{32}Pd_{62}Se_6$. The key equilibrium in Figure 2.26 is the three-phase field $ZnSe–Zn_{32}Pd_{62}Se_6–Pd_{17}Se_{15}$.

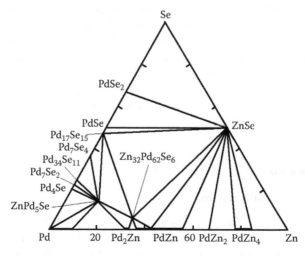

FIGURE 2.26

Isothermal section of the Zn–Pd–Se ternary system at 340°C. (From Goesmann, F. et al., *J. Cryst. Growth*, 184/185, 406, 1998.)

References

Abdullayev A.G., Kerimova T.G., Kyazumov M.G., Khidirov A.Sh. An electron diffraction study of the $ZnIn_2Se_4$ crystal structure. A novel phase, *Thin Solid Films*, **190**(2), 309–315 (1989).

Abrikosov N.H., Bankina V.F., Efimova B.A., Krotova E.F. Investigation of $Cu_{2-x}Se$ doping with zinc [in Russian], in: *Legir. poluprovodn. materialy*, Moscow, Russia: Nauka, pp. 73–75 (1985).

Aliev K.A. Synthesis and properties of $ZnLa_2Se_4$ and $ZnLa_2Te_4$ [in Russian], in: *4th Resp. konf. "Fiz.-khim. analiz i neorgan. materialoved."*, Baku, 114–117 (1998).

Asadov Yu.G., Dzhafarov K.M., Asadova S.Yu. X-ray investigations of cation substitution in the Cu_2Se, *Neorgan. materialy*, **36**(5), 542–544 (2000).

Atroshchenko L.V., Voronkin E.F., Galkin S.N., Lalayants A.I., Rybalka I.A., Ryzhikov V.D., Fedorov A.G. Influence of the tellurium impurity on the ZnSe crystal structure [in Russian], *Neorgan. materialy*, **40**(6), 656–659 (2004).

Beun J.A., Nitsche R., Lichensteiger M. Optical and electrical properties of ternary chalcogenides, *Physica*, **27**(5), 448–452 (1961).

Busch G., Mooser E., Pearson W.B. Neue halbleitende Verbindungen mit diamantähnlicher Struktur, *Helv. Phys. Acta*, **29**(3), 192–193 (1956).

Cemič L., Neuhaus A. Untersuchungen zum Mischbarkeits, Zustands und Strukturverhalten im System ZnSe–MnSe bie Drücken bis 125 kbar und Temperaturen von 500–1100°C, *High Temp.-High Pressures*, **6**(2), 203–215 (1974).

Charifi Z., Hassan El Haj F., Baaziz H., Khosravizadeh Sh., Hashemifar S.J., Akbarzadeh H. Structural and electronic properties of the wide-gap $Zn_{1-x}Mg_xS, Zn_{1-x}Mg_xSe$ and $Zn_{1-x}Mg_xTe$ ternary alloys, *J. Phys.: Condens. Matter*, **17**(44), 7077–7088 (2005).

Cobb S.D., Andrews R.N., Szofran F.R., Lehoczky S.L. Characterization of directionally solidified $Hg_{1-x}Zn_xSe$ semiconducting alloys, *J. Cryst. Growth*, **110**(3), 415–422 (1991).

Debska U., Giriat W., Harrison H.R., Yoder-Short D.R. RF-heated Bridgman growth of $(ZnSe)_{1-x}(MnSe)_x$ in self-sealing graphite crucibles, *J. Cryst. Growth*, **70**(1–2), 399–402 (1984).

Demianiuk M. Growth of $Zn_{1-x}Mn_xSe$ and $Zn_{1-x}Fe_xSe$ mixed crystals, *Mater. Res. Bull.*, **25**(3), 337–342 (1990).

Dohnke I., Muehlberg M., Neumann W. ZnSe single-crystal growth with SnSe as solvent, *J. Cryst. Growth*, **198–199**, 287–291 (1999).

Dvoretskov G.A., Vishniakov A.V., Kovtunenko P.V. Solubility isotherm of gallium in cadmium and zinc selenides [in Russian], in: *Legir. poluprovodn.*, Moscow, Russia: Nauka, pp. 67–71 (1982).

Edwards J.G., Burckel P., Norwisz J.P. Effusion reactions in the $ZnSe-ZnIn_2Se_4$ system, *Thermochim. Acta*, **340–341**, 323–339 (1999).

Errandonea D., Kumar R.S., Manjón F.J., Ursaki V.V., Tiginyanu I.M. High-pressure x-ray diffraction study on the structure and phase transitions of the defect-stannite $ZnGa_2Se_4$ and defect chalcopyrite $CdGa_2S_4$, *J. Appl. Phys.*, **104**(6), 063524_1–063524_9 (2008).

Fadeev A.V., Van K.V., Gaspariants N.R. Distribution coefficient of Ag, In and Cu in ZnSe [in Russian], *Izv. AN SSSR. Neorgan. materialy*, **13**(10), 1920–1921 (1977).

Falke U., Cichos A., Firszt F., Męczyńska H., Dłużewski P., Paszkowicz W., Lenzner J., Hietschold M. Structural investigations of polytypes in $Zn_{1-x}Mg_xSe$ by transmission electron microscopy and cathodoluminescence, *J. Cryst. Growth*, **184/185**, 1015–1020 (1998).

Fedorov A.G., Zagoruiko Yu.A., Fedorenko O.A., Kovalenko N.O. X-ray characterization of ZnSe single crystals doped with Mg, *Semicond. Phys. Quant. Electron. Optoelectron.*, **4**(2), 118–122 (2001).

Fedorov V.A., Ganshin V.A., Korkishko Yu.N. Determination of the point of the zincblende-to-wurtzite structural phase transition in cadmium selenide crystals, *Phys. Status Solidi (a)*, **126**(1), K5–K7 (1991).

Firszt F., Łęgowski S., Męczyńska H., Szatkowski J., Paszkowicz W., Godwod K. Growth and characterisation of $Zn_{1-x}Be_xSe$ mixed crystals, *J. Cryst. Growth*, **184/185**, 1335–1337 (1998).

Firszt F., Mechynska H., Sekulska B., Szatkowski J., Paszkowicz W., Kachniarz J. Composition dependence of the unit cell dimensions and the energy gap in $Zn_{1-x}Mg_xSe$ crystals, *Semicond. Sci. Technol.*, **10**(2), 197–200 (1995).

Galiulin E.A., Odin I.N., Novoselova A.V. The $ZnSe–SnSe$, $ZnSe–SnSe_2$, $CdSe–SnSe_2$ systems [in Russian], *Zhurn. neorgan. khimii*, **27**(1), 266–268 (1982).

Gamidov R.S., Kulieva T.Z. X-ray investigation of the thallium and zinc chalcoaluminates [in Russian], in: *Issled. v obl. neorgan. i fiz. khim. i ih rol' v khim. promyshlennosti*, Baku, Azerbaijan: AzINTI, pp. 32–34 (1969).

Gan'shin V.A., Korkishko Yu.N., Fedorov V.A. The $ZnSe–CdSe$ phase diagram in the subsolidus region [in Russian], *Zhurn. neorgan. khimii*, **38**(3), 533–537 (1993).

Gastaldi L., Simeone M.G., Viticoli S. Structure refinement of $ZnIn_2Se_4$, *J. Solid State Chem.*, **66**(2), 251–255 (1987).

Gavaleshko N.P., Gorley P.N., Paranchich S.Yu., Frasuniak V.M., Homiak V.V. Phase diagrams of $CdSe–HgSe$, $ZnSe–HgSe$, $MgTe–HgTe$ quasibinary systems [in Russian], *Izv. AN SSSR. Neorgan. materialy*, **10**(2), 327–329 (1983).

Gocke E., Schramm W., Dolscheid P., Schöllhorn R. Molybdenum cluster chalcogenides Mo_6X_8: Electrochemical intercalation of closed shell ions Zn^{2+}, Cd^{2+}, and Na^+, *J. Solid State Chem.*, **70**(1), 71–81 (1987).

Goesmann F., Studnitzky T., Schmid-Feltzer R., Pisch A. Palladium thin film contacts on p-type ZnSe: Adjustment of electrical properties by reaction diffusion, *J. Cryst. Growth*, **184/185**, 406–410 (1998).

Golovey M.I., Olekseyuk I.D., Voroshilov Yu.V. Chemical bond and solubility in the systems $A^{II}_3B^V_2–A^{II}C^{VI}$ [in Russian], in: *Khim. svyaz' v poluprovodn. i polumetallah*. Minsk, Belarus: Nauka i Tekhnika Publishing, pp. 233–239 (1972).

Goriunova N.A. Substitutional solid solutions in the compounds with zinc blend structure [in Russian], in: *Voprosy teorii i issled. poluprovodn. i protsessov poluprovodn. metallurgii*. Moscow, Russia: Izd. AN SSSR, pp. 29–37 (1955).

Goriunova N.A., Kotovich V.A., Frank-Kamenetski V.A. About simultaneous crystallization of hexagonal cadmium selenide with ZnSe, InAs and In_2Se_3 [in Russian], *Zhurn. tehn. phiziki*, **25**(14), 2419–2421 (1955a).

Goriunova N.A., Kotovich V.A., Frank-Kamenetski V.A. X-ray investigation of isomorphism for some gallium and indium compounds [in Russian], *Dokl. AN SSSR*, **103**(4), 659–662 (1955b).

Guseinov F.H., Babanly M.B., Kuliev A.A. Phase equilibria in the $Tl_2Te_3–Zn(Cd)$ Te systems [in Russian], *Izv. vuzov. Khimia i khim. tehnologia*, **24**(10), 1245–1248 (1981).

Guseinov F.H., Babanly M.B., Kuliev A.A. Phase equilibria in the TlX–Cd(Zn)X (X – S, Se, Te) systems [in Russian], *Izv. AN SSSR. Neorgan. materialy*, 18(5), 759–763 (1982).

Haeuseler H., Himmrich M. Zur Kenntnis des quaternären Systems Zn–In–S–Se. Der Schnitt $ZnIn_2S_4$–$ZnIn_2Se_4$–In_2S_3–In_2Se_3, *Z. anorg. und allg. Chem.*, 535(4), 13–19 (1986).

Hahn H., Frank G., Klingler W., Störger A.D., Störger G. Über ternäre Chalkogenide des Aluminiums, Galliums und Indiums mit Zink, Cadmium und Quecksilber, *Z. anorg. und allg. Chem.*, 279(5/6), 241–270 (1955).

Hahn H., Lorent C. Untersuchungen über ternäre Chalkogenide. Über ternäre Sulfide und Selenide des Germaniums mit Zink, Cadmium und Quecksilber, *Naturwissenschaften*, 45(24), 621–622 (1958).

Hahn H., Schröder K.F. Über die Structur des $ZnCr_2Se_4$ und $CdCr_2Se_4$, *Z. anorg. und allg. Chem.*, 269(3), 135–140 (1952).

Harif Ya.L., Vishniakov A.V. Phase equilibria in the Zn–Cu–Se system [in Russian], *Izv. AN SSSR. Neorgan. materialy*, 11(7), 1202–1205 (1975).

Hawthorne F.C., Ercit T.S., Groat L.A. Structures of zinc selenite and copper selenite, *Acta Crystallogr.*, C42(10), 1285–1287 (1986).

Huan G., Greenblatt M. New $A_xNb_6Se_8$ (A = Na, K, Rb, Cu Ag, Zn, Cd, Pb) phases with the Nb_3Te_4 structure, *Mater. Res. Bull.*, 22(4), 505–512 (1987).

Il'ichev N.N., Shapkin P.V., Gulyamova E.S., Kulevskiy L.A., Nasibov A.S. Determination of the Fe^{2+} diffusion coefficient in single crystalline ZnSe [in Russian], *Neorgan. materialy*, 46(2), 149–153 (2010).

Jonker B.T., Krebs J.J., Prinz G.A. $Zn_{1-x}Co_xSe$. A new diluted magnetic semiconductor growth by molecular beam epitaxy, *Appl. Phys. Lett.*, 53(5), 450–452 (1988).

Juza R., Rabenau A., Pascher G. Über teste Lösungen in den Systemen ZnS/MnS, ZnSe/MnSe und ZnTe/MnTe, *Z. anorg. und allg. Chem.*, 285(1/2), 61–69 (1956).

Kaldis E., Krausbauer L., Widmer R. Cd_4SiS_6 and Cd_4SiSe_6, new ternary compounds, *J. Electrochem. Soc.*, 114(10), 1074–1076 (1967).

Kalinkin I.P., Aleksandrova L.A., Sanitarov V.A. Thermodynamics of zinc selenotelluride solid solutions [in Russian], *Elektronnaya tehnika. Ser. 6. Materialy*, 10(195), 12–14 (1984).

Kimura S. Phase equilibria in the systems Zn+S+Sn, Zn+Se+Sn and Cd+S+Sn, *J. Chem. Thermodyn.*, 3(1), 7–17 (1971).

Kimura S., Panish M.B. Phase equilibria in the systems Zn+S+Bi, Zn+Se+Bi and Cd+S+Bi, *J. Chem. Thermodyn.*, 2(1), 77–86 (1970).

Kirovskaya I.A., Budanova E.M. Obtaining and properties of the $Zn_xCd_{1-x}Se$ solid solutions [in Russian], *Neorgan. materialy*, 37(8), 913–916 (2001).

Kohn K., Inoue K., Horie O., Akimoto S.-I. Crystal chemistry of $MSeO_3$ and $MTeO_3$ (M = Mg, Mn, Co, Ni, Cu and Zn), *J. Solid State Chem.*, 18(1), 27–37 (1976).

Komar V.K. Thermodynamic aspects of carbon incorporation into ZnSe and CdSe crystals, *Funct. Mater.*, 5(4), 513–516 (1998).

Koren' N.N., Matias E.E., Shrubova E.F., Pashkovski O.I. About synthesis of Zn_2GeSe_4 and $ZnGeS_3$ ternary compounds [in Russian], *Izv. AN SSSR. Neorgan. materialy*, 20(11), 1924–1925 (1984).

Kot M.V., Simashkevich A.V. Structure and electrical properties of the ZnSe–HgSe system [in Russian], *Izv. AN SSSR. Ser. fiz.*, 28(6), 1065–1068 (1964).

Kot M.V., Tyrziu V.G., Simashkevich A.V., Maronchuk Yu.E., Mshenski V.A. Dependence of energy activation in thin films from molar composition for some $A^{II}B^{VI}$–$A^{II}B^{VI}$ systems [in Russian], *Fizika tverdogo tela*, 4(6), 1535–1541 (1962).

Kovach A.P., Potoriy M.V., Voroshilov Yu.V., Tovt V.V. Physico-chemical investigation of the Cd(Zn)S(Se)–P$_2$S$_4$(Se$_4$) systems [in Russian], *Ukr. khim. zhurn.*, **63**(2), 92–96 (1997).

Krämer V., Hirth W., Hofherr W., Trash H.-P. Phase studies in the systems Ag$_2$Te–Ga$_2$Te$_3$, ZnSe–In$_2$Se$_3$ and ZnS–Ga$_2$S$_3$, *Thermochim. Acta*, **112**(1), 88–94 (1987).

Krucheanu E., Nikulesku D., Nistor N. Solid solutions in the ZnSe–HgSe and ZnTe– HgTe pseudobinary systems [in Russian], *Rev. Phys. Acad. RPR*, **8**(4), 379–382 (1963).

Kulakov M.P., Baliakina I.V. Wurtzite-sphalerite correlation in the subsolidus region of the CdSe–ZnSe system [in Russian], *Kristallografiya*, **35**(6), 1479–1482 (1990).

Kulakov M.P., Baliakina I.V., Kolesnikov N.N. Phase diagram and crystallization in the CdSe–ZnSe system [in Russian], *Izv. AN SSSR. Neorgan. materialy*, **25**(10), 1637–1640 (1989).

Kulakov M.P., Fadeev A.V. Oxidation of mechanically polished zinc selenide in the air [in Russian], *Izv. AN SSSR. Neorgan. materialy*, **19**(3), 347–351 (1983).

Kulakov M.P., Fadeev A.V. Growth of zinc selenide crystals doping with magnesium [in Russian], *Izv. AN SSSR. Neorgan. materialy*, **22**(3), 392–394 (1986).

Kulakov M.P., Kulakovski V.D., Savchenko I.B., Fadeev A.V. About phase transition in the crystals of zinc selenide [in Russian], *Fizika tverdogo tela*, **18**(3), 909–911 (1976).

Kumar S., Maurya T.K., Auluck S. Optical properties and critical points in ordered Be$_x$Zn$_{1-x}$Se alloys, *J. Alloys Compd.*, **480**(2), 717–722 (2009).

Larach S., Schrader R.E., Stocker C.F. Anomalous variation of band gap with composition in zinc sulfo- and seleno-tellurides, *Phys. Rev.*, **108**(3), 587–589 (1957).

Leute V., Plate H. The phase diagram of the semiconductor alloy Zn$_k$Hg$_{(1-k)}$Se$_l$Te$_{(1-l)}$, *Ber. Bunsenges. Phys. Chem.*, **93**(7), 757–763 (1989).

Leute V., Wulff B. The phase diagram of the quasiternary system (Zn$_k$Cd$_{1-k}$)(Se$_l$Te$_{1-l}$), *Ber. Bunsenges. Phys. Chem.*, **96**(2), 119–128 (1992).

Marsh R.E., Robinson W.R. On the structure of ZnIn$_2$Se$_4$, *J. Solid State Chem.*, **73**(2), 591–592 (1988).

Marugin V.V., Odin I.N., Novoselova A.V. Physico-chemical investigation of the zinc chalcogenides—bismuth chalcogenides systems [in Russian], *Zhurn. neorgan. khimii*, **28**(8), 2104–2107 (1983).

Masaharu A., Masaoko W., Hiroshi N. Growth of Zn and Cd chalcogenides from solutions using Te as a solvent, *J. Fac. Eng. Univ. Tokyo*, **A**(18), 38–39 (1980).

Masako W., Ken-ichi S., Masaharu A. Solution growth of ZnS, ZnSe, CdS and their mixed compounds using tellurium as a solvent, *Jpn. J. Appl. Phys.*, **18**(5), 869–872 (1979).

Medvedev Yu.V., Berchenko N.N. Peculiarities of the interface between solid solutions based on mercury chalcogenides and native oxides [in Russian], *Zhurn. neorgan. khimii*, **38**(12), 1940–1945 (1993).

Meenakshi S., Vijyakumar V., Godwal B.K., Eifler A., Orgzall I., Tkachev S., Hochheimer H.D. High pressure X-ray diffraction study of CdAl$_2$Se$_4$ and Raman study of AAl$_2$Se$_4$ (A = Hg, Zn) and CdAl$_2$X$_4$ (X = Se, S), *J. Phys. Chem. Solids*, **67**(8), 1660–1667 (2006).

Meunier G., Bertaud M. Cristallochimie du sélénium (+IV): II. Structure cristalline de ZnSe$_2$O$_5$, *Acta Crystallogr.*, **B30**(12), 2840–2843 (1974).

Mishra S., Ganguli B. Electronic and structural properties of AAl$_2$Se$_4$ (A = Ag, Cu, Cd, Zn) chalcopyrite semiconductors, *J. Solid State Chem.*, **184**(7), 1614–1621 (2011).

Mizetskaya I.B., Oleinik G.S., Tomashik V.N. Phase diagrams of the CdSe–ZnSe and ZnTe–ZnSe quasibinary systems [in Russian], *Ukr. khim. zhurn.*, **44**(2), 163–164 (1978).

Mizetskaya I.B., Oleinik G.S., Trishchuk L.I. Phase diagrams of the $Cu_{2-x}S-ZnS$ and $Cu_{2-x}Se-ZnSe$ systems [in Russian], *Izv. AN SSSR. Neorgan. materialy*, **18**(11), 1792–1794 (1982).

Nasar A. Investigation of the thermodynamic properties of Te-saturated ZnSe–ZnTe solid solutions, *J. Alloys Compd.*, **217**(2), 167–175 (1995).

Odin I.N. Physico-chemical analysis of ternary and ternary mutual systems, containing cadmium, zinc, silicon, bismuth chalcogenides and properties of ingots in these systems [in Russian], *Zhurn. neorgan. khimii*, **41**(6), 941–953 (1996).

Odin I.N., Marugin V.V. Phase diagrams and electrophysical properties of ingots in the Bi_2X_3-ZnX (X – S, Se, Te) systems [in Russian], *Zhurn. neorgan. khimii*, **36**(7), 1842–1846 (1991).

Oleinik G.S., Mizetski P.A., Nizkova A.I. Interaction between lead and zinc chalcogenides [in Russian], *Izv. AN SSSR. Neorgan. materialy*, **18**(5), 873–874 (1982).

Oleinik G.S., Tomashik V.N., Mizetskaya I.B. CdTe + ZnSe $\Leftrightarrow$ CdSe + ZnTe ternary mutual system [in Russian], *Izv. AN SSSR. Neorgan. materialy*, **14**(3), 441–443 (1978).

Olekseyuk I., Parasyuk O., Piskach L., Horhut H., Zmiy O., Krykhovets' O., Sysa L., Kadykalo E., Strok O., Marchuk O., Halka V. *Quasiternary chalcogenide systems* [in Ukrainian], Vol. 1, Luts'k, Ukraine: Vezha Publishing, 168 p. (1999).

Olekseyuk I.D., Parasyuk O.V., Sysa L.V. The $ZnSe-Ga_2Se_3$ system, *Pol. J. Chem.*, **70**(9), 1111–1113 (1996).

Olekseyuk I.D., Stoika I.M. The Zn–As–Se system [in Russian], *Zhurn. neorgan. khimii*, **23**(9), 2496–2501 (1978).

Olekseyuk I.D., Stoyka I.M., Gerasimenko V.S. Phase equilibria and properties of solid solutions single crystals in the $Zn_3As_2-ZnSe(ZnTe)$ systems [in Russian], *Izv. AN SSSR. Neorgan. materialy*, **14**(8), 1383–1388 (1978).

Pajaczkowska A. Physicochemical properties and crystal growth of $A^{II}B^{VI}-MnB^{VI}$ systems, *Progr. Crystal Growth Caract.*, **1**(3), 289–326 (1978).

Parasyuk O.V., Olekseyuk I.D., Mazurets I.I., Piskach L.V. Phase equilibria in the quasiternary $ZnSe-Ga_2Se_3-SnSe_2$ system, *J. Alloys Compd.*, **379**(1–2), 143–147 (2004).

Paszkowicz W., Dłużewski P., Spolnik Z.M., Firszt F., Męczyńska H. Formation of 4H and 8H polytypes in bulk $Zn_{1-x}Mg_xSe$ crystals, *J. Alloys Compd.*, **286**(1–2), 224–235 (1999).

Paszkowicz W., Dlużewski P., Spolnik Z., Godwod K., Domagała J., Firszt F., Sekulska B., Szatkowski J. Structure of bulk ZnSe based solid solutions containing Mg and Be, *Electron. Technol.*, **31**(2), 208–212 (1998).

Paszkowicz W., Spolnik Z., Firszt F., Męczyńska H. H-polytype formation in $Zn_{1-x}Mg_xSe$, *Acta Crystallogr. A*, **52**, Suppl., 319 (1996).

Paszkowicz W., Szuszkiewicz W., Dynowska E., Domagała J.Z., Firszt F., Męczyńska H., Łęgowski S., Lathe C. High-pressure structural and optical properties of wurtzite-type $Zn_{1-x}Mg_xSe$, *J. Alloys Compd.*, **371**(1–2), 168–171 (2004).

Raccah P.M., Bouchard R.J., Wold A. Crystallographic study of chromium spinels, *J. Appl. Phys.*, **37**(3), 1436–1437 (1966).

Radautsan S.I., Ivanova R.A. Formation of the solid solutions based on $A^{II}B^{IV}C_3^{VI}$ compounds [in Russian], *Izv. AN MSSR*, **10**(88), 64–70 (1961).

Range K.-J., Becker W., Weiss A. Über Hochdruckphasen des $CdAl_2S_4$, $HgAl_2S_4$, $ZnAl_2Se_4$, $CdAl_2Se_4$ und $HgAl_2Se_4$ mit Spinellstruktur, *Z. Naturforsch.*, **23B**(7), 1009 (1968).

Rigan M.Yu., Stasiuk N.P. Some questions of charge synthesis and crystal growth of ZnGa$_2$S$_4$ and ZnGa$_2$Se$_4$ [in Russian], in: *Poluch. i svoistva slozhn. poluprovodn., Uzhgorod. gos. un-t*, Kyiv, Ukraine: Nauk. Dumka Publishing, pp. 62–69 (1991).

Rovnyi A.G., Krylov V.S., Gershun A.S. About growing of ZnSe$_x$Te$_{1-x}$ single crystals [in Russian], in: *Obtaining and Properties of A^{II}B^{VI} and A^{IV}B^{VI} Semiconductor Compounds and Solid Solutions on Its Base*. Tez. dokl. M.: MISIS, pp. 2, 242 (1977).

Rubenstein M. Solubilities of some II-VI compounds in bismuth, *J. Electrochem. Soc.*, **113**(6), 623–624 (1966).

Rubenstein M. Solution growth of some II-VI compounds using tin as a solvent, *J. Cryst. Growth*, **3–4**, 309–312 (1968).

Savickas R., Baršauskas K. ZnSe–CdSe ir CdSe–CdTe trejinių sistemų struktūros klausimu [in Lithuanian], *Tr. Kaunas. polytehn. in-ta*, **14**(1), 15–18 (1960).

Shalimova K.V., Botnev A.F., Dmitriev V.A., Kognovitskaya N.Z., Starostin V.V. Crystal structure of the solid solutions in the ZnSe–CdSe system [in Russian], *Kristallografiya*, **14**(4), 629–633 (1969).

Sirota N.N., Yanovich V.D. Identify periods and root-mean-square ion displacements for zinc cadmium selenide solid solutions [in Russian], *Dokl. AN SSSR*, **204**(3), 583–585 (1972).

Skums V.F., Pan'ko E.P., Vecher A.A. Influence of high pressures on the electrical resistivity of cadmium and zinc chalcogenide solid solutions, *Inorg. Mater.* (Engl. Trans.), **28**(4), 572–577 (1992), transl. from *Neorgan. Mater.*, **28**(4), 745–751 (1992).

Su C.-H., Feth S., Volz M.P., Matyi R., George M.A., Chattopadyay K., Burger A., Lehoczky S.L. Vapor growth and characterization of Cr-doped ZnSe crystals, *J. Cryst. Growth*, **207**(1–2), 35–42 (1999).

Trah H.P., Krämer V. Crystal structure of zinc indium selenide, ZnIn$_2$Se$_4$, *Z. Kristallogr.*, **173**(3–4), 199–203 (1985).

Trishchuk L.I., Oleinik G.S., Mizetskaya I.B. Phase equilibria in the Ag$_2$Se–ZnSe and Ag$_2$Se–CdSe systems [in Russian], *Izv. AN SSSR. Neorgan. materialy*, **18**(11), 1795–1797 (1982).

Trishchuk L.I., Oleinik G.S., Mizetskaya I.B. Thermal analysis determination of A^{II}B^{VI} solid solubility in A$^I_2$B^{VI}, *Thermochim. Acta*, **92**, 611–613 (1985).

Vasiliev M.G., Novikova E.M. Investigation of equilibria in the Me–ZnSe, Sn–GaAs, ZnSe–GaAs polythermal sections of Me–GaAs–ZnSe ternary systems, where Me—Ga, Sn [in Russian], Deposited in VINITI, No 971–77Dep. (1977).

Veliev R.K., Mamedov K.K., Kerimov I.G., Mehtiev M.I., Masimov E.A. Heat capacity and thermodynamic properties of ZnGa$_2$Se$_4$ and ZnIn$_2$Se$_4$ semiconductors at low temperatures [in Russian], *Zhurn. fiz. khimii*, **50**(3), 746–748 (1976).

Vishniakov A.V., Dvoretskov G.A., Zubkovskaya V.N., Tyurin O.A., Kovtunenko P.V. Phase equilibria in the systems formed by the II-VI compounds and the elements of I and III groups of the periodic table [in Russian], *Tr. Mosk. Khim.-Tekhnol. In-t*, (120), 87–103 (1981).

Vishniakov A.V., Harif Ya.L. Solubility of Cu$_2$Se in ZnSe and CdSe [in Russian], *Izv. AN SSSR. Neorgan. materialy*, **8**(2), 217–219 (1972).

Wagner P., Lorenz M.R. Solubility of ZnSe and ZnTe in Ga and In, *J. Phys. Chem. Solids*, **27**(11/12), 1749–1752 (1966).

Yoder-Short D.R., Debska U., Furdyna J.K. Lattice parameters of $Zn_{1-x}Mn_xSe$ and tetrahedral bond lengths in $A^{II}_{1-x}Mn_xB^{VI}$ alloys, *J. Appl. Phys.*, **58**(11), 4056–4060 (1985).

Zagoruiko Yu.A., Kovalenko N.O. Fedorenko O.A., Fedorov A.G., Mateychenko P.V. $Zn_{1-x}Mg_xSe$ single crystals as a functional material for optoelectronics, *Funct. Mater.*, **12**(4), 731–734 (2005).

Zherdev Yu.V., Ormont B.F. About dependence of energy gap from structure and composition in the ZnSe–CdSe system [in Russian], *Zhurn. neorgan. khimii*, **5**(1), 239 (1960a).

Zherdev Yu.V., Ormont B.F. About dependence of energy gap from structure and composition in the ZnSe–CdSe system [in Russian], *Zhurn. neorgan. khimii*, **5**(8), 1796–1799 (1960b).

Zhukov E.G., Poluliak E.S., Varnakova E.S., Fedorov V.A. Investigation of chalcogenide spinel stability at the heating in the air [in Russian], *Neorgan. materialy*, **33**(8), 939–941 (1997).

Zuyev A.P., Kulakov M.P., Fadeev A.V., Kireiko V.V. Oxygen contents in the ZnS, ZnSe and CdS crystals, obtained from the melt [in Russian], *Izv. AN SSSR. Neorgan. materialy*, **17**(7), 1159–1161 (1981).

3

Systems Based on ZnTe

3.1 Zinc–Copper–Tellurium

ZnTe–Cu$_{1.95}$Te: The phase diagram belongs to the eutectic type (Figure 3.1) (Trishchuk et al. 1984, 2008). The eutectic composition and temperature are 53 mol. % ZnTe and 1038°C, respectively. The solubility of ZnTe in Cu$_{1.95}$Te at the eutectic temperature reaches 51 mol. % and decreases to 10 mol. % at 700°C and 1.5 mol. % at 545°C (Trishchuk et al. 1984, 1985, 2008). The maximum solubility of Cu$_{1.95}$Te in ZnTe is not higher than 3 mol. %.

According to the data of Asadov et al. (1992), the ingots with the composition (ZnTe)$_{0.2}$(Cu$_{1.95}$Te)$_{0.8}$ are single phase at the annealing at 100°C for 50 h. Single crystals of such composition were grown using the Bridgman method.

This system was investigated using DTA and metallography (Trishchuk et al. 1984, 2008).

3.2 Zinc–Silver–Tellurium

ZnTe–Ag$_2$Te: The phase diagram belongs to the eutectic type of phase diagram with limited solubility (Figure 3.2) (Trishchuk 2009, Trishchuk et al. 1986, 2008). The eutectic composition and temperature are 32 mol. % ZnTe and 880°C, respectively. Solid solutions based on γ-Ag$_2$Te contain up to 29 mol. % ZnTe (Trishchuk et al. 1985, 1986, 2008, 2009) and undergo a eutectoid transformation at 713°C (eutectoid composition is 11 mol. % ZnTe) (Trishchuk 2009, Trishchuk et al. 1986, 2008). The solubility of ZnTe in β-Ag$_2$Te reaches 9 mol. %, and the solubility of ZnTe in α-Ag$_2$Te is insignificant. The solubility of Ag$_2$Te in ZnTe at 880°C and 700°C is equal correspondingly to 2 and 0.5 mol. %.

This system was investigated using DTA and metallography. The ingots were annealed at 870°C during 20 h (Trishchuk 2009, Trishchuk et al. 1986, 2008).

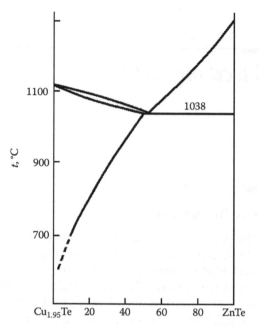

FIGURE 3.1
ZnTe–Cu$_{1.95}$Te phase diagram. (From Trishchuk, L.I. et al., *Izv. AN SSSR. Neorgan. materialy*, 20(9), 1486, 1984; Trishchuk, L.I. et al., *Chem. Met. Alloys*, 1(1), 58, 2008.)

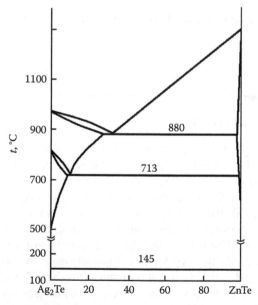

FIGURE 3.2
ZnTe–Ag$_2$Te phase diagram. (From Trishchuk, L.I. et al., *Ukr. khim. zhurn.*, 52(8), 799, 1986; Trishchuk, L.I. et al., *Chem. Met. Alloys*, 1(1), 58, 2008; Trishchuk, L.I., *Optoelektronika i poluprovodn. Tekhnika*, (44), 99, 2009.)

3.3 Zinc–Beryllium–Tellurium

ZnTe–BeTe: Bulk $Zn_{1-x}Be_xTe$ crystals were grown in the range extending up to $x = 0.12$ (Paszkowicz et al. 2002). Their structural, optical, and photothermal properties were characterized using x-ray diffraction (XRD), photoluminescence, and photoacoustic methods. The lattice constants are found to follow a linear dependence on x. Energy gap and thermal diffusivity were determined from photoacoustic spectra. It was shown that the energy gap increases with rising beryllium content.

3.4 Zinc–Magnesium–Tellurium

ZnTe–MgTe: The phase diagram is shown in Figure 3.3 (Parker et al. 1971). Solid solutions are formed in this system. They crystallize in hexagonal and cubic (within the interval of 0–53 mol. % MgTe) structure. Two phases with hexagonal structure exist in the $Zn_{0.18}Mg_{0.82}Te$ ingot, and one of them is similar to MgTe. Two-phase region (hexagonal + cubic) is not investigated. MgTe undergoes polymorphous transformation at 1170°C ± 5°C. Concentration dependence of energy gap changes linearly, but a hump exists at 53 mol. % MgTe, that is, line inclination for cubic solid solutions differs from one for hexagonal solid solutions.

The lattice parameters of the $Zn_{1-x}Mg_xTe$ solid solutions do not change linearly, and a small deviation from Vegard's law with lattice bowing equal to –2 pm has been observed (Charifi et al. 2005). The bulk modulus of these solid solutions decreases on increasing the Mg concentration, which is due

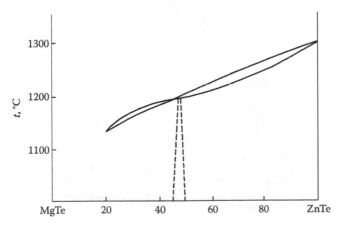

FIGURE 3.3
ZnTe–MgTe phase diagram. (From Parker, S.G. et al., *J. Electrochem. Soc.*, 118(6), 979, 1971. With permission.)

to the increase in the ionicity of bonds. A significant deviation of the bulk modulus from linear concentration dependence was found.

This system was investigated using the differential thermal analysis (DTA) and XRD (Parker et al. 1971).

3.5 Zinc–Cadmium–Tellurium

The phase equilibria data for the ZnTe–CdTe–Te section were used in Radautsan and Maksimova (1976) and Haloui et al. (1997) to determine the liquidus surface. Projections of the Zn–Cd–Te ternary liquidus isotherms on composition triangle according to Haloui et al. (1997) are shown in Figure 3.4. The eutectic troughs vary smoothly between the binary Zn–Te and Cd–Te liquidus curves, but there is a steep slope along the equiatomic ZnTe–CdTe composition line. The thermodynamic calculations give the composition of the solid for each point of the liquidus surface. The isoconcentration curves are shown in Figure 3.5 for the Te-rich region (Laugier 1973).

The 265°C isothermal section of the Zn–Cd–Te ternary system just at the temperature of the ternary eutectic equilibrium $L_E \Leftrightarrow ZnTe + Cd + Zn$ (Haloui et al. 1997) is shown in Figure 3.6.

Steininger et al. (1970) investigated the liquidus isopleths at constant Cd/Zn ratios of 4.0, 1.0, and 0.25 and at constant Te concentrations of 0.10, 0.30, 0.60,

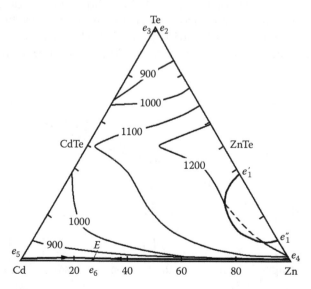

FIGURE 3.4

Liquidus surface of the Zn–Cd–Te ternary system. (From Haloui, A. et al., *J. Alloys Compd.*, 260(1–2), 179, 1997.)

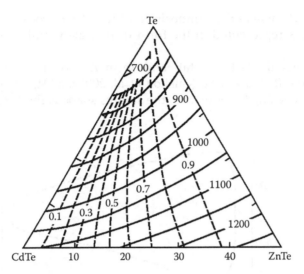

FIGURE 3.5
Calculated ternary phase diagram of the Zn–Cd–Te ternary system in the Te-rich part: solid lines, liquidus isotherms, and dash lines, isoconcentration curves in the solid solution. (From Laugier, A., *Rev. Phys. Appl.*, 8(3), 259, 1973.)

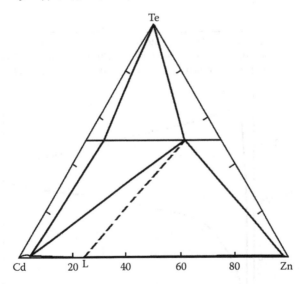

FIGURE 3.6
Isothermal section of the Zn–Cd–Te ternary system at 265°C. (From Haloui, A. et al., *J. Alloys Compd.*, 260(1–2), 179, 1997.)

0.70, 0.80, and 0.90. Some vertical sections (Cd + Zn)–Te from the Te-rich side were calculated by Litvak and Charykov (1991). Five isopleths of the Zn–Cd–Te ternary system were experimentally studied in Haloui et al. (1997).

The *T-x-y* projection of the *p-T-x-y* phase diagram for the Zn–Cd–Te system and the *x-y* diagram accompanied with the *T-x-y* isobar was constructed in Nipan (2004), where the composition of the phases coexisting at a fixed temperature is represented in the form of triangle parallel to the base of the prism.

The solubility limits for metals and Te in $Zn_xCd_{1-x}Te_{1\pm\delta}$ solid solutions at $x = 0.05$, 0.1, 0.15, and 0.90 (Guskov et al. 2003a, 2004) are presented in Figure 3.7. According to these data, the solidus shape is the same as that for

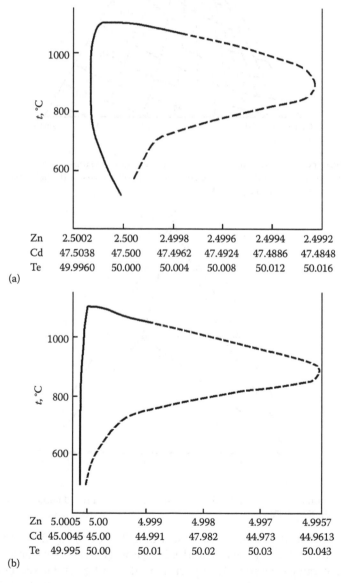

(a)

Zn	2.5002	2.500	2.4998	2.4996	2.4994	2.4992
Cd	47.5038	47.500	47.4962	47.4924	47.4886	47.4848
Te	49.9960	50.000	50.004	50.008	50.012	50.016

(b)

Zn	5.0005	5.00	4.999	4.998	4.997	4.9957
Cd	45.0045	45.00	44.991	47.982	44.973	44.9613
Te	49.995	50.00	50.01	50.02	50.03	50.043

FIGURE 3.7
T-x projection of the $Zn_xCd_{1-x}Te$ solid solutions for $x = 0.05$ (a), 0.1 (b).

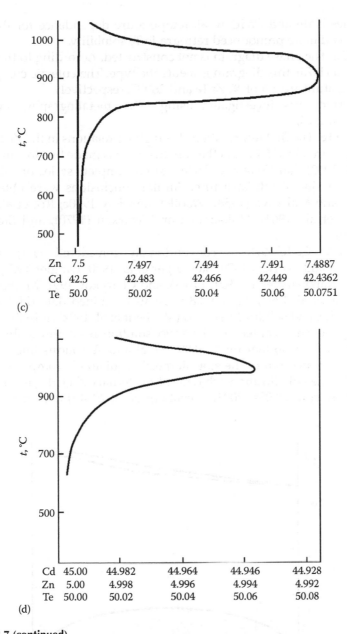

Zn	7.5	7.497	7.494	7.491	7.4887
Cd	42.5	42.483	42.466	42.449	42.4362
Te	50.0	50.02	50.04	50.06	50.0751

(c)

Cd	45.00	44.982	44.964	44.946	44.928
Zn	5.00	4.998	4.996	4.994	4.992
Te	50.00	50.02	50.04	50.06	50.08

(d)

FIGURE 3.7 (continued)
T-x projection of the $Zn_xCd_{1-x}Te$ solid solutions for $x = 0.15$ (c), and 0.90 (d). (From Guskov, V.N. et al., *Dokl. Phys. Chem.*, 391(4–6), 203, 2003, transl. from *Dokl. Ros. Akad. Nauk*, 391(6), 788, 2003a; Guskov, V.N. et al., *J. Alloys Compd.*, 371(1–2), 118, 2004.)

the binaries ZnTe and CdTe: weak temperature dependence for the metal solubility and more pronounced retrograde Te solubility.

ZnTe–Cd: The phase diagram is not constructed. According to the data of Maksimova (1973), this diagram is a eutectic type. The eutectic composition and temperature are 2 mol. % ZnTe and 280°C, respectively.

This system was investigated using DTA, metallography, and XRD (Maksimova 1973).

ZnTe–CdTe: The first information about phase relations in the ZnTe–CdTe system appeared in 1953 when the conclusion was made by Goriunova and Fedorova (1953) that CdTe and ZnTe form a complete series of solid solutions with a zinc-blende structure. Similar conclusions were obtained in Kolomiets and Mal'kova (1958), Woolley and Ray 1960c, Kot et al. (1962), Gromakov et al. (1964), Maksimova and Tsurkan (1970), and Steininger et al. (1970).

The section CdTe–ZnTe is conventionally considered as quasibinary (Maksimova and Tsurkan 1970, Steininger et al. 1970, Laugier 1973, Patrick et al. 1988, Katayama et al. 1991, Litvak and Charykov 1991, Marbeuf et al. 1992, Haloui et al. 1997). Figure 3.8 presents combined data (Haloui et al. 1997) (liquidus and solidus lines) and (Marbeuf et al. 1992) (miscibility gap). Increasing the ZnTe content in the solid solution results in a shift of the solidus toward Te, so that for $x=0.85$, metal and Te solidus lines are both found to be located out of the stoichiometric equiatomic composition. This is shown by the volatilization behavior of the ternary $Zn_xCd_{1-x}Te$ solid solution (Guskov et al. 2003a, 2004, Greenberg et al. 2004b) and of the binary

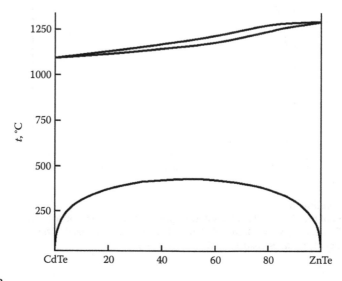

FIGURE 3.8
ZnTe–CdTe phase diagram. (From Marbeuf, A. et al., *J. Cryst. Growth*, 117(1–4), 10, 1992; Haloui, A. et al., *J. Alloys Compd.*, 260(1–2), 179, 1997.)

non-stoichiometric compound $ZnTe_{1+x}$ (Guskov et al. 2003a). The phase diagram of the ZnTe–CdTe system can be described by the model of regular solutions (Ilegems and Pearson 1975, Zabdyr 1984).

The cubic parameter of the $Zn_xCd_{1-x}Te$ solid solutions varies linearly according to Vegard's law (Woolley and Ray 1960c, Maksimova and Tsurkan 1970, Webb et al. 1987, Schenk et al. 1996, Haloui et al. 1997, Rabadanov and Simonov 2006). At high pressures, $Zn_xCd_{1-x}Te$ solid solutions undergo phase transitions to a cubic structure of NaCl type (Webb et al. 1987, Skums et al. 1992). The pressure that corresponds to the beginning of the phase transition is a function of composition according to a linear function $p_1 (GPa) = 3.76 + 8.91x$ (at $1 < x < 0.36$) (Skums et al. 1992). Maksimova and Tsurkan (1970) registered additional heat effects on cooling around 1100°C on $Zn_xCd_{1-x}Te$ samples with $x \leq 0.40$ that might be considered as polymorphic transition of unclear nature.

As undoubtedly shown in a number of works (Feldman et al. 1987, Wei et al. 1990, Marbeuf et al. 1992, Ruault et al. 1994, Edelman et al. 2002, Guskov et al. 2003b, Voronin and Pentin 2005, Pentin et al. 2006), the $Zn_xCd_{1-x}Te$ solid solution separates at low temperatures; however, there are discrepancies concerning the coordinates of the miscibility gap. The temperature of 428°C predicted in Marbeuf et al. (1992) seems to be the most probable. The direct TEM study of Marbeuf et al. (1992) on layers grown at 365°C has revealed such a phase separation. This conclusion is consistent with the thermodynamic data of Goncharuk et al. (1990), Gupta et al. (1994), Shamsuddin et al. (1994), and Goncharuk and Sidorko (1996): the components ZnTe and CdTe are completely miscible over the range 470°C–575°C.

The miscibility gap for the $Zn_xCd_{1-x}Te$ solid solution is under question up to now. Predicted theoretically (Wei et al. 1990, Marbeuf et al. 1992) and determined experimentally both in crystals (Ruault et al. 1994, Greenberg et al. 2003, Alikhanian et al. 2004) and in layers (Feldman et al. 1987, Marbeuf et al. 1992), the phase separation has been objected by the earlier work of Schenk et al. (1996). There it was noted that neither electron diffraction patterns nor images by diffraction contrast gave any indication of a phase separation or ordered structures in the $Zn_{0.24}Cd_{0.76}Te$ and $Zn_{0.83}Cd_{0.17}Te$ solid solutions. Recently, low-temperature annealing results (Guskov et al. 2003b) confirmed partly the possibility of such decomposition. Phase separation was also seen by TEM in specimens made from slowly cooled bulk $Zn_{0.04}Cd_{0.96}Te$ crystals (Ruault et al. 1994).

The need to produce $Zn_xCd_{1-x}Te$ solid solutions homogeneously has stimulated fundamental investigations on the Zn–Cd–Te phase equilibria with vapor (Vydyanath et al. 1992, Yu and Brebrick 1992, Sang et al. 2000, Alikhanian et al. 2002) as ZnTe and CdTe are volatile components and both are characterized by having notable homogeneity ranges. Descriptions of the phase equilibria involving $Zn_xCd_{1-x}Te$ cannot be constrained to one coordinate of composition, for example, as p-T-x diagram of the ZnTe–CdTe quasibinary system. In solid solutions that more correctly can be presented as $(Zn_xCd_{1-x})_{0.5+\delta}Te_{0.5-\delta}$, the ratio $(Zn+Cd)/Te$ varies and deviates from the stoichiometry. To stay in agreement with the Gibbs phase rule, it is necessary to construct a p-T-x-y phase diagram for the Zn–Cd–Te system (y being the second coordinate of the composition). While the binaries, ZnTe and CdTe, have been extensively studied long ago, the first p-T-x-y equilibria data for this ternary system appeared only recently in Guskov et al. (2003b) followed by the investigations of Alikhanian et al. (2003a,b), Greenberg et al. (2003, 2004a,b), Guskov et al. (2003a, 2004), Takahashi and Mochizuki (2003), and Nipan (2004).

The solid solutions $Zn_xCd_{1-x}Te$ were found to exhibit both negative deviation from ideality (Zabdyr 1984) and a positive one [according to the experimental data of Goncharuk et al. (1990), Mezhuev et al. (1990), Katayama et al. (1991), Gupta et al. (1994), Sidorko and Goncharuk (1995), and Goncharuk and Sidorko (1996)] that is related with the tendency of these solid solutions to decompose with decreasing temperature. At 627°C, the Gibbs free energy of mixing for the ZnTe–CdTe solid solutions decreases in comparison with values for the ideal ones. At 450°C, the excess integral Gibbs energy and the enthalpy of the solid-solution formation are positive in the entire concentration range (Goncharuk et al. 1990, Sidorko and Goncharuk 1995, Goncharuk and Sidorko 1996), depending on the interaction of the metallic sublattice components. Thus, thermodynamic stability of the solid solutions decreases with decreasing temperature.

The activities of both Zn and Cd over $Zn_xCd_{1-x}Te$ solid solutions at different temperatures and compositions exhibit high negative deviations from Raoult's law (Shamsuddin et al. 1994). Both a_{Cd} and a_{Zn} have been found to increase with temperature, and the change in activities is more pronounced at higher temperatures. The continuous variation of the activity and activity coefficients of both the components and the smooth gradual variation of the stability and excess stability with composition indicate that over the temperature range 470°C–575°C, a single-phase field exists throughout the entire range of composition (Gupta et al. 1994, Shamsuddin et al. 1994).

The activities of binary components CdTe and ZnTe reported in Alikhanian et al. (2002, 2003a,b, 2004) need a careful consideration since at 627°C, the activities of CdTe were found to have negative deviation from ideal behavior whereas those of ZnTe exhibited positive deviation, which is not consistent with the Gibbs–Duhem relationship. At 507°C, the data spread of the CdTe activity does not allow to exclude the possibility that there is the separation of the phases in the range from 40 to 75 mol. % CdTe.

The equilibrium vapor pressures of Cd (Zn) over ZnTe–CdTe solutions increase with increasing CdTe (ZnTe) concentration at any temperature (Shamsuddin et al. 1994, Guskov et al. 2004). It was observed that the equilibrium vapor pressure of Cd is larger than that of Zn over the solid solutions at

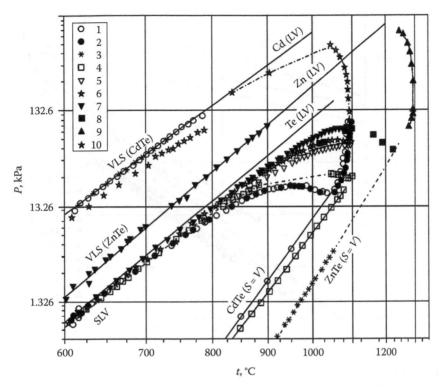

FIGURE 3.9

p-T phase diagram of the $Zn_xCd_{1-x}Te$ solid solutions at $x = 1$ (1), 0.95 (2), 0.85 (3), 0.5 (4), 0.2 (5), 0.1 (6), 0 (7–10). (From Guskov, V.N. et al., *J. Alloys Compd.*, 371(1–2), 118, 2004.)

all temperatures (Figure 3.9). The sublimation in this system is incongruent, and no azeotropic composition was observed in the solid–vapor equilibrium (Greenberg et al. 2003, 2004a). The main vapor species observed in the ZnTe–CdTe system are Cd and Zn atoms and Te_2 molecules (Alikhanian et al. 2002, 2003a,b, 2004, Greenberg et al. 2003).

Sang et al. (2000) determined that p_{Cd} and p_{Zn} over $Cd_{0.8}Zn_{0.2}Te$ melt at 1162°C that are about 690 and 84 Pa, respectively. The p_{Cd}-T projection of the $Zn_{0.05}Cd_{0.95}Te_{1\pm\delta}$ sublimation range and p-T projection of the p-T-x-y phase diagram for the Zn–Cd–Te system are shown in Figures 3.10 and 3.11, respectively.

Concentration dependence of the energy gap of the $Zn_xCd_{1-x}Te$ solid solutions is linear (Kolomiets and Mal'kova 1958, Kot et al. 1962).

This system was investigated using DTA, metallography, differential scanning calorimetry (DSC), electron probe microanalyzer (EPMA), and XRD (Maksimova and Tsurkan 1970, Steininger et al. 1970, Haloui et al. 1997, Katayama et al. 1991).

The full description of the Zn–Cd–Te ternary system is given in Tomashik and Shcherbak (2006).

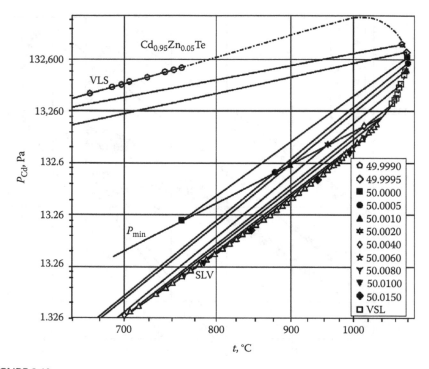

FIGURE 3.10
p_{Cd}-T projection of the $Zn_{0.05}Cd_{0.95}Te_{1\pm\delta}$ sublimation range and isopleths of Cd partial pressures (numbers in the legend correspond to the composition in at. % Te). (From Greenberg, J.H. et al., *J. Cryst. Growth*, 270, 69, 2004a.)

3.6 Zinc–Mercury–Tellurium

Ternary liquidus temperature in the Zn–Hg–Te system was measured in a transparent furnace using a modified direct observation technique (Smith et al. 1987, Andruhiv et al. 1992, 1993) and calculated using a model of completely associated solutions (Andruhiv et al. 1992, 1993). There is a good agreement between experimental and calculated data. Liquidus temperature data for $(Hg_{1-z}Zn_z)_{1-y}Te_y$ are plotted in Figure 3.12.

ZnTe–HgTe: The phase diagram is shown in Figure 3.13 (Woolley 1960, Krucheanu et al. 1964, Marbeuf et al. 1992). At the interaction of ZnTe and HgTe, solid solutions with sphalerite structure are formed over the entire range of concentrations. Krucheanu et al. (1963, 1964) obtained two-phase ingots after melting previously annealed at 650°C specimens.

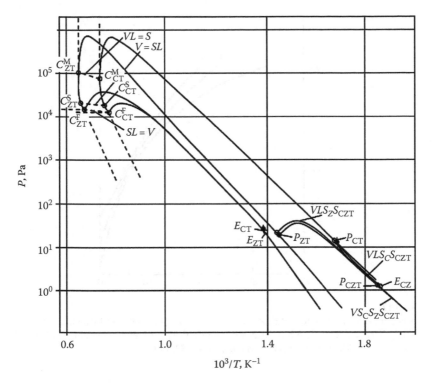

FIGURE 3.11
p-T projection of the $Zn_xCd_{1-x}Te_{1\pm\delta}$ solid solutions. (From Nipan, G.D. et al., *J. Alloys Compd.*, 371(1–2), 160, 2004.)

A miscibility gap is predicted in this system (Wei et al. 1990, Marbeuf et al. 1992, Yu and Brebrick 1992, Voronin and Pentin 2005, Pentin et al. 2006). Critical temperature and composition of the solid-solution decomposition are 285°C and 33.2 mol. % HgTe (Voronin and Pentin 2005, Pentin et al. 2006) [T_C=329°C (Marbeuf et al. 1992), T_C=182°C and x_C=44 mol. % HgTe (Wei et al. 1990), T_C=264°C and x_C=36 mol. % HgTe (Leute and Plate 1989), T_C=7°C (Andruhiv et al. 1993), T_C=−43.7°C (Yu and Brebrick 1992), T_C=121°C (Ohtani et al. 1992), T_C=481°C (Laugier 1973)].

The solid solutions $Zn_xHg_{1-x}Te$ were found to exhibit a positive deviation from ideality (Lukashenko et al. 1991, Sidorko and Goncharuk 1995). At 380°C, the excess integral Gibbs energy and the enthalpy of the solid-solution formation are positive in the entire concentration range (Lukashenko et al. 1991, Sidorko and Goncharuk 1995), depending on the interaction of the metallic sublattice components. Thus, thermodynamic stability of the solid solutions decreases with decreasing temperature. A cluster theory based on the quasi-chemical approximation has been applied by Patrick et al. (1988) to study the local correlation, bond-length distribution, and ZnTe–HgTe phase diagram.

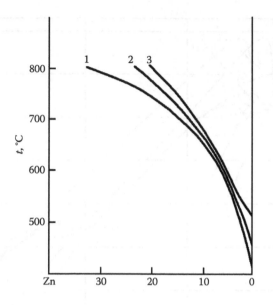

FIGURE 3.12
Liquidus temperature data for $(Hg_{1-z}Zn_z)_{1-y}Te_y$ at $y = 0.75$ (1), 0.8 (2), and 0.85 (3). (From Smith, E.J. et al., *J. Vac. Sci. Technol.*, A5(5), 3043, 1987; Andruhiv, A.M. et al., *Pis'ma v ZhTF*, 18(13), 10, 1992; Andruhiv, A.M. et al., *Neorgan. Materialy*, 29(4), 492, 1993.)

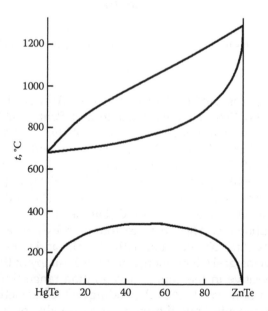

FIGURE 3.13
ZnTe–HgTe phase diagram. (From Woolley, J.C. and Ray, B., *J. Phys. Chem. Solids*, 13(1/2), 151, 1960c; Krucheanu, E. et al., *Rev. Roum. Phys.* [in Russian], 9(5), 499, 1964; Marbeuf, A. et al., *J. Cryst. Growth*, 117(1–4), 10, 1992.)

Concentration dependence of the lattice parameters can be described by the Vegard's law (Woolley and Ray 1960c, Danilyuk and Kot 1964, Krucheanu et al. 1964, Toulouse et al. 1987, Leute and Plate 1989). Energy gap at room temperature changes linearly with composition (Danilyuk and Kot 1964). For each composition, E_g follows very well a linear variation with temperature. At 80 K, concentration dependence of energy gap for $Zn_xHg_{1-x}Te$ solid solutions can be described by the following equation (Toulouse et al. 1987):

$$E_g(eV) = -0.258 + 2.676x - 0.629x^2 + 0.533x^3.$$

According to the data of Niculescu and Dziuba (1969), the alloys containing up to 10 mol. % ZnTe are semimetals, and at higher ZnTe contents, they have semiconductor properties.

Isothermal interdiffusion experiments between ZnTe and HgTe have been performed by Pobla et al. (1986) within the temperature interval from 400°C to 600°C. The interdiffusion coefficient D has been deduced from composition profiles.

The Bridgman–Stockbarger method and the slow zone recrystallization have been used for the preparation of $Zn_xHg_{1-x}Te$ solid solutions (Cruceanu and Nistor 1966).

This system was investigated using DTA and XRD (Woolley and Ray 1960c, Krucheanu et al. 1964). Phase diagram was calculated using the model of regular associated solutions. The solid/liquid equilibria were calculated with the assumption of an ideal behaving melt (Leute and Plate 1989).

3.7 Zinc–Aluminum–Tellurium

According to the data of Gamidov and Kulieva (1969), $ZnAlTe_2$ exists in the Zn–Al–Te ternary system. This compound crystallizes in a tetragonal structure of the chalcopyrite type with lattice parameters $a = 598.5$ and $c = 1179.0$ pm.

$ZnTe–Al_2Te_3$: The phase diagram is not constructed. $ZnAl_2Te_4$ compound was obtained in this system, which crystallizes in a tetragonal structure of chalcopyrite type with lattice parameters $a = 509.4$ and $c = 1203$ pm ($c = 601.5$ pm) [$a = 842.0$ and $c = 2418.2$ pm (Schwer and Krämer 1988)] and calculation and experimental density 4.965 and 4.91 g cm^{-3}, respectively (Hahn et al. 1955).

$ZnAl_2Te_4$ was obtained by annealing of mixtures, containing ZnTe, Al, and Te in stoichiometric ratio at 800°C for 12–24 h (Hahn et al. 1955).

3.8 Zinc–Gallium–Tellurium

ZnTe–Ga: The phase diagram is not constructed. According to the data of Wagner and Lorenz (1966), this diagram is a eutectic type with degenerated eutectic from the Ga-rich side. Liquidus temperatures of the ZnTe–Ga system are given in Table 3.1.

Liquidus temperatures were determined by a visual observation of a solid-phase appearance and disappearance for heating and cooling of ingots (accuracy of measurement is ±10°C). Compositions of the ingots were investigated using XRD (Wagner and Lorenz 1966).

ZnTe–GaTe: The phase diagram is not constructed. According to the data of Guseinov (1969), $ZnGaTe_2$ compound is formed in this system, which crystallizes in a tetragonal structure with the lattice parameters $a = 1141$ and $c = 661$ pm and calculation and experimental density 6.05 and 6.03 g cm^{-3}, respectively, and energy gap $E_g = 1.2$ eV.

ZnTe–Ga$_2$Te$_3$: The phase diagram is shown in Figure 3.14 (Burlaku et al. 1977). The peritectic temperature is 850°C. $ZnGa_2Te_4$ is formed in this system. It melts incongruently at 880°C ± 5°C and crystallizes in a tetragonal structure with the lattice parameters $a = 836$ and $c = 4799$ pm (Schwer and Krämer 1988) [$a = 836.4$ and $c = 4799$ pm (Burlaku et al. 1977), $a = 592.5$ and $c = 1185$ pm (Hahn et al. 1955)] and calculation and experimental density 5.674 and 5.57 g cm^{-3}, respectively (Hahn et al. 1955), and energy gap $E_g = 1.22$ eV (Goriunova et al. 1955a). Low-temperature modification of $ZnGa_2Te_4$ is formed after annealing at 320°C for 300 h and crystallizes also in the tetragonal structure with lattice parameters $a = 593.7$ and $c = 1187$ pm (Hahn et al. 1955).

Large regions of solid solutions based on ZnTe and Ga$_2$Te$_3$ exist in this system (Goriunova 1955, Goriunova et al. 1955b, Burlaku et al. 1977). At 765°C, the solubility of Ga$_2$Te$_3$ in 3ZnTe is equal to 55 mol. %, and the solubility of 3ZnTe in Ga$_2$Te$_3$ reaches 27 mol. % (Woolley and Ray 1960a).

This system was investigated using DTA, metallography, and XRD. The ingots were annealed at temperatures near to solidus temperatures for 600–700 h (Burlaku et al. 1977).

TABLE 3.1

Liquidus Temperatures in the ZnTe–Ga System

x_{ZnTe}, mol. %	t, °C	x_{ZnTe}, mol. %	t, °C
0.134	527	3.50	876
0.378	613	6.77	963
1.08	728	12.6	1041
1.77	793	35.0	1141

Source: Wagner, P. and Lorenz, M.R., *J. Phys. Chem. Solids*, 27(11/12), 1749, 1966.

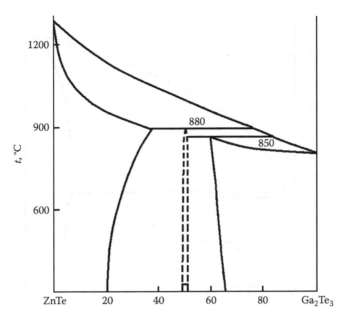

FIGURE 3.14
ZnTe–Ga$_2$Te$_3$ phase diagram. (From Burlaku, G.G. et al., *Izv. AN SSSR. Neorgan. materialy*, 13(5), 820, 1977.)

3.9 Zinc–Indium–Tellurium

ZnTe–In: The phase diagram is not constructed. According to the data of Wagner and Lorenz (1966), this diagram is a eutectic type with degenerated eutectic from the In-rich side. Liquidus temperatures of the ZnTe–In system are given in Table 3.2.

Liquidus temperatures were determined by visual observation of a solid-phase appearance and disappearance for heating and cooling of ingots

TABLE 3.2

Liquidus Temperatures in the ZnTe–In System

x_{ZnTe}, mol. %	t, °C	x_{ZnTe}, mol. %	t, °C
0.25	408	5.00	735
0.314	445	5.03	745
0.36	457	10.01	827
0.532	491	25.0	970
0.999	547	50.1	1107
2.01	625	100	1295

Source: Wagner, P. and Lorenz, M.R., *J. Phys. Chem. Solids*, 27(11/12), 1749, 1966.

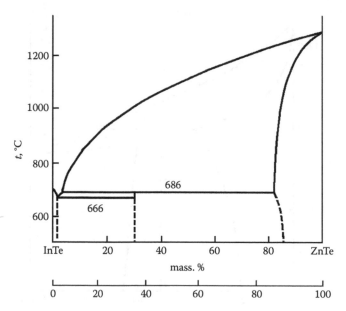

FIGURE 3.15
ZnTe–InTe phase diagram. (From Svechnikova, T.E. et al., *Izv. AN SSSR. Neorgan. materialy*, 7(6), 939, 1971.)

(accuracy of measurement is ±10°C). Compositions of the ingots were investigated using XRD (Wagner and Lorenz 1966).

ZnTe–InTe: The phase diagram is shown in Figure 3.15 (Svechnikova et al. 1971). The eutectic composition and temperature are 3.1 mol. % (2.5 mass. %) and 666°C, respectively. Ternary compound, containing approximately 35 mol. % (30 mass. %) ZnTe, is formed in this system. It melts incongruently at 686°C. The peritectic composition is 5 mol. % (4 mass. %) ZnTe.

The solubility of InTe in ZnTe at 500°C reaches 12.7 mol. % (15.5 mass. %) (Svechnikova et al. 1971).

According to the data of Guseinov (1969), $ZnInTe_2$ compound is formed in this system that crystallizes in a tetragonal structure with lattice parameters $a = 1218$ and $c = 600$ pm, experimental density 6.39 g cm^{-3}, and energy gap $E_g = 0.39$ eV.

This system was investigated using DTA, metallography, XRD, and measuring of microhardness (Svechnikova et al. 1971).

ZnTe–In$_2$Te$_3$: The phase diagram is shown in Figure 3.16 (O'Kane and Mason 1965). There are four peritectic transformations in this system: the first (β-phase) at 50 mol. % In$_2$Te$_3$ and 795°C, the second (γ-phase) at approximately 75 mol. % In$_2$Te$_3$ and 708°C, the third (δ-phase) at approximately 92 mol. % In$_2$Te$_3$ and 692°C, and the fourth (ε-phase) at approximately 98 mol. % In$_2$Te$_3$ and 670°C.

β-phase is a ZnIn$_2$Te$_4$ compound (Woolley and Ray 1960b) and crystallizes in tetragonal structure of a chalcopyrite type with lattice parameters $a = 611.0$ and $c = 1222$ pm and calculation and experimental density 5.825 and 5.82 g cm^{-3} (Hahn et al. 1955) and energy gap $E_g = 1.4$ eV (Busch et al. 1956).

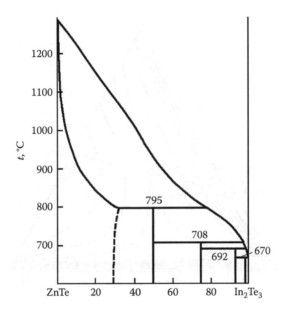

FIGURE 3.16

ZnTe–In$_2$Te$_3$ phase diagram. (From O'Kane, D.F. and Mason, D.R., *Trans. Metallurg. Soc. AIME*, 233(6), 1189, 1965.)

α-, δ-, and ε-phases crystallize in disordered zinc-blende structure (O'Kane and Mason 1965).

The ingots containing 30 and 50 mol. % In$_2$Te$_3$ have phase transition at 420°C (O'Kane and Mason 1965).

This system was investigated using DTA. The ingots were annealed at 600°C for 1 week (O'Kane and Mason 1965).

3.10 Zinc–Thallium–Tellurium

The field of ZnTe primary crystallization occupies almost all Zn–Tl–Te ternary system, the fields of Tl$_2$Te and Tl$_5$Te$_3$ form narrow bands along the Tl–Te binary system, and the fields of Zn, Tl, and Te are degenerated (Figure 3.17) (Babanly et al. 1986). Two immiscibility regions exist in this system. The intersection of e_4–Tl eutectic line with immiscibility region leads to nonvariant equilibrium (point *M* in Figure 3.17): L$_1$ ⇔ L$_2$ + α(Tl$_2$Te) + ZnTe (372°C).

Ternary eutectic between α(Tl$_2$Te), γ(Tl$_5$Te$_3$), and ZnTe is degenerated near e_4 eutectic (binary eutectic in the ZnTe–Tl$_2$Te system at 377°C).

Ternary compounds were not found in the Zn–Tl–Te ternary system (Figure 3.18) (Asadov et al. 1983, Babanly et al. 1986).

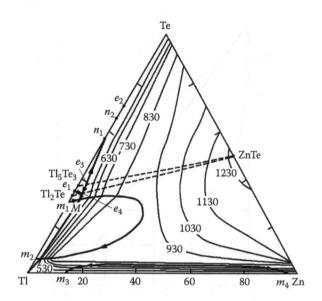

FIGURE 3.17
Liquidus surface of the Zn–Tl–Te ternary system. (From Babanly, M.B. et al., *Izv. AN SSSR. Neorgan. materialy*, 22(11), 1822, 1986.)

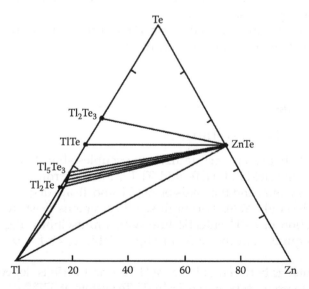

FIGURE 3.18
Isothermal section of the Zn–Tl–Te ternary system at room temperature. (From Babanly, M.B. et al., *Izv. AN SSSR. Neorgan. materialy*, 22(11), 1822, 1986.)

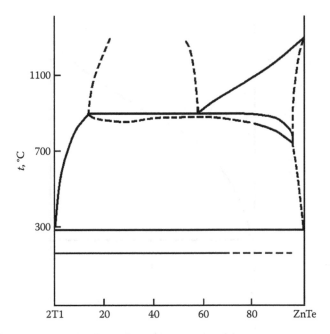

FIGURE 3.19
Phase relations in the ZnTe–2Tl system. (From Babanly, M.B. et al., *Izv. AN SSSR. Neorgan. materialy*, 22(11), 1822, 1986.)

ZnTe–Tl: According to the data of Babanly et al. (1986), this section is a nonquasibinary section of Zn–Tl–Te ternary system (Figure 3.19). These nonquasibinarity authors explain the existence of invariant equilibrium $L_1 \Leftrightarrow L_2 + \beta(ZnTe)$ and appreciable retrograde solubility based on ZnTe. Two phases ZnTe and Tl exist in the solid state.

This system was investigated using DTA, XRD, and measuring of microhardness and emf of concentration chains. The ingots were annealed for 400–600 h (Babanly et al. 1986).

ZnTe–TlTe: This section is a nonquasibinary section of Zn–Tl–Te ternary system (Figure 3.20) (Guseinov et al. 1982). Thermal effects at 360°C correspond to secondary crystallization of γ-phase based on Tl_5Te_3 and at 300°C to the four-phase equilibrium $L + \gamma(Tl_5Te_3) \Leftrightarrow ZnTe + TlTe$. The mutual solubility of ZnTe and TlTe is less than 1 mol. %.

This system was investigated using DTA, XRD, and measuring of microhardness. The ingots were annealed at temperatures 20°C–30°C lower than solidus temperature for 400–500 h (Guseinov et al. 1982).

ZnTe–Tl₂Te: The phase diagram is shown in Figure 3.21 (Sztuba et al. 2006). $ZnTl_{18}Te_{10}$ compound that melts congruently at 440°C ± 0.5°C is formed in this system. This compound forms a eutectic with ZnTe, the coordinates of which are 22.8 mol. % ZnTe and 429°C ± 0.5°C. Another eutectic composition is degenerated from Tl_2Te side. It is necessary to note that the ZnTe melting temperature was not correct in the author version and was revised.

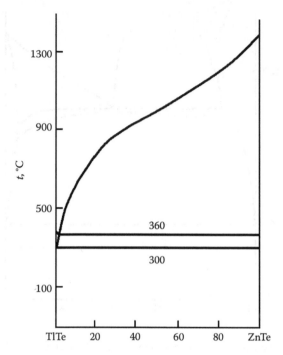

FIGURE 3.20

Phase relations in the ZnTe–TlTe system. (From Guseinov, F.H. et al., *Izv. AN SSSR. Neorgan. materialy*, 18(5), 759, 1982.)

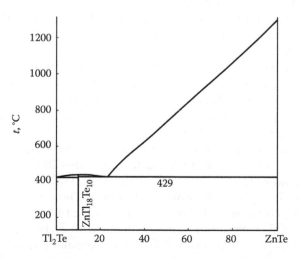

FIGURE 3.21

ZnTe–Tl$_2$Te phase diagram. (From Sztuba, Z. et al., *Calphad*, 30(4), 421, 2006.)

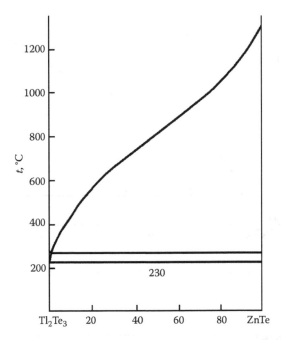

FIGURE 3.22
Phase relations in the ZnTe–Tl$_2$Te$_3$ system. (From Guseinov, F.H. et al., *Izv. vuzov. khimia i khim. tehnologia*, 24(10), 1245, 1981b.)

According to the data of Guseinov et al. (1981a), the phase diagram of this system is of a simple eutectic type, with the eutectic coordinates being 377°C and 8 mol. % ZnTe without forming a ternary compound. Mutual solubility of ZnTe and Tl$_2$Te is insignificant.

This system was investigated using DTA, XRD, and measuring of microhardness (Guseinov et al. 1981a, Sztuba et al. 2006).

ZnTe–Tl$_2$Te$_3$: According to the data of Guseinov et al. (1981b), the ZnTe–Tl$_2$Te$_3$ section is a nonquasibinary section of the Zn–Tl–Te ternary system (Figure 3.22). Incongruently melting of Tl$_2$Te$_3$ leads to four-phase peritectic reaction at 235°C: L + TlTe ⇔ ZnTe + Tl$_2$Te$_3$. The mixture of ZnTe and Tl$_2$Te$_3$ exists in the solid state. ZnTl$_2$Te$_4$ ternary compound was not found in this system. Mutual solubility of ZnTe and Tl$_2$Te$_3$ is insignificant.

This system was investigated using DTA, XRD, and measuring of microhardness. The ingots were annealed at 230°C for 400 h (Guseinov et al. 1981b).

ZnTe–Tl$_5$Te$_3$: The phase diagram is a eutectic type (Figure 3.23) (Babanly et al. 1986). The eutectic composition and temperature are 5 mol. % 4ZnTe (the system is regarded as 4ZnTe–Tl$_5$Te$_3$) and 427°C, respectively.

This system was investigated using DTA, XRD, and measuring of microhardness and emf of concentration chains. The ingots were annealed for 400–600 h (Babanly et al. 1986).

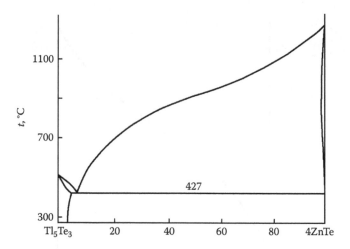

FIGURE 3.23
4ZnTe–Tl$_5$Te$_3$ phase diagram. (From Babanly, M.B. et al., *Izv. AN SSSR. Neorgan. materialy*, 22(11), 1822, 1986.)

3.11 Zinc–Lanthanum–Tellurium

ZnTe–La$_2$Te$_3$: The phase diagram is not constructed. ZnLa$_2$Te$_4$ is formed in this system (Aliev 1998). It melts incongruently at 1262°C. This compound is semiconductor with energy gap 2.25 eV and calculation density 6.35 g cm^{-3}.

ZnLa$_2$Te$_4$ was obtained by the interaction of ZnTe and La$_2$Te$_3$ at 800°C–900°C with next annealing at 730°C for 150 h, cooling up to 530°C and the annealing at this temperature during 250 h (Aliev 1998). Single crystals of this compound were grown using chemical transport reactions using iodine as the transport agent.

3.12 Zinc–Silicon–Tellurium

The field of Si primary crystallization occupies almost all Zn–Si–Te ternary system (Figure 3.24) (Odin 1996). There are also the fields of ZnTe, Si$_2$Te$_3$, and Zn (degenerated field) primary crystallization in this system. Three- and four-phase equilibria in the Zn–Si–Te ternary system with liquid participation are given in Table 3.3 (Odin 1996). Ternary compounds are not found in this system (Kaldis et al. 1967, Odin 1996).

ZnTe–Si: The phase diagram is not constructed. According to the data of Odin (1996), this diagram is a eutectic type. The eutectic composition and temperature are 5 ± 1 at. % Si and 1267°C ± 10°C. Mutual solubility of ZnTe and Si is insignificant.

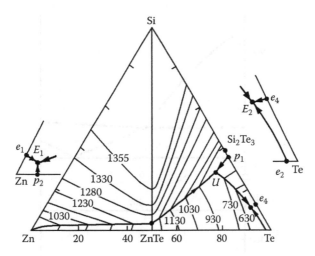

FIGURE 3.24

Liquidus surface of the Zn–Si–Te ternary system. (From Odin, I.N., *Zhurn. Neorgan. Khimii*, 41(6), 941, 1996.)

TABLE 3.3

Three- and Four-Phase Equilibria with Liquid Participation in the Zn–Si–Te Ternary System

Symbol	Reaction	$t, °C$
e_1	$L \Leftrightarrow Zn + Si$	419
e_2	$L \Leftrightarrow ZnTe + Te$	450
e_4	$L \Leftrightarrow Te + \beta\text{-}Si_2Te_3$	414
p_1	$L + Si \Leftrightarrow \beta\text{-}Si_2Te_3$	886
p_2	$ZnTe + L \Leftrightarrow Zn$	422
E_1	$L \Leftrightarrow ZnTe + Zn + Si$	418
E_2	$L \Leftrightarrow \beta\text{-}Si_2Te_3 + ZnTe + Te$	407
U	$L + Si \Leftrightarrow ZnTe + \beta\text{-}Si_2Te_3$	818

Source: Odin, I.N., *Zhurn. neorgan. khimii*, 41(6), 941, 1996.

This system was investigated using DTA, metallography, and XRD. The ingots were annealed at temperatures 20°C lower than temperatures of nonvariant equilibria with liquid participation (Odin 1996).

ZnTe–Si₂Te₃: This section is a nonquasibinary section of Zn–Si–Te ternary system (Figure 3.25) (Odin 1996). Silicon primarily crystallizes from the Si₂Te₃-rich side. Crystallization ends at 818°C by the next peritectic reaction: $L + Si \Leftrightarrow ZnTe + \beta\text{-}Si_2Te_3$.

Si₂Te₃ has polymorphous transformation at 406°C–409°C. The solubility of Si₂Te₃ in ZnTe is not higher than 0.1 mol. %, and the solubility of ZnTe in Si₂Te₃ is equal to 0.1–0.2 mol. %.

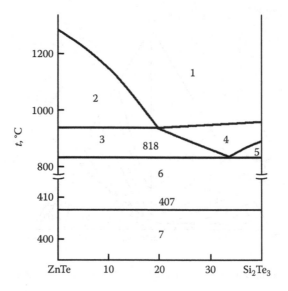

FIGURE 3.25
Phase relations in the $ZnTe–Si_2Te_3$ system: 1, L; 2, L+ZnTe; 3, L+ZnTe+Si; 4, L+Si; 5, L+Si+β-Si_2Te_3; 6, ZnTe+β-Si_2Te_3; and 7, ZnTe+α-Si_2Te_3. (From Odin, I.N., *Zhurn. neorgan. khimii*, 41(6), 941, 1996.)

This system was investigated using DTA, metallography, and XRD. The ingots were annealed at the temperature 20°C lower than temperatures of nonvariant equilibria with liquid participation (Odin 1996).

3.13 Zinc–Germanium–Tellurium

Ternary compounds in the Zn–Ge–Te system were not found (Kaldis et al. 1967).

ZnTe–GeTe: The phase diagram is a eutectic type (Figure 3.26) (Glazov et al. 1970, 1972, 1975). The eutectic temperature is 705°C. Interaction of ZnTe with low-temperature GeTe modification is described by the eutectoid phase diagram. The eutectoid transformation takes place at 380°C. Maximum mutual solubility of ZnTe and GeTe is not higher than 5 mol. % (Glazov et al. 1972). According to the data of Kutsia and Stavrianidis (1983), the solubility of ZnTe in GeTe at 530°C and 230°C is equal correspondingly 2.5 and 2 mol. %.

This system was investigated using DTA, metallography, and XRD (Glazov et al. 1970, 1972, 1975, Kutsia and Stavrianidis 1983). The ingots were annealed at 530°C, 430°C, 330°C, and 230°C for 200, 400, 600, and 800 h, respectively (Kutsia and Stavrianidis 1983).

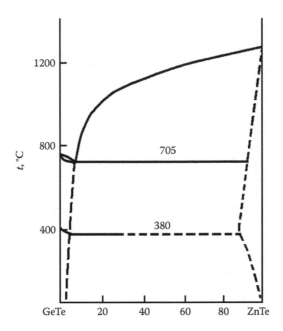

FIGURE 3.26
ZnTe–GeTe phase diagram. (From Glazov, V.M. et al., *Izv. AN SSSR. Neorgan. materialy*, 6(3), 569, 1970; Glazov, V.M. et al., *Izv. vuzov. Ser. Tsvetnaya metallurgia*, (5), 116, 1972.)

3.14 Zinc–Tin–Tellurium

ZnTe–Sn: The phase diagram is not constructed. The ZnTe solubility in liquid tin was studied (Figure 3.27) (Rubenstein 1968).

Liquidus temperatures were determined by high-temperature filtration (Rubenstein 1968).

ZnTe–SnTe: The phase diagram is a eutectic type (Figure 3.28) (Sultanova et al. 1974). The eutectic composition and temperature are 10 mol. % ZnTe and 770°C, respectively. The solubility of ZnTe in SnTe is not higher than 2 mol. %, and the solubility of SnTe in ZnTe corresponds to 1 mol. %.

This system was investigated using DTA, metallography, and XRD. The ingots were annealed at 350°C for 500 h (Sultanova et al. 1974).

3.15 Zinc–Lead–Tellurium

Liquidus surface of the Zn–Pb–Te ternary system includes five fields of primary crystallization (Figure 3.29) (Movsum-zade et al. 1986). The field of ZnTe primary crystallization occupies almost all liquidus surface.

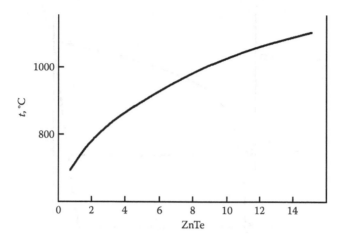

FIGURE 3.27
Temperature dependence of ZnTe solubility in liquid Sn. (From Rubenstein, M., *J. Cryst. Growth,* 3–4, 309, 1968.)

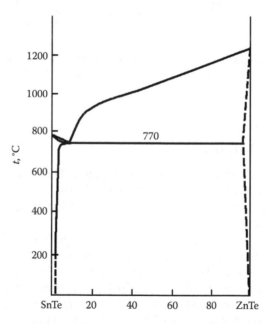

FIGURE 3.28
ZnTe–SnTe phase diagram. (From Sultanova, N.R. et al., *Izv. AN SSSR. Neorgan. materialy,* 10(8), 1418, 1974.)

The immiscibility region from the Zn–Pb binary system penetrates deeply in the ternary system and occupies the great part of the ZnTe field. The ternary eutectics E_1, E_2, and E_3 crystallize at 400°C, 300°C, and 310°C, respectively.

ZnTe–Pb: The phase diagram is a eutectic type (Figure 3.30) (Movsumzade et al. 1986). The eutectic composition and temperature are 3 mol. % ZnTe

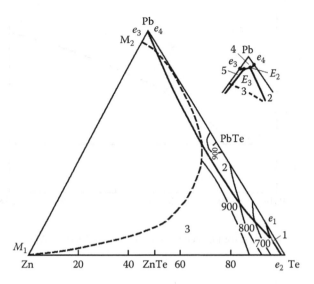

FIGURE 3.29
Liquidus surface of the Zn–Pb–Te ternary system: Te (1), PbTe (2), ZnTe (3), Pb (4), and Zn (5) primary crystallization fields. (From Movsum-zade, A.A. et al., *Zhurn. neorgan. khimii*, 31(1), 198, 1986.)

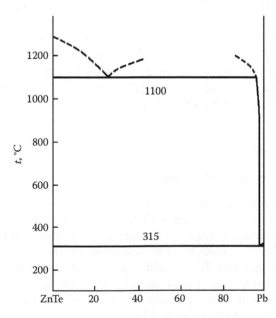

FIGURE 3.30
ZnTe–Pb phase diagram. (From Movsum-zade, A.A. et al., *Zhurn. neorgan. khimii*, 31(1), 198, 1986.)

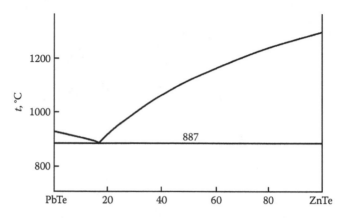

FIGURE 3.31
ZnTe–PbTe phase diagram. (From Grytsiv, V.I. et al., *Izv. AN SSSR. Neorgan. materialy*, 16(3), 543, 1980.)

and 315°C, respectively. An immiscibility region within the interval from 25 to 95 at. % Pb exists in this system at 1100°C.

This system was investigated using DTA, metallography, XRD, and measuring of microhardness. The ingots were annealed at a temperature 50°C lower than solidus temperature for 200 h (Movsum-zade et al. 1986).

ZnTe–PbTe: The phase diagram is a eutectic type (Figure 3.31) (Grytsiv et al. 1980, Movsum-zade et al. 1986). The eutectic composition and temperature are 17 mol. % ZnTe and 887°C ± 3°C, respectively (Grytsiv et al. 1980) [10 mol. % ZnTe and 875°C (Movsum-zade et al. 1986)]. The solubility of ZnTe in PbTe at 800°C, 720°C, and 250°C is equal correspondingly to 1.5, 1, and 1 mol. % (Rosenberg et al. 1964, Sealy and Crocker 1973).

This system was investigated using DTA, metallography, XRD, and measuring of microhardness (Grytsiv et al. 1980, Movsum-zade et al. 1986).

3.16 Zinc–Arsenic–Tellurium

There are seven fields of primary crystallization on the liquidus surface of the Zn–As–Te ternary system (Figure 3.32) (Olekseyuk and Stoika 1977). The field of ZnTe crystallization occupies the biggest part of this surface. The phases of Zn_3As_2, Zn, $ZnAs_2$, As, As_2Te_3, and Te also primarily crystallize; moreover, the fields of Zn, As, and Te crystallization are degenerated. Ternary compounds were not found in this system. Information about four-phase equilibria in the Zn–As–Te ternary system is given in Table 3.4. Ternary eutectic E_1 is degenerated from the Zn-rich side.

Isothermal section of the Zn–As–Te ternary system at 340°C is shown in Figure 3.33 (Olekseyuk and Stoika 1977). The solubility of different

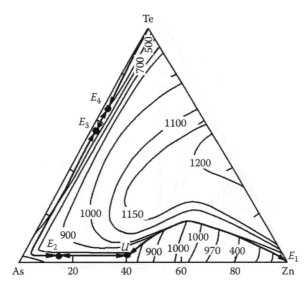

FIGURE 3.32
Liquidus surface of the Zn–As–Te ternary system. (From Olekseyuk, I.D. and Stoika, I.M., *Zhurn. neorgan. khimii*, 22(7), 1916, 1977.)

TABLE 3.4

Nonvariant Equilibria in the Zn–As–Te Ternary System

Symbol	Reaction	$t, °C$
E_1	$L \Leftrightarrow ZnTe + Zn_3As_2 + Zn$	417
E_2	$L \Leftrightarrow ZnTe + ZnAs_2 + As$	714
E_3	$L \Leftrightarrow ZnTe + As + As_2Te_3$	355
E_4	$L \Leftrightarrow ZnTe + Te + As_2Te_3$	345
U	$L + ZnTe \Leftrightarrow ZnAs_2 + Zn_3As_2$	758

Source: Olekseyuk, I.D. and Stoika, I.M., *Zhurn. neorgan. khimii*, 22(7), 1916, 1977.

phases in α-Zn_3As_2 at 340°C is equal to 0.4 mol. % Te, 4.85 mol. % ZnTe, 11.36 mol. % $Zn_{0.6}Te_{0.4}$, 4.8 mol. % $Zn_{0.8}Te_{0.2}$, and 3.84 mol. % $Zn_{0.9}Te_{0.1}$, and the solubility in β-Zn_3As_2 is within the intervals of 7.18–33.56 mol. % ZnTe and 16.3–26.96 mol. % $Zn_{0.6}Te_{0.4}$.

ZnTe–ZnAs₂: The phase diagram is a eutectic type (Figure 3.34) (Olekseyuk and Stoika 1977). The eutectic composition and temperature are 3.9 mol. % ZnTe and 763°C.

This system was investigated using DTA, metallography, XRD, and measuring of microhardness. The ingots were annealed at temperatures 20°C–40°C lower than solidus temperatures for 200–1500 h (Olekseyuk and Stoika 1977).

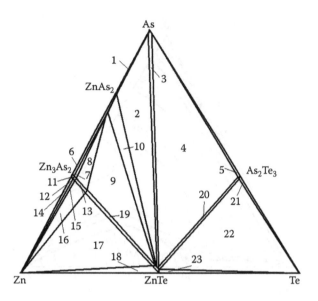

FIGURE 3.33
Isothermal section of the Zn–As–Te ternary system at 340°C: 1, $ZnAs_2 + As$; 2, $ZnTe + ZnAs_2 + As$; 3, $ZnTe + As$; 4, $ZnTe + As + As_2Te_3$; 5, $As + As_2Te_3$; 6, $\alpha\text{-}Zn_3As_2 + ZnAs_2$; 7, $\alpha\text{-}Zn_3As_2 + \beta\text{-}Zn_3As_2 + ZnAs_2$; 8, $\beta\text{-}Zn_3As_2 + ZnAs_2$; 9, $ZnAs_2 + \beta\text{-}Zn_3As_2 + ZnTe$; 10, $ZnAs_2 + ZnTe$; 11, $\alpha\text{-}Zn_3As_2$; 12, $\alpha\text{-}Zn_3As_2 + \beta\text{-}Zn_3As_2$; 13, $\beta\text{-}Zn_3As_2$; 14, $\alpha\text{-}Zn_3As_2 + \beta\text{-}Zn_3As_2 + Zn$; 15, $\alpha\text{-}Zn_3As_2 + Zn$; 16, $\beta\text{-}Zn_3As_2 + Zn$; 17, $ZnTe + \beta\text{-}Zn_3As_2 + Zn$; 18, $Zn + ZnTe$; 19, $\beta\text{-}Zn_3As_2 + ZnTe$; 20, $ZnTe + As_2Te_3$; 21, $Te + As_2Te_3$; 22, $ZnTe + Te + As_2Te_3$; and 23, $ZnTe + Zn$. (From Olekseyuk, I.D. and Stoika, I.M., *Zhurn. neorgan. khimii*, 22(7), 1916, 1977.)

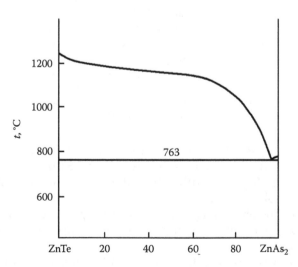

FIGURE 3.34
$ZnTe–ZnAs_2$ phase diagram. (From Olekseyuk, I.D. and Stoika, I.M., *Zhurn. neorgan. khimii*, 22(7), 1916, 1977.)

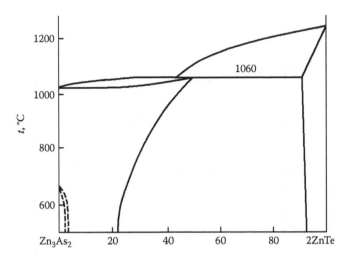

FIGURE 3.35

2ZnTe–Zn$_3$As$_2$ phase diagram. (From Olekseyuk, I.D. et al., *Izv. AN SSSR. Neorgan. materialy,* 7(5), 747, 1971; Golovey, M.I. et al., Chemical bond and solubility in the systems A$^{II}_3$B$^V_2$–A^{II}C^{VI} [in Russian], in *Khim. svyaz' v poluprovodn. i polumetallah.* Minsk, Nauka i Tekhnika Publish., pp. 233–239, 1972; Golovey, M.I. et al., *Izv. AN SSSR. Neorgan. materialy,* 9(6), 930, 1973; Olekseyuk, I.D. et al., *Izv. AN SSSR. Neorgan. materialy,* 14(8), 1383, 1978.)

ZnTe–Zn$_3$As$_2$: The phase diagram is a peritectic type (Figure 3.35) (Olekseyuk et al. 1971, Golovey et al. 1972, 1973, Olekseyuk et al. 1978). The peritectic composition and temperature are 45 mol. % 2ZnTe and 1060°C, respectively. Large regions of solid solutions based on ZnTe and Zn$_3$As$_2$ exist in this system (Kradinova et al. 1969, Goriunova et al. 1970, Olekseyuk et al. 1971, 1978, Golovey et al. 1972, 1973, Olekseyuk and Stoika 1977). The solubility of 2ZnTe in Zn$_3$As$_2$ increases from 21.5 mol. % at 500°C to 50 mol. % at 1060°C. The solubility of Zn$_3$As$_2$ is equal to 10 mol. % at 1060°C and practically does not change at the decreasing temperature. Solid solutions based on Zn$_3$As$_2$ have polymorphous transformation (Olekseyuk et al. 1971, 1978, Golovey et al. 1972, 1973).

This system was investigated using DTA, metallography, XRD, and measuring of microhardness. The ingots were annealed at the temperatures 20°C–40°C lower than the solidus temperatures for 200–1500 h (Kradinova et al. 1969, Goriunova et al. 1970, Olekseyuk et al. 1971, 1978, Golovey et al. 1972, 1973, Olekseyuk and Stoika 1977).

ZnTe–As: The phase diagram is a eutectic type (Figure 3.36) (Olekseyuk and Stoika 1977). The eutectic composition and temperature are 2.85 mol. % ZnTe and 795°C, respectively.

This system was investigated using DTA, metallography, XRD, and measuring of microhardness. The ingots were annealed at temperatures 20°C–40°C lower than the solidus temperatures for 200–1500 h (Olekseyuk and Stoika 1977).

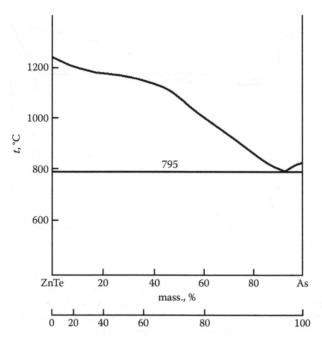

FIGURE 3.36
ZnTe–As phase diagram. (From Olekseyuk, I.D. and Stoika, I.M., *Zhurn. neorgan. khimii*, 22(7), 1916, 1977.)

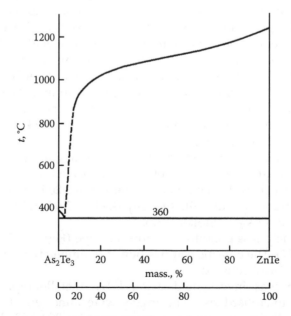

FIGURE 3.37
ZnTe–As$_2$Te$_3$ phase diagram. (From Olekseyuk, I.D. et al., *Izv. AN SSSR. Neorgan. materialy*, 11(11), 2081, 1975.)

ZnTe–As₂Te₃: The phase diagram is a eutectic type (Figure 3.37) (Olekseyuk et al. 1975). The eutectic composition and temperature are 6.6 mol. % ZnTe and 360°C (Olekseyuk and Stoika 1977) [approximately 2.5 mass. % ZnTe and 355°C ± 5°C (Olekseyuk et al. 1975)].

This system was investigated using DTA, metallography, XRD, and measuring of microhardness. The ingots were annealed at temperatures 20°C–40°C lower than the solidus temperatures for 200–1500 h (Olekseyuk et al. 1975, Olekseyuk and Stoika 1977).

3.17 Zinc–Antimony–Tellurium

ZnTe–Zn₃Sb₂: The phase diagram is a eutectic type (Figure 3.38) (Svechnikova et al. 1971). The eutectic composition and temperature are 2.2 mol. % (1 mass. %) ZnTe and 565°C. Thermal effects at 448°C and 408°C correspond to polymorphous transformations of Zn₃Sb₂. The solubility of ZnTe in Zn₃Sb₂ is not higher than 2.2 mol. % (1 mass. %), and the solubility of Zn₃Sb₂ in ZnTe reaches 6.7 mol. % (1.6 mass. %). Zn₄Sb₃ was found in cast alloys, which indicates that crystallization in this system is nonequilibrium.

This system was investigated using DTA, metallography, and XRD (Svechnikova et al. 1971).

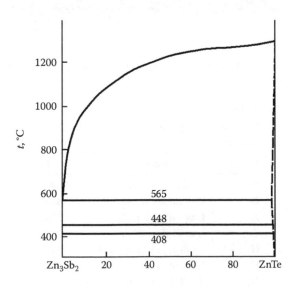

FIGURE 3.38
ZnTe–Zn₃Sb₂ phase diagram. (From Svechnikova, T.E. et al., *Izv. AN SSSR. Neorgan. materialy*, 7(6), 939, 1971.)

3.18 Zinc–Bismuth–Tellurium

There are eight fields of primary crystallization on the liquidus surface of the Zn–Bi–Te ternary system (Figure 3.39) (Marugin et al. 1984). The field of ZnTe crystallization occupies almost all liquidus surface. Two immiscibility regions exist in this system. One of them adjoins to the Zn–Bi binary system and forms narrow band in the Zn–Bi–Te ternary system, and the other is located inside the triangle along the ZnTe–Bi quasibinary system. Nonvariant equilibria in this ternary system are given in Table 3.5.

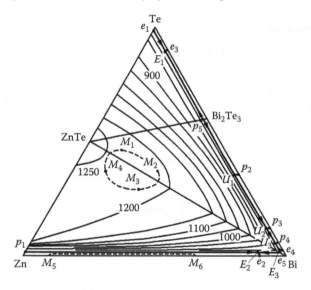

FIGURE 3.39

Liquidus surface of the Zn–Bi–Te ternary system. (From Marugin, V.V. et al., *Zhurn. neorgan. khimii*, 29(6), 1561, 1984.)

TABLE 3.5

Nonvariant Equilibria in the Zn–Bi–Te
Ternary System

Symbol	Reaction	t, °C
p_5	$L + ZnTe \Leftrightarrow Bi_2Te_3$	600
E_1	$L \Leftrightarrow ZnTe + Te + Bi_2Te_3$	410
E_2	$L \Leftrightarrow ZnTe + Zn + Bi$	250
E_3	$L \Leftrightarrow Bi_{14}Te_6 + Bi + ZnTe$	260
U_1	$L + Bi_2Te_3 \Leftrightarrow ZnTe + BiTe$	545
U_2	$L + BiTe \Leftrightarrow ZnTe + Bi_2Te$	415
U_3	$L + Bi_2Te \Leftrightarrow ZnTe + Bi_{14}Te_6$	310

Source: Marugin, V.V. et al., *Zhurn. neorgan. khimii*, 29(6), 1561, 1984.

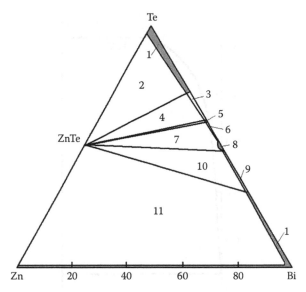

FIGURE 3.40
Isothermal section of the Zn–Bi–Te ternary system at 500°C: 1, L; 2, ZnTe+L; 3, Bi_2Te_3+L; 4, ZnTe+Bi_2Te_3+L; 5, Bi_2Te_3; 6, Bi_2Te_3+BiTe; 7, ZnTe+Bi_2Te_3+BiTe; 8, BiTe; 9, BiTe+L; 10, ZnTe+BiTe+L; and 11, ZnTe+L. (From Marugin, V.V. et al., *Zhurn. neorgan. khimii*, 29(6), 1561, 1984.)

Isothermal section of the Zn–Bi–Te ternary system at 500°C is shown in Figure 3.40 (Marugin et al. 1984). Liquid regions adjoin to the Zn–Bi binary system and Te-rich side.

ZnTe–Bi: The phase diagram is a eutectic type (Figure 3.41) (Rubenstein 1966, Marugin et al. 1984). The eutectic is degenerated from the Bi-rich side and crystallizes at 270°C±3°C. An immiscibility region exists in this system within the interval of 50–80 mol. % ZnTe. The monotectic temperature is 1235°C. Mutual solubility of ZnTe and Bi is insignificant.

This system was investigated using DTA, metallography, and XRD. The ingots were annealed at a temperature 50°C lower than the temperatures of corresponding nonvariant equilibria for 700 h (Marugin et al. 1984). Temperature dependence of ZnTe solubility in liquid Bi was determined by high-temperature filtration. The mixtures of ZnTe and Bi were annealed at low temperature for 16 h and at 1300°C for 6 h. The quantity of ZnTe dissolved in Bi was determined by weighing of the ingots before and after its treatment (Rubenstein 1966).

ZnTe–BiTe: This section is a nonquasibinary section of the Zn–Bi–Te ternary system (Figure 3.42) (Marugin et al. 1983, 1984). ZnTe and Bi_2Te_3 primarily crystallize in this system. At 570°C, secondary crystallization of ZnSe and Bi_2Se_3 takes place, and at 550°C, peritectic equilibrium L+Bi_2Te_3 ⇔ ZnTe+BiTe exists in this system. Crystallization ends at the temperature of ternary peritectic (545°C). The solubility of ZnTe in BiTe at 500°C reaches 2 mol. %, and the solubility of BiTe in ZnTe at the same temperature is less than 0.5 mol. %.

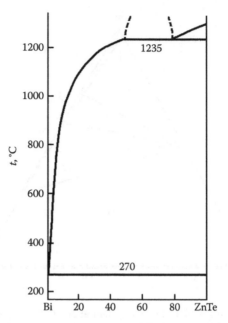

FIGURE 3.41
ZnTe–Bi phase diagram. (From Marugin, V.V. et al., *Zhurn. neorgan. khimii*, 29(6), 1561, 1984.)

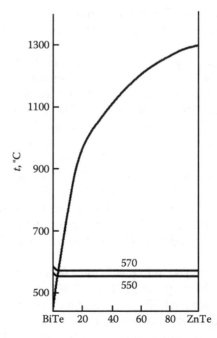

FIGURE 3.42
Phase relations in the ZnTe–BiTe system. (From Marugin, V.V. et al., *Zhurn. neorgan. khimii*, 28(8), 2104, 1983.)

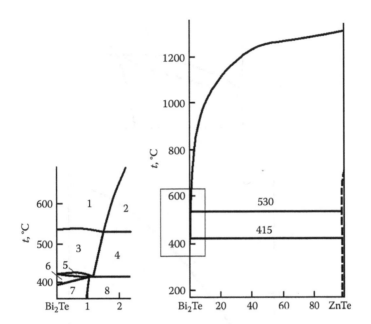

FIGURE 3.43
Phase relations in the ZnTe–Bi$_2$Te system: 1, L; 2, ZnTe + L; 3, BiTe + L; 4, ZnTe + BiTe + L; 5, Bi$_2$Te + BiTe + L; 6, Bi$_2$Te + L; 7, Bi$_2$Te; and 8, Bi$_2$Te + ZnTe. (From Marugin, V.V. et al., *Zhurn. neorgan. khimii*, 29(6), 1561, 1984.)

This system was investigated using DTA, metallography, and XRD. The ingots were annealed at a temperature 50°C lower than temperatures of corresponding nonvariant equilibria for 700 h (Marugin et al. 1984) [at 500°C for 500 h (Marugin et al. 1983)].

ZnTe–Bi$_2$Te: This section is a nonquasibinary section of the Zn–Bi–Te ternary system (Figure 3.43) (Marugin et al. 1984). ZnTe and BiTe primarily crystallize in this system. Crystallization ends at the temperature of ternary peritectic (415°C). The solubility of ZnTe in Bi$_2$Te and Bi$_2$Te in ZnTe is equal to 0.5 and 1 mol. %, respectively.

This system was investigated using DTA, metallography, and XRD. The ingots were annealed at a temperature 50°C lower than temperatures of corresponding nonvariant equilibria for 700 h (Marugin et al. 1984).

ZnTe–Bi$_2$Te$_3$: The phase diagram is a peritectic type (Figure 3.44) (Marugin et al. 1983, 1984, Odin and Marugin 1991). The peritectic composition and temperature are 4 mol. % ZnTe and 600°C ± 5°C. The solubility of ZnTe in Bi$_2$Te$_3$ reaches 5 mol. %, and the solubility of Bi$_2$Te$_3$ is less than 0.5 mol. %. As Bi$_2$Te$_3$ has maximum melting temperature (585°C) at the composition with 40.05 at. % Bi, this system is characterized by small deviations from quasibinarity (Odin and Marugin 1991).

This system was investigated using DTA, metallography, and XRD (Marugin et al. 1983, 1984, Odin and Marugin 1991). The ingots were annealed at 500°C for 720 h (Odin and Marugin 1991) [at 500°C for 500 h

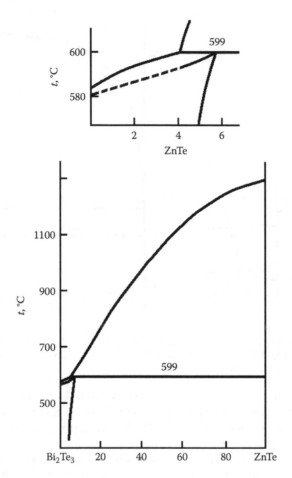

FIGURE 3.44
ZnTe–Bi$_2$Te$_3$ phase diagram. (From Odin, I.N. and Marugin, V.V., *Zhurn. neorgan. khimii*, 36(7), 1842, 1991.)

and at a temperature 50°C lower than the temperatures of corresponding nonvariant equilibria for 700 h (Marugin et al. 1984)].

3.19 Zinc–Oxygen–Tellurium

Isothermal section of the Zn–O–Te ternary system at room temperature, constructing according to thermodynamic calculations, is shown in Figure 3.45 (Medvedev et al. 1993).

ZnO–TeO$_2$: The phase diagram is not constructed. ZnTeO$_3$, Zn$_2$Te$_3$O$_8$, and ZnTe$_6$O$_{13}$ ternary compounds exist in this system. ZnTeO$_3$ melts at 780°C,

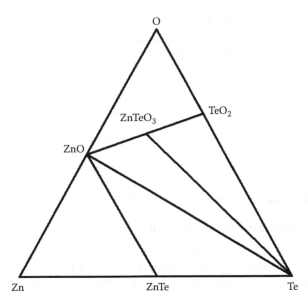

FIGURE 3.45
Isothermal section of the Zn–O–Te ternary system at room temperature. (From Medvedev, Yu.V. and Berchenko, N.N., *Zhurn. neorgan. khimii*, 38(12), 1940, 1993.)

is stable at a heating up to 1000°C, and crystallizes in an orthorhombic structure with the lattice parameters $a = 736$, $b = 638$, and $c = 1232$ pm (Hanke 1967) and experimental density 5.31 ± 0.01 g cm^{-3} (Markovskii and Pron' 1968). The heat of formation of amorphous and crystalline $ZnTeO_3$ is equal to 689.6 ± 0.4 and 717.4 ± 0.8 kJ mol^{-1} correspondingly (Pron' and Markovskii 1969). This compound has been obtained by the precipitation from water solutions at the mixing of Na_2TeO_3 and $ZnSO_4$ (Markovskii and Pron' 1968).

$Zn_2Te_3O_8$ crystallizes in a monoclinic structure with the lattice parameters $a = 1271$, $b = 521$, $c = 1182$ pm, and $\beta = 100.0°$ (Hanke 1966).

According to the data of Nawash et al. (2007), single crystals of the $ZnTe_6O_{13}$ ternary compounds have been grown using multiple heating and cooling of the $ZnO + TeO_2$ mixture. This compound crystallizes in a tetragonal structure with the lattice parameters at 86 K $a = 1012.83$ and $c = 1894.8$ pm.

3.20 Zinc–Chlorine–Tellurium

ZnTe–ZnCl₂: The phase diagram is not constructed. The temperature dependence of ZnTe solubility in liquid $ZnCl_2$ was determined using isothermal saturation with subsequent chemical analysis of salt phase (Table 3.6) (Rodionov et al. 1972).

TABLE 3.6

Temperature Dependence of ZnTe
Solubility in Liquid $ZnCl_2$

t, °C	x_{ZnTe}	t, °C	x_{ZnTe}
430	0.9	660	10.7
535	3.4	690	14.6
580	4.8	820	25.8

Source: Rodionov, Yu.I. et al., *Zhurn.*
neorgan. khimii, 17(3), 846, 1972.

3.21 Zinc–Manganese–Tellurium

ZnTe–MnTe: The phase equilibria in this system have been discussed in Pajaczkowska (1978) base on papers that have appeared up to 1977. The phase diagram is not constructed. Solid solutions $Zn_{1-x}Mn_xTe$ are formed in this system for values of x below about 0.75 (Furdyna et al. 1983) [below 0.8 (Juza et al. 1956)]. No solubility of ZnTe in MnTe was detected (Juza et al. 1956). This material crystallizes in the sphalerite structure, and the lattice parameter satisfies Vegard's law and is given by $a(x) = 610.3 + 23.7x$ pm (Furdyna et al. 1983).

The samples of $Zn_{1-x}Mn_xTe$ were prepared either by the Bridgman method or by sintering using in both instances individual elements as starting materials (Furdyna et al. 1983). The ingots were annealed at 800°C–900°C for 24 h or at 1000°C for 12 h and then after grinding at 600°C (Juza et al. 1956). The crystal growth of the $Zn_{1-x}Mn_xTe$ solid solutions is given in Pajaczkowska (1978).

References

Aliev K.A. Synthesis and properties of $ZnLa_2Se_4$ and $ZnLa_2Te_4$ [in Russian], in: *4th Resp. konf. Fiz.-khim. analiz i neorgan. materialoved*. Baku, Azerbaijan, pp. 114–117 (1998).

Alikhanian A.S., Guskov V.N., Greenberg J.H., Fiederle M., Benz K.W. Mass spectrometric study of the CdTe–ZnTe system, *J. Alloys Compd.*, **371**(1–2), 82–85 (2004).

Alikhanian A.S., Guskov V.N., Natarovskii A.M. Thermodynamic properties of ZnTe–CdTe solid solution at 899 K [in Russian], *Zhurn. fiz. khimii*, **77**(10), 1817–1822 (2003a).

Alikhanian A.S., Guskov V.N., Natarovskii A.M., Kovalenko V.V. Thermodynamic properties of ZnTe–CdTe solid solutions, *Inorg. Mater. (Engl. Trans.)*, **39**(3), 234–239 (2003), transl. from *Neorgan. Mater.*, **39**(3), 298–304 (2003b).

Alikhanian A.S., Guskov V.N., Natarovsky A.M., Greenberg J.H., Fiederle M., Benz K.W. Mass spectrometric study of the CdTe–ZnTe system, *J. Cryst. Growth*, **240**(1–2), 73–79 (2002).

Andruhiv A.M., Litvak A.M., Mironov K.E. ZnTe–HgTe–Te and ZnTe–CdTe–HgTe–Te phase diagrams [in Russian], *Pis'ma v ZhTF*, **18**(13), 10–16 (1992).

Andruhiv A.M., Litvak A.M., Mironov K.E. Phase equilibria in the Zn–Hg–Te system and elastic stress at epitaxial growth [in Russian], *Neorgan. materialy*, **29**(4), 492–498 (1993).

Asadov M.M., Babanly M.B., Guseinov F.H., Kuliev A.A. Application of electromoving forces for the determination of physico-chemical properties of Zn(Cd)–Tl–Te ternary system [in Russian], *Zhurn. fiz. khimii*, **57**(11), 2865–2867 (1983).

Asadov Yu.G., Dzhafarov K.M., Gasymov G.B., Babaev A.G. Structural transformation in $Cu_{1.55}Zn_{0.20}Te$ [in Russian], *Neorgan. materialy*, **28**(3), 531–535 (1992).

Babanly M.B., Guseinov F.H., Kuliev A.A., Bashirov E.A. Phase equilibria in the Tl–Zn–Te system [in Russian], *Izv. AN SSSR. Neorgan. materialy*, **22**(11), 1822–1826 (1986).

Burlaku G.G., Markus M.M., Tyrziu V.G. The ZnTe–Ca_2Te_3 phase diagram [in Russian], *Izv. AN SSSR. Neorgan. materialy*, **13**(5), 820–823 (1977).

Busch G., Mooser E., Pearson W.B. Neue halbleitende Verbindungen mit diamantähnlicher Struktur, *Helv. Phys. Acta*, **29**(3), 192–193 (1956).

Charifi Z., Hassan El Haj F., Baaziz H., Khosravizadeh Sh., Hashemifar S.J., Akbarzadeh H. Structural and electronic properties of the wide-gap $Zn_{1-x}Mg_xS$, $Zn_{1-x}Mg_xSe$ and $Zn_{1-x}Mg_xTe$ ternary alloys, *J. Phys.: Condens. Matter*, **17**(44). 7077–7088 (2005).

Cruceanu E., Nistor N. Preparation of homogeneous HgTe–ZnTe alloys, *J. Electrochem. Soc.*, **113**(9), 955–956 (1966).

Danilyuk S.A., Kot M.V. Structure and electrical properties of the HgTe–ZnTe system [in Russian], *Izv. AN SSSR. Ser. fiz.*, **28**(6), 1073–1076 (1964).

Edelman F., Stolyarova S., Chack A., Zakharov N., Werner P., Beserman R., Weil R., Nemirovsky Y. Spontaneous ordering in thin polycrystalline $Cd_{1-x}Zn_xTe$ films during annealing, *Phys. Status Solidi (b)*, **229**(1), 141–144 (2002).

Feldman R.D., Austin R.F., Fuoss P.H., Dayem A.H., Westerwirk E.H., Nakamura S., Boone T., Menendez A., Pinchuk A., Valladers J.P., Brennan S. Phase separation in $Cd_{1-x}Zn_xTe$ grown by molecular-beam epitaxy, *J. Vac. Sci. Technol. B*, **B5**(3), 690–693 (1987).

Furdyna J.K., Giriat W., Mitchell D.F., Sproule G.I. The dependence of the lattice parameter and density of $Zn_{1-x}Mn_xTe$ on composition, *J. Solid State Chem.*, **46**(3), 349–352 (1983).

Gamidov R.S., Kulieva T.Z. X-ray investigation of the thallium and zinc chalcoaluminates [in Russian], in: *Issled. v obl. neorgan. i fiz. khim. i ih rol' v khim. promyshlennosti*, Baku, Azerbaijan: AzINTI, pp. 32–34 (1969).

Glazov V.M., Nagiev V.A., Nuriev R.S. Investigation and analysis of interaction in the quasibinary systems based on zinc, cadmium and mercury tellurides with germanium telluride [in Russian], *Izv. vuzov. Ser. Tsvetnaya metallurgia*, (5), 116–120 (1972).

Glazov V.M., Nagiev V.A., Nuriev R.S. Intermolecular interaction and thermodynamics of GeTe–$A^{II}Te$ melts [in Russian], in: *Termodinamicheskiye svoistva metallicheskih splavov*, Baku, Azerbaijan: Elm, pp. 376–379 (1975).

Glazov V.M., Nagiev V.A., Zargarova M.I. Investigation of phase equilibria in the GeTe–$A^{II}Te$ systems [in Russian], *Izv. AN SSSR. Neorgan. materialy*, **6**(3), 569–571 (1970).

Golovey M.I., Olekseyuk I.D., Shpyrko G.N., Voroshylov Yu.V. Chemical interaction in the $(Zn_3As_2)_{1-x}-(2ZnTe)_{1-x}$ system [in Russian], *Izv. AN SSSR. Neorgan. materialy*, **9**(6), 930–934 (1973).

Golovey M.I., Olekseyuk I.D., Voroshilov Yu.V. Chemical bond and solubility in the systems $A^{II}_3B^V_2-A^{II}C^{VI}$ [in Russian], in: *Khim. svyaz' v poluprovodn. i polumetallah.* Minsk, Belarus: Nauka i Tekhnika Publishing, pp. 233–239 (1972).

Goncharuk L.V., Lukashenko G.M., Budennaya L.D., Dubrovin I.V. Thermodynamic properties of quasibinary solid solutions ZnTe–CdTe [in Russian], *Zhurn. fiz. khimii*, **64**(7), 1954–1957 (1990).

Goncharuk L.V., Sidorko V.R. Thermodynamic properties of some solid solutions formed by $A^{II}B^{IV}$ and $A^{II}B^V$ semiconductor compounds, *Powder Metall. Met. Cer.*, **35**(7/8), 392–396 (1996), transl. from *Poroshkovaya Metallurg.*, **35**(7–8), 79–84 (1996).

Goriunova N.A. Substitutional solid solutions in the compounds with zinc blend structure [in Russian], in: *Voprosy teorii i issled. poluprovodn. i protsessov poluprovodn. metallurgii.* Moscow, Russia: Izd. AN SSSR, pp. 29–37 (1955).

Goriunova N.A., Fedorova N.N. About isomorphism of compounds with covalent bond [in Russian], *Dokl. AN SSSR*, **90**(6), 1039–1041 (1953).

Goriunova N.A., Grigor'eva V.S., Konovalenko B.M., Ryvkin S.M. Photoelectrical properties of some compounds with zinc blend structure [in Russian], *Zhurn. tehn. fiziki*, **25**(10), 1675–1682 (1955a).

Goriunova N.A., Kotovich V.A., Frank-Kamenetski V.A. X-ray investigation of isomorphism for some gallium and indium compounds [in Russian], *Dokl. AN SSSR*, **103**(4), 659–662 (1955b).

Goriunova N.A., Olekseyuk I.D., Golovey M.I. Solid solutions in the $(ZnTe)_{2x}-(Zn_3As_2)_{1-x}$ system [in Russian], in: *Poluprovodn. soed. i ih tverdyie rastvory*, Kishinev, Moldova: RIO AN MSSR, pp. 182–187 (1970).

Greenberg J.H., Guskov V.N., Alikhanyan A.S. Solid-vapor equilibrium in quasibinary CdTe–ZnTe, *Cryst. Res. Technol.*, **38**(7–8), 598–603 (2003).

Greenberg J.H., Guskov V.N., Fiederle M., Benz K.-W. Vapor pressure scanning of non-stoichiometry in $Cd_{0.95}Zn_{0.05}Te_{1+\delta}$, *J. Cryst. Growth*, **270**, 69–76 (2004a).

Greenberg J.H., Guskov V.N., Fiederle M., Benz K.-W. Experimental study of non-stoichiometry in $Cd_{1-x}Zn_xTe_{1\pm\delta}$, *J. Electron. Mater.*, **33**(6), 719–723 (2004b).

Gromakov S.D., Zoroatskaya I.V., Latypov Z.M., Chvala M.A., Eidel'man E.A., Badygina L.I., Zaripova L.G. About investigation of phase diagrams of semiconductor systems [in Russian], *Zhurn. neorgan. khimii*, **9**(10), 2485–2487 (1964).

Grytsiv V.I., Tomashik V.N., Oleinik G.S., Tomashik Z.F. Investigation of the PbTe–ZnTe system [in Russian], *Izv. AN SSSR. Neorgan. materialy*, **16**(3), 543–544 (1980).

Gupta T.K.S.P., Nasar A., Shamsuddin M., Ramachandrarao P. Thermodynamic investigations of liquid-Te-saturated CdTe–ZnTe solid solutions, *Thermochim. Acta*, **247**(2), 415–429 (1994).

Guseinov F.H., Babanly M.B., Kuliev A.A. Phase equilibria in the $Tl_2Te-Zn(Cd)Te$ systems [in Russian], *Izv. AN SSSR. Neorgan. materialy*, **17**(1), 31–33 (1981a).

Guseinov F.H., Babanly M.B., Kuliev A.A. Phase equilibria in the $Tl_2Te_3-Zn(Cd)Te$ systems [in Russian], *Izv. vuzov. Ser. Khimia i khim. tehnologia*, **24**(10), 1245–1248 (1981b).

Guseinov F.H., Babanly M.B., Kuliev A.A. Phase equilibria in the TlX–Cd(Zn) X (X – S, Se, Te) systems [in Russian], *Izv. AN SSSR. Neorgan. materialy*, **18**(5), 759–763 (1982).

Guseinov G.D. Searching and physical investigation of new semiconductors-analogues [in Russian], *Avtoref. dis. … doct. fiz.-mat. nauk.* Baku, Azerbaijan, 82pp (1969).

Guskov V.N., Greenberg J.H., Fiederle M., Benz K.-W. Vapour pressure investigation of CdZnTe, *J. Alloys Compd.*, **371**(1–2), 118–121 (2004).

Guskov V.N., Izotov A.D., Natarovskii A.M. Nonstoichiometry of the $Cd_{1-x}Zn_xTe_y$ solid solution, *Dokl. Phys. Chem.*, **391**(4–6), 203–205 (2003), transl. from *Dokl. Ros. Akad. Nauk*, **391**(6), 788–790 (2003a).

Guskov V.N., Nipan G.D., Kol'tsova T.N. p-T-x phase equilibria in the Cd–Zn–Te system, *Inorg. Mater. (Engl. Trans.)*, **39**(4), 336–341 (2003), transl. from *Neorgan. Mater.*, **39**(4), 415–421 (2003b).

Hahn H., Frank G., Klingler W., Störger A.D., Störger G. Über ternäre Chalkogenide des Aluminiums, Galliums und Indiums mit Zink, Cadmium und Quecksilber, *Z. anorg. und allg. Chem.*, **279**(5/6), 241–270 (1955).

Haloui A., Feutelais Y., Legendre B. Experimental study of the ternary system Cd–Te–Zn, *J. Alloys Compd.*, **260**(1–2), 179–192 (1997).

Hanke K. Die Kristallstruktur von $Zn_2Te_3O_8$, *Naturwissenschaften*, **53**(11), 273 (1966).

Hanke K. Zinktellurit: Kristallstruktur und Beziehungen zu einigen Seleniten, *Naturwissenschaften*, **54**(8), 1999 (1967).

Ilegems M., Pearson G.L. Phase studies in III-V, II-VI and IV-VI compound semiconductor alloy systems, *Ann. Rev. Mater. Sci.*, **5**, 345–371 (1975).

Juza R., Rabenau A., Pascher G. Über teste Lösungen in den Systemen ZnS/MnS, ZnSe/MnSe und ZnTe/MnTe, *Z. anorg. und allg. Chem.*, **285**(1/2), 61–69 (1956).

Kaldis E., Krausbauer L., Widmer R. Cd_4SiS_6 and Cd_4SiSe_6, new ternary compounds, *J. Electrochem. Soc.*, **114**(10), 1074–1076 (1967).

Katayama I., Inomoto T., Kozuka Z., Iida T. Activity measurement of Zn in ZnTe–CdTe solid solutions by EMF method, *Mater. Trans., JIM*, **32**(2), 169–173 (1991).

Kolomiets B.T., Mal'kova A.A. Properties and structure of ternary semiconductor systems. IV. Electrical and photoelectrical properties of substitution solid solutions in the ZnTe–CdTe system [in Russian], *Zhurn. tehn. fiziki*, **28**(8), 1662–1669 (1958).

Kot M.V., Tyrziu V.G., Simashkevich A.V., Maronchuk Yu.E., Mshenski V.A. Dependence of energy activation in thin films from molar composition for some $A^{II}B^{VI}$–$A^{II}B^{VI}$ systems [in Russian], *Fizika tverdogo tela*, **4**(6), 1535–1541 (1962).

Kradinova L.V., Vaipolin A.A., Goriunova N.A. Solid solutions in the $A_3^{II}B_2^V$–$A^{II}B^{VI}$ systems [in Russian], in: *Khim. sviaz' v kristallah*, Minsk, Belarus: Nauka i tehnika, pp. 417–422 (1969).

Krucheanu E., Nikulesku D., Nistor N. Solid solutions in the ZnSe–HgSe and ZnTe–HgTe pseudobinary systems [in Russian], *Rev. phys. Acad. RPR*, **8**(4), 379–382 (1963).

Krucheanu E., Nikulesku D., Nistor N. Investigation of ZnTe–HgTe pseudobinary system [in Russian], *Rev. Roum. Phys.*, **9**(5), 499–506 (1964).

Kutsia N.M., Stavrianidis S.A. Solubility of zinc and cadmium tellurides in germanium telluride [in Russian], *Izv. AN SSSR. Neorgan. materialy*, **19**(8), 1302–1303 (1983).

Laugier A. Thermodynamics and phase diagram calculations in II-VI and IV-VI ternary systems using an associated solution model, *Rev. Phys. Appl.*, **8**(3), 259–270 (1973).

Leute V., Plate H. The phase diagram of the semiconductor alloy $Zn_kHg_{(1-k)}Se_lTe_{(1-l)}$, *Ber. Bunsenges. Phys. Chem.*, **93**(7), 757–763 (1989).

Litvak A.M., Charykov N.A. Thermodynamic modeling in condensed A^2B^6-phases. Model of fully associated solutions, *J. Appl. Chem. USSR (Engl. Transl.)*, **64**(8), 1488–1495 (1991), transl. from *Zhurn. Prikl. Khimii*, **64**(8), 1633–1640 (1991).

Lukashenko G.M., Goncharuk L.V., Budennaya L.D., Dubrovin I.V. Thermodynamic properties of solid solution based on zinc and mercury tellurides [in Russian], *Zhurn. fiz. khimii*, **65**(11), 3141–3144 (1991).

Maksimova O.G. Obtaining and investigation of solid solutions in the Zn–Cd–Te system [in Russian], *Avtoref. dis. … kahd. fiz.-mat. nauk*, Kishinev, Moldova, 19pp (1973).

Maksimova O.G., Tsurkan A.E. The $Zn_xCd_{1-x}Te$ phase diagram [in Russian], in: *Issled. slozhnyh poluprovodnikov*, Kishinev, Moldova: RIO AN MSSR, pp. 145–153 (1970).

Marbeuf A., Druilhe R., Triboulet R., Patriarche G. Thermodynamic analysis of Zn–Cd–Te, Zn–Hg–Te and Cd–Hg–Te: Phase separation in $Zn_xCd_{1-x}Te$ and $Zn_xHg_{1-x}Te$, *J. Cryst. Growth*, **117**(1–4), 10–15 (1992).

Markovskii L.Ya., Pron' G.F. Synthesis and some properties of the zinc, cadmium and mercury tellurites [in Russian], *Zhurn. neorgan. khimii*, **13**(10), 2640–2644 (1968).

Marugin V.V., Odin I.N., Novoselova A.V. Physico-chemical investigation of the zinc chalcogenides—bismuth chalcogenides systems [in Russian], *Zhurn. neorgan. khimii*, **28**(8), 2104–2107 (1983).

Marugin V.V., Odin I.N. Novoselova A.V. Phase equilibria in the Zn–Bi–Te system [in Russian], *Zhurn. neorgan. khimii*, **29**(6), 1561–1565 (1984).

Medvedev Yu.V., Berchenko N.N. Pecularities of the interface between solid solutions based on mercury chalcogenides and native oxides [in Russian], *Zhurn. neorgan. khimii*, **38**(12), 1940–1945 (1993).

Mezhuev O.M., Vishniakov A.V., Zubkovskaya V.N. Obtaining and thermodynamic properties of $Zn_xCd_{1-x}Te$ solid solutions [in Russian], *Izv. AN SSSR. Neorgan. materialy*, **26**(4), 732–734 (1990).

Movsum-zade A.A., Allazov M.R., Suleimanova A.U., Seidova N.A. The Zn–Pb–Te system [in Russian], *Zhurn. neorgan. khimii*, **31**(1) 198–201 (1986).

Nawash J.M., Twamlye B., Lynn K.G. $ZnTe_6O_{13}$, a new $ZnO–TeO_2$ phase, *Acta Crystallogr.*, **C63**(8), i66–i68 (2007).

Niculescu D., Dziuba E.Z. Galvanomagnetic properties of $Zn_xHg_{1-x}Te$, *Phys. Status Solidi*, **31**(2), 465–470 (1969).

Nipan G.D. *p-T-x-y* phase diagram of the Cd–Zn–Te system, *J. Alloys Compd.*, **371**(1–2), 160–163 (2004).

O'Kane D.F., Mason D.R. The phase diagram of pseudo-binary system $ZnTe–In_2Te_3$, *Trans. Metallurg. Soc. AIME*, **233**(6), 1189–1191 (1965).

Odin I.N. Physico-chemical analysis of ternary and ternary mutual systems, containing cadmium, zinc, silicon, bismuth chalcogenides and properties of ingots in these systems [in Russian], *Zhurn. neorgan. khimii*, **41**(6), 941–953 (1996).

Odin I.N., Marugin V.V. Phase diagrams and electrophysical properties of ingots in the $Bi_2X_3–ZnX$ (X – S, Se, Te) systems [in Russian], *Zhurn. neorgan. khimii*, **36**(7), 1842–1846 (1991).

Ohtani H., Kojima K., Ishida K., Nishizawa T. Miscibility gap in II-VI semiconductor systems, *J. Alloys Compd.*, **182**(1), 103–114 (1992).

Olekseyuk I.D., Golovey M.I., Goriunova N.A. The $(Zn_3As_2)_{1-x}–(2ZnTe)_x$ system [in Russian], *Izv. AN SSSR. Neorgan. materialy*, **7**(5), 747–752 (1971).

Olekseyuk I.D., Golovey M.I., Stoika I.M., Yatskovich I.I. Phase diagram of the $As_2Te_3–ZnTe$ system [in Russian], *Izv. AN SSSR. Neorgan. materialy*, **11**(11), 2081–2082 (1975).

Olekseyuk I.D., Stoika I.M. Phase interaction in the Zn–As–Te system [in Russian], *Zhurn. neorgan. khimii*, **22**(7), 1916–1924 (1977).

Olekseyuk I.D., Stoyka I.M., Gerasimenko V.S. Phase equilibria and properties of solid solutions single crystals in the Zn_3As_2–ZnSe(ZnTe) systems, *Izv. AN SSSR. Neorgan. materialy*, **14**(8), 1383–1388 (1978).

Pajaczkowska A. Physicochemical properties and crystal growth of $A^{II}B^{VI}$–MnB^{VI} systems, *Progr. Crystal Growth Caract.*, **1**(3), 289–326 (1978).

Parker S.G., Reinberg A.R., Pinnell J.E., Holton W.C. Preparation and properties of $Mg_xZn_{1-x}Te$, *J. Electrochem. Soc.*, **118**(6), 979–983 (1971).

Paszkowicz W., Firszt F., Łęgowski S., Męczyñska H., Zakrewski J., Marczak M. Structural, optical and thermal properties of bulk $Zn_{1-x}Be_xTe$ crystals, *Phys. Status Solidi*, **B229**(1), 57–62 (2002).

Patrick R.S., Chen A.-B., Sher A., Berding M.A. Phase diagrams and microscopic structures of (Hg,Cd)Te, (Hg,Zn)Te and (Cd,Zn)Te alloys, *J. Vac. Sci. Technol.*, **6**(4), 2643–2649 (1988).

Pentin I.V., Grosheva A.A., Kozhemyakina N.V. The miscibility gap in cadmium, mercury and zinc telluride systems: Theoretical description, *Calphad*, **30**(2), 191–195 (2006).

Pobla C., Granger R., Rolland S., Triboulet R. Interdiffusion between HgTe and ZnTe, *J. Cryst. Growth*, **79**(1–3), Pt.1, 515–518 (1986).

Pron' G.F., Markovskii L.Ya. Heats of formation of the zinc, cadmium and mercury tellurites [in Russian], *Zhurn. neorgan. khimii*, **14**(4), 880–882 (1969).

Rabadanov M.Kh., Simonov V.I. Atomic structure of $Cd_{1-x}Zn_xTe$ solid solution single crystals and structural prerequisites of their ferroelectricity [in Russian], *Kristallografiya*, **51**(5), 830–843 (2006).

Radautsan S.I., Maksimova O.G. Phase interaction in the Te–ZnTe–CdTe and Ga–ZnTe–CdTe systems and growth of the $Zn_xCd_{1-x}Te$ crystals from solution in melt [in Russian], in: *Poluprovodn. materialy i ih primenienie.* Kishinev, Moldova: Shtiintsa, pp. 3–12 (1976).

Rodionov Yu.I., Klokman V.R., Miakishev K.G. Solubility of $A^{II}B^{VI}$, $A^{IV}B^{VI}$ and $A^{V}B^{VI}$ semiconductors compounds in the melts of halogenides [in Russian], *Zhurn. neorgan. khimii*, **17**(3), 846–851 (1972).

Rosenberg A.J., Grierson R., Woolley J.C., Nikolič P. Solid solutions of CdTe and InTe in PbTe and SnTe. I. Crystal chemistry, *Trans. Metallurg. Soc. AIME*, **230**(2), 342–350 (1964).

Ruault M.-O., Kaitasov O., Triboulet R., Crestou J., Gasgnier M. Electron microscopy observation of phase separation in bulk $Cd_{0.96}Zn_{0.04}Te$ crystals, *J. Cryst. Growth*, **143**(1–2), 40–45 (1994).

Rubenstein M. Solubilities of some II–VI compounds in bismuth, *J. Electrochem. Soc.*, **113**(6), 623–624 (1966).

Rubenstein M. Solution growth of some II–VI compounds using tin as a solvent, *J. Cryst. Growth*, **3–4**, 309–312 (1968).

Sang W., Qian Y., Shi W., Wang L., Yang J., Liu D. Equilibrium partial pressures and crystal growth of $Cd_{1-x}Zn_xTe$, *J. Cryst. Growth*, **214/215**, 30–34 (2000).

Schenk M., Haehnert I., Duong L.T.H., Niebsch H.-H. Validity of the lattice-parameter Vegard-rule in $Cd_{1-x}Zn_xTe$ solid solutions, *Cryst. Res. Technol.*, **31**(5), 665–672 (1996).

Schwer H., Krämer V. Neue Überstrukturen von AB_2Z_4-Defekt-Tetraheder-Verbindungen, *Z. Kristallogr.*, **182**(1–4), 245–246 (1988).

Sealy B.J., Crocker A.J. A comparison of phase equilibria in some II–IV–VI compounds based on PbTe, *J. Mater. Sci.*, **8**(12), 1731–1736 (1973).

Shamsuddin M., Gupta T.K.S.P., Nasar A., Ramachandrarao P. Thermodynamic behaviour of zinc and cadmium in liquid-Te-saturated CdTe–ZnTe solid solutions, *Thermochim. Acta*, **246**, 213–227 (1994).

Sidorko V.R., Goncharuk L.V. Thermodynamic properties of the solid solutions ZnTe–CdTe, ZnTe–HgTe and CdTe–HgTe, *J. Alloys Compd.*, **228**(1), 13–15 (1995).

Skums V.F., Pan'ko E.P., Vecher A.A. Influence of high pressures on the electrical resistivity of cadmium and zinc chalcogenide solid solutions, *Inorg. Mater. (Engl. Trans.)*, **28**(4), 572–577 (1992), transl. from *Neorgan. Mater.*, **28**(4), 745–751 (1992).

Smith E.J., Tung T., Sen S., Konkel W.H., James J.B., Harper V.B., Zuck B.F., Cole R.A. Epitaxial growth, characterization, and phase diagram of HgZnTe, *J. Vac. Sci. Technol.*, **A5**(5), 3043–3047 (1987).

Steininger J., Strauss A.J., Brebrick R.E. Phase diagram of the Zn–Cd–Te ternary system, *J. Electrochem Soc.*, **117**(10), 1305–1309 (1970).

Sultanova N.R., Nasirov Ya.N., Zargarova M.I., Pirzade M.M. Thermoelectric properties of SnTe–ZnTe solid solutions [in Russian], *Izv. AN SSSR. Neorgan. materialy*, **10**(8), 1418–1420 (1974).

Svechnikova T.E., Belaya A.D., Zemskov V.S. The InTe–ZnTe and Zn_3Sb_2–ZnTe systems [in Russian], *Izv. AN SSSR. Neorgan. materialy*, **7**(6), 939–942 (1971).

Sztuba Z., Mucha I., Gawel W. Phase studies on the quasi-binary thallium(I) telluride—zinc telluride system, *Calphad*, **30**(4), 421–424 (2006).

Takahashi J., Mochizuki K. Melt growth and stoichiometry control of $(Cd_{1-x}Zn_x)_{1+y}$Te single crystals, *Mat. Sci. Semicon. Proc.*, **6**, 453–456 (2003).

Tomashik V., Shcherbak L. Cadmium–tellurium–zinc, in: *Landolt–Börnstein New Series. Group IV: Physical Chemistry. Vol. 11. Ternary Alloy Systems. Phase Diagrams, Crystallographic and Thermodynamic Data. Subvolum C. Non-Ferrous Metal Systems. Pt. 1. Selected Semiconductor Systems*. Berlin, Heidelberg, Germany: Springer-Verlag, pp. 269–287 (2006).

Toulouse B., Granger R., Rolland S., Triboulet R. Band gap in $Hg_{1-x}Zn_x$Te solid solutions, *J. Phys. (France)*, **48**(2), 247–251 (1987).

Trishchuk L.I. Growth of the zinc and cadmium tellurides single crystals by the crystallization from the solution-melt [in Ukrainian], *Optoelektronika i poluprovodn. tekhnika*, (44), 99–106 (2009).

Trishchuk L.I., Oleinik G.S., Mizetskaya I.B. Phase equilibria in the Cu_{2-x}Te–ZnTe and Cu_{2-x}Te–CdTe systems [in Russian], *Izv. AN SSSR. Neorgan. materialy*, **20**(9), 1486–1489 (1984).

Trishchuk L.I., Oleinik G.S., Mizetskaya I.B. Thermal analysis determination of $A^{II}B^{VI}$ solid solubility in $A^I_2B^{VI}$, *Thermochim. Acta*, **92**, 611–613 (1985).

Trishchuk L.I., Oleinik G.S., Mizetskaya I.B. Physico-chemical investigation of interaction in the Ag_2Te–ZnTe and Ag_2Te–CdTe systems [in Russian], *Ukr. khim. zhurn.*, **52**(8), 799–803 (1986).

Trishchuk L.I., Oliynyk G.S., Tomashyk V.M. Phase diagrams of the $Cu_{1.95}(Ag_2)$Te–ZnTe(CdTe) quasibinary systems and liquidus surfaces of the $Cu_{1.95}(Ag_2)$Te–ZnTe–CdTe quasiternary systems, *Chem. Met. Alloys.* **1**(1), 58–61 (2008).

Voronin G.F., Pentin I.V. Decomposition of the solid solution based on cadmium, mercury and zinc tellurides [in Russian], *Zhurn. fiz. khimii*, **79**(10), 1771–1778 (2005).

Vydyanath H.R., Elsworth J.A., Risher R.F. Vapor phase equilibria in the $Cd_{1-x}Zn_x$Te alloy system, *J. Electron. Mater.*, **22**(8), 1067–1071 (1992).

Wagner P., Lorenz M.R. Solubility of ZnSe and ZnTe in Ga and In, *J. Phys. Chem. Solids*, **27**(11/12), 1749–1752 (1966).

Webb A.W., Qadri S.B., Carpetner E.R., Skelton Jr. E.F. Effects of pressure on $Cd_{1-x}Zn_xTe$ alloys ($0 \leq x < 0.5$), *J. Appl. Phys.*, **61**(7), 2492–2494 (1987).

Wei S.-H., Ferreira L.G., Zunger A. First-principles calculation of temperature-composition phase diagrams of semiconductor alloys, *Phys. Rev. B*, **41**(12), 8240–8269 (1990).

Woolley J.C., Ray B. Effects of solid solutions of Ga_2Te_3 with $A^{II}B^{VI}$ tellurides, *J. Phys. Chem. Solids*, **16**(1/2), 102–106 (1960a).

Woolley J.C., Ray B. Effects of solid solutions of In_2Te_3 with $A^{II}B^{VI}$ tellurides, *J. Phys. Chem. Solids*, **15**(1/2), 27–32 (1960b).

Woolley J.C., Ray B. Solid solution in $A^{II}B^{VI}$ tellurides, *J. Phys. Chem. Solids*, **13**(1/2), 151–153 (1960c).

Yu T.C., Brebrick R.F. The Hg–Cd–Zn–Te phase diagram, *J. Phase Equilibria*, **13**(5), 476–496 (1992).

Zabdyr L.A. Thermodynamics and phase diagram of pseudobinary ZnTe–CdTe system, *J. Electrochem. Soc.*, **131**(9), 2157–2160 (1984).

4

Systems Based on CdS

4.1 Cadmium–Lithium–Sulfur

CdS–Li: The lattice constant of Li-doped CdS single crystals should be smaller than those of undoped crystals (Yoshizawa 1968). But reasons why the growth rate and lattice constant of the <0001> direction (*c*) are smaller than the ones of other directions are not clear.

CdS powder containing small amount of $LiSO_4$ was sublimated at 1100°C in nitrogen atmosphere, and the vapor was carried into a lower-temperature zone at 800°C–1000°C by flow of N_2 gas, where hexagonal-type crystals grow (Yoshizawa 1968). The crystal habits of Li-doped CdS single crystals are affected by the Li concentration.

4.2 Cadmium–Sodium–Sulfur

CdS–Na_2S: The phase diagram belongs to the eutectic type with peritectic transformations (Figure 4.1) (Polyviannyi et al. 1981). The eutectic compositions and temperatures are 45 and 52 mol. % CdS and 805±5°C and 790±5°C, respectively. Three compounds were found out in this system: Na_6CdS_4 ($3Na_2S$ CdS) and Na_2CdS_2 (Na_2S CdS) melt incongruently at 850±5°C and 900±5°C, respectively, and Na_4CdS_3 ($2Na_2S$ CdS) is melted congruently at 840±5°C (Polyviannyi et al. 1981, Polyvyanny and Lata 1985).

Na_6CdS_4 crystallizes in a hexagonal structure with the lattice parameters $a = 912.4$ and $c = 697.3$ pm and has the polymorph transformations at 690±5°C and 725±5°C.

According to the data of Axtell and Kanatzidis (1996), $Na_6Cd_7S_{10}$ also exists in the CdS–Na_2S system. It crystallizes in a monoclinic structure with the lattice parameters $a = 2656.4 \pm 0.3$, $b = 423.85 \pm 0.04$, $c = 1052.3 \pm 0.2$ pm, and $\beta = 108.48 \pm 0.01°$. This compound is a semiconductor with a room temperature band gap of 2.62 eV.

$Na_6Cd_7S_{10}$ could be prepared by the reaction of Na_2S, Cd, and S at the heating at 800°C for 3 days in evacuated quartz tube with the next cooling

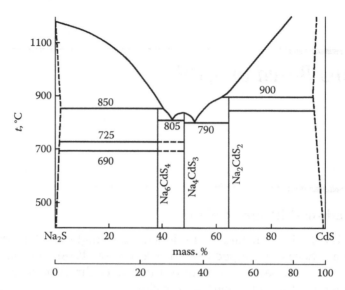

FIGURE 4.1

The CdS–Na$_2$S phase diagram. (From Polyviannyi, I.R. et al., *Zhurn. neorgan. khimii*, 26(4), 1038, 1981.)

at a rate of 10°C/h to 400°C and 20°C/h to room temperature (Axtell and Kanatzidis 1996). This compound is stable in air and in water and is insoluble in common organic solvents.

This system was investigated using differential thermal analysis (DTA), x-ray diffraction (XRD), and metallography (Polyviannyi et al. 1981).

CdS–Na$_2$S$_3$: According to the data of Mellikov et al. (1980), the phase diagram is a eutectic type.

CdS–Na$_2$S$_4$: According to the data of Mellikov et al. (1980), the phase diagram is a eutectic type. The eutectic is degenerated from the Na$_2$S$_4$ side, contains 0.05 mol. % (0.04 mass. %) CdS, and crystallizes at 238°C.

This system was investigated using DTA.

CdS–Na$_2$S$_5$: According to the data of Mellikov et al. (1980), the phase diagram is a eutectic type. The eutectic is degenerate from the Na$_2$S$_5$ side, contains 0.2 mol. % (0.15 mass. %) CdS, and crystallizes at 248°C.

This system was investigated using DTA.

4.3 Cadmium–Potassium–Sulfur

CdS–K$_2$S: The phase diagram is not constructed. K$_2$Cd$_2$S$_3$ and K$_2$Cd$_3$S$_4$ ternary compounds are formed in this system (Axtell et al. 1993, 1996). K$_2$Cd$_2$S$_3$ crystallizes in a hexagonal structure with lattice parameters $a = 1451.7$ and

$c = 691.2$ pm and calculation density 3.36 g cm^{-3}. This compound is a semi-conductor with energy gap 2.89 eV.

K$_2$Cd$_3$S$_4$ melts congruently at 816°C and crystallizes in an orthorhombic structure with lattice parameters $a = 1388.0 \pm 0.4$, $b = 1024.7 \pm 0.3$, and $c = 660.8 \pm 0.1$ pm and calculation density 3.84 g cm^{-3} and energy gap 2.75 eV (Axtell et al. 1996).

K$_2$Cd$_2$S$_3$ was obtained by a caking of K$_2$S, Cd, and S in the vacuum at 600°C (Axtell et al. 1993). K$_2$Cd$_3$S$_4$ could be synthesized by the interaction of K$_2$S, Cd, and S in the evacuated quartz tube (Axtell et al. 1996). The mixture must be heated at 600°C for 2 days, followed by 5°C/h cooling to 50°C. This compound was also prepared by a direct combination reaction at higher temperature. The mixture of CdS and K$_2$S was heated in the evacuated quartz tube at 800°C for 2 days, followed by cooling at 1°C/h to 750°C and 25°C/h to 50°C. K$_2$Cd$_3$S$_4$ is air and water stable and is insoluble in common organic solvents (Axtell et al. 1996).

4.4 Cadmium–Rubidium–Sulfur

CdS–Rb$_2$S: The phase diagram is not constructed. Rb$_2$Cd$_3$S$_4$ ternary compound is formed in this system (Axtell et al. 1996). It melts congruently at 912°C and crystallizes in an orthorhombic structure with lattice parameters $a = 1421.7 \pm 0.6$, $b = 1052.5 \pm 0.4$, and $c = 655.8 \pm 0.2$ pm at 151 K and calculation density 4.31 g cm^{-3} and energy gap 2.92 eV.

Rb$_2$Cd$_3$S$_4$ was obtained by the interaction of Rb$_2$S, Cd, and S in the evacuated quartz tube (Axtell et al. 1996). The mixture was heated at 800°C for 3 days, followed by 10°C/h cooling to 400°C, 20°C/h cooling to 50°C, and quenching to room temperature. This compound is air and water stable and is insoluble in common organic solvents.

4.5 Cadmium–Copper–Sulfur

Phase equilibria in the region CdS–Cd–Cu$_2$S–Cu of the Cd–Cu–S ternary system is shown in Figure 4.2 (Vishniakov et al. 1982). According to Charbonnier (1973), the crystals with lattice parameters $a = b = 2570 \pm 2$ and $c = 2015 \pm 2$ are formed in this system. However, the mixture of primary sulfides was detected at the breakage of such crystals.

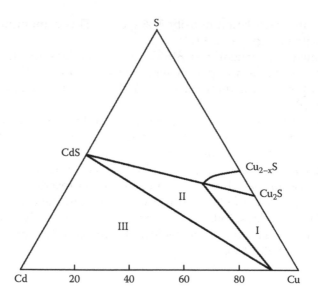

FIGURE 4.2
The phase equilibria in the region CdS–Cd–Cu$_2$S–Cu of the Cd–Cu–S ternary system: I—the equilibria of the phase based on Cu$_2$S with Cu–Cd solutions; II—the equilibria of solid solutions based on Cu$_2$S and CdS with Cu–Cd metallic phase; III—the equilibria of the phase based on CdS with Cu–Cd solutions. (From Vishniakov, A.V. et al., *Zhurn. neorgan. khimii*, 27(7), 1820, 1982.)

At room temperature, the high-temperature bcc$_1$ and bcc$_2$ phases coexist metastably with rhombic analyte (Cu$_{1.75}$S) and monoclinic djurleite (Cu$_{1.96}$S) in the Cu$_{1.70}$Cd$_{0.05}$S crystal (Asadov and Baykulov 2005). At $28 \pm 0.5°C$, the analyte transforms into bcc$_1$ phase and at $79 \pm 0.5°C$ djurleite turns into bcc$_2$ phase and these transformations are reversible.

CdS–Cu: According to the metallography, the solubility of Cu in CdS does not exceed $4.5 \cdot 10^{-2}$ at. % ($2 \cdot 10^{-2}$ mass. %) (Dreeben 1964). Copper is distributed along the dislocations at higher concentrations.

Cu solubility x_{Cu} (cm^{-3}) does not depend on the cadmium vapor pressure, and temperature dependence can be express as follows (Aarna et al. 1976, 1982): $x_{Cu} = 3.0 \cdot 10^{24}$ exp (E/kT), where E—the activation energy ($E = 0.94$ eV).

Cu$_2$S is formed at the CdS single crystals doping by the copper (Aarna et al. 1976).

CdS–CuS: The attempts to synthesize ternary sulfides in the CdS–CuS system had not given any results (Charbonnier 1973). The ingots were prepared from chemical elements or binary compounds by the annealing of mixtures at 800°C for 15 days and at 1000°C for 8 days with next treatment at 1350°C for 6 days. The CdS content in the ingots was 10, 50, and

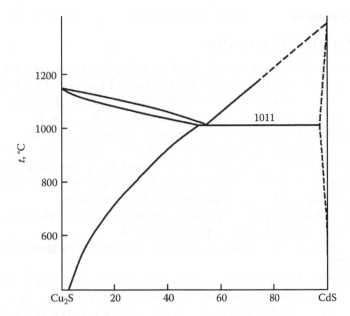

FIGURE 4.3

The CdS–Cu$_2$S phase diagram. (From Mihalev, A.A. et al., *Sb. nauch. tr./VNII liuminoforov i osobo chistyh veshchestv*, (3), 85, 1970; Vishniakov, A.V. et al., *Zhurn. neorgan. khimii*, 25(5), 1358, 1980; Mizetskaya, I.B. et al., *Izv. AN SSSR. Neorgan. materialy*, 18(4), 684, 1982.)

90 mol. % in the different series of experiments. According to the data of XRD, the mixture of copper and cadmium sulfides is always present in the ingots.

CdS–Cu$_2$S: The phase diagram is a eutectic type with limited solubility in the solid state (Figure 4.3) (Mihalev et al. 1970, Vishniakov et al. 1980, Mizetskaya et al. 1982, Trishchuk et al. 1985). The eutectic composition and temperature are 53 mol. % CdS and 1011°C (Mizetskaya et al. 1982, Trishchuk et al. 1985) [50 ± 5 mol. % CdS and 980 ± 10°C (Vishniakov et al. 1981)]. The solid solutions based on γ-Cu$_2$S contain 53 mol. % CdS at the eutectic temperature. The solubility of Cu$_2$S in CdS at 395°C constitutes 2.5 mol. % (Trishchuk et al. 1985) (1.7 and 2.6 mol. % at 900°C and 1000°C, respectively (Bundel' et al. 1970).

This system was investigated using DTA, XRD, and metallography (Vishniakov et al. 1980, Mizetskaya et al. 1982, Vishniakov et al. 1982, Trishchuk et al. 1985). The solubility of Cu$_2$S in CdS was determined using diffusion saturation (Mihalev et al. 1970). The ingots for the crystal structure studying were obtained by the annealing of CdS + Cu$_{1.8}$S mixture, containing 10 at. % Cu at 700°C for 15 days (Charbonnier 1973).

4.6 Cadmium–Silver–Sulfur

The field of CdS primary crystallization occupies the largest part of the Cd–Ag–S liquidus surface (Figure 4.4) (Tulva and Koppel 1976). The limited solubility area of components is within CdS–Ag–Ag$_2$S subsystem. The ternary peritectic contains 60 at. % Ag, 32 at. % S, and 8 at. % Cd.

The attempts to synthesize the ternary sulfides in the CdS–CuS system at the Cd contents not higher than 50 at. % had not given any results (Charbonnier 1973). The ingots were prepared from chemical elements or binary compounds by the annealing of mixtures at 800°C for 15 days and at 1000°C for 8 days with next treatment at 1350°C for 6 days.

The phase relations in the Cd–Ag–S ternary system are shown schematically in Figure 4.5 (Vydyanath and Kröger 1975).

The liquidus surface was constructed using DTA method (Tulva and Koppel 1976).

CdS–Ag: The phase diagram is a eutectic type (Figure 4.6) (Tulva et al. 1973). The eutectic composition and temperature are 5.26 at. % Ag and 920±5°C. The solubility of Ag in CdS is insignificant, and the addition of 2.7·10^{-2} at. % (2·10^{-2} mass. %) Ag leads to the formation of Ag$_2$S phase (Dreeben 1964). The discordance of Dreeben (1964) data with conclusion of Tulva et al. (1973) about quasibinary of this system can be explained by the particularities of CdS single crystal obtained from the vapor phase.

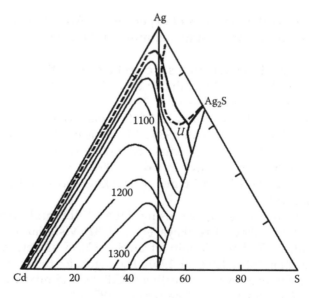

FIGURE 4.4
The Cd–Ag–S liquidus surface. (From Tulva, L. and Koppel, H., *Izv. AN ESSR. Ser. Khimia, Geologia*, 25(1), 32, 1976.

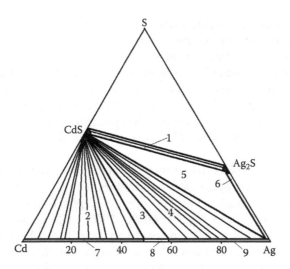

FIGURE 4.5
Schematic ternary diagram for the Cd–Ag–S system at 800°C: 1, $CdS + \gamma\text{-}Ag_2S$; 2, $CdS + L$; 3, $CdS + Ag_{1-x}Cd_x + L$; 4, $CdS + Ag_{1-x}Cd_x$; 5, $CdS + \gamma\text{-}Ag_2S + Ag$; 6, $\gamma\text{-}Ag_2S + Ag$; 7, L; 8, $Ag_{1-x}Cd_x + L$; and 9, $Ag_{1-x}Cd_x$. (From Vydyanath, H.R. and Kröger, F.A., *J. Phys. Chem. Solids*, 36(6), 509, 1975.)

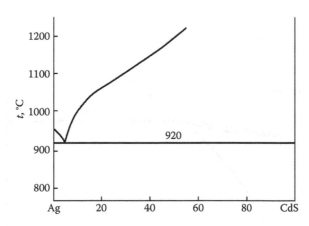

FIGURE 4.6
The CdS–Ag phase diagram. (From Tulva, L. et al., *Izv. AN ESSR. Ser. Khimia, Geologia*, 22(1), 46, 1973.)

The temperature dependence of Ag solubility (cm^{-3}) in CdS can be expressed as follows (Aarna et al. 1976, 1982): $x_{Ag} = 1.6 \cdot 10^{23} \exp (E/kT)$, where E is the activation energy $(E = -0.80 \text{ eV})$.

Gleize and Cabane-Brouty (1975) determine the solubility of Ag in CdS at 700°C using radioactive isotopes. The solubility lays within the limits from $4.6 \cdot 10^{17} \text{ cm}^{-3}$ $(2.41 \cdot 10^{-3} \text{ at. \%})$ to $3.06 \cdot 10^{18} \text{ cm}^{-3}$ $(16.07 \cdot 10^{-3} \text{ at. \%})$ at different deviations from stoichiometry.

This system was investigated using DTA and metallography, measuring of electroconductivity and using radioactive isotopes (Dreeben 1964, Tulva et al. 1973, Chaus et al. 1975, Aarna et al. 1976, 1982). Ag-doped crystals were made by its diffusion into plate-shaped samples of undoped CdS with thickness 0.5–1 mm cut from a boule (Vydyanath and Kröger 1975).

CdS–Ag$_2$S: The phase diagram is a peritectic type (Figure 4.7) (Tulva et al. 1973). The peritectic composition and temperature are 33 mol. % CdS and 893°C. The thermal effects at 175°C correspond to the Ag$_2$S polymorphous transition.

In the obtained crystals, there were detected two disposed side-by-side elementary cells with aggregate parameters a and c ($a_1 = a_2 = 538 \pm 2$ pm, $b_1 = 781 \pm 2$, and $b_2 = 383 \pm 2$ pm and $c_1 = c_2 = 688 \pm 2$ pm) (Dreeben 1964).

The solubility of CdS in Ag$_2$S at 600°C, 700°C, 800°C, and 893°C constitutes 11, 18, 27, and 38 mol. %, respectively, and the solubility of Ag$_2$S in CdS at 893°C is not higher than 0.2 mol. % (Tulva et al. 1973). According to Vishniakov and Iofis (1974), the solubility of Ag$_2$S in CdS at 600°C, 700°C, and 790°C constitutes 0.055, 0.12, and 0.27 mol. %.

This system was investigated using DTA, XRD, and measuring of electroconductivity (Tulva et al. 1973). The solubility of Ag$_2$S in CdS was studied by the saturation of thin films (Vishniakov and Iofis 1974). The crystals

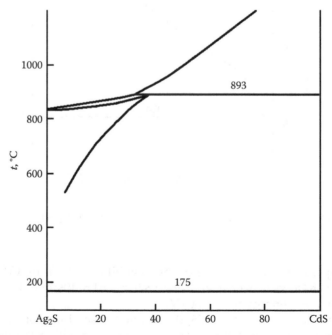

FIGURE 4.7
The CdS–Ag$_2$S phase diagram. (From Tulva, L. et al., *Izv. AN ESSR. Ser. Khimia, Geologia*, 22(1), 46, 1973.)

for structure investigations were obtained by the recrystallization of CdS–Ag$_2$S mixture containing 10 at. % Cd at 700°C in H$_2$S flow for 15 days (Dreeben 1964).

4.7 Cadmium–Gold–Sulfur

CdS–Au: The temperature (x_{Au}) and pressure dependences of the Au solubility in CdS are given in Figure 4.8 (Nebauer 1968). From the Arrhenius relation $x_{Au} = x_0 \cdot \exp(-\Delta H_s/kT)$ of Figure 4.8a, the parameters $x_0 = 5.6 \cdot 10^{20}$ cm^{-3} and $\Delta H_s = 0.47$ eV were obtained.

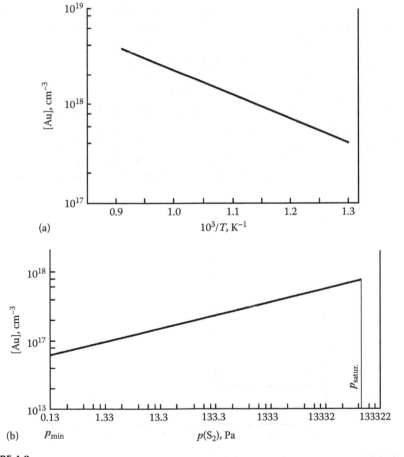

(a)

(b)

FIGURE 4.8
Temperature (a) and pressure (b) dependences of Au solubility in CdS. (From Nebauer, E., *Phys. Status Solidi*, 29(1), 269, 1968.)

According to the data of Dreeben (1964), a second phase precipitates at the addition of $1.4 \cdot 10^{-2}$ at. % ($2 \cdot 10^{-2}$ mass. %) Au in CdS.

The solubility of Au in CdS was determined from diffusion experiments (Nebauer 1968).

4.8 Cadmium–Magnesium–Sulfur

CdS–MgS: The phase diagram is not constructed. At 500°C, a wide region of solid solutions based on CdS with a strained sphalerite structure and solid solution based on MgS with NaCl-type structure exists in this system (Figure 4.9) (Kobayashi et al. 1980). Solid solutions with a rock-salt structure are formed within the interval of 50–100 mol. % MgS at 600°C–800°C and 2–3 GPa (20–30 kbar) (Kobayashi et al. 1979, Susa et al. 1980). Dependence of CdS in MgS solubility on pressure at 800°C is shown in Figure 4.10 (Kobayashi et al. 1980). According to the data of Atroshchenko et al. (1986), solid solutions with wurtzite structure based on CdS reach 33 mol. % MgS, and Dmitrenko et al. (1990) indicate that the solubility of MgS in CdS is equal to 15–17 and 17–19 mol. % at 800°C and 900°C, respectively.

Sysoev and Obuhova (1973) indicate that at small additions of MgS to CdS (12–22 mol. % or 5–10 mass. % MgS), the stressless crystals are formed, but with increasing of MgS content, the inner stresses in crystals increase, and at 39–52 mol. % (20–30 mass. %) MgS, two separate phases exist in the specimens. The c/a ratio decreases at addition of MgS to CdS. The wurtzite phase

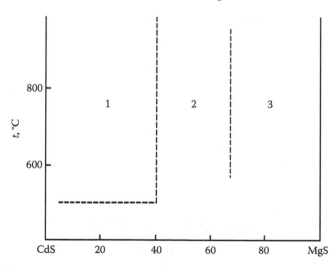

FIGURE 4.9

Phase relations in the CdS–MgS system at 2 GPa: 1, $Cd_{1-x}Mg_xS$; 2, $Cd_{1-x}Mg_xS + Cd_xMg_{1-x}S$; and 3, $Cd_xMg_{1-x}S$. (From Kobayashi, T. et al., *J. Solid State Chem.*, 33(2), 203, 1980.)

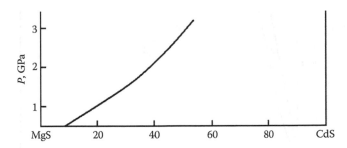

FIGURE 4.10
Dependence of CdS in MgS solubility on pressure at 800°C. (From Kobayashi, T. et al., *J. Solid State Chem.*, 33(2), 203, 1980.)

with minimum value of c/a is obtained at 23.7 mol. % (10.8 mass. %) MgS. The microhardness of $Cd_{1-x}Mg_xS$ single crystals on the basic and lateral faces almost does not change and has the significances, which are typical for CdS.

The solid solutions based on CdS are characterized using a broad XRD pattern, and they have a strained wurtzite-type structure (Kobayashi et al. 1976, 1980, Susa et al. 1980).

Single crystals of solid solutions with wurtzite structure were obtained up to 40 mol. % MgS (Gaisinskii et al. 1981) [32 mol. % MgS (Atroshchenko et al. 1979), 10 mol. % MgS (Atroshchenko et al. 1977, Gaisinskii et al. 1977)].

This system was investigated using metallography, XRD, and measuring of microhardness (Sysoev and Obuhova 1973, Gaisinskii et al. 1977, Atroshchenko et al. 1979, Kobayashi et al. 1979, 1980). The ingots were stand at given pressures for 2 h (Kobayashi et al. 1976, 1980, Susa et al. 1980).

4.9 Cadmium–Calcium–Sulfur

CdS–CaS: The phase diagram is not constructed. There is a wide region of solid solutions based on CaS with NaCl-type structure, two-phase region, and solid solutions based on CdS with a strained sphalerite structure in this system (Figure 4.11) (Kobayashi et al. 1980). According to the data of Kobayashi et al. (1979) and Susa et al. (1980), continuous solid solutions are formed in this system at 600°C–800°C and 2–3 GPa (20–30 kbar). Lattice parameters change approximately linearly from composition. Dependence of CdS solubility in CaS from pressure at 800°C is shown on Figure 4.12 (Kobayashi et al. 1980).

Addition of CaS stabilizes CdS with a rock-salt structure, which exists at high pressures and temperatures (Kobayashi et al. 1976). Stabilizing CdS modification is inconvertible at room temperature and atmospheric pressure. The solubility of CaS in CdS at 127°C and pressure more than 1 GPa (>10 kbar) constitutes 12 mol. %.

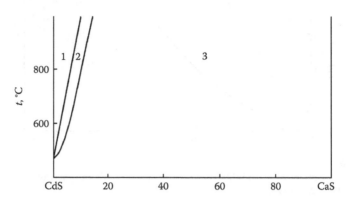

FIGURE 4.11
Phase relations in the CdS–CaS system at 2 GPa: 1, $Cd_{1-x}Ca_xS$; 2, $Cd_{1-x}Ca_xS + Cd_xCa_{1-x}S$; and 3, $Cd_xCa_{1-x}S$. (From Kobayashi, T. et al., *J. Solid State Chem.*, 33(2), 203, 1980.)

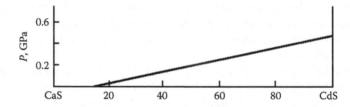

FIGURE 4.12
Dependence of CdS in CaS solubility on pressure at 800°C. (From Kobayashi, T. et al., *J. Solid State Chem.*, 33(2), 203, 1980.)

This system was investigated using XRD. The ingots were annealed at the given pressures and temperatures for 1–4 h (Kobayashi et al. 1976, 1979, 1980, Susa et al. 1980).

4.10 Cadmium–Strontium–Sulfur

CdS–SrS: The phase diagram is not constructed. There is a narrow field of solid solutions with a rock-salt structure, limited by two two-phase regions (Figure 4.13) (Kobayashi et al. 1980). At 2–3 GPa (20–30 kbar) and 600°C–800°C, the solid solutions based on CdS reach 24 mol. % SrS, and ones based on SrS contain 2 mol. % CdS (Kobayashi et al. 1979, Susa et al. 1980). Dependence of CdS solubility in SrS from pressure at 800°C is shown in Figure 4.14 (Kobayashi et al. 1980).

Addition of SrS stabilizes CdS with a rock-salt structure, which exists at high pressures and temperatures (Kobayashi et al. 1976). Stabilizing CdS modification is inconvertible at room temperature and atmospheric pressure.

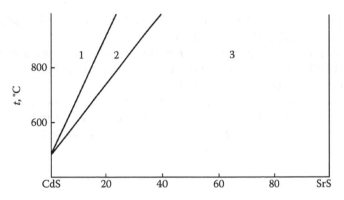

FIGURE 4.13
Phase relations in the CdS–SrS system at 2 GPa: 1, $Cd_{1-x}Sr_xS$ with a strained sphalerite structure + $Cd_{1-x}Sr_xS$ with NaCl-type structure; 2, $Cd_{1-x}Sr_xS$ with NaCl-type structure; and 3, $Cd_{1-x}Sr_xS$ with NaCl-type structure + $Cd_xSr_{1-x}S$. (From Kobayashi, T. et al., *J. Solid State Chem.*, 33(2), 203, 1980.)

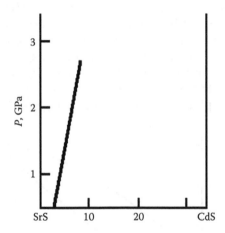

FIGURE 4.14
Dependence of CdS in SrS solubility on pressure at 800°C. (From Kobayashi, T. et al., *J. Solid State Chem.*, 33(2), 203, 1980.)

This system was investigated using XRD. The ingots were annealed at the given pressures and temperatures for 1–4 h (Kobayashi et al. 1976, 1979, 1980, Susa et al. 1980).

4.11 Cadmium–Barium–Sulfur

CdS–BaS: The phase diagram is not constructed. According to the date of Iglesias et al. (1974), $BaCdS_2$ and Ba_2CdS_3 were synthesized in this system. Both compounds crystallize in an orthorhombic structure with lattice parameters $a = 727.81$, $b = 416.70$, and $c = 1391.89$ pm for $BaCdS_2$ and $a = 891.45$, $b = 433.56$, and $c = 1724.39$ pm for Ba_2CdS_3.

$BaCdS_2$ and Ba_2CdS_3 were obtained by heating of mixtures containing CdS and BaS at 900°C for 3 weeks for the first compound and at 1000°C–1200°C for the second one (Iglesias et al. 1974).

Susa et al. (1980) investigated this system under the pressure of 2 GPa and 800°C and showed that an unknown phase is formed at these conditions. These data confirm the results of Iglesias et al. (1974) about the formation of ternary compounds in the CdS–BaS system.

4.12 Cadmium–Mercury–Sulfur

CdS–HgS: The phase diagram is a peritectic type (Figure 4.15) (Kulakov 1976). The ingots containing more than 1.5 mol. % HgS consist of phases based on cubic α-HgS and hexagonal α-CdS, or they are mixtures of these two phases. These ingots are stable during long keeping. Only the ingots containing less than 1.5 mol. % CdS included hexagonal β-HgS (cinnabar) after 6-month keeping at room temperature. There was determined a small endothermic effect at the heating of ingots, containing 0.8 mol. % CdS. This effect coincides with phase transformation of HgS [α-β-transformation for

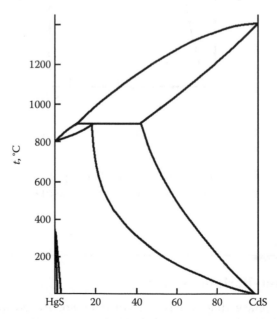

FIGURE 4.15
The CdS–HgS phase diagram. (From Kulakov, M.P., *Zhurn. neorgan. khimii*, 21(2), 513, 1976.)

HgS takes place at 344°C (Kulakov 1975)]. The limit solubility of CdS in cinnabar reaches 0.9 ± 0.1 mol. % (Kulakov 1976).

Immiscibility gap in the solid state, determined using DTA and EPMA, exists within the intervals of 18–42 mol. % CdS at 900°C and 35–67 mol. % at 400°C and approaches zero at next temperature decreasing. The next phases were found in this system at room temperature: hexagonal solid solutions based on CdS (100–45 mol. % CdS), a mixture of hexagonal solid solutions based on CdS, cubic solid solutions based on HgS (45–33 mol. % CdS), cubic solid solutions based on HgS (33–10 mol. % CdS), a mixture of cubic and hexagonal solid solutions based on HgS (10–5 mol. % CdS), and hexagonal solid solutions based on HgS (5–0 mol. % CdS) (Charbonnier 1973).

Solid solutions containing up to 43 mol. % HgS crystallize in wurtzite structure with constant lattice parameters $a = 413$ and $c = 674$ pm. The mixture of the hexagonal and cubic phases exists in this system within the interval of 43–60 mol. % HgS, and a cubic phase crystallizes at the concentrations more than 60 mol. % HgS (Kremheller et al. 1960). Solid solutions, obtained by coprecipitation from water solutions, contain 34.2 mol. % CdS (Rudniev and Dzhumayev 1966).

At 350°C and 10 MPa, $Cd_xHg_{1-x}S$ solid solutions are formed for 3–5 days (Kremheller et al. 1960).

According to the data of Chen (2011), Cd_4HgS_5 ternary compound is formed in this system. It crystallizes in an orthorhombic structure with lattice parameters $a = 1256.61 \pm 0.05$, $b = 725.51 \pm 0.05$, and $c = 1075.20 \pm 0.07$ pm and calculation density 5.492 g cm^{-3}. To obtain this compound, the mixture of CdS, HgS, and S was loaded into a silica tube, which was heated to 200°C in 6 h from room temperature and kept for 24 h, then heated to 450°C in 6 h and kept for 15 days, followed by cooling to 100°C at a rate of 6°C/h to promote crystal growth, then cooled to 35°C in 5 h and power off.

This system was investigated using DTA and XRD (Kremheller et al. 1960, Rudniev and Dzhumayev 1966, Charbonnier 1973, Kulakov 1976). The ingots were annealed at 300°C, 500°C, and 600°C for accordingly 1800, 750, and 200 h (Kulakov 1976).

4.13 Cadmium–Boron–Sulfur

Liquidus surface of the $CdS-B_2S_3-S$ subsystem of the Cd–B–S ternary system is shown in Figure 4.16 (Odin et al. 2001). Practically, all liquidus surface is occupied by the field of CdS primary crystallization, and the fields of BS_2, B_2S_3, and $Cd_2B_2S_5$ primary crystallization occupy narrow bands near B_2S_3-S subsystem. Ternary eutectics E_1 and E_2 crystallize at 375°C and 118°C, respectively. The immiscibility region occupies practically all fields of CdS primary crystallization.

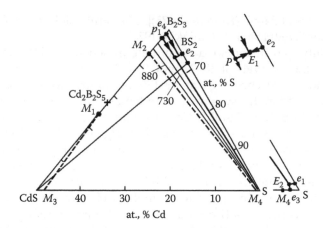

FIGURE 4.16
Liquidus surface of the CdS–B$_2$S$_3$–S subsystem. (From Odin, I.N. et al., *Zhurn. neorgan. khimii*, 46(7), 1210, 2001.)

CdS–B: This system is a quasibinary section of the Cd–B–S ternary system (Odin et al. 2001). According to the data of the atomic absorption analysis, the doping CdS contains 0.79–1.31 at. % B (0.06–0.10 mass. % B) (Odin et al. 1999b).

CdS–BS$_2$: The phase diagram is a eutectic type (Figure 4.17) (Odin et al. 2001). The eutectic composition and temperature are 1 mol. % CdS and 412 ± 3°C, respectively. CdS dissolves less than 0.4 mol. % BS$_2$.

This system was investigated using DTA, metallography, and XRD (Odin et al. 2001).

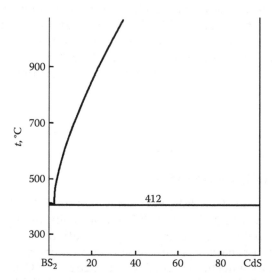

FIGURE 4.17
The CdS–BS$_2$ phase diagram. (From Odin, I.N. et al., *Zhurn. neorgan. khimii*, 46(7), 1210, 2001.)

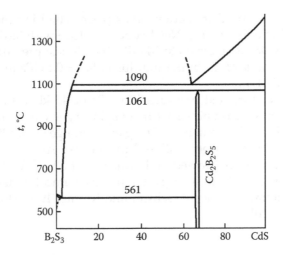

FIGURE 4.18
The CdS–B$_2$S$_3$ phase diagram. (From Odin, I.N. et al., *Zhurn. neorgan. khimii*, 44(12), 2061, 1999b.)

CdS–B$_2$S$_3$: The phase diagram is a eutectic type (Figure 4.18) (Odin et al. 1999c). The eutectic composition and temperature are 3±0.5 mol. % CdS and 561±5°C, respectively. Immiscibility region exists in this system at 1090±8°C. Cd$_2$B$_2$S$_5$ ternary compound is formed at the interaction of CdS and B$_2$S$_3$ (Odin et al. 1999b,c). It melts incongruently at 1061°C, and a composition of 8±1 mol. % CdS corresponds to the peritectic point (Odin et al. 1999c). The homogeneity region of Cd$_2$B$_2$S$_5$ is within the interval from 66.1±0.4 to 67.8±0.3 mol. % CdS, and the energy gap is equal to 2.55 eV.

The solubility of B$_2$S$_3$ in CdS reaches 0.5 mol. %, and the solubility of CdS in B$_2$S$_3$ is not higher than 0.4 mol. % (Odin et al. 1999c, 2001).

This system was investigated using DTA, metallography, and XRD (Odin et al. 1999c).

4.14 Cadmium–Aluminum–Sulfur

CdS–Al$_2$S$_3$: The phase diagram is not constructed. CdAl$_2$S$_4$ is formed in this system. It crystallizes in a tetragonal structure with lattice parameters $a = 555.23$ and $c = 1010.31$ pm (Schwer and Krämer 1990) [$a = 556.0$ and $c = 1032.0$ pm (Meenakshi et al. 2010), $a = 555.3$ and $c = 1030$ pm (Hahn et al. 1955), $a = 556$ and $c = 1032$ pm (Range et al. 1968)], calculation and experimental density 3.062 [3.06 (Range et al. 1968)] and 3.04 g cm^{-3} (Hahn et al. 1955), and energy gap $E_g \approx 4.2$ eV (Moldovyan 1991). The ambient tetragonal structure is retained in CdAl$_2$S$_4$ up to 9 GPa (Meenakshi et al. 2010).

At 10.3 GPa, a structural transition takes place. The high-pressure phase could be fitted to a disordered NaCl-type structure with lattice parameter $a = 473.56$ pm, which is stable up to 34 GPa, the highest pressure of the measurements. The value of the bulk modulus is 44.6 ± 0.1 GPa in the chalcopyrite phase of $CdAl_2S_4$.

At high temperature and pressure (4.5 GPa or 45 kbar and 400°C), this compound can exist in a cubic structure of spinel type with lattice parameter $a = 1024$ pm and calculation density 3.65 g cm^{-3} (Range et al. 1968).

$CdAl_2S_4$ was obtained from CdS and powderlike Al and S in equimolar ratio. The mixtures were pressed and annealed at 800°C for 12–24 h (Hahn et al. 1955). Its single crystals were grown using chemical transport reactions with iodine as a transport agent (Schwer and Krämer 1990, Moldovyan 1991, Meenakshi et al. 2010).

4.15 Cadmium–Gallium–Sulfur

CdS–GaS: The phase diagram is a eutectic type (Figure 4.19) (Odin et al. 2005). The eutectic composition and temperature are 32 ± 2 mol. % CdS and 835 ± 8°C, respectively. The solubility of GaS in CdS is equal to 1 mol. %, and CdS dissolves 0.5 mol. % GaS.

The ingots were annealed at 630°C during 1000 h. This system was investigated using DTA, metallography, and XRD (Odin et al. 2005).

CdS–Ga$_2$S$_3$: The results of this system investigation are contradictory. The most reliable phase diagram is given in Olekseyuk et al. (2001) (Figure 4.20). Three intermediate phases are formed in this system. The most stable of them is the phase $CdGa_2S_4$, which melts congruently at 978°C [985 ± 5°C (Radautsan

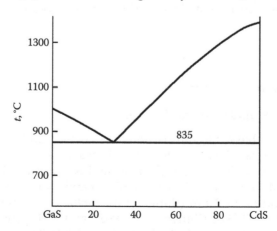

FIGURE 4.19

The CdS–GaS phase diagram. (From Odin, I.N. et al., *Zhurn. neorgan. khimii*, 50(4), 714, 2005.)

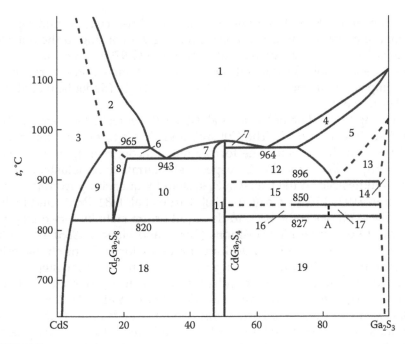

FIGURE 4.20

The CdS–Ga$_2$S$_3$ phase diagram: 1, L; 2, L+(CdS); 3, (CdS); 4, L+(β-Ga$_2$S$_3$); 5, (β-Ga$_2$S$_3$); 6, L+(Cd$_5$Ga$_2$S$_8$); 7, L+(CdGa$_2$S$_4$); 8, (Cd$_5$Ga$_2$S$_8$); 9, (CdS)+(Cd$_5$Ga$_2$S$_8$); 10, (Cd$_5$Ga$_2$S$_8$)+(CdGa$_2$S$_4$); 11, (CdGa$_2$S$_4$); 12, (CdGa$_2$S$_4$)+(β-Ga$_2$S$_3$); 13, (β-Ga$_2$S$_3$)+(α-Ga$_2$S$_3$); 14, (α-Ga$_2$S$_3$); 15, (CdGa$_2$S$_4$)+(α-Ga$_2$S$_3$); 16, (CdGa$_2$S$_4$)+A; 17, A+(α-Ga$_2$S$_3$); 18, (CdS)+(CdGa$_2$S$_4$); and 19, (CdGa$_2$S$_4$)+(α-Ga$_2$S$_3$). (From Olekseyuk, I.D. et al., *J. Alloys Compd.*, 325(1–2), 167, 2001.)

et al. 1982), 984°C (Voroshilov et al. 1979), and 1147°C (Zhitar' et al. 1975)]. The homogeneity range of CdGa$_2$S$_4$ is 3 mol. % (47–50 mol. % Ga$_2$S$_3$). The other two phases are metastable and exist in the defined temperature interval. The congruently melting of Cd$_5$Ga$_2$S$_8$ has not been verified. It forms incongruently at 965°C and decomposes eutectoidally at 820°C. The third intermediate phase A exists in the concentration interval between 80 and 85 vol. % Ga$_2$S$_3$. The temperature interval of its existence is not large (827°C–850°C). Invariant reactions at 943°C and 964°C correspond to the crystallization of the eutectics, which contain 33 and 63 mol. % Ga$_2$S$_3$. The highest solid solubility is observed for β-Ga$_2$S$_3$, which extends to 28 mol. % CdS at 964°C. The solid solution based on α-Ga$_2$S$_3$ decomposes at 896°C according to the eutectoid reaction.

There are also other results concerning the phase equilibria in the CdS–Ga$_2$S$_3$ system. According to the data of Radautsan et al. (1982), the phase diagram belongs to a eutectic type. The eutectic temperatures are 940°C and 970°C. The CdGa$_2$S$_4$ ternary compound is formed in this system. The maximum solubility of Ga$_2$S$_3$ in CdS is equal to 21 mol. % at 940°C and decreases with temperature decreasing. Ga$_2$S$_3$ crystallizes in two hexagonal and one cubic structure. Solid solutions based on hexagonal modifications of Ga_2S_3 contain 18 (δ-phase) and 6 mol. % CdS (γ-phase). δ- and γ-phases decompose at 880°C and 840°C correspondingly according to the eutectoid reactions. Homogeneity region of $CdGa_2S_4$ at 940°C–970°C is within the interval of 47.5–50.5 mol. % CdS, and the lattice constants a and c vary linearly with composition in the range 50.0–52.5 mol. % Ga_2S_3 (Radautsan et al. 1982, Derid et al. 1985).

According to the data of Pardo et al. (1984), the eutectic temperature from the CdS-rich side is 913°C. The $CdGa_2S_4$ and $Cd_3Ga_2S_6$ ternary compounds are formed in this system. The former melts incongruently and the latter congruently at 930°C and 960°C, respectively. Tetrahedral metastable phases can be synthesized in the CdS–Ga_2S_3 system by quenching from the liquid state for $n = 0.03$–0.30 [$n = Cd/(Cd + Ga)$] (Pardo et al. 1984, Pardo and Flahaut 1985). These phases come back to the equilibrium state by a two-step process with two exothermic transitions (Pardo and Flahaut 1985). In the first step, other tetrahedral nonequilibrium phases are formed. The characteristic phases of the equilibrium phase diagram appear in the second step.

The $CdGa_2S_4$ ternary compound crystallizes in a tetragonal structure of the chalcopyrite type (Hahn et al. 1955, Beun et al. 1961, Donika et al. 1968, Guseinov et al. 1980, Pardo et al. 1984, Gastaldi et al. 1985) with lattice parameters $a = 554.6 \pm 0.1$ and $c = 1016.9 \pm 0.2$ pm for the crystals obtained using the Bridgman method and $a = 554.65 \pm 0.02$ and $c = 1016.14 \pm 0.05$ pm for the crystals obtained by the chemical vapor deposition method (Bodnar' et al. 2004) [$a = 555.3$ and $c = 1017.2$ pm (Gastaldi et al. 1985), $a = 555.5$ and $c = 1019.0$ pm (Guseinov et al. 1980), $a = 556.6$ and $c = 1005$ pm (Voroshilov et al. 1979), $a = 556.6$ and $c = 1006$ pm (Hahn et al. 1955, Donika et al. 1968), $a = 557$ and $c = 1001$ pm (Mamedov et al. 1972], calculation and experimental density 4.032 [4.07 (Pardo et al. 1984)] and 3.97 g cm⁻³ (Hahn et al. 1955, Donika et al. 1968) [4.03 and 3.94 g cm⁻³ (Voroshilov et al. 1979)], respectively, and energy gap $E_g = 2.93 \pm 0.2$ eV (Zhitar' et al. 1975) [$E_g = 3.40$–3.44 eV (Beun et al. 1961)].

At a compression, $CdGa_2S_4$ exhibits a defect tetragonal stannite-type structure up to 17 GPa (Errandonea et al. 2008). Beyond this pressure up to 23.0 GPa, a pressure-induced phase transition takes place. High-pressure phase has been characterized as a defect-cubic NaCl-type structure.

High-temperature metastable modification of this compound crystallizes in an orthorhombic structure with lattice parameters $a = 1288$, $b = 780$, and $c = 613$ pm and calculation density 3.90 g cm⁻³ (Pardo et al. 1984). The treatment of $CdGa_2S_4$ in the Cd melt at 750°C for 20 h leads to its decomposition into CdS, containing 0.7 mass. % Ga and gallium as inclusions of second phase (Georgobiani et al. 1985). This compound decomposes in the oxygen atmosphere at 500°C (Wu et al. 1988).

The absolute value of entropy and a variation of enthalpy under standard conditions of $CdGa_2S_4$ have been calculated using experimental measurement heat capacity by Mamedov et al. (1972): $S^0_{298} = 226.1$ J/mol K and $H^0_{298} - H^0_0 = 32.12$ kJ/mol.

A part of the CdS–Ga$_2$S$_3$ phase diagram from the CdS-rich side was constructed according to the determination of Ga equilibrium concentration in the CdS crystals at the different annealing temperature (Jones and Mykura 1980). According to these data, the solubility of Ga$_2$S$_3$ in CdS at low temperature is negligible.

The interaction of CdS and Ga$_2$S$_3$ at coprecipitation from water solutions of chlorides using Na$_2$S leads to the formation of 3CdS Ga$_2$S$_3$ [Cd$_3$Ga$_2$S$_6$: this compound was obtained at the investigation of phase diagram by Pardo et al. (1984)], CdS Ga$_2$S$_3$ (CdGa$_2$S$_4$), and CdS 3Ga$_2$S$_3$ (CdGa$_6$S$_{10}$) (Chaus and Sheka 1967, Chaus et al. 1975). CdS Ga$_2$S$_3$ is the most stable from these compounds. The mutual solubility of CdS and Ga$_2$S$_3$ is negligible (Hahn et al. 1955, Chaus and Sheka 1967, Chaus et al. 1975).

This system was investigated using DTA and XRD (Radautsan et al. 1982, Pardo et al. 1984, Derid et al. 1985). The CdGa$_2$S$_4$ compound was obtained by the annealing of CdS and Ga$_2$S$_3$ at 900°C for 12–24 h (Hahn et al., 1955) or by the melting of CdS + Ga$_2$S$_3$ mixtures at 1200°C under sulfur pressure 1.013 · 10^3 kPa (10 atm) (Springford 1963) or by using chemical transport reactions (Donika et al. 1968). Single crystals of this compound were grown using chemical transport reactions (Beun et al. 1961, Zhitar' et al. 1975, Guseinov et al. 1980, Gastaldi et al. 1985, Georgobiani et al. 1985, Wu et al. 1988) or by using an oriented crystallization (Zhitar' et al. 1975, Voroshilov et al. 1979). The single crystals of different compositions were grown from the melts of the component elements (Derid et al. 1985).

4.16 Cadmium–Indium–Sulfur

Liquidus surface of the Cd–In–S system is shown in Figure 4.21 (Kas'k and Koppel 1976). Practically all Cd–CdS–In subsystem is occupied by the field of CdS primary crystallization. The fields of Cd and In primary crystallizations are degenerated. The field of CdS primary crystallization occupies also a majority of CdS–CdIn$_2$S$_4$–In subsystem. Primary crystallization of CdIn$_2$S$_4$ occupies a small area of this subsystem, and the field of In crystallization is degenerated. The immiscibility region from In–S binary system reaches CdS–CdIn$_2$S$_4$–In subsystem and occupies a part of CdS and CdIn$_2$S$_4$ crystallization field.

In the Cd–In–S ternary system, sharp maximum of solubility in direction to CdIn$_2$S$_4$ exists in the solid state from the CdS-rich side achieving 15 mol. % CdIn$_2$S$_4$. The solubility in all other directions is much lower. CdS is in equilibrium with Cd, In, CdIn$_2$S$_4$, and two compounds, which are forming as a result of the solid-state interaction (Kas'k and Koppel 1976).

CdS–In: The phase diagram is a eutectic type (Figure 4.22) (Kas'k and Koppel 1974). Eutectic is degenerated from the In-rich side and crystallizes

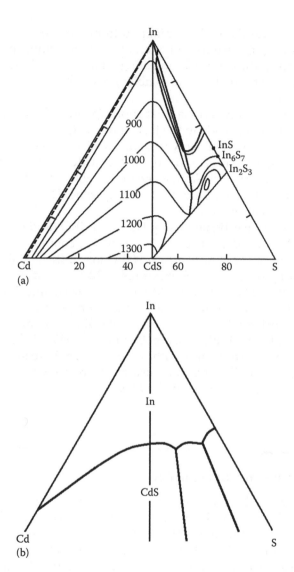

FIGURE 4.21
Liquidus surface (a) and scheme of the liquidus surface near In (b) of the Cd–In–S ternary system. (From Kas'k, R. and Koppel, H., *Izv. AN ESSR. Ser. Khimia, Geologia*, 25(1), 28, 1976.)

at $155 \pm 3°C$. The solubility of In in CdS is not higher than 1 mol. % (Kas'k and Koppel 1974). The solubility of CdS in liquid indium can be described as follows: $\ln x_{CdS} = E/RT + 4.373$, where E—activation energy ($E = -62.3$ kJ/mol) [75Bue].

This system was investigated using DTA (Kas'k and Koppel 1974). The solubility of CdS in liquid In was determined by visual observation of solid-phase appearance and disappearance for heating and cooling of ingots [75Bue].

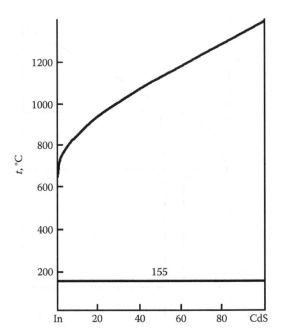

FIGURE 4.22
The CdS–In phase diagram. (From Kas'k, R. and Koppel, H., *Izv. AN ESSR. Ser. Khimia, Geologia*, 23(3), 214, 1974.)

CdS–InS: This section is a nonquasibinary section of the Cd–In–S ternary system (Figure 4.23) (Kas'k and Koppel 1975). Four-phase peritectic interaction L + In_6S_7 ⇔ InS + $CdIn_3S_5$(?) takes place at 640 ± 5°C. At the temperatures below 690°C, immiscibility exists in this system within the interval of 22–31 mol. % CdS.

$CdInS_2$ compound (Guseinov et al. 1969) was not found (Kas'k and Koppel 1975).

CdS–In₂S₃: The most reliable phase diagram is shown in Figure 4.24 (Kozer et al. 2009). Eutectic composition and temperature are 35 mol. % In_2S_3 and 1082°C, respectively. $CdIn_2S_4$ and $Cd_5In_2S_8$ ternary compounds are formed in this system. $CdIn_2S_4$ melts congruently at 1125°C (Goriunova 1968) [1087°C (Suchow and Stemple 1964), 1105°C (Czaja and Krausbauer 1969)], and $Cd_5In_2S_8$ forms peritectically at 1117°C, exists within the temperature interval 932°C–1117°C, and has a polymorphous transition at 992°C (Kozer et al. 2009).

Solid solutions with spinel structure are formed in the In_2S_3–$CdIn_2S_4$ subsystem (Donika et al. 1967, Tezlevan and Zhitar' 1970, Val'kovskaya et al. 1970, Kas'k and Koppel 1973, Donika et al. 1980). Concentration dependence of lattice parameter (Kshirsagar et al. 1982), energy gap, and density is almost linear (Radautsan et al. 1969, Tezlevan and Zhitar' 1970, Val'kovskaya et al. 1971), and microhardness (Val'kovskaya et al. 1970) of forming solid solutions does not change.

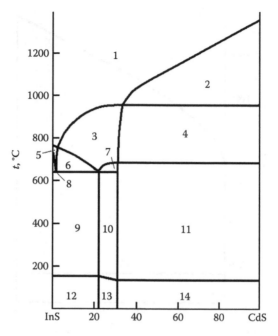

FIGURE 4.23
Phase relations in the CdS–InS system: 1, L; 2, L + CdS; 3, L + CdIn₃S₅; 4, L + CdS + CdIn₃S₅; 5, L + In₆S₇; 6, L + CdIn₃S₅ + In₆S₇; 7, L₁ + L₂ + CdIn₃S₅; 8, L + InS + In₆S₇; 9, L + CdIn₃S₅ + InS; 10, L + CdIn₃S₅; 11, L + CdS + CdIn₃S₅; 12, InS CdIn₃S₅ + In; 13, CdIn₃S₅ + In; and 14, CdIn₃S₅ + CdS + In. (From Kas'k, R. and Koppel, H., *Izv. AN ESSR. Ser. Khimia, Geologia*, 24(3), 210, 1975.)

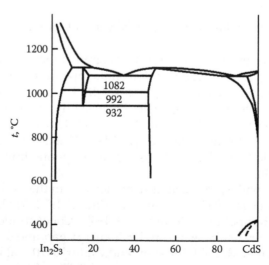

FIGURE 4.24
The CdS–In₂S₃ phase diagram. (From Kozer, V.R. et al., *J. Alloys Compd.*, 480(2), 360, 2009.)

According to the data of Kas'k and Koppel (1973), two immiscibility regions exist in the $CdIn_2S_4$–CdS subsystem within the intervals of 22–45 and 65–85 mol. % CdS at 1100°C and 1120°C, respectively. These data were not confirmed by Kozer et al. (2009).

$CdIn_2S_4$ crystallizes in a cubic structure of spinel type (Hahn and Klingler 1950, Hahn et al. 1955, Beun et al. 1960, Donika et al. 1967, Aresti and Congiu 1973, Sobolev 1976) with lattice parameter $a = 1079.7 \pm 0.7$ pm (Hahn and Klingler 1950) [$a = 1086$ pm (Mamedov et al. 1972), $a = 1082.2$ pm (Donika et al. 1967)], calculation and experimental density 4.934 and 4.93 g cm^{-3}, respectively (Hahn and Klingler 1950), and energy gap $E_g = 2.3$ eV (Sobolev 1976). At 130°C, this compound undergoes second-order transition (Guseinov and Mamedov 1994). The treatment of $CdIn_2S_4$ single crystals in the Cd melt at 750°C for 20 h leads to its decomposition with the formation of CdS, containing 0.7 mass. % In, and indium in the form of second-phase inclusions (Georgobiani et al. 1985). The absolute value of entropy and a variation of enthalpy under standard conditions of $CdIn_2S_4$ have been calculated using experimental measurement heat capacity by Mamedov et al. (1972): $S_{298}^0 = 242.1$ J/mol K and $H_{298}^0 - H_0^0 = 34.05$ kJ/mol.

At 980°C using metallography, there was revealed appearance of a new phase, which is connected evidently with $Cd_2In_2S_5$ formation in the solid state. This compound is stable at temperatures below 1010°C, and its homogeneity range is elongated in the direction to CdS (γ-phase). At 958 ± 5°C, $Cd_2In_2S_5$ has apparently a phase transformation. At 950°C, there are other thermal effects for the ingots, containing more than 50 mol. % CdS, with maximum at 87–88 mol. % CdS. A new phase with supposed formula of $Cd_8In_2S_{11}$ was revealed on the microstructure of ingots, annealing at temperatures below 950°C. At 1010°C, there are thermal effects with maximum at 50 mol. % $CdIn_2S_4$ (Kas'k and Koppel 1973). The existence of the $Cd_2In_2S_5$ and $Cd_8In_2S_{11}$ ternary compounds was not confirmed by the new investigations of Odin et al. (2004) and Kozer et al. (2009).

There were found $Cd_4In_2S_7$ (4CdS In_2S_3), $CdIn_2S_4$ (CdS In_2S_3), and $CdIn_8S_{13}$ (CdS $4In_2S_3$) ternary compounds at coprecipitation of CdS and In_2S_3 from water solution (Chaus et al. 1975). $Cd_4In_2S_7$ and $CdIn_8S_{13}$ have not been determined by the method of physicochemical analysis.

The solubility of CdS in $CdIn_2S_4$ is maximum at 1090°C and equal to 14 mol. % CdS, and the solubility of $CdIn_2S_4$ reaches 15 mol. % at the eutectic temperature and decreases to 9 and 2.5 mol. % at 1010°C and 950°C, respectively. The solubility of CdS in In_2S_3 at 400°C reaches 12 mol. % (Odin et al. 2004) (according to the data of Suchow and Stemple (1964), the solubility of CdS in In_2S_3 reaches 54 mol. %).

This system was investigated using DTA, XRD, and metallography (Kas'k and Koppel 1973, Odin et al. 2004, Kozer et al. 2009). $CdIn_2S_4$ was obtained by the heating of chemical element mixture at 1110°C for 8 h (Kas'k and Koppel 1973), by the melting of CdS and In_2S_3 in stoichiometric ratio at 1200°C–1250°C with next crystallization using Bridgman method (Kas'k and Koppel 1973, Endo 1976, Larionkina and Nani 1976), by using chemical transport reactions

(Beun et al. 1960, Donika et al. 1967, Valov and Payonchkovska 1970, Endo 1976, Larionkina and Nani 1976) by annealing of CdS and In_2S_3 mixtures at 900°C (Hahn and Klingler 1950) [1150°C (Koelmans and Grimmeiss 1959)] for 12 h, or by the melting of CdS and In_2S_3 in stoichiometric ratio at 1050°C and pressure of sulfur vapor 1 MPa (10 atm) (Hahn et al. 1955). Single crystals of $CdIn_2S_4$ and solid solutions were grown using chemical transport reactions (Donika et al. 1967, Georgobiani et al. 1985).

4.17 Cadmium–Thallium–Sulfur

The field of CdS primary crystallization occupies almost all liquidus surface of Cd–Tl–S ternary system, and the small field of Tl_2S primary crystallization is situated nearly Tl–S binary system as a narrow band (Figure 4.25) (Babanly et al. 1983, 1986). Primary crystallization fields of other phases are degenerated. Therefore, four-phase invariant equilibria are also degenerated from the side of invariant equilibria in the Tl–S binary system.

There are two immiscibility gaps in this ternary system. The first of them occupies a large region in the subsystem $CdS–Cd–Tl–Tl_2S$, and the second is situated in the sulfur corner of the Cd–Tl–S ternary system. The second immiscibility gap was constructed by interpolation by reason of high sulfur vapor pressure (Babanly et al. 1983, 1986).

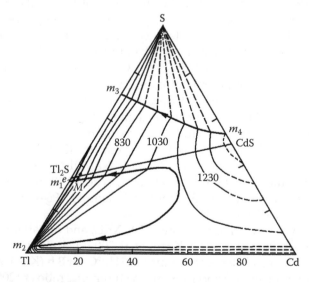

FIGURE 4.25

The liquidus surface of Cd–Tl–S ternary system. (From Babanly, M.B. et al., *Zhurn. neorgan. khimii*, 31(10), 2634, 1986.)

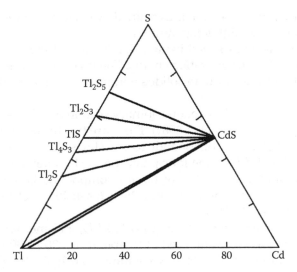

FIGURE 4.26
The isothermal section of Cd–Tl–S ternary system at room temperature. (From Babanly, M.B. et al., *Zhurn. neorgan. khimii*, 31(10), 2634, 1986.)

At room temperature, ternary compounds and solid solutions (excepting α′-solid solutions) do not form in the Cd–Tl–S ternary system (Figure 4.26) (Babanly et al. 1983, 1986). Glass formation is absent in the Cd–Tl–S ternary system (Babanly et al. 1983).

CdS–Tl: This section is a nonquasibinary section of Cd–Tl–S ternary system (Figure 4.27) (Babanly et al. 1983, 1986). There is a large immiscibility region in this section. Below the solidus temperature, this section is stable and the ingots consist of CdS and Tl. According to the data Babanly et al. (1986),

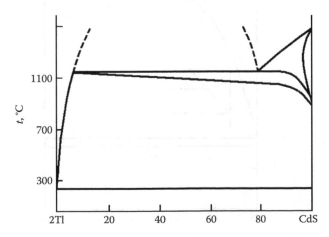

FIGURE 4.27
The phase relation in the CdS–2Tl system. (From Babanly, M.B. et al., *Zhurn. neorgan. khimii*, 31(10), 2634, 1986.)

nonquasibinarity of this section is explained by existence of retrograde solubility based on CdS at high temperatures.

This system was investigated using DTA, XRD, and measuring of microhardness and emf of concentration chains. The ingots were annealed at temperatures 20°C–40°C lower than solidus temperature for 400–600 h (Babanly et al. 1986).

CdS–TlS: This section is a nonquasibinary section of Cd–Tl–S ternary system (Figure 4.28) (Guseinov et al. 1982). There are two horizontal lines on this section. The first at 260°C corresponds to the Tl_4S_3 secondary crystallization and the second at 235°C to the four-phase peritectic interaction $L + Tl_4S_3 \Leftrightarrow TlS + CdS$. The mutual solubility of binary compounds is not higher than 1 mol. %. $CdTlS_2$ compound (Guseinov 1969, 60Gus2) is not formed in this system (Guseinov et al. 1982).

This system was investigated using DTA, XRD, and measuring of microhardness. The ingots were annealed at temperatures 20°C–30°C lower than solidus temperature for 400–500 h (Guseinov et al. 1982). Single crystals of $Cd_xTl_{1-x}S$ solid solutions ($x < 0.01$) were grown using the Bridgman method (Asadov 1984).

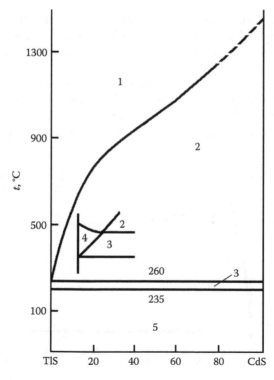

FIGURE 4.28
Phase relations in the CdS–TlS system: 1, L; 2, L+CdS; 3, L+CdS+Tl_4S_3; 4, L+Tl_4S_3; and 5, TlS+CdS. (From Guseinov, F.H. et al., *Izv. AN SSSR. Neorgan. materialy*, 18(5), 759, 1982.)

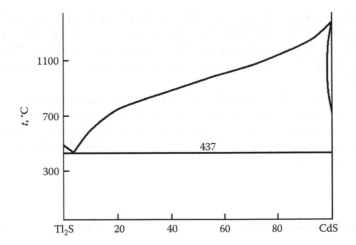

FIGURE 4.29

The CdS–Tl$_2$S phase diagram. (From Babanly, M.B. et al., *Zhurn. neorgan. khimii*, 31(10), 2634, 1986.)

CdS–Tl$_2$S: The phase diagram is a eutectic type (Figure 4.29) (Babanly et al. 1986). The eutectic composition and temperature are 5 mol. % CdS and 437°C, respectively. Solid solutions based on CdS and Tl$_2$S are not higher than 1 mol. % (Babanly et al. 1983, 1986).

This system was investigated using DTA, XRD, and measuring of micro-hardness and emf of concentration chains. The ingots were annealed at temperatures 20°C–40°C lower than solidus temperature for 400–600 h (Babanly et al. 1986). Single crystals of the Cd$_x$Tl$_{2-x}$S solid solutions ($x < 0.01$) were grown using the Bridgman method (Asadov 1984).

CdS–Tl$_4$S$_3$: This section is a nonquasibinary section of Cd–Tl–S ternary system because Tl$_4$S$_3$ melts incongruently (Figure 4.30) (Babanly et al.

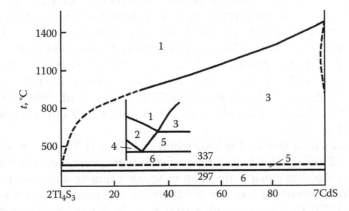

FIGURE 4.30

Phase relations in the 7CdS–2Tl$_4$S$_3$ system: 1, L; 2, L+Tl$_2$S; 3, L+CdS; 4, L+Tl$_2$S+Tl$_4$S$_3$; 5, L+Tl$_2$S+CdS; and 6, Tl$_4$S$_3$+CdS. (From Guseinov, F.H. et al., *Azerb. khim. zhurn.*, (4), 108, 1985; Babanly, M.B. et al., *Zhurn. neorgan. khimii*, 31(10), 2634, 1986.)

1983, 1986, Guseinov et al. 1985). The field of Tl_2S primary crystallization is degenerated from the Tl_4S_3 side. The secondary crystallization of Tl_2S takes place at 337°C, and peritectic transformation $L + Tl_2S \Leftrightarrow Tl_4S_3 + CdS$ occurs at 297°C. All ingots are two phase in the solid state and consist of CdS and Tl_4S_3.

This system was investigated using DTA, XRD, and measuring of microhardness and emf of concentration chains (Guseinov et al. 1985, Babanly et al. 1986). The ingots were annealed at temperatures 20°C–40°C lower than solidus temperature for 400–600 h (Babanly et al. 1986) [at 260° C for 300 h (Guseinov et al. 1985)].

4.18 Cadmium–Scandium–Sulfur

CdS–Sc$_2$S$_3$: The phase diagram is not constructed. $CdSc_2S_4$ is formed in this system, which crystallizes in a cubic structure of spinel type with lattice parameter $a = 1073.3$ pm and energy gap $E_g = 2.3$ eV (Yim et al. 1973).

$CdSc_2S_4$ was obtained by the interaction of CdS and Sc_2S_3 in $CS_2 + He$ flow at 1000°C–1200°C for 3–6 h. Its single crystals were grown using chemical transport reactions (Yim et al. 1973).

4.19 Cadmium–Yttrium–Sulfur

CdS–Y$_2$S$_3$: The phase diagram is not constructed. CdY_2S_4 and CdY_4S_7 ternary compounds are formed in this system. CdY_2S_4 crystallizes in a cubic structure of spinel type with lattice parameter $a = 1121.6 \pm 0.3$ pm [$a = 1117.2$ pm (Yim et al. 1973)] and calculation density 3.94 g cm^{-3} (Tomas et al. 1986). CdY_4S_7 crystallizes in a monoclinic structure with lattice parameters $a = 1270$, $b = 383$, $c = 1149$ pm, and $\beta = 106°$ (Flahaut et al. 1962) [$a = 1282$, $b = 381$, $c = 1147$ pm, and $\beta = 105.72°$ (Aliev et al. 1986, 1997)] and calculation and experimental density 4.28 and 4.15 g cm^{-3}, respectively (Flahaut et al. 1962).

CdY_2S_4 was obtained by the interaction of CdS and Y_2S_3 in $CS_2 + He$ flow at 1000°C–1200°C for 3–6 h. Its single crystals were prepared by heating CdS and Y_2S_3 in stoichiometric ratio in an evacuated quartz ampoule with KBr, at 950°C for 1 month, or were grown by using chemical transport reactions (Yim et al. 1973). CdY_4S_7 was synthesized at the high temperature and fusion or by the sulfurization of Cd and Y at 800°C–950°C using a mixture of carbon disulfide and hydrogen sulfide (Aliev et al. 1986).

4.20 Cadmium–Lanthanum–Sulfur

CdS–La$_2$S$_3$: The phase diagram is shown in Figure 4.31 (Aliev et al. 1997). The eutectic from the CdS side crystallizes at 1337°C. CdLa$_2$S$_4$ and CdLa$_4$S$_7$ ternary compounds are formed in this system. CdLa$_2$S$_4$ melts congruently at 1598°C and crystallizes in a cubic structure of Th$_3$P$_4$ type with lattice parameter $a = 872$ pm (Agayev et al. 1989a, Aliev et al. 1997) [$a = 870.33$ pm (Yim et al. 1973)], calculation and experimental density 5.19 and 5.12 g cm^{-3}, respectively, and energy gap $E_g = 1.30$ eV (Agayev et al. 1989a, Aliev et al. 1997) [$E_g = 2.6$ eV (Yim et al. 1973)]. CdLa$_4$S$_7$ melts incongruently at 1627°C.

Wide regions of solid solutions are formed on the La$_2$S$_3$ base. The limit solubility of La$_2$S$_3$ in CdS reaches 1 mol. % at room temperature (Aliev et al. 1997).

This system was investigated using DTA, metallography, XRD, and measuring of microhardness (Aliev et al. 1997). CdLa$_2$S$_4$ was obtained by the interaction of CdS and La$_2$S$_3$ in CS$_2$ + He flow at 1000°C–1200°C for 3–6 h (Yim et al. 1973). Its single crystals were grown using chemical transport reactions (Yim et al. 1973, Agayev et al. 1989a, Aliev et al. 1997).

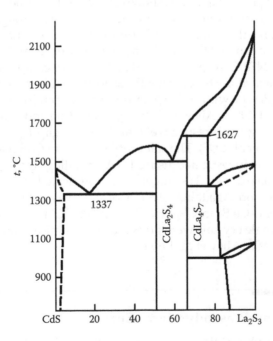

FIGURE 4.31
The CdS–La$_2$S$_3$ phase diagram. (From Aliev, O.M., *Neorgan. materialy*, 33(11), 1327, 1997.)

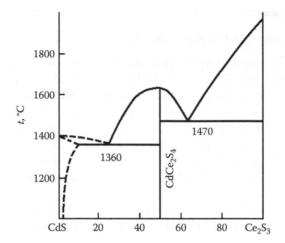

FIGURE 4.32
The $CdS-Ce_2S_3$ phase diagram. (From Aliev, O.M. and Azadaliev, R.A., *Zhurn. neorgan. khimii*, 25(11), 3160, 1980.)

4.21 Cadmium–Cerium–Sulfur

$CdS-Ce_2S_3$: The phase diagram is shown in Figure 4.32 (Aliev and Azadaliev 1980). The melting point of CdS is wrong in this phase diagram; therefore, it was correct and the liquidus and solidus lines from the CdS side were redrawn. Eutectic compositions and temperatures are 24 and 63 mol. % Ce_2S_3 and 1360°C and 1470°C, respectively. $CdCe_2S_4$ ternary compounds are formed in this system, which melts congruently at 1630°C and crystallizes in a cubic structure of Th_3P_4 type with lattice parameter $a = 869$ pm (Aliev and Azadaliev 1980, Agayev et al. 1989a, Aliev et al. 1997), calculation and experimental density 5.27 and 5.24 g cm^{-3}, respectively, and energy gap $E_g = 1.35$ eV (Agayev et al. 1989a, Aliev et al. 1997). The solubility of Ce_2S_3 in CdS reaches 3 mol%.

This system was investigated using DTA, metallography, XRD, and measuring of microhardness. The ingots were annealed at 700°C–750°C for 3 months (Aliev and Azadaliev 1980). $CdCe_2S_4$ was obtained by the interaction of CdS and Ce_2S_3 in $CS_2 + He$ flow at 1000°C–1200°C for 3–6 h (Yim et al. 1973). Its single crystals were grown using chemical transport reactions (Agayev et al. 1989a, Aliev et al. 1997).

4.22 Cadmium–Praseodymium–Sulfur

$CdS-Pr_2S_3$: The phase diagram is not constructed. $CdPr_2S_4$ ternary compound is formed in this system. It melts congruently at 1500°C and crystallizes in a cubic structure of Th_3P_4 type with lattice parameter $a = 866$ pm

(Agayev et al. 1989a, Aliev et al. 1997) [according to the data of Yim et al. (1973), it crystallizes in a trigonal structure], calculation and experimental density 5.36 and 5.34 g cm^{-3}, respectively, and energy gap $E_g = 1.43$ eV (Agayev et al. 1989a, Aliev et al. 1997) [$E_g = 2.1$ eV (Yim et al. 1973)].

CdPr$_2$S$_4$ was obtained by the interaction of CdS and Pr$_2$S$_3$ in CS$_2$ + He flow at 1000°C–1200°C for 3–6 h (Yim et al. 1973). Its single crystals were grown using chemical transport reactions (Agayev et al. 1989a, Aliev et al. 1997).

4.23 Cadmium–Neodymium–Sulfur

CdS–Nd$_2$S$_3$: The phase diagram is not constructed. CdNd$_2$S$_4$ ternary compound is formed in this system. It melts congruently at 1600°C and crystallizes in a cubic structure of Th$_3$P$_4$ type with lattice parameter $a = 864$ pm (Agayev et al. 1989a, Aliev et al. 1997) (according to the data of Yim et al. 1973, CdNd$_2$S$_4$ crystallizes in a trigonal structure with lattice parameters $a = 460$ and $c = 800$ pm), calculation and experimental density 5.47 and 5.40 g cm^{-3}, respectively, and energy gap $E_g = 1.48$ eV (Agayev et al. 1989a, Aliev et al. 1997).

CdNd$_2$S$_4$ was obtained by the interaction of CdS and Nd$_2$S$_3$ in CS$_2$ + He flow at 1000°C–1200°C for 3–6 h (Yim et al. 1973). Its single crystals were grown using chemical transport reactions (Agayev et al. 1989a, Aliev et al. 1997).

4.24 Cadmium–Samarium–Sulfur

CdS–Sm$_2$S$_3$: The phase diagram is not constructed. CdSm$_2$S$_4$ ternary compound is formed in this system. It melts congruently at 1590°C and crystallizes in a cubic structure of Th$_3$P$_4$ type with lattice parameter $a = 862$ pm, calculation and experimental density 5.63 g cm^{-3}, and energy gap $E_g = 1.48$ eV (Agayev et al. 1989a, Aliev et al. 1997).

CdSm$_2$S$_4$ was obtained by the interaction of CdS and Sm$_2$S$_3$ in CS$_2$ + He flow at 1000°C–1200°C for 3–6 h (Yim et al. 1973). Its single crystals were grown using chemical transport reactions (Agayev et al. 1989a, Aliev et al. 1997).

4.25 Cadmium–Gadolinium–Sulfur

CdS–Gd$_2$S$_3$: The phase diagram is shown in Figure 4.33 (Agayev et al. 1984). The eutectic temperature from CdS side is 1350°C. CdGd$_2$S$_4$ and CdGd$_4$S$_7$ ternary compounds are formed in this system. CdGd$_2$S$_4$ melts congruently at

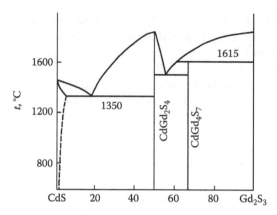

FIGURE 4.33

The CdS–Gd$_2$S$_3$ phase diagram. (From Agayev, A.B. et al., *Azerb. khim. zhurn.*, (4), 105, 1984.)

1870°C (Agayev et al. 1984) and crystallizes in a cubic [monoclinic (Yim et al. 1973)] structure of Th$_3$P$_4$ type with lattice parameter $a = 860$ pm [$a = 836.8$ pm (Ben-Dor and Shilo 1980)], calculation and experimental density 5.76 g cm^{-3}, and energy gap $E_g = 1.15$ eV (Agayev et al. 1989a, Aliev et al. 1997). According to the data of Agayev et al. (1989a) and Aliev et al. (1997), CdGd$_2$S$_4$ melts at 1620°C [Ben-Dor and Shilo (1980) give for this compound a composition of Cd$_{0.5}$Gd$_{2.3}$S$_4$], and CdGd$_4$S$_7$ melts incongruently at 1615°C.

This system was investigated using DTA, metallography, XRD, and measuring of microhardness (Agayev et al. 1984, Aliev et al. 1997). The ingots were annealed at 530°C–730°C. CdGd$_2$S$_4$ was obtained by the interaction of CdS and Gd$_2$S$_3$ in CS$_2$ + He flow at 1000°C–1200°C for 3–6 h (Yim et al. 1973). Its single crystals were grown using chemical transport reactions (Agayev et al. 1989a, Aliev et al. 1997).

4.26 Cadmium–Terbium–Sulfur

CdS–Tb$_2$S$_3$: The phase diagram is not constructed. CdTb$_2$S$_4$ ternary compound is formed in this system. It melts congruently at 1370°C and crystallizes in a cubic structure of Th$_3$P$_4$ type with lattice parameter $a = 858$ pm [$a = 834.3$ pm (Ben-Dor and Shilo 1980)], calculation and experimental density 5.89 and 5.80 g cm^{-3}, respectively, and energy gap $E_g = 1.18$ eV (Agayev et al. 1989a, Aliev et al. 1997). Ben-Dor and Shilo (1980) gives for this compound a composition Cd$_{0.4}$Tb$_{2.4}$S$_4$.

CdTb$_2$S$_4$ was obtained by the interaction of CdS and Tb$_2$S$_3$ in CS$_2$ + He flow at 1000°C–1200°C for 3–6 h (Yim et al. 1973). Its single crystals were grown using chemical transport reactions (Agayev et al. 1989a, Aliev et al. 1997).

4.27 Cadmium–Dysprosium–Sulfur

CdS–Dy$_2$S$_3$: The phase diagram is not constructed. CdDy$_2$S$_4$ and CdDy$_4$S$_7$ ternary compounds are formed in this system. CdDy$_2$S$_4$ melts congruently at 1590°C and crystallizes in a cubic structure of spinel type with lattice parameter $a = 1126$ pm, calculation and experimental density 5.28 and 5.22 g cm^{-3}, respectively, and energy gap $E_g = 1.20$ eV (Agayev et al. 1989a, Aliev et al. 1997) [$E_g = 2.5$ eV (Yim et al. 1973)].

CdDy$_4$S$_7$ crystallizes in a monoclinic structure with lattice parameters $a = 1286$, $b = 384$, $c = 1150$ pm, and $\beta = 105.64°$ (Aliev et al. 1986, Aliev et al. 1997).

CdDy$_2$S$_4$ was obtained by the interaction of CdS and Dy$_2$S$_3$ in CS$_2$ + He flow at 1000°C–1200°C for 3–6 h (Yim et al. 1973). CdDy$_4$S$_7$ was synthesized by the high temperature and fusion or by the sulfurization of Cd and Dy at 800°C–950°C using a mixture of carbon disulfide and hydrogen sulfide (Aliev et al. 1986). Single crystals of CdDy$_2$S$_4$ and CdDy$_4$S$_7$ were grown using chemical transport reactions (Agayev et al. 1989a, Aliev et al. 1997).

4.28 Cadmium–Holmium–Sulfur

CdS–Ho$_2$S$_3$: The phase diagram is not constructed. CdHo$_2$S$_4$ and CdHo$_4$S$_7$ ternary compounds are formed in this system. CdHo$_2$S$_4$ melts congruently at 1570°C and crystallizes in a cubic structure of spinel type with lattice parameter $a = 1124$ pm (Fujii et al. 1972, Agayev et al. 1989a, Aliev et al. 1997) ($a = 1117 \pm 1$ pm (Bakker et al. 1982), $a = 1116.74$ pm (Yim et al. 1973), $a = 1116.7$ pm (Ben-Dor and Shilo 1980), calculation and experimental density 5.35 and 5.28 g cm^{-3}, respectively, and energy gap $E_g = 1.26$ eV (Aliev et al. 1997). At a temperature above 900°C, the spinel-type structure of CdHo$_2$S$_4$ becomes unstable and gradually transforms into a rock-salt-type structure (Bakker et al. 1982). Small domains of the high-temperature structure appear to occur also in samples that were annealed well below 900°C for a long period.

CdHo$_4$S$_7$ crystallizes in a monoclinic structure with lattice parameters $a = 1275$, $b = 381$, $c = 1145$ pm, and $\beta = 105.60°$ (Aliev et al. 1986, Agayev et al. 1989a).

CdHo$_2$S$_4$ was obtained by the interaction of stoichiometric quantities of chemical elements at 800°C–900°C (Fujii et al. 1972, Ben-Dor and Shilo 1980) or by the interaction of CdS and Ho$_2$S$_3$ in CS$_2$ + He flow at 1000°C–1200°C for 3–6 h (Yim et al. 1973). Equimolar mixtures of CdS and Ho$_2$S$_3$ were heated between 900°C and 1100°C for 2 h and air quenched (Bakker et al. 1982). Also, a mixture was heated to 900°C and annealed at 700°C for 2 weeks. Finally, one sample was kept for 24 h at 900°C and air quenched.

CdHo$_4$S$_7$ was synthesized by the high temperature and fusion or by the sulfurization of Cd and Ho at 800°C–950°C using a mixture of carbon disulfide and hydrogen sulfide (Aliev et al. 1986). Single crystals of CdHo$_2$S$_4$ and CdHo$_4$S$_7$ were grown using chemical transport reactions (Agayev et al. 1989a, Aliev et al. 1997).

4.29 Cadmium–Erbium–Sulfur

CdS–Er$_2$S$_3$: The phase diagram is shown in Figure 4.34 (Agayev et al. 1989b). The eutectic compositions and temperatures are 24 and 57.5 mol. % Er$_2$S$_3$ and 1200°C and 1250°C, respectively. CdEr$_2$S$_4$ and CdEr$_4$S$_7$ ternary compounds are formed in this system. CdEr$_2$S$_4$ melts congruently at 1470°C and crystallizes in a cubic structure of spinel type with lattice parameter $a = 1119$ pm (Agayev et al. 1989a, Aliev et al. 1997) [$a = 1160$ pm (Agayev et al. 1989b), $a = 1119.2$ (Fujii et al. 1972), $a = 1112.8$ (Ben-Dor and Shilo 1980), $a = 1113.47$ pm (Yim et al. 1973), $a = 1110$ pm (Tomas et al. 1978)], calculation and experimental density 5.47 and 5.41 g cm^{-3}, respectively, and energy gap $E_g = 1.30$ eV (Agayev et al. 1989a, Aliev et al. 1997) $E_g = 1.8$ eV (Yim et al. 1973).

CdEr$_4$S$_7$ melts incongruently at 1350°C and crystallizes in a monoclinic structure of Y$_5$S$_7$ type with lattice parameters $a = 1268$, $b = 379$, $c = 1140$ pm, and $\beta = 105.50°$ (Aliev et al. 1986, Agayev et al. 1989b, Aliev et al. 1997).

This system was investigated using DTA, metallography, XRD, and measuring of microhardness and density. The ingots were annealed at 1110°C for 240 h (Agayev et al. 1989b).

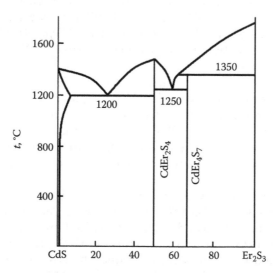

FIGURE 4.34

The CdS–Er$_2$S$_3$ phase diagram. (From Agayev, A.B. et al., *Zhurn. neorgan. khimii*, 34(4), 1065 1989b.)

CdEr$_2$S$_4$ was obtained by the interaction of stoichiometric quantities of chemical elements at 800°C–900°C (Fujii et al. 1972, Ben-Dor and Shilo 1980), by the interaction of binary compounds at 1250°C (Walker and Ward 1984), or by the interaction of CdS and Er$_2$S$_3$ in CS$_2$+He flow at 1000°C–1200°C for 3–6 h (Yim et al. 1973). CdEr$_4$S$_7$ was synthesized by the high temperature and fusion or by the sulfurization of Cd and Er at 800°C–950°C using a mixture of carbon disulfide and hydrogen sulfide (Aliev et al. 1986). Single crystals of CdEr$_2$S$_4$ and CdEr$_4$S$_7$ were grown using chemical transport reactions (Agayev et al. 1989a,b, Aliev et al. 1997).

4.30 Cadmium–Thulium–Sulfur

CdS–Tm$_2$S$_3$: The phase diagram is not constructed. CdTm$_2$S$_4$ and CdTm$_4$S$_7$ ternary compounds are formed in this system. CdTm$_2$S$_4$ crystallizes in a cubic structure of spinel type with lattice parameter $a = 1110$ pm (Pokrzywnicki and Czopnik 1975, Agayev et al. 1989a, Aliev et al. 1997) [$a = 1109.7$ pm (Ben-Dor and Shilo 1980), $a = 1109$ (Yim et al. 1973), $a = 1108.5$ (Tomas et al. 1985)], calculation and experimental density 5.64 and 5.63 g cm^{-3}, respectively (Agayev et al. 1989a, Aliev et al. 1997), and energy gap $E_g = 2.4$ eV (Yim et al. 1973).

CdTm$_4$S$_7$ crystallizes in a monoclinic structure with lattice parameters $a = 1264$, $b = 376$, $c = 1139$ pm, and $\beta = 105.54°$ (Aliev et al. 1986, Aliev et al. 1997).

CdTm$_2$S$_4$ was obtained by the interaction of stoichiometric quantities of binary compounds at 800°C–850°C for 10 days (Pokrzywnicki and Czopnik 1975, Ben-Dor and Shilo 1980, Tomas et al. 1985) or by the interaction of CdS and Tm$_2$S$_3$ in CS$_2$+He flow at 1000°C–1200°C for 3–6 h (Yim et al. 1973). CdTm$_4$S$_7$ was synthesized by the high temperature and fusion or by the sulfurization of Cd and Tm at 800°C–950°C using a mixture of carbon disulfide and hydrogen sulfide (Aliev et al. 1986). Single crystals of CdTm$_2$S$_4$ and CdTm$_4$S$_7$ were grown using chemical transport reactions (Agayev et al. 1989a, Aliev et al. 1997).

4.31 Cadmium–Ytterbium–Sulfur

CdS–YbS: The phase diagram is not constructed. Solid solutions Cd$_{1-x}$Yb$_x$S were obtained in this system within the interval of $0.05 < x < 0.5$ (Susa et al. 1975).

CdS–Yb$_2$S$_3$: The phase diagram is not constructed. CdYb$_2$S$_4$ and CdYb$_4$S$_7$ ternary compounds are formed in this system. CdYb$_2$S$_4$ melts congruently at 1600°C and crystallizes in a cubic structure of spinel type with lattice parameter $a = 1106.84$ pm (Yim et al. 1973) [$a = 1104$ pm (Agayev et al. 1989a, Aliev et al. 1997), $a = 1106.6$ pm (Pokrzywnicki 1975), $a = 1104.3$ pm (Ben-Dor and Shilo 1980), $a = 1105.5$ pm (Tomas et al. 1985)], calculation and experimental density 5.81 and 5.79 g cm^{-3}, respectively, and energy gap $E_g = 1.38$ eV (Agayev et al. 1989a, Aliev et al. 1997) [$E_g = 2.5$ eV (Yim et al. 1973)].

CdYb$_4$S$_7$ crystallizes in a monoclinic structure with lattice parameters $a = 1260$, $b = 374$, $c = 1135$ pm, and $\beta = 105.48°$ (Aliev et al. 1986, 1997).

CdYb$_2$S$_4$ was obtained by the interaction of CdS and Yb$_2$S$_3$ in stoichiometric ratio at 820°C–880°C for 10 days (Pokrzywnicki 1975) or by the interaction of CdS and Yb$_2$S$_3$ in CS$_2$ + He flow at 1000°C–1200°C for 3–6 h (Yim et al. 1973). CdYb$_4$S$_7$ was synthesized by the high temperature and fusion or by the sulfurization of Cd and Yb at 800°C–950°C using a mixture of carbon disulfide and hydrogen sulfide (Aliev et al. 1986). Single crystals of CdYb$_2$S$_4$ and CdYb$_4$S$_7$ were grown using chemical transport reactions (Agayev et al. 1989a, Aliev et al. 1997).

4.32 Cadmium–Lutecium–Sulfur

CdS–Lu$_2$S$_3$: The phase diagram is not constructed. CdLu$_2$S$_4$ and CdLu$_4$S$_7$ ternary compounds are formed in this system. CdLu$_2$S$_4$ melts congruently at 1640°C (Aliev et al. 1997) and crystallizes in a cubic structure of spinel type with lattice parameter $a = 1094.5$ pm (Fujii et al. 1972) [$a = 1104.5$ pm (Yim et al. 1973), $a = 1100$ pm (Aliev et al. 1997)] and calculation and experimental density 5.91 and 5.85 g cm^{-3}, respectively (Aliev et al. 1997).

CdLu$_4$S$_7$ crystallizes in a monoclinic structure with lattice parameters $a = 1254$, $b = 370$, $c = 1130$ pm, and $\beta = 105.40°$ (Aliev et al. 1997).

CdLu$_2$S$_4$ was obtained by the interaction of stoichiometric quantities of chemical elements at 800°C–900°C for 1 week (Fujii et al. 1972) or by the interaction of CdS and Lu$_2$S$_3$ in CS$_2$ + He flow at 1000°C–1200°C for 3–6 h (Yim et al. 1973). Single crystals of CdLu$_2$S$_4$ and CdLu$_4$S$_7$ were grown using chemical transport reactions (Aliev et al. 1997).

4.33 Cadmium–Silicon–Sulfur

The field of Si primary crystallization occupies almost all liquidus surface of the Cd–Si–S ternary system (Figure 4.35) (Odin 1996). There are also the fields of SiS, α- and β-Cd$_4$SiS$_6$, CdS, and SiS$_2$ primary crystallization and degenerated fields of Cd and β-S primary crystallization. Information about

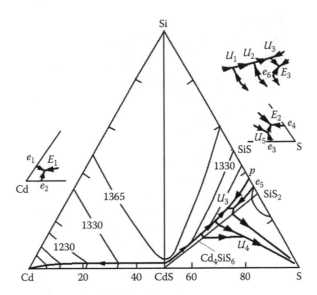

FIGURE 4.35
Liquidus surface of the Cd–Si–S ternary system. (From Odin, I.N., *Zhurn. neorgan. khimii*, 41(6), 941, 1996.)

three- and four-phase equilibria in the Cd–Si–S ternary system is given in Table 4.1, and a part of isothermal section at 730°C of this system is shown in Figure 4.36.

CdS–Si: The phase diagram is not constructed. According to the data of Odin (1996), it is a eutectic type with eutectic temperature of 1337 ± 10°C. The mutual solubility of CdS and Si is insignificant.

This system was investigated using DTA, metallography, and XRD. The ingots were annealed at temperatures 20°C lower than the temperature of invariant equilibrium, including liquid, for 1000 h (Odin 1996).

CdS–SiS$_2$: The phase diagram is shown in Figure 4.37 (Odin et al. 1985). The eutectic composition and temperature are 40 mol. % CdS and 986 ± 6°C, respectively, and the peritectic composition is 78 mol. % CdS. Cd_4SiS_6 is formed in this system. It melts incongruently at 1235 ± 8°C (Odin et al. 1985) and crystallizes in a monoclinic structure with lattice parameters $a = 1231.0 \pm 0.6$, $b = 704.1 \pm 0.4$, $c = 1233.6 \pm 0.6$ pm, and $\beta = 110.38 \pm 0.05°$ (Krebs von and Mandt 1972) [$a = 1234$, $b = 708.9$, $c = 1235$ pm, and $\beta = 110°44'$ (Serment et al. 1968); $a = 1199$, $b = 702$, $c = 1212$ pm, and $\beta = 110.20°$ (Susa and Steinfink 1971); $\beta = 110.5°$ (Kaldis et al. 1967)] and calculation and experimental density 4.446 and 4.41 ± 0.01 g cm^{-3} (Krebs von and Mandt 1972). At 1085 ± 7°C, polymorphous transition of Cd_4SiS_6 takes place (Odin et al. 1985).

Solid-solution regions based on CdS and SiS$_2$ do not exceed 0.7 mol. % SiS$_2$ and 1 mol. % CdS, respectively. Homogeneity region of Cd_4SiS_6 is within the interval 79.4–80.0 mol. % CdS (Odin et al. 1985).

TABLE 4.1

Three- and Four-Phase Equilibria
in the Cd–Si–S Ternary System

Symbol	Reaction	t, °C
e_1	L $\Leftrightarrow$ Cd + Si	320
e_2	L $\Leftrightarrow$ Cd + CdS	319
e_3	L $\Leftrightarrow$ CdS + β-S	117
e_4	L $\Leftrightarrow$ SiS$_2$ + β-S	115
e_5	L $\Leftrightarrow$ SiS$_2$ + SiS	1030
p_1	L + Si $\Leftrightarrow$ SiS	1202
E_1	L $\Leftrightarrow$ Cd + Si + CdS	317
E_2	L $\Leftrightarrow$ α-Cd$_4$SiS$_6$ + SiS$_2$ + β-S	113
E_3	L $\Leftrightarrow$ α-Cd$_4$SiS$_6$ + SiS + SiS$_2$	1004
U_1	L + CdS $\Leftrightarrow$ β-Cd$_4$SiS$_6$ + Si	1197
U_2	L + β-Cd$_4$SiS$_6$ $\Leftrightarrow$ Si + α-Cd$_4$SiS$_6$	1056
U_3	L + Si $\Leftrightarrow$ α-Cd$_4$SiS$_6$ + SiS	1043
U_4	L + β-Cd$_4$SiS$_6$ $\Leftrightarrow$ α-Cd$_4$SiS$_6$ + CdS	1036
U_5	L + CdS $\Leftrightarrow$ α-Cd$_4$SiS$_6$ + β-S	116

Source: Odin, I.N., *Zhurn. neorgan. khimii*, 41(6), 941,
              1996.

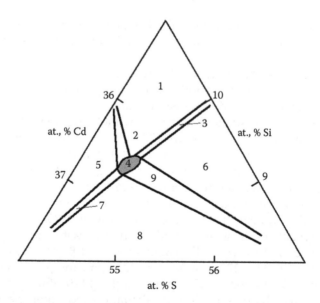

FIGURE 4.36

Part of the Cd–Si–S isothermal section at 730°C: 1, α-Cd$_4$SiS$_6$ + Si + SiS$_2$; 2, α-Cd$_4$SiS$_6$ + Si;
3, α-Cd$_4$SiS$_6$ + SiS$_2$; 5, α-Cd$_4$SiS$_6$ + Cd + Si; 6, α-Cd$_4$SiS$_6$ + SiS$_2$ + L; 7, α-Cd$_4$SiS$_6$ + CdS;
8, α-Cd$_4$SiS$_6$ + CdS + L; and 9, α-Cd$_4$SiS$_6$ + L. (From Odin, I.N., *Zhurn. neorgan. khimii*, 41(6), 941,
1996.)

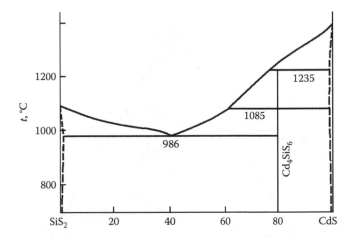

FIGURE 4.37

The CdS–SiS$_2$ phase diagram. (From Odin, I.N. et al., *Zhurn. neorgan. khimii*, 30(1), 207, 1985.)

This system was investigated using DTA and XRD. The ingots were annealed at 700°C for 720 h (Odin et al. 1985). Cd$_4$SiS$_6$ was obtained by annealing of mixtures of Cd, Si, and S in stoichiometric ratio at 200°C–400°C for 2–4 h and then at 600°C–800°C for 12 h (Susa and Steinfink 1971b).

4.34 Cadmium–Germanium–Sulfur

Eight fields of primary crystallizations are on the liquidus surface of the Cd–Ge–S ternary system (Figure 4.38) (Movsum-zade et al. 1987). The field of Cd crystallization is very small, and the field of sulfur crystallization is degenerated. Ternary eutectic E_4 is degenerated from the sulfur-rich side. Considerable part of the liquidus surface is occupied by immiscibility region. Information about four-phase equilibria in the Cd–Ge–S ternary system is given in Table 4.2.

According to the data of Odin et al. (1983), the Cd$_4$GeS$_5$ compound (Movsum-zade et al. 1987) was not found in the Cd–Ge–S ternary system.

Limits of α-Cd$_4$GeS$_6$ homogeneity region at 544°C are shown in Figure 4.39 (Odin and Grin'ko 1991a). This region is strongly elongated along the CdS–GeS$_2$ binary system.

CdS–Ge: The phase diagram is a eutectic type (Figure 4.40) (Movsum-zade et al. 1987). The eutectic composition and temperature are 53.84 at. % Ge and 900°C, respectively. At 925°C, there is an immiscibility region within the interval of 66.66–92.30 at. % Ge.

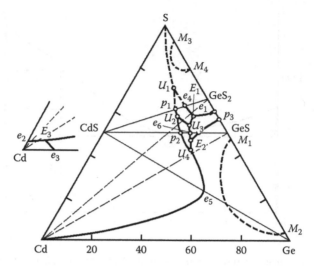

FIGURE 4.38

Projection of liquidus surface of Cd–Ge–S ternary system. (From Movsum-zade, A.A. et al., *Zhurn. neorgan. khimii*, 32(4), 1025, 1987.)

TABLE 4.2

Nonvariant Equilibria in the Cd–Ge–S
Ternary System

Symbol	Reaction	t, °C
E_1	$L \Leftrightarrow Cd_4GeS_6 + GeS + GeS_2$	525
E_2	$L \Leftrightarrow GeS + Ge + Cd_4GeS_5$	530
E_3	$L \Leftrightarrow Cd + CdS + Ge$	300
U_1	$L + CdS \Leftrightarrow GeS_2 + Cd_4GeS_5$	—
U_2	$L + CdS \Leftrightarrow Cd_4GeS_6 + Cd_4GeS_5$	770
U_3	$L + Cd_4GeS_5 \Leftrightarrow Cd_4GeS_6 + GeS$	540
U_4	$L + CdS \Leftrightarrow Cd_4GeS_5 + Ge$	575

Source: Movsum-zade, A.A. et al., *Zhurn. neorgan. khimii*, 32(4), 1025, 1987.

The Ge content in the CdS-doped single crystals reaches 0.5 at. % (0.25 mass. %) (Odin et al. 1999a) (GeS was used as a dopant).

This system was investigated using DTA, metallography, XRD, and measuring of microhardness. The ingots were annealed at temperatures 50°C–100°C lower than the eutectic temperature (Movsum-zade et al. 1987).

CdS–GeS: This section is a nonquasibinary section of Cd–Ge–S ternary system (Figure 4.41) (Odin et al. 1983). CdS, α-Cd_4GeS_6, and Ge primary crystallize in this system. Thermal effects at 925°C correspond to joint crystallization of CdS and α-Cd_4GeS_6. Nonvariant equilibrium

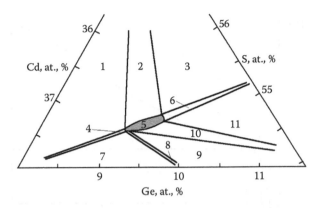

FIGURE 4.39
Limits of α-Cd$_4$GeS$_6$ homogeneity region at 544°C in the Cd–Ge–S ternary system: 1, α-Cd$_4$GeS$_6$+CdS+L; 2, α-Cd$_4$GeS$_6$+L; 3, α-Cd$_4$GeS$_6$+GeS$_2$+L; 4, α-Cd$_4$GeS$_6$+CdS; 5, α-Cd$_4$GeS$_6$; 6, α-Cd$_4$GeS$_6$+GeS$_2$; 7, α-Cd$_4$GeS$_6$+Ge+CdS; 8, α-Cd$_4$GeS$_6$+Ge; 9, α-Cd$_4$GeS$_6$+GeS+Ge; 10, α-Cd$_4$GeS$_6$+GeS; and 11, α-Cd$_4$GeS$_6$+GeS$_2$+GeS. (From Odin, I.N. and Grin'ko, V.V., *Zhurn. neorgan. khimii*, 36(5), 1332, 1991.)

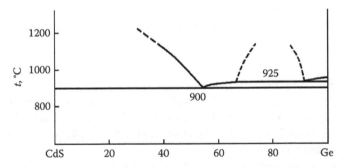

FIGURE 4.40
The CdS–Ge phase diagram. (From Movsum-zade, A.A. et al., *Zhurn. neorgan. khimii*, 32(4), 1025 1987.)

L+CdS $\Leftrightarrow$ α-Cd$_4$GeS$_6$+Ge takes place at 725°C. Within the interval of 0.5–66.6 mol. % CdS, crystallization ends in the ternary peritectic at 625°C. The mutual solubility of CdS and GeS is not higher than 0.5 mol. % (Odin et al. 1983). The solubility of GeS in CdS reached 30 mol. % when solid solutions were synthesized under 2–3 GPa and 600°C–700°C for 1–4 h (Kobayashi et al. 1979).

The Cd$_4$GeS$_5$ compound, which according to the data of Movsum-zade et al. (1987) melts incongruently at 605°C, was not found (Odin et al. 1983).

This system was investigated using DTA, metallography, XRD, and measuring of microhardness (Odin et al. 1983, Movsum-zade et al. 1987). The ingots were annealed at 540°C for 1200 h (Odin et al. 1983) or at the temperatures 50–100°C lower than eutectic temperature (Movsum-zade et al. 1987).

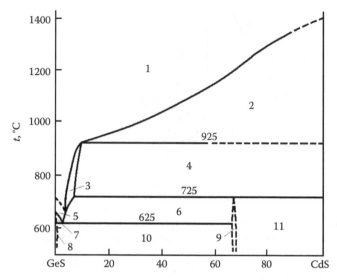

FIGURE 4.41

Phase relations in the CdS–GeS system: 1, L; 2, L + CdS; 3, L + α-Cd$_4$GeS$_6$; 4, L + CdS + α-Cd$_4$GeS$_6$; 5, L + Ge; 6, L + Ge + α-Cd$_4$GeS$_6$; 7, L + Ge + GeS; 8, Ge + GeS; 9, Ge + α-Cd$_4$GeS$_6$; 10, GeS + Ge + α-Cd$_4$GeS$_6$; and 11, α-Cd$_4$GeS$_6$ + Ge + CdS. (From Odin, I.N. et al., *Zhurn. neorgan. khimii*, 28(9), 2362, 1983.)

CdS–GeS$_2$: The phase diagram is shown in Figure 4.42 (Odin et al. 1983). The peritectic composition is 68 mol. % CdS, and eutectic composition and temperature are 31 ± 3 mol. % CdS and 803 ± 5°C, respectively (Odin et al. 1983) [20 mol. % CdS and 800°C (Barnier et al. 1990), 26.47 at. % Ge and 700°C (Movsum-zade et al. 1987)].

The Cd$_4$GeS$_6$ ternary compound is formed in this system. It melts incongruently at 1064 ± 7°C (Odin et al. 1983) [1020°C (Barnier et al. 1990), 775°C (Movsum-zade et al. 1987)] and crystallizes in a monoclinic structure with lattice parameters $a = 1233.8 ± 0.2$, $b = 709.3 ± 0.2$, $c = 1238.1 ± 0.3$ pm, and $\beta = 110.097 ± 0.008°$ (Parasyuk et al. 2005) [$a = 1232.6$, $b = 708.8$, $c = 1236.5$ pm, and $\beta = 110.08°$ (Julien-Pouzol and Jaulmes 1995); $a = 1239.5$, $b = 710.7$, $c = 1234.9$ pm, and $\beta = 110.75°$ (Motria et al. 1986); $a = 1234.6$, $b = 708.4$, $c = 1237.8$ pm, and $\beta = 110.20°$ (Susa and Steinfink 1971); $a = 1235$, $b = 708$, $c = 1238$ pm, and $\beta = 110.2°$ (Susa and Steinfink 1971); $a = 1236$, $b = 710.7$, $c = 1238$ pm, and $\beta = 110°08'$ (Serment et al. 1968)] and calculation and experimental density 4.67 [4.66 Motria et al. (1986)] and 4.70 [4.57 (Serment et al. 1968, Motria et al. 1986)] g.cm^{-3} (Julien-Pouzol and Jaulmes 1995). The structure of Cd$_4$GeS$_6$ can be described also using a pseudohexagonal structure with lattice parameters $a = 700$ and $c = 3450$ pm (Susa and Steinfink 1971). This compound has a polymorphous transition at 1034 ± 7°C (Odin et al. 1983) and decomposes at the heating on the air up to 410°C (Motria et al. 1987). Its high-temperature modification was not obtained (Odin et al. 1983).

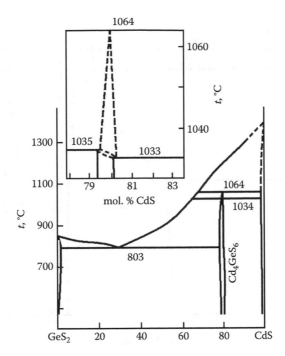

FIGURE 4.42
The CdS–GeS$_2$ phase diagram. (From Odin, I.N. et al., *Zhurn. neorgan. khimii*, 28(9), 2362, 1983.)

The solubility of GeS$_2$ in CdS at the peritectic temperature is equal to 1.5 mol. % and decreases to 0.5–1.0 mol. % at 540°C. Solid solutions based on GeS$_2$ contain less than 2 mol. % CdS at the eutectic temperature [31.27 at. % Ge at 700°C (Movsum-zade et al. 1987)] and approximately 1.5 mol. % CdS at 540°C (Odin et al. 1983). Homogeneity region of Cd$_4$GeS$_6$ at 540°C is equal approximately to 1 mol. % (Odin et al. 1983). Cd$_2$GeS$_4$ was not found in this system (Odin et al. 1983, Movsum-zade et al. 1987). A glass region is found in the CdS–GeS$_2$ system within the interval from 70 to 100 mol. % GeS$_2$ (Barnier et al. 1990).

Vapor composition in the CdS–GeS$_2$ system corresponds to GeS$_2$ compound, that is, it lies on this section (Odin and Grin'ko 1991a,b). The vapor above Cd$_4$GeS$_6$ consists mainly from GeS and S$_2$ in ratio 1:0.5, and the Cd concentration in vapor is 0.01 mass. % (Motria et al. 1987, Odin and Grin'ko 1991a,b). The limit of α-Cd$_4$GeS$_6$ homogeneity region from the CdS side does not change with temperature and from the GeS$_2$-side changes at temperatures above eutectic temperature (Figure 4.43) (Odin and Grin'ko 1991a). (T–x)$_p$ sections of the CdS–GeS$_2$ system at p = 267 and 667 hPa are shown in Figure 4.44, and (p–x)$_T$ section at 850°C is shown in Figure 4.45 (Odin and Grin'ko 1991a,b).

On p–T projection of CdS–GeS$_2$ phase diagram (Figure 4.46), the lines represent two-phase [AB—S(GeS$_2$)/V, BC—L(GeS$_2$)/V, BD—L(GeS$_2$)/S(GeS$_2$)] and

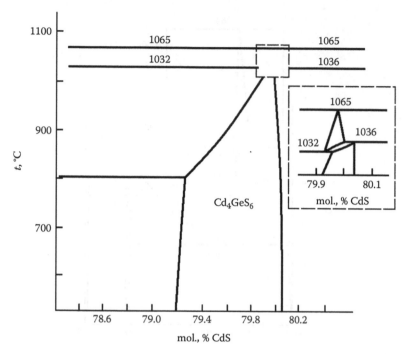

FIGURE 4.43

T–x projection of CdS–GeS$_2$ phase diagram close to Cd$_4$GeS$_6$. (From Odin, I.N. and Grin'ko, V.V., *Zhurn. neorgan. khimii*, 36(5), 1332, 1991).

three-phase equilibria [ME—V/S(GeS$_2$)/S(α-Cd$_4$GeS$_6$), EF—S(GeS$_2$)/L/S(α-Cd$_4$GeS$_6$), EB—S(GeS$_2$)/L/V, EQ—V/L/S(α-Cd$_4$GeS$_6$), NG—V/S(α-Cd$_4$GeS$_6$)/S(CdS)] (Odin and Grin'ko 1991b). The lines, representing equilibria of condensed phases [L(GeS$_2$)/S(GeS$_2$), S(GeS$_2$)/L/S(α-Cd$_4$GeS$_6$)], are practically vertical. The parameters for four-phase equilibria in the CdS–GeS$_2$ system are given in Table 4.3.

This system was investigated using DTA, metallography, XRD, measuring of microhardness, and using tensimeter (Odin et al. 1983, Movsum-zade et al. 1987, Odin and Grin'ko 1991a,b). The ingots were annealed at 540°C for 1200 h (Odin et al. 1983) [at the temperatures 50°C–100°C lower than eutectic temperature (Movsum-zade et al. 1987) and at 630°C for 500 h (Odin and Grin'ko 1991b)].

Cd$_4$GeS$_6$ was obtained from CdS and GeS$_2$ at 1000°C (Julien-Pouzol and Jaulmes 1995); by interaction of Ge, S, and CdS at 800°C–855°C using vibrational mixing Motria et al. (1986); by heating of Cd, Ge, and, S mixtures in stoichiometric ratio at 200°C–400°C for 2–4 h and then at 600°C–800°C for 12 h (Susa and Steinfink 1971); by heating of mixtures from CdS, Ge,

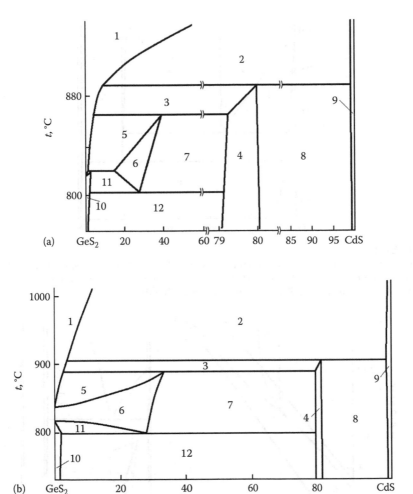

FIGURE 4.44

$(T–x)_p$ sections of CdS–GeS$_2$ phase diagram at 267 (a) and 667 hPa (b) [91Odi1, 91Odi2]: 1, V; 2, V+S(CdS); 3, V+S(α-Cd$_4$GeS$_6$); 4, S(α-Cd$_4$GeS$_6$); 5, V+L; 6, L; 7, S(α-Cd$_4$GeS$_6$)+L; 8, S(α-Cd$_4$GeS$_6$)+S(CdS); 9, S(CdS); 10, S(GeS$_2$); 11, S(GeS$_2$)+L; and 12, S(GeS$_2$)+S(α-Cd$_4$GeS$_6$). (From Odin, I.N. and Grin'ko, V.V., *Zhurn. neorgan. khimii*, 36(5), 1332, 1991a; Odin, I.N. and Grin'ko, V.V., *Zhurn. neorgan. khimii*, 36(4), 1056, 1991b.)

and S at 850°C (Kaldis and Widmer 1965); or by coprecipitation from the water solutions with next long-time heating at 600°C (Kislinskaya 1974). Its single crystals were grown using chemical transport reaction (Kaldis and Widmer 1965, Odin et al. 1983, Motria et al. 1986). Glasses in the CdS–GeS$_2$ system were obtained by quenching the melt at 1000°C (Barnier et al. 1990).

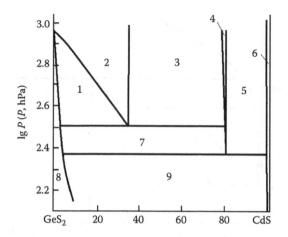

FIGURE 4.45
$(p-x)_T$ section of CdS–GeS$_2$ phase diagram at 850°C: 1, L+V; 2, L; 3, L+S(α-Cd$_4$GeS$_6$); 4, S(α-Cd$_4$GeS$_6$); 5, S(α-Cd$_4$GeS$_6$)+(CdS); 6, S(CdS); 7, V S(α-Cd$_4$GeS$_6$); 8, V; and 9, V+S(CdS). (From Odin, I.N. and Grin'ko, V.V., *Zhurn. neorgan. khimii*, 36(4), 1056, 1991b.)

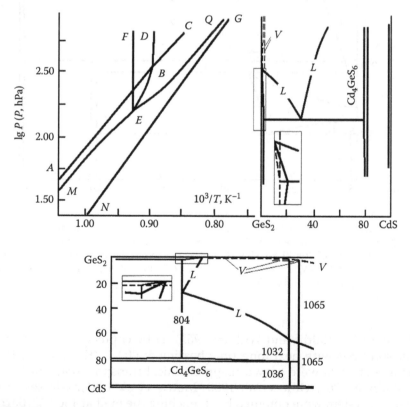

FIGURE 4.46
P_{total}–T–x phase diagram of the CdS–GeS$_2$ system (explanations are in the text). (From Odin, I.N. and Grin'ko, V.V., *Zhurn. neorgan. khimii*, 36(4), 1056, 1991b.)

TABLE 4.3

Parameters for Four-Phase Equilibria in the CdS–GeS$_2$ System

Phases in Equilibrium	Phase Compositions, mol. % CdS					t, °C	p, hPa
	S(CdS)	S(α)	S(β)	S(GeS$_2$)	L		
S(GeS$_2$), S(α), L, V	—	79.3	—	1.0	31.0	804	132.5
S(α), S(β), L, V	—	79.95	79.9	—	66.0	1032	1360
S(α), S(β), S(CdS), V	99.0	80.05	80.0	—	—	1036	1350
S(β), S(CdS), L, V	99.0	—	80.0	—	68.0	1065	—

Source: Odin, I.N. and Grin'ko, V.V., *Zhurn. neorgan. khimii*, 36(4), 1056, 1991b.
Note: S(α) and S(β) correspond to S(α-Cd$_4$GeS$_6$) and S(β-Cd$_4$GeS$_6$).

4.35 Cadmium–Tin–Sulfur

Eleven fields of primary crystallizations are on the liquidus surface of the Cd–Sn–S ternary system (Figure 4.47) (Zargarova et al. 1985). The field of CdS primary crystallization occupies considerable part of liquidus surface, and the field of sulfur crystallization is degenerated. Information about four-phase equilibria in the Cd–Sn–S ternary system is given in Table 4.4 (Zargarova et al. 1985).

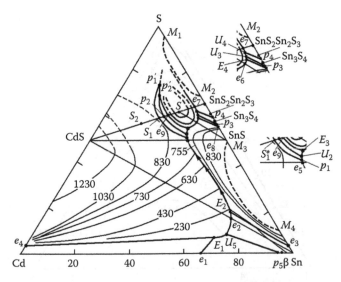

FIGURE 4.47
Liquidus surface of Cd–Sn–S ternary system (ternary compounds CdSn$_2$S$_5$, CdSnS$_3$, and Cd$_2$SnS$_4$ are indicated as S, S$_1$, and S$_2$, respectively). (From Zargarova, M.I. et al., *Zhurn. neorgan. khimii*, 30(5), 1279, 1985.)

TABLE 4.4

Nonvariant Equilibria in the Cd–Sn–S
Ternary System

Symbol	Reaction	t, °C
E_1	$L \Leftrightarrow CdS + Cd + \beta\ (Sn)$	142
E_2	$L \Leftrightarrow CdS + SnS + Sn$	182
E_3	$L \Leftrightarrow CdSnS_3 + SnS + CdSn_2S_5$	477
E_4	$L \Leftrightarrow CdSn_2S_5 + SnS + Sn_3S_4$	537
U_1	$L + CdS \Leftrightarrow Cd_2SnS_4 + Sn_3S_4$	587
U_2	$L + Cd_2SnS_4 \Leftrightarrow CdSnS_3 + SnS$	557
U_3	$L + Sn_2S_3 \Leftrightarrow Sn_3S_4 + CdSn_2S_5$	592
U_4	$L + SnS_2 \Leftrightarrow Sn_2S_3 + CdSn_2S_5$	647
U_5	$L + Sn \Leftrightarrow CdS + \beta\ (Sn)$	175

Source: Zargarova, M.I. et al., *Zhurn. neorgan. khimii*, 30(5), 1279, 1985.

Kimura (1971) determined liquidus temperatures of Cd–Sn–S ternary system from the Sn-rich side and constructed a part of Cd–Sn–S liquidus surface on the Sn-rich side, which is shown in Figure 4.48. According to the data of Kimura (1971), and Buehler and Bachmann (1976), a border between the fields of CdS and SnS primary crystallization passes across the two-phase region.

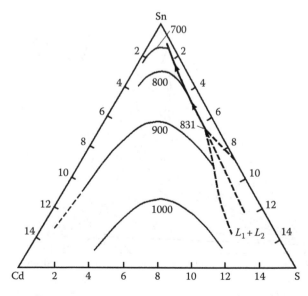

FIGURE 4.48
Part of the Cd–Sn–S liquidus surface on the Sn-rich side. (From Kimura, S., *J. Chem. Thermodyn.*, 3(1), 7, 1971.)

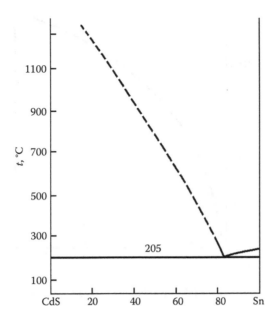

FIGURE 4.49
The CdS–Sn phase diagram. (From Zargarova, M.I. et al., *Zhurn. neorgan. khimii*, 30(5), 1279, 1985.)

Liquidus temperatures were determined by visual observation of solid-phase appearance and disappearance at the heating and cooling of ingots (accuracy of measurement is $\pm 5°C$) (Kimura 1971).

CdS–Sn: The phase diagram is a eutectic type (Figure 4.49) (Alieva et al. 1983, Zargarova et al. 1985). The eutectic composition and temperature are 82 at. % Sn [70 at. % Sn (Alieva et al. 1983)] and 205°C, respectively (Zargarova et al. 1985). The mutual solubility of CdS and Sn is insignificant.

According to the data of Rubenstein (1968) and Kimura (1971), the eutectic is degenerated from the Sn-rich side (Figure 4.50). The solubility of CdS in liquid tin at 900°C–1250°C can be described as follows (Buehler and Bachmann 1976): $\ln x_{CdS} = E/RT + 3.795$, where E—activation energy ($E = -70$ kJ/mol).

The Sn content in the CdS-doped single crystals reaches 0.11–0.15 at. % (0.09–0.12 mass. %) (Odin et al. 1999a) (SnS was used as a dopant).

This system was investigated using DTA, metallography, XRD, and measuring of microhardness (Alieva et al. 1983, Zargarova et al. 1985). The ingots were annealed. Liquidus temperatures were determined by visual observation of solid-phase appearance and disappearance for heating and cooling of ingots (Kimura 1971, Buehler and Bachmann 1976) and by the method of high-temperature filtration (Rubenstein 1968).

CdS–SnS: The phase diagram is a eutectic type (Figure 4.51) (Galiulin et al. 1981). The eutectic composition and temperature are 13 mol. % CdS and 831°C, respectively (Galiulin et al. 1981) [27 mol. % CdS and 750°C (Zargarova et al. 1985)]. In the solid state, there is a phase transition of solid solutions based on SnS. At 650°C, the mutual solubility of CdS and SnS is not higher

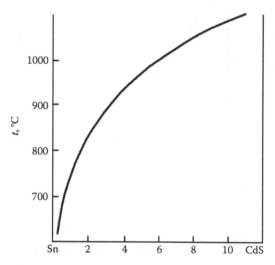

FIGURE 4.50
Temperature dependence of CdS solubility in liquid Sn. (From Kimura, S., *J. Chem. Thermodyn.*, 3(1), 7, 1971.)

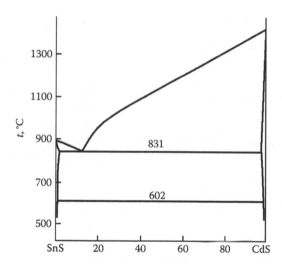

FIGURE 4.51
The CdS–SnS phase diagram. (From Galiulin, E.A. et al., *Zhurn. neorgan. khimii*, 26(7), 1881, 1981.)

than 1 mol. % (Galiulin et al. 1981) (according to the data of Zargarova et al. 1985, the solubility of CdS in SnS at room temperature reaches 4 mol. %). The solubility of SnS in CdS reaches 30 mol. % when solid solutions were synthesized under 2–3 GPa and 600°C–800°C for 1–4 h (Kobayashi et al. 1979).

This system was investigated using DTA, metallography, XRD, and measuring of microhardness (Galiulin et al. 1981, Zargarova et al. 1985).

The ingots were annealed at 650°C for 250 h (Galiulin et al. 1981) [at 150°C for 200 h (Zargarova et al. 1985)].

CdS–SnS₂: The results of this system investigation are contradictory. According to the data of Piskach et al. (1998), the phase diagram is of a eutectic type (Figure 4.52a). The eutectic point coordinates are 22 mol. % CdS and 770°C. Intermediate phases have not been found in this system.

According to the data of Zargarova et al. (1985), CdSnS₃, CdSn₂S₅, and Cd₂SnS₄ ternary compounds are formed in this system (Figure 4.52b). CdSn₂S₅ melts congruently at 800°C, and CdSnS₃ and Cd₂SnS₄ melt incongruently

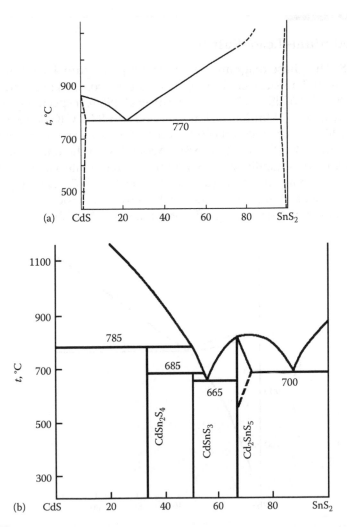

FIGURE 4.52
The CdS–SnS₂ phase diagram: (a) (From Piskach, L.V. et al., *J. Alloys Compd.*, 279(2), 142, 1998.) and (b) (From Zargarova, M.I. et al., *Zhurn. neorgan. khimii*, 30(5), 1279, 1985.)

correspondingly at 685°C and 785°C. The eutectic compositions and temperatures are 55 and 88 mol. % SnS_2 and 665°C and 700°C, respectively.

Kislinskaya (1974) noted that the limited solid solutions form at the heating up to 400°C of CdS and SnS_2 mixtures obtained by coprecipitation from water solutions.

This system was investigated using DTA, metallography, XRD, and measuring of microhardness. The ingots were annealed at 450°C for 200 h (Zargarova et al. 1985).

4.36 Cadmium–Lead–Sulfur

CdS–PbS: The phase diagram is a eutectic type (Figure 4.53) (Calawa et al. 1972, Stetiu 1973, Oleinik et al. 1983). The eutectic composition and temperature are 38 mol. % CdS and 1050°C (Oleinik et al. 1983), respectively [40 ± 2 mol. % CdS and 1054°C (Odin 2001), 42 mol. % CdS and 1052°C (Calawa et al. 1972)]. The solubility of CdS in PbS at 600°C and 1000°C is equal correspondingly to 4 and 31.5 mol. % (Oleinik et al. 1983). According to the data of Bethke and Barton (1971), the solubility of CdS in PbS within the interval of 400°C–930°C can be described as follows: $\log x_{CdS} = 5.216 \log T - 14.677$. Lattice parameters of solid solutions change according to Vegard's law (Sood et al. 1978). The solubility of PbS in CdS is insignificant (Calawa et al. 1972).

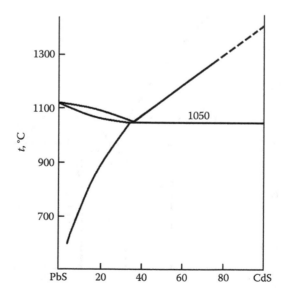

FIGURE 4.53

The CdS–PbS phase diagram. (From Oleinik, G.S. et al., *Izv. AN SSSR. Neorgan. materialy*, 19(11), 1799, 1983.)

The solid solutions over the entire range of concentrations form in this system when they are synthesized under 2–3 GPa and 600°C–800°C for 1–4 h (Kobayashi et al. 1979, Susa et al. 1980). Supersaturated and metastable $Cd_xPb_{1-x}S$ solid solutions at $0.03 < x < 0.18$ could be synthesized at low temperature (Maskayeva et al. 2003). The higher-temperature limit of the stability of the $Cd_xPb_{1-x}S$ solid-solution-deposited films is 132°C–137°C.

Deshmukh et al. (1994) employed a modified chemical deposition for the preparation of thin film $(CdS)_x(PbS)_{1-x}$ composites with $0.2 \leq x \leq 0.8$. $CdSO_4$, $Pb(CH_3COO)_2$, and thiourea were used as the basic source materials. It was shown that these composites include hexagonal and cubic CdS, cubic CdO and PbS, tetragonal PbO and PbO_2, and free elemental Cd and S. The content of hexagonal CdS was increased, while the content of cubic PbS was decreased as x increased from 0.2 to 0.8.

This system was investigated using DTA, metallography, and XRD (Calawa et al. 1972, Stetiu 1973, Oleinik et al. 1983, Deshmukh et al. 1994). The solubility of CdS in PbS was determined by saturation of PbS from mixtures CdS + PbS or Cd + Pb + S (Bethke and Barton 1971). The ingots were annealed at 1000°C for 10 days (Calawa et al. 1972).

4.37 Cadmium–Phosphorus–Sulfur

Isothermal section of the Cd–P–S ternary system at 500°C is shown in Figure 4.54 (Potoriy et al. 1999). $CdPS_4$ ternary compound was not obtained in this system.

CdS–"P_2S_3": The phase diagram is not constructed. CdP_2S_4 was obtained in this system, which melts at 917 ± 5°C and crystallizes in a trigonal structure with lattice parameters $a = 621.1$, $b = 1075.7$, and $c = 1961$ pm; experimental density 3.3 g cm^{-3}; and energy gap $E_g = 3.76$ eV (Golovey et al. 1973).

Brusilovets and Fedoruk (1976) doubted in the correctness of the composition determination for this compound, and according to the data of Potoriy et al. (1999), such compounds do not exist in this system.

CdS–"P_2S_4": The phase diagram is a eutectic type (Figure 4.55) (Kovach et al. 1997, Potoriy et al. 1999). The eutectic composition and temperature are 6 mol. % "P_2S_4" and 860 ± 5°C, respectively. $Cd_2P_2S_6$ ($CdPS_3$) was obtained in this system, which melts incongruently at 895 ± 5°C (Kovach et al. 1997, Potoriy et al. 1999) and crystallizes in a monoclinic structure with lattice parameters $a = 619.5 \pm 0.3$, $b = 1067.4 \pm 0.7$, $c = 687.4 \pm 0.4$ pm, and $\beta = 107.21 \pm 0.08°$ (Kovach et al. 1997, Potoriy et al. 1999) [$a = 617$, $b = 1067$, $c = 682$ pm, and $\beta = 107.1°$ (Klinger et al. 1970, 1973) and $a = 621.8$, $b = 1076.3$, $c = 686.7$ pm, and $\beta = 107.68°$ (Brec et al. 1980, Ouvrard et al. 1985)] and calculation and experimental density 3.71 and 3.49 [3.67 (Tovt et al. 1998)] g cm^{-3}, respectively (Klinger et al. 1970, Klinger et al. 1973). According to the data of Covino et al. (1985),

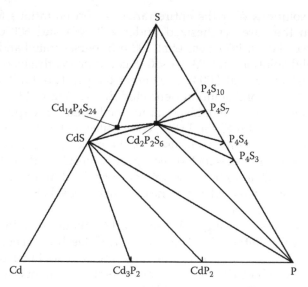

FIGURE 4.54
Isothermal section of the Cd–P–S ternary system at 500°C. (Potoriy, M.V. et al., *Neorgan. materialy*, 35(11), 1297, 1999.)

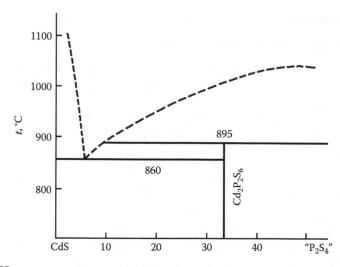

FIGURE 4.55
The CdS–"P_2S_4" phase diagram. (From Kovach, A.P. et al., *Ukr. khim. zhurn.*, 63(2), 92, 1997; Potoriy, M.V. et al., *Neorgan. materialy*, 35(11), 1297, 1999.)

$CdPS_3$ crystallizes in a hexagonal structure with lattice parameters $a = 1078$ and $c = 1963$ pm. At the temperatures above 500°C, $CdPS_3$ decomposes with the formation of CdS and volatile phosphorus sulfides (Golovey et al. 1973, Covino et al. 1985). Brusilovets and Fedoruk (1976) do not confirm the existence of $Cd_2P_2S_6$ compound.

Single crystals of $Cd_2P_2S_6$ compound were grown using chemical transport reactions (Nitsche and Wild 1970, Kovach et al. 1997, Tovt et al. 1998, Potoriy et al. 1999).

The ingots were annealed at 500°C–530°C during 2 weeks, and this system was investigated using DTA and XRD (Kovach et al. 1997, Potoriy et al. 1999).

CdS–P₂S₅: The phase diagram is not constructed. $Cd_{14}P_4S_{24}$ (Grieshaber et al. 1976, Hozhainov et al. 1976), CdP_2S_6 (cadmium metathiophosphate) (Hozhainov et al. 1976), and $Cd_3(PS_4)_2$ ($Cd_3P_2S_8$) (Soklakov and Nechaeva 1970) compounds were obtained in this system. $Cd_{14}P_4S_{24}$ melts incongruently at 390°C (Grieshaber et al. (1976) and crystallizes in a monoclinic structure with lattice parameters $a = 1224.1 \pm 0.9$, $b = 1227.4 \pm 0.9$, $c = 699.6 \pm 0.6$ pm, and $\beta = 110.53 \pm 0.05°$ (Potoriy et al. 1999) [$a = 1216 \pm 1$, $b = 1226 \pm 1$, $c = 698 \pm 1$ pm, and $\beta = 110.55 \pm 0.03°$ (Grieshaber et al. 1976)] and calculation and experimental density 4.200 and 4.14 g cm⁻³, respectively (Grieshaber et al. 1976). Around 385°C, a reversible solid–solid transition into a cubic $Cd_{14}P_2S_{24}$ takes place.

CdP_2S_6 and $Cd_3P_2S_8$ crystallize in a cubic structure with lattice parameter $a = 924$ pm (Hozhainov et al. 1976) and $a = 1308 \pm 4$ pm (Soklakov and Nechaeva 1970), respectively. According to the data of Potoriy et al. (1999), $Cd_3P_2S_8$ does not exist in this system: the ingot of such composition is three phase and contains $Cd_2P_2S_6$, $Cd_{14}P_4S_{24}$, and sulfur. Brusilovets and Fedoruk (1976) do not confirm the existence of CdP_2S_6 compound.

$Cd_{14}P_2S_{24}$ single crystals were prepared by sinter–annealing (Grieshaber et al. 1976). This compound is perfectly stable in air to about 480°C, when decomposition, coupled with oxidation, sets in.

This system was investigated using DTA and XRD (Hozhainov et al. 1976).

4.38 Cadmium–Arsenic–Sulfur

Six nonvariant equilibria are on the liquidus surface of Cd–As–S ternary system (Table 4.5) (Olekseyuk et al. 1976). The fields of primary crystallization in this system are shown in Figure 4.56. All phases, forming eight fields of primary crystallization, are in equilibria with CdS, which occupies majority of Cd–As–S system. Ternary compounds do not form in this system. The quasibinary sections in this system are $CdS–Cd_3As_2$, $CdS–CdAs_2$, CdS–As, $CdS–As_2S_2$, and $CdS–As_2S_3$ (Olekseyuk et al. 1976).

Glass region in the Cd–As–S ternary system is shown in Figure 4.57 (Olekseyuk et al. 1976, Olekseyuk et al. 1977). The specimens for determination of this region were quenched from 400°C–450°C.

CdS–CdAs₂: The phase diagram is a eutectic type (Figure 4.58) (Ruut and Koppel 1973). The eutectic composition and temperature are 1 mol. % CdS and 616°C [621°C (Ruut and Koppel 1973)], respectively (Olekseyuk et al. 1976). The mutual solubility of CdS and $CdAs_2$ is insignificant.

TABLE 4.5

Nonvariant Equilibria in the Cd–As–S
Ternary System

Symbol	Reaction	t, °C
E_1	$L \Leftrightarrow CdS + Cd_3As_2 + Cd$	320
U	$L + CdS \Leftrightarrow Cd_3As_2 + CdAs_2$	614
E_2	$L \Leftrightarrow CdS + CdAs_2 + As$	610
E_3	$L \Leftrightarrow CdS + As_2S_2 + As$	285
E_4	$L \Leftrightarrow CdS + As_2S_3 + As_2S_2$	270
E_5	$L \Leftrightarrow CdS + As_2S_3 + S$	100

Source: Olekseyuk, I.D. et al., *Zhurn. neorgan. khimii*, 21(12), 3382, 1976.

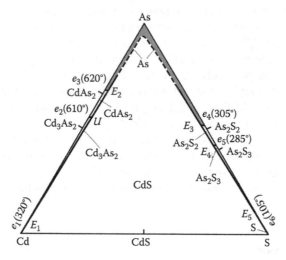

FIGURE 4.56
Liquidus surface of the Cd–As–S ternary system. (From Olekseyuk, I.D. et al., *Zhurn. neorgan. khimii*, 21(12), 3382, 1976.)

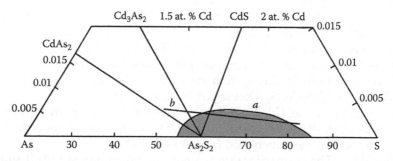

FIGURE 4.57
Glass region in the Cd–As–S ternary system (the ingots were quenched from 400°C to 450°C) (a) and the liquidus isotherm at 400°C (b) in the Cd–As–S ternary system. (From Olekseyuk, I.D. et al., *Zhurn. neorgan. khimii*, 21(12), 3382, 1976.)

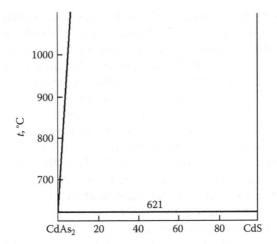

FIGURE 4.58
The CdS–CdAs$_2$ phase diagram. (From Ruut, A. and Koppel, H., *Izv. AN ESSR. Ser. Khimia, Geologia*, 22(2), 137, 1973.)

This system was investigated using DTA, XRD, and measuring of micro-hardness. The ingots were annealed at 300°C for 840 h (Ruut and Koppel 1973, Olekseyuk et al. 1976).

CdS–Cd$_3$As$_2$: The phase diagram is a peritectic type (Figure 4.59) (Ruut and Koppel 1973). The peritectic composition and temperature are 4 mol. % CdS and 745 ± 5°C [740 ± 5°C (Olekseyuk et al. 1976)], respectively (Ruut and Koppel 1973). At 602 ± 2°C, an α-solid solution transforms according to a peritectoid reaction into an α′-solid solution based on low-temperature modification of Cd$_3$As$_2$ (this transformation is not shown on the phase diagram).

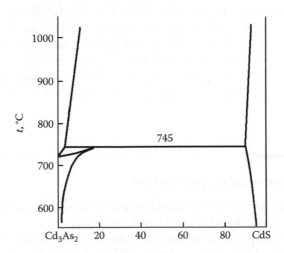

FIGURE 4.59
The CdS–Cd$_3$As$_2$ phase diagram. (From Ruut, A. and Koppel, H., *Izv. AN ESSR. Ser. Khimia, Geologia*, 22(2), 137, 1973.)

Region of α'-solid solution reaches 15 and 1.5 mol. % CdS at 600°C and 240°C, respectively (Olekseyuk et al. 1976) [according to the data of Golovey et al. (1972), solid solutions were not determined in this system]. The solubility of CdS in Cd_3As_2 at peritectic temperature is equal to 18 mol. % (Ruut and Koppel 1973, Olekseyuk et al. 1976) and decreases to 4.5 and 3.5 mol. % [2 mol. % (Olekseyuk et al. 1976)] at 700°C and 600°C, respectively (Ruut and Koppel 1973). The solubility of Cd_3As_2 in CdS at peritectic temperature reaches 9.4 mol. % and decreases to 7.5, 6, and less than 0.5 mol. % at 950°C, 620°C, and 240°C, respectively (Ruut and Koppel 1973, Olekseyuk et al. 1976).

According to the data of Golovey et al. (1974), within the interval of $0 \leq x \leq 0.06$ [the system is considered as $(2CdS)_x(Cd_3As_2)_{1-x}$], the ingots have a polymorphous transformation, and its temperature increases from 580°C for Cd_3As_2 to 595°C for the ingot with $x = 0.02$.

Single crystals of the solid solutions based on CdS were grown using chemical transport reactions (Golovey and Rigan 1977). This system was investigated using DTA, XRD, and measuring of microhardness. The ingots were annealed at 300°C for 840 h (Ruut and Koppel 1973, Golovey et al. 1974, Olekseyuk et al. 1976).

CdS–As: The phase diagram is not constructed. According to the data of Olekseyuk et al. (1976), it is a eutectic type. The eutectic composition and temperature are nearly 2 mol. % CdS and 814°C, respectively.

This system was investigated using DTA, XRD, and measuring of microhardness. The ingots were annealed at 300°C for 840 h (Olekseyuk et al. 1976).

CdS–As$_2$S$_2$: The phase diagram is not constructed. According to the data of Olekseyuk et al. (1976), it is a eutectic type. The eutectic composition and temperature are 1.5 mol. % CdS and 302 ± 2°C, respectively.

This system was investigated using DTA, XRD, and measuring of microhardness. The ingots were annealed at 300°C for 840 h (Olekseyuk et al. 1976).

CdS–As$_2$S$_3$: The phase diagram is not constructed. According to the data of Olekseyuk et al. (1976), it is a eutectic type. The eutectic composition and temperature are 0.5 mol. % CdS and 300 ± 5°C, respectively.

This system was investigated using DTA, XRD, and measuring of microhardness. The ingots were annealed at 300°C for 840 h (Olekseyuk et al. 1976).

4.39 Cadmium–Antimony–Sulfur

Liquidus surface of the Cd–Sb–S is characterized by the two big immiscibility regions (Figure 4.60) (Odin and Chukichev 2000).

CdS–CdSb: The phase diagram is a eutectic type (Figure 4.61) (Odin and Chukichev 2000). The eutectic contains 0.5 mol. % CdS and crystallizes at 455 ± 2°C. The mutual solubility of CdS and CdSb is insignificant.

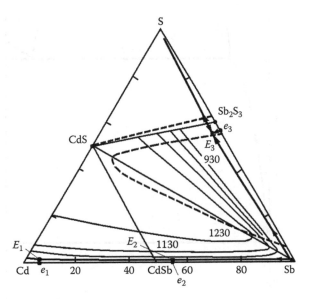

FIGURE 4.60
Liquidus surface of Cd–Sb–S ternary system. (From Odin, I.N. and Chukichev, M.V., *Zhurn. neorgan. khimii*, 45(2), 255, 2000.)

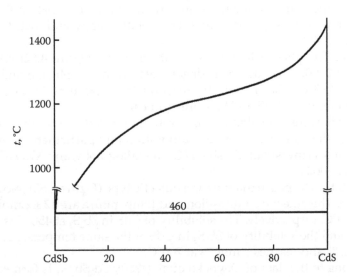

FIGURE 4.61
The CdS–CdSb phase diagram. (From Odin, I.N. and Chukichev, M.V., *Zhurn. neorgan. khimii*, 45(2), 255, 2000.)

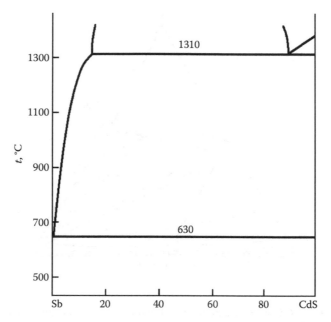

FIGURE 4.62
The CdS–Sb phase diagram. (From Odin, I.N. and Chukichev, M.V., *Zhurn. neorgan. khimii*, 45(2), 255, 2000.)

The ingots were annealed at temperatures for 20°C lower than the temperature of the nonvariant equilibrium with liquid participation for 1000 h. The system was investigated using DTA, metallography, and XRD (Odin and Chukichev 2000).

CdS–Sb: The phase diagram is a eutectic type (Figure 4.62) (Odin and Chukichev 2000). The eutectic is degenerated from the Sb side and crystallizes at 630±3°C. The monotectic reaction takes place at 1310±10°C. The mutual solubility of CdS and Sb is insignificant.

The ingots were annealed at temperatures for 20°C lower than the temperature of the nonvariant equilibrium with liquid participation for 1000 h. The system was investigated using DTA, metallography, and XRD (Odin and Chukichev 2000).

CdS–Sb₂S₃: The phase diagram is a eutectic type (Figure 4.63) (Novoselova et al. 1978). The eutectic composition and temperature are 12±2 mol. % CdS and 537±3°C, respectively. The solubility of CdS in Sb_2S_3 at 450°C is equal to 4 mol. %, and the solubility of Sb_2S_3 in CdS at the same temperature is equal to 1.5 mol. % and reaches 2 mol. % at 700°C.

According to the data of Deyneko et al. (1980), $Cd_2Sb_6S_{11}$ is formed in this system. It crystallizes in an orthorhombic structure with lattice parameters $a = 391.7$, $b = 959.9$, and $c = 1249.7$ pm.

This system was investigated using DTA and XRD. The ingots were annealed at 700°C for 240 h (Novoselova et al. 1978).

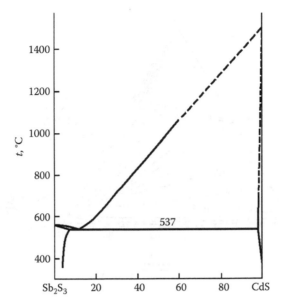

FIGURE 4.63
The CdS–Sb$_2$S$_3$ phase diagram. (From Novoselova, A.V. et al., *Zhurn. neorgan. khimii*, 23(12), 3321, 1978.)

4.40 Cadmium–Bismuth–Sulfur

Liquidus surface of the Cd–Bi–S ternary system is shown in Figure 4.64 (Odin and Chukichev 2000). It characterizes by two immiscibility regions: one of them is elongated along the CdS–Bi quasibinary system, and another occupies practically all S corner of the concentration triangle.

Kimura and Panish (1970) determined liquidus temperatures of Cd–Bi–S ternary system and constructed a part of Cd–Bi–S liquidus surface on the Bi-rich side, which is shown in Figure 4.65. The field of CdS primary crystallization occupies the majority of liquidus surface.

CdS–Bi: The phase diagram is a eutectic type (Figure 4.66) (Odin and Chukichev 2000). The eutectic is degenerated from the Bi side and crystallizes at $282 \pm 1°C$. The monotectic reaction takes place at $1305 \pm 10°C$. The mutual solubility of CdS and Sb is insignificant.

Temperature dependence of CdS solubility in liquid Bi was determined by Rubenstein (1966) and Buehler and Bachmann (1976), and according to the data of Buehler and Bachmann (1976) at $750°C–1000°C$, it can be described as follows: $\ln x_{CdS} = E/RT + 5.928$, where E—activation energy ($E = -98.4$ kJ/mol).

The ingots were annealed at temperatures for 20°C lower than the temperature of the nonvariant equilibrium with liquid participation for 1000 h.

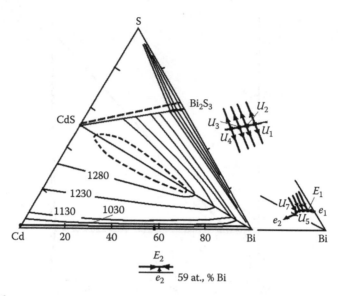

FIGURE 4.64
Liquidus surface of the Cd–Bi–S ternary system. (From Odin, I.N. and Chukichev, M.V., *Zhurn. neorgan. khimii*, 45(2), 255, 2000.)

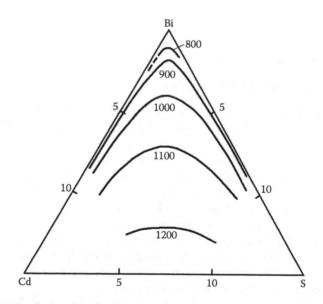

FIGURE 4.65
Part of the Cd–Bi–S liquidus surface on the Bi-rich side. (From Kimura, S. and Panish, M.B., *J. Chem. Thermodyn.*, 2(1), 77, 1970.)

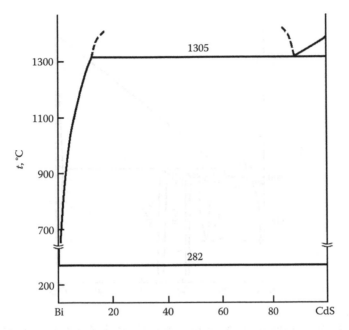

FIGURE 4.66
The CdS–Bi phase diagram. (From Odin, I.N. and Chukichev, M.V., *Zhurn. neorgan. khimii*, 45(2), 255, 2000.)

The system was investigated using DTA, metallography, and XRD (Odin and Chukichev 2000). Temperature dependence of CdS solubility in liquid Bi was determined by visual observation of solid-phase appearance and disappearance at the heating and cooling of ingots (Buehler and Bachmann 1976) and by high-temperature filtration (Rubenstein 1966). The mixtures of ZnS and Bi were annealed at low temperature for 16 h and at 1300°C for 6 h. The quantity of ZnS dissolved in Bi was determined by weighting of the ingots before and after its treatment (Rubenstein 1966).

CdS–Bi$_2$S$_3$: The phase diagram is shown in Figure 4.67 (Novoselova et al. 1978). CdBi$_2$S$_4$ (CdS Bi$_2$S$_3$), Cd$_2$Bi$_{10}$S$_{17}$ (2CdS 5Bi$_2$S$_3$), and Cd$_3$Bi$_{10}$S$_{18}$ (3CdS 5Bi$_2$S$_3$) are formed in this system. CdBi$_2$S$_4$ melts incongruently at 843°C. Its homogeneity range is within the interval of 45–50 mol. % CdS (Novoselova et al. 1978) [48–50 mol. % CdS at 700°C (Brouty et al. 1975)].

The compounds Cd$_3$Bi$_{10}$S$_{18}$ and Cd$_2$Bi$_{10}$S$_{17}$ contain 35–40 and 27–32 mol. % CdS and also melt incongruently at 817°C and 808°C, respectively (Brouty et al. 1975, Novoselova et al. 1978). The annealing of Cd$_2$Bi$_{10}$S$_{17}$ at 450°C leads to its decomposition with the formation of Bi$_2$S$_3$ and Cd$_3$Bi$_{10}$S$_{18}$ (Novoselova et al. 1978). A solid solution based on Bi$_2$S$_3$ is formed peritectically at 782°C. The energy gaps of Cd$_2$Bi$_{10}$S$_{17}$, Cd$_3$Bi$_{10}$S$_{18}$, and CdBi$_2$S$_4$ are equal to 1.26, 1.69, and 1.83 eV, respectively (Odin and Chukichev 2000).

The solubility of Bi$_2$S$_3$ in CdS and CdS in Bi$_2$S$_3$ is equal correspondingly to 0.75 and 2 mol. % (Novoselova et al. 1978) [up to 1 and 5 mol. % in the case of

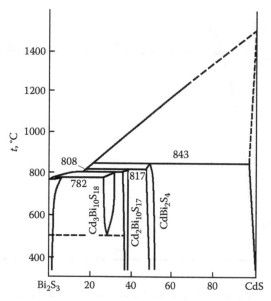

FIGURE 4.67

The CdS–Bi$_2$S$_3$ phase diagram. (From Novoselova, A.V. et al., *Zhurn. neorgan. khimii*, 23(12), 3321, 1978.)

the CdS–Bi$_2$S$_3$ films (Semenov et al. 2000a)]. All ternary compounds crystallize in a pseudoorthorhombic structure like the Bi$_2$S$_3$ structure (Brouty et al. 1975).

According to the data of Choe et al. (1997), four compounds are formed in the CdS–Bi$_2$S$_3$ system: CdBi$_2$S$_4$ (I), CdBi$_4$S$_7$ (II), Cd$_{2.8}$Bi$_{8.1}$S$_{15}$ (III), and Cd$_2$Bi$_6$S$_{11}$ (IV). These compounds crystallize in a monoclinic structure with lattice parameters: I—$a = 1309.5$, $b = 397.9$, $c = 1461.1$ pm, and $\beta = 116.50°$; II—$a = 1311.2$, $b = 400.32$, $c = 1176.5$ pm, and $\beta = 105.21°$; III—$a = 1311.1$, $b = 398.85$, $c = 2471$ pm, and $\beta = 97.84°$; and IV—$a = 1311.2$, $b = 399.7$, $c = 3584$ pm, and $\beta = 90.39°$.

The ingots were annealed at temperatures for 20°C lower than the temperature of the nonvariant equilibrium with liquid participation for 1000 h. The system was investigated using DTA, metallography, and XRD (Novoselova et al. 1978, Odin and Chukichev 2000).

4.41 Cadmium–Oxygen–Sulfur

The following four equilibrium reactions were confirmed to exist in the Cd–O–S ternary system in the temperature range of 630°C–830°C at the total pressure of 10^2 kPa (1 atm) (Fukatsu et al. 1988):

$$CdS + 2O_2 = CdSO_4$$

$$2CdS + 3O_2 = 2CdO + 2SO_2$$

$$6CdO + 2SO_2 + O_2 = 2(2CdO\ CdSO_4)$$

$$(2CdO\ CdSO_4) + 2SO_2 + O_2 = 3CdSO_4$$

The 2CdS 2CdO CdSO$_4$ ternary phase was synthesized in this system. A log $P(O_2)$–log $P(S_2)$ potential diagram was drawn at 730°C and is shown in Figure 4.68 (Fukatsu et al. 1988). It should be noted that the stable region of the new phase is limited to a very small area surrounded by the phases CdSO$_4$, 2CdO CdSO$_4$, CdO, and CdS.

At 1100°C, the solubility of oxygen in CdS changes for three degrees, and the maximum solubility is equal to 10^{21} cm^{-3} at deviation from stoichiometry in the Cd-rich side (Morozova et al. 1993). Abrupt decreasing of oxygen concentration coincides with stoichiometry composition, and oxygen concentration reduces to 10^{19} cm^{-3} at the changing of CdS composition in the S-rich side. Oxygen concentration in CdS cannot be significantly lower than 10^{18} cm^{-3} at the real conditions of single crystal growing.

2CdS 2CdO CdSO$_4$ could be obtained at 101.3 kPa (1 atm) total pressure by keeping material made only from the elements Cd, S, and O under the controlled temperature and the oxygen partial pressure corresponding to the region where the new phase was found to be stable (Fukatsu et al. 1988).

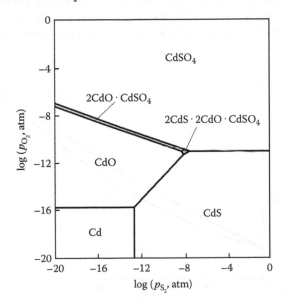

FIGURE 4.68
log $P(S_2)$–log $P(O_2)$ potential diagram of Cd–O–S ternary system at 730°C. (From Fukatsu, N. et al., *J. Electrochem. Soc.*, 135(4), 997, 1988. With permission.)

CdS–CdO: The phase diagram is not constructed. The next reaction takes place at the interaction of CdS and CdO (Goriayev et al. 1973): $CdS + 2CdO = 3Cd + SO_2$. The reaction rate is insignificant at 600°C and appreciably increases at 700°C, and at 800°C, the interaction is completed for 30 min.

CdS single crystals, growing using the Bridgman method, contain $(1-2) \cdot 10^{-2}$ mol. % CdO (Zuyev et al. 1981).

4.42 Cadmium–Selenium–Sulfur

CdS–CdSe: The phase diagram is shown in Figure 4.69 (Tomashik et al. 1979). Solid solutions are formed in this system over the entire range of concentrations. The concentration dependence of lattice parameters changes according to the Vegard's law (Hansevarov et al. 1958).

Using mass spectrometry, Karasev et al. (1972) determined that vapor composition, leaving effusion chamber, coincides with solid-solution compositions at 13, 40, 60, and 80 mol. % CdS. Such behavior of solid solutions indicates on ideality of CdS–CdSe system. At 727°C, total vapor pressure in this system increases with increasing of CdSe contents (Lopatin et al. 1975).

Kinetic and thermodynamic properties of the evaporation of single crystalline CdS_xSe_{1-x} ($0 \le x \le 1.0$) were investigated in the temperature range of 571°C–808°C (Lam and Munir 1980). The total equilibrium pressure over such solid solutions appears to be independent of x, and similar conclusions can be drawn regarding other calculated thermodynamic properties (ΔH_{298}^0 and ΔS_{298}^0).

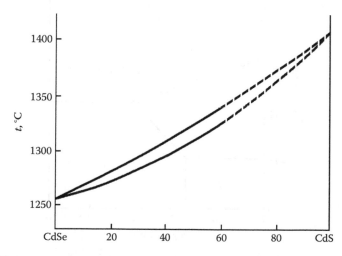

FIGURE 4.69
The CdS–CdSe phase diagram. (From Tomashik, V.N. et al., *Izv. AN SSSR. Neorgan. materialy*, 15(2), 202, 1979.)

The formation of cadmium sulfoselenide from CdS and Se has been studied using DSC (Kulkarni and Garn 1986). It was shown that the formation of CdS_xSe_{1-x} solid solutions begins following the melting of Se. The reaction of CdS and Se starts at ~265°C and can be considered practically complete at 380°C.

This system was investigated using DTA and XRD (Tomashik et al. 1979). Solid solutions were obtained from chemical element in vapor phase (Vitrihovski and Mizetskaya 1959a) or from binary compounds by melting and caking (Budionnaya et al. 1976, Tomashik et al. 1979) or by interaction of CdS and Se at 250°C–350°C (Cini and Melandri 1971). The liquidus and solidus were extrapolated to the melting point of CdS [1405 ± 10°C (Sysoev et al. 1967)]. Vapor pressure determinations were made using the torsion method (Lam and Munir 1980). The ingots were annealed at 900°C for 100 h (Tomashik et al. 1979).

4.43 Cadmium–Tellurium–Sulfur

CdS–CdTe: The phase diagram is shown in Figure 4.70 (Ohata et al. 1973, Budionnaya et al. 1978). Solid solutions with limited mutual solubility are formed in this system (Vitrihovski and Mizetskaya 1959b, Budionnaya et al. 1972, 1978, Ohata et al. 1973, Sanitarov et al. 1980). The azeotrope

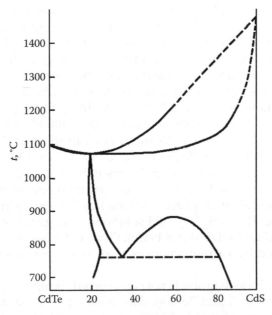

FIGURE 4.70
The CdS–CdTe phase diagram. (From Ohata, K. et al., *Jap. J. Appl. Phys.*, 12(8), 1198, 1973.)

composition and temperature are 80 mol. % CdTe and 1071°C (Ohata et al. 1973). The two-phase region is very narrow at 1000°C. Composition dependence of lattice parameters both in the cubic and hexagonal region changes according to the Vegard's law (Ohata et al. 1973, Budionnaya et al. 1978).

According to the data of Nunoue et al. (1990), the phase boundaries of sphalerite/(sphalerite + wurtzite) are in good agreement with the result reported by Ohata et al. (1973). However, the concentration of CdS at the phase boundaries of (sphalerite + wurtzite)/wurtzite is 4 mol. % less than obtained by Ohata et al. (1973). Vapor species above the CdS–CdTe system at around 700°C are Cd, Te_2, S_2, and STe (Nunoue et al. 1990).

The mixture of $CdTe_{0.87}S_{0.13}$ and $CdTe_{0.04}S_{0.96}$ coexists at a sintering temperature of 630°C (Saraie et al. 1977). At 800°C, solid solutions, containing up to 20 mol. % CdS (Budionnaya et al. 1978) [22.5 mol. % CdS (Tai et al. 1975)], crystallize in sphalerite structure; at 1025°C, they crystallize in wurtzite structure within the interval of 22.5–100 mol. % CdS (Tai et al. 1975). A microheterogeneous model of the solid-solution cluster structure could be used for the characterization of their physicochemical properties (Sanitarov et al. 1980).

Concentration dependence of energy gap has a minimum of 22.5 mol. % CdS (Tai et al. 1976).

A condensation diagram of thin-film solid solutions of the CdS–CdTe system was constructed by Vityuk et al. (1982a) over a wide range of condensation temperatures (50°C–550°C) and molecular vapor flux densities incident on the substrate (10^{20}–10^{23} m^{-2} s^{-1}). Thermodynamic unstable films of compositions $0.1 < x < 0.9$ ($CdTe_{1-x}S_x$) were shown to disintegrate at temperatures above 300°C forming a two-phase system (Vityuk et al. 1982a,b). In films obtained under maximum nonequilibrium conditions, the disintegration process proceeds at a greater rate and to a greater extent.

This system was investigated using DTA, XRD (Ohata et al. 1973, Saraie et al. 1977, Budionnaya et al. 1978), and isothermal evaporation method (Nunoue et al. 1990). The ingots were annealed at 1000°C for 100 h and then at 650°C–720°C for 1000–1500 h (Nunoue et al. 1990), at 1080°C for 5 weeks (Cruceanu and Niculescu 1965), at 800°C–850°C for 100 h (Budionnaya et al. 1972, 1978), or at 1000°C for 5–7 days (Ohata et al. 1973). In addition, the ingots were annealed for 5 weeks at temperatures below solidus temperature for determination phase relations in solid state (Ohata et al. 1973). Single crystals of the $CdS_{1-x}Te_x$ solid solutions were grown by the sublimation (Vitrihovski and Mizetskaya 1959b). Films of CdS–CdTe solid solutions were synthesized by vacuum condensation in a quasi-closed cell (Vityuk et al. 1982a).

CdS–Te: The phase diagram is not constructed. Temperature dependence of CdS solubility in liquid Te was determined by the saturation method within the interval of 600°C–1000°C (Figure 4.71) (Masako et al. 1979, Masaharu et al. 1980).

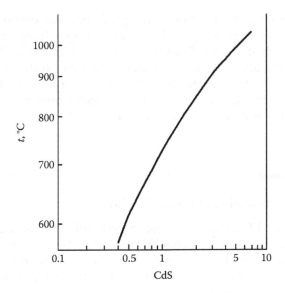

FIGURE 4.71
Temperature dependence of CdS solubility in liquid Te. (From Masako, W. et al., *Jap. J. Appl. Phys.*, 18(5), 869, 1979; Masaharu, A. et al., *J. Fac. Eng. Univ. Tokyo*, A(18), 38, 1980.)

4.44 Cadmium–Chromium–Sulfur

CdS–CrS: The phase diagram is not constructed. The solubility of CrS in CdS reaches 30 mol. % when solid solutions were synthesized under 2 GPa and 600°C (Susa et al. 1980). Phase with NiAs structure was found in this system at 50 mol. % CrS.

CdS–Cr$_2$S$_3$: The phase diagram is not constructed. CdCr$_2$S$_4$ is formed in this system (Koelmans and Grimmeiss 1959, Czaja 1967, Pinch and Berger 1968, Czaja and Krausbauer 1969, Boruhovich et al. 1974, Samal' and Sadovnikova 1974, Semenov et al. 2000b). It melts at 1105°C (Czaja 1967, Czaja and Krausbauer 1969) and crystallizes in a cubic structure of spinel type with lattice parameter $a = 1024.3$ pm and experimental density 4.23 g cm^{-3} (Boruhovich et al. 1974), and energy gap $E_g = 2.2$ eV (Koelmans and Grimmeiss 1959, Czaja and Krausbauer 1969). At 750°C–800°C, the CdCr$_2$S$_4$ formation runs practically to the end for 50–60 h. According to the data of Tret'yakov et al. (1972), CdCr$_2$S$_{4-x}$ has a narrow gap of homogeneity with $0 \leq x \leq 0.19$, and the lattice parameter does not change at the x changing. It was indicated by Pleshchev et al. (1978) that lattice parameter of this compound increases when at the annealing the partial chalcogen pressure increases.

At obtaining the films in the CdS–Cr$_2$S$_3$ system, the solubility of Cr$_2$S$_3$ in CdS is equal approximately to 5 mol. %, and Cr$_2$S$_3$ dissolves about 10 mol. % CdS (Semenov et al. 2000b).

CdCr$_2$S$_4$ was obtained by the alloying of chemical elements or binary sulfides (Boruhovich et al. 1974, Pinch and Berger 1968), by solid–phase interaction of CdS and CrCl$_3$ (Boruhovich et al. 1974), or by alloying of CdS and Cr$_2$S$_3$ in equimolar ratio under vapor pressure of sulfur up to 5·10^2 kPa (5 atm) (Koelmans and Grimmeiss 1959). Single crystals of CdCr$_2$S$_4$ were grown using the Bridgman method under vapor pressure of sulfur 30.39–50.65 kPa (0.3–0.5 atm) (Czaja and Krausbauer 1969).

4.45 Cadmium–Molybdenum–Sulfur

Ternary compound Cd$_x$Mo$_{15}$S$_{19}$ (Tarascon and Hull 1986) and Chevrel phase CdMo$_6$S$_8$ (Gocke et al. 1987) are formed in the Cd–Mo–S ternary system. CdMo$_6$S$_8$ crystallizes in a rhombohedral structure with lattice parameters $a = 651.7$ pm and $\alpha = 92.77°$ (in a hexagonal structure with lattice parameters $a = 943.7$ and $c = 1072.9$ pm) (Gocke et al. 1987).

Cd$_x$Mo$_{15}$S$_{19}$ was prepared at low temperatures by diffusion of Cd into the Mo$_{15}$S$_{19}$ matrix (Tarascon and Hull 1986). This phase is stable at room temperature in air. CdMo$_6$S$_8$ was obtained via intercalation of Cd^{2+} in aqueous electrolytes into Mo$_6$S$_8$ (Gocke et al. 1987).

4.46 Cadmium–Chlorine–Sulfur

CdS–CdCl$_2$: The phase diagram is a eutectic type (Figure 4.72) (Kulakov and Sokolovskaya 1971, Mellikov et al. 1980, Oleinik et al. 1986). The eutectic composition and temperature are 23 mol. % CdS [26.4 mol. % CdS (Mellikov et al. 1980)] and 525°C, respectively (Oleinik et al. 1986) [40.8 mol. % CdS and 522°C (Kulakov and Sokolovskaya 1971)] and 26.4 mol. % CdS and 500 ± 5°C (Bidnaya et al. 1962, Andreev and Loginova 1970). The solubility of CdS in CdCl$_2$ is not higher than 0.64 mol. % (0.5 mass. %), and the solubility of CdCl$_2$ in CdS is insignificant (Kulakov and Sokolovskaya 1971, Oleinik et al. 1986). According to the data of Bidnaya et al. (1962), Andreev and Loginova (1970), and Loginova and Andreev (1970), the solubility of CdS in CdCl$_2$ at 500°C is equal to 15.3 mol. % (12.5 mass. %), which is improbable.

At the minimum CdS total pressure, its solubility (mol. parts) in the CdCl$_2$ melt within the interval of 650°C–1100°C and can be described as follows (Mellikov et al. 1980):

$$x_{CdS} = 5.16 \exp\left[(-0.23 \pm 0.01)\ eV/kT\right].$$

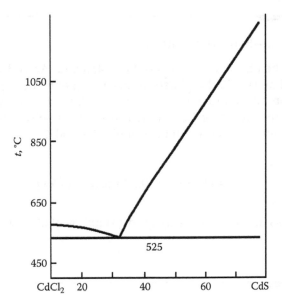

FIGURE 4.72
The CdS–CdCl₂ phase diagram. (From Oleinik, G.S. et al., *Izv. AN SSSR. Neorgan. materialy,* 22(1), 164, 1986.)

Analytical expression for solidus can be described in the following equation (Altosaar et al. 1983):

$$x_{Cl} \, (cm^{-3}) \; = \; 4 \cdot 10^{20} \exp \left(-0.32 \; eV/kT \right).$$

According to the data of Hiye et al. (1982), the solubility of chlorine in CdS crystals in the presence of CdCl₂ melt within the interval of 600°C–1100°C is expressed as follows:

$$x_{Cl} \, (cm^{-3}) \; = \; 3 \cdot 10^{21} p_{Cd}^{-1/8} \exp \left(-0.6 \; eV/kT \right),$$

and at temperatures below CdCl₂ condensation temperature,

$$x_{Cl} \, (cm^{-3}) \; = \; 2.1 \cdot 10^{19} \exp \left(0.05 \; eV/kT \right) p\left(CdCl_2 \right)^{1/2} p_{Cd}^{-1/8}.$$

This system was investigated using DTA and XRD (Bidnaya et al. 1962, Andreev and Loginova 1970, Kulakov and Sokolovskaya 1971, Mellikov et al. 1980, Oleinik et al. 1986).

4.47 Cadmium–Bromine–Sulfur

CdS–CdBr$_2$: The phase diagram is not constructed. According to the data of Mellikov et al. (1980), the phase diagram is a eutectic type. The eutectic composition and temperature are 2.24 mol. % (1.2 mass. %) CdS and 559°C. At the minimum of CdS total pressure, its solubility (mol. parts) in the CdBr$_2$ melt within the interval of 650°C–1100°C can be described as follows:

$$x_{CdS} = 11.2 \exp \left[(-0.30 \pm 0.02) \text{ eV} / kT \right].$$

This system was investigated using DTA (Mellikov et al. 1980).

4.48 Cadmium–Iodine–Sulfur

CdS–CdI$_2$: The phase diagram is a eutectic type (Figure 4.73) (Odin 2001). The eutectic composition and temperature are 22 ± 1 mol. % CdS and 357°C [2.5 mol. % CdS and 381°C (Mellikov et al. 1980)], respectively. The solubility of CdI$_2$ in CdS reaches 0.2 mol. % (Odin 2001). At the minimum of CdS total

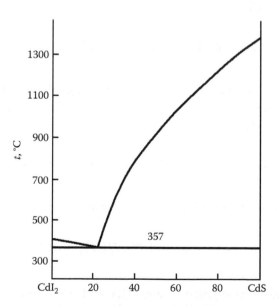

FIGURE 4.73
The CdS–CdI$_2$ phase diagram. (From Odin, I.N., *Zhurn. neorgan. khimii*, 46(10), 1733, 2001.)

pressure, its solubility (mol. parts) in the CdI_2 melt within the interval of 650°C–1100°C can be described as follows:

$$x_{Cds} = 9.6 \exp [(-0.31 \pm 0.03) \text{ eV/kT}].$$

This system was investigated using DTA, metallography, and XRD (Mellikov et al. 1980, Odin 2001). The ingots were annealed at 500°C during 1000 h (Odin 2001).

CdS–I₂: The phase diagram is not constructed. Equilibrium constant for the next reaction

$$CdS + I_2 = CdI_2 + \frac{1}{2}S_2$$

was calculated using temperature dependence of total vapor pressure (Gordeev and Karelin 1969). In the temperature range of 680°C–803°C, it can be described as follows:

$$\log K = \frac{-5071}{T} + 4.94.$$

4.49 Cadmium–Manganese–Sulfur

CdS–MnS: The phase equilibria in this system have been discussed in Pajaczkowska (1978) on the base of papers, which have appeared up to 1977. The phase diagram is a eutectic type (Figure 4.74) (Cooke 1968). The eutectic temperature is 1365 ± 10°C. The difference between the liquidus and solidus temperatures from the CdS-rich side is within the limits of 5°C. The solubility of MnS in CdS is equal to 49 mol. % at 800°C and 44 mol. % at 1385°C. The solubility of CdS in MnS increases from 11.5 mol. % at 800°C to 37 mol. % at 1390°C (Cooke 1968). Solid solutions $Cd_{1-x}Mn_xS$ ($x \leq 0.43$) crystallize in the wurtzite cell types (Rodic et al. 1996).

According to the data of Wiedemeier and Kahn (1968), the solubility of CdS in MnS at 600°C, 700°C, 800°C, and 1000°C is equal correspondingly to 6, 10, 13.5, and 17.5 mol. %, and the solubility of MnS in CdS at the same temperatures constitutes 44.3, 48.0, 50.0, and 53.0 mol. %, respectively.

Miller et al. (1966) investigated the phase relations in the CdS–MnS system in the temperature range of 100°C–600°C under pressures up to 4 GPa (40 kbar). The solubility of MnS in CdS weakly depends from temperature at atmospheric pressure and is equal to 45 mol. % at 100°C–400°C. The solubility of CdS in MnS increases from 4 mol. % at 100°C to 9 mol. % at 500°C.

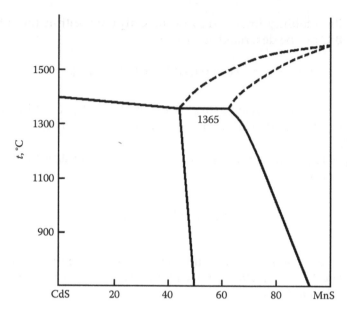

FIGURE 4.74
The CdS–MnS phase diagram. (From Cooke, W.R., *J. Amer. Ceram. Soc.*, 51(9), 518, 1968.)

The solubility of MnS in CdS at coprecipitation from water solutions is approximately equal to 0.1 mol. % (Paič 1971, Paič and Despotovič 1971).

Energy gap of $Cd_{1-x}Mn_xS$ solid solutions decreases with increasing of Mn concentration up to $x \approx 0.03$ and then increases (Ma and Giriat 1986). It changes linearly within the interval of $0.15 \leq x \leq 0.35$.

This system was investigated using DTA and XRD (Cooke 1968). The ingots were baked at 1140°C–1175°C for 18–24 h and then annealed at 800°C and 1000°C for 96 h (at 600°C and 700°C for 144 h) (Wiedemeier and Kahn 1968). Single crystals of $Cd_{1-x}Mn_xS$ solid solutions with $x \leq 0.35$ were grown using chemical transport reactions (Ma and Giriat 1986). The crystal growth of the $Cd_{1-x}Mn_xS$ solid solutions is also given in Pajaczkowska (1978).

4.50 Cadmium–Iron–Sulfur

CdS–FeS: The phase diagram is not constructed. There were obtained mixtures containing the phases with a strained wurtzite and NiAs-type structures at the annealing of CdS and FeS in different molar ratios under 2 GPa and 800°C (Susa et al. 1980).

Nanocrystals of $Cd_{1-x}Fe_xS$ ($x = 0-0.05$) have been synthesized by aqueous solution precipitation method at room temperature (Tripathi et al. 2008). The structure of these nanocrystals is found to be single-phase hexagonal.

The lattice parameters a and c are equal to 4.142 and 6.724, 4.136 and 6.713, and 4.100 and 6.693 A for CdS, $Cd_{0.99}Fe_{0.01}S$, and $Cd_{0.95}Fe_{0.05}S$, respectively.

4.51 Cadmium–Cobalt–Sulfur

CdS–CoS: The phase diagram is not constructed. The solubility of CoS in CdS at 1000°C reaches 20 mol. %, and the solubility of CdS in CoS is insignificant (Becker and Lutz 1978).

This system was investigated using XRD. The ingots were annealed at 600°C for 10 days and then at 500°C, 600°C, 700°C, 800°C, 900°C, and 1000°C for 40–6 days (Becker and Lutz 1978).

References

Aarna H.A., Kukk P.L., Koh P.R., Soorm T.V. The solubility of copper and silver in the cadmium sulfide [in Russian], *Tr. Tallin. polytehn. in-ta*, (404), 3–8 (1976).

Aarna H.A., Voogne M.P., Kukk P.L. The solubility of copper and silver in the cadmium sulfide single crystals [in Russian], *Legir. poluprovodn.*, Moscow: Nauka, 63–66 (1982).

Agayev A.B., Azadaliev R.A., Alieva U.R. Physico-chemical investigation of the CdS–Gd_2S_3 system [in Russian], *Azerb. khim. zhurn.*, (4), 105–107 (1984).

Agayev A.B., Rustamov P.G., Aliev O.M., Azadaliev R.A. Synthesis and properties of cadmium-lanthanoid chalcogenides [in Russian], *Izv. AN SSSR. Neorgan. materialy*, **25**(2), 294–298 (1989a).

Agayev A.B., Rustamov P.G., Aliev O.M., Azadaliev R.A. The CdS–Er_2S_3 system [in Russian], *Zhurn. neorgan. khimii*, **34**(4), 1065–1068 (1989b).

Aliev O.M., Agayev A.B., Azadaliev R.A. Synthesis and properties of $CdLn_2S_4(Se_4,Te_4)$ and $CdLn_4S_7$ [in Russian], *Neorgan. materialy*, **33**(11), 1327–1332 (1997).

Aliev O.M., Azadaliev R.A. The CdS–Ce_2S_3 system [in Russian], *Zhurn. neorgan. khimii*, **25**(11), 3160–3162 (1980).

Aliev O.M., Khasaev G.G., Kurbanov T.Kh. Synthesis and physico-chemical study of the $Me^{2+}Ln_4S_7(Se_7)$ type compounds, *Bull. Soc. Chim. France*, (1), 26–28 (1986).

Alieva Sh.B., Movsum-zade A.A., Allazov M.R. The CdS–Sn and SnS–Cd systems [in Russian], *Zhurn. neorgan. khimii*, **28**(7), 1788–1789 (1983).

Altosaar M.E., Kukk P.L., Hiye Ya.V. Solubility of chlorine in CdS and structure of CdS:Cl point defects [in Russian], *Zhurn. neorgan. khimii*, **28**(1), 69–75 (1983).

Andreev Yu.V., Loginova M.V. Investigation of $CdCl_2$–CdS phase diagram [in Russian], *Zhurn. neorgan. khimii*, **15**(12), 3377–3379 (1970).

Aresti A., Congiu A. Transport properties of $CdIn_2S_4$ single crystals, *Phys. Status Solidi (a)*, **16**(1), K55-K58 (1973).

Asadov M.M. Single crystals of $Tl_{2-x}Cd_xS$ and $Cd_{1-x}Tl_xS$ ($x < 0.01$) solid solutions [in Russian], *Zhurn. neorgan. khimii*, **29**(10), 2708–2709 (1984).

Asadov Yu.G., Baykulov R.B., Diffractometric study of structure transformations in $Cu_{1.70}Cd_{0.05}S$ crystals [in Russian], *Kristallographiya*, **50**(3), 434–439 (2005).

Atroshchenko L.V., Kukol V.V., Lakin E.E., Nosachev B.G., Sysoev L.A. Growth from the melt and investigation of the crystal structure, defect and phase transformation of pure, dopped and mixed $A^{II}B^{VI}$ single crystals [in Russian], *Rost i legirovanie poluprovodn. kristallov i plenok*, Pt.2, 298–301 (1977).

Atroshchenko L.V., Obuhova N.F., Pisarevski Y.V., Sil'vestrova I.M., Sysoev L.A. The Regularity of properties change for ZnS–MgS, CdS–MgS and CdSe–MgSe solid solution single crystals [in Russian], in: *Fiz. slozhn. poluprovodn. soedin.* Kishinev, Moldova: Shtiinza, 189–198 (1979).

Atroshchenko L.V., Obuhova N.F., Sysoev L.A. Solid solutions with wurtzite type structure in the systems based on $A^{II}B^{VI}$ [in Russian], *Izv. AN SSSR. Neorgan. materialy*, **22**(1), 162–163 (1986).

Axtell E.A. (III), Kanatzidis M.G. Synthesis and characterization of $Na_6Cd_7S_{10}$: a new framework sulphide with 1-D channels containing 12- and 16-member rings and a sulphide anion in an umbrella-like geometry, *Chem. Mater.*, **8**(7), 1350–1352 (1996).

Axtell E.A. (III), Liao J.-H., Pikramenou Z., Kanatzidis M.G. Dimensional reduction in II-VI Materials: $A_2Cd_3Q_4$ (A=K, Q=S, Se, Te; A=Rb, Q=S, Se), novel ternary low-dimensional cadmium chalcogenides produced by incorporation of A_2Q in CdQ, *Chem. Eur. J.*, **2**(6), 656–666 (1996).

Axtell E.A., Liao Ju-H., Pikramenou Z., Park Y., Kanatzidis M.G. $K_2Cd_2S_3$ vs CdS: Can the properties of quantum-sized CdQ semiconductors be emulated by bulk alkali metal ternary A/Cd/Q phases (Q=chalcogen)? *J. Am. Chem. Soc.*, **115**(25), 12191–12192 (1993).

Babanly M.B., Guseinov F.H., Kuliev A.A., Bashirov E.A. Liquidus surface of Tl–Cd–S system [in Russian], *Zhurn. neorgan. khimii*, **31**(10), 2634–2638 (1986).

Babanly M.B., Kulieva N.A., Guseynov F.H., Kurbanov A.A. Phase equilibria and glass formation in the Tl–Cd(Ge)–S and Hg–As–Se ternary system [in Russian], in: *Neorgan. soyedin.—sintez i svoystva*, Baku, 114–118 (1983).

Bakker M., Vollebregt F.H.A., Plug C.M. The transition from spinel-type to rocksalt-type structure in the compound $CdHo_2S_4$, as observed by electron microscopy/diffraction, *J. Solid State Chem.*, **42**(1), 11–21 (1982).

Barnier S., Guittard M., Julien C. Glass formation and structural studies of chalcogenide glasses in the $CdS–Ga_2S_3–GeS_2$ system, *Mater. Sci. Eng.*, **B7**(3), 209–214 (1990).

Becker W., Lutz H. D. Phase studies in the systems CoS–MnS, CoS–ZnS and CoS–CdS, *Mater Res. Bull.*, **13**(9), 907–911 (1978).

Ben-Dor L., Shilo I. Structure and magnetic properties of sulfides of the type $CdRe_2S_4$ and $Mg(Gd_xYb_{1-x})_2S_4$, *J. Solid State Chem.*, **35**(2), 278–285 (1980).

Bethke P.M., Barton P.B. Sub-solidus relations in the system PbS–CdS, *Amer. Mineralogist*, **56**(11–12), 2034–2039 (1971).

Beun J. A., Nitsche R., Lichtensteiger M. L. Photoconductivity in ternary sulfides, *Physica*, **26**(8), 647–649 (1960).

Beun J.A., Nitsche R., Lichtensteiger M. Optical and electrical properties of ternary chalcogenides, *Physica*, **27**(5) 448–452 (1961).

Bidnaya D.S., Obuhovski Ya.A., Sysoiev L.A. Searching of the methods for the growth of CdS crystals from solutions, *Zhurn. neorgan. khimii*, 7(12), 2671–2673 (1962).

Bodnar' I.V., Rud' V.Yu., Rud' Yu.V. Growth and properties of the $CdGa_2S_4$ compound single crystals [in Russian], *Neorgan. materialy*, **40**(2), 144–148 (2004).

Boruhovich A.S., Marunia M.S., Lobachevskaya N.I., Bamburov V.G., Gel'd P.V. Thermal capacity of CdCr$_2$S$_4$ ferromagnetic spinel [in Russian], *Fizika tverdogo tela*, **16**(7), 2084–2086 (1974).

Brec R., Ouvrard G., Louisy A., Rouxel J. Proprietés structurales de phases M^{2+}PX$_3$ (X = S, Se), *Ann. chim. (France)*, **5**(6), 499–512 (1980).

Brouty C., Spinat P., Herpin P. Mise en évidence et étude radiocristallographique de nouvelles phases dans le système pseudobinaire Bi$_2$S$_3$–CdS, *Rev. chim. miner.*, **12**(1), 60–68 (1975).

Brusilovets A.I., Fedoruk T.I. Investigation of interaction in the CdSe–P$_2$Se$_5$, CdS–P$_2$S$_5$ systems [in Russian], *Zhurn. neorgan. khimii*, **21**(6), 1648–1651 (1976).

Budionnaya L.D., Mitina L.A., Mizetskaya I.B. Investigation of cadmium sulfide interaction at crystallization from the melt in the CdS–CdTe semiconductor system [in Russian], in: *Problemy fiziki soiedineniy A^{II}B^{VI}*, Vil'nius, Vil'nius. un-t, vol. 1, 240–244 (1972).

Budionnaya L.D., Mizetskaya I.B., Nuzhnaya T.P., Sharkina E.V. Formation of solid solutions at the interaction of cadmium telluride and sulfide [in Russian], *Kvantovaya elektronika*, (15), 97–106 (1978).

Budionnaya L.D., Mizetskaya I.B., Sharkina E.V. Interaction of CdS and CdSe at the caking [in Russian], *Poluprovodn. tehnika i mikroelektron.*, (23), 81–84 (1976).

Buehler E., Bachmann K. J. Solubilities of InP and CdS in Cd, Sn, In, Bi and Pb, *J. Cryst. Growth*, **35**(1), 60–64 (1976).

Bundel' A.A., Vishniakov A.V., Zubkovskaya V.N. The solubility of Cu$_2$S in ZnS and CdS [in Russian], *Izv. AN SSSR. Neorgan. materialy*, **6**(7), 1248–1251 (1970).

Calawa A. R., Mroczkowski J. A., Harman T. C. Preparation and properties of Pb$_{1-x}$Cd$_x$S, *J. Electron. Mater.*, **1**(1), 191–201 (1972).

Charbonnier M. Contribution à l'étude des sulfures simples et mixtes des métaux des groupes I B (Cu–Ag) et II B (Zn–Cd–Hg): These doct. *sci. phys. Univ. Claude-Bernard.—Lyon*, 164 p. (1973).

Chaus I.S., Koval' L.B., Sheka I.A. Interaction of indium, gallium and cadmium sulfides at coprecipitation from hydrochloric solutions [in Russian], *Ukr. khim. zhurn.*, **41**(9), 914–917 (1975).

Chaus I.S., Sheka I.A. Interaction of gallium and cadmium sulfides in hydrochloric solutions [in Russian], *Izv. AN SSSR. Neorgan. materialy*, **3**(2), 324–328 (1967).

Chen W. A novel metal chalcogenide—HgCdS$_4$, *Zhurn. neorgan. khimii*, **56**(3), 383–386 (2011).

Choe W., Lee S., O'Konnel P., Corey A. Synthesis and structure of new Cd–Bi–S homologous series: A study in intergrowth and the control of twinning patterns, *Chem. Mater.*, **9**(9), 2025–2030 (1997).

Cini L., Melandri L. Formation of the solid solution CdS/CdSe from CdS and Se and its thermal stability: A thermogravimetric investigation, *J. Therm. Anal.*, **3**(2), 131–136 (1971).

Cooke W.R. The CdS–MnS and CdSe–MnSe phase diagrams, *J. Amer. Ceram. Soc.*, **51**(9), 518–519 (1968).

Covino J., Dragovich P., Lowe-Ma C.K., Kubin R.F., Schwartz R.W. Synthesis and characterization of stoichiometric CdPS$_3$, *Mater. Res. Bull.*, **20**(9), 1099–1107 (1985).

Cruceanu E., Niculescu D. Solutions solides dans les systèmes CdS–CdTe et HgSe–CdSe, *C. r. Acad. sci.*, **261**(4), 935–938 (1965).

Czaja W. Ordnung im Thiospinel CdIn$_2$S$_4$, *Helv. physica acta*, **40**(4), 352 (1967).

Czaja W., Krausbauer L. Photoluminescence of $CdIn_2S_4$ and mixed crystals with In_2S_4 as related to their structural properties, *Phys. status solidi*, **33**(1), 191–199 (1969).

Derid Yu.O., Georgobiani A.N., Gruzintzev A.N., Radautsan S.I., Tiginyanu I.M. On the influence of nonstoichiometry on luminescent properties of $CdGa_2S_4$ single crystals, *Gryst. Res. Technol.*, **20**(6), 857–861 (1985).

Deshmukh L.P., More B.M., Holikatti S.G., Hankare P.P. Preparation and properties of $(CdS)_x(PbS)_{1-x}$ thin-film composites, *Bull. Mater. Sci.*, **17**(5), 455–463 (1994).

Deyneko I.P., Egorov-Tismenko Yu.K., Spitsyna V.D., Simonov M.A., Belov N.V. Crystal structure of the $Cd_2Sb_6S_{11}$ [in Russian], *Dokl. AN SSSR*, **254**(4), 877–881 (1980).

Dmitrenko A.O., Abramova L.V., Bukesov S.A. Solid solutions in the ZnS–CdS–MgS system [in Russian], *Izv. AN SSSR. Neorgan. materialy*, **26**(12), 2483–2487 (1990).

Donika F.G., Radautsan S.I., Tezlevan V.E. X-ray investigation of the ingots in the $(CdS)_{3x}-(In_2S_3)_{1-x}$ system [in Russian], in: *Materialy dokl. na 3 nauchno-tehnich. konf. Kishin. politeh. in-ta*. Koshinev: Kishin. politehn. in-t, 119–120 (1967).

Donika F.G., Zhitar' V.F., Radautsan S.I. Semiconductors of the $ZnS-In_2S_3$ system [in Russian], *Kishinev, Shtiintsa Publish.*, 148 p. (1980).

Donika F.G., Zhitar' V.F., Radautsan S.I., Semiletov S.A., Donika T.V. Investigation of the structure and some physical properties of the alloys in the $(CdS)_{3x}-(Ga_2S_3)_{1-x}$ system [in Russian], *Tr. Kishin. politehn. in-ta*, (112), 3–12 (1968).

Dreeben A. Precipitation of impurities in large single crystals of CdS, *J. Electrochem. Soc.*, **111**(2), 174–179 (1964).

Endo S. Transport properties of $CdIn_2S_4$ single crystals, *J. Phys. Chem. Solids*, **37**(2), 201–209 (1976).

Errandonea D., Kumar R.S., Manjón F.J., Ursaki V.V., Tiginyanu I.M. High-pressure X-ray diffraction study on the structure and phase transitions of the defect-stannite $ZnGa_2Se_4$ and defect chalcopyrite $CdGa_2S_4$, *J. Appl. Phys.*, **104**(6), 063524_1–063524_9 (2008).

Flahaut J., Domange L., Patrie M. Vue d'ensemble sur les systèmes formés par le sulfure d'yttrium avec les sulfures des éléments divalents, *Bull. soc. chim. France*, **35**(1), 159–163 (1962).

Fujii H., Okamoto T., Kamigaichi T. Crystallographic and magnetic studies of rare earth chalcogenide spinels, *J. Phys. Soc. Jap.*, **32**(5), 1432 (1972).

Fukatsu N., Saito A., Shimizu N., Ohashi T. Phase equilibria in the system Cd–S–O and the stability region of a new phase $2CdS·2CdO\ CdSO_4$, *J. Electrochem. Soc.*, **135**(4), 997–1003 (1988).

Gaisinskii V.B., Dmitriev Yu.N., Sysoev L.A. Formation mechanism of solid solutions $Cd_{1-x}Mg_xS$, *Sov. Phys. Crystallogr.*, **26**(1), 114 (1981), translated from *Krystallografiya*, **26**(1), 206 (1981).

Gaisinskii V.B., Sysoev L.A., Fainer M.S. Piezoelectric properties of CdS–MgS single crystals [in Russian], in: *Fiz. i kim. kristallov*, Khar'kov, 162–164 (1977).

Galiulin E.A., Odin I.N., Novoselova A.V. Physico-chemical investigation of the CdS–SnS and CdSe–SnSe systems [in Russian], *Zhurn. neorgan. khimii*, **26**(7), 1881–1883 (1981).

Gastaldi L., Simeone L.G., Viticoli S. Cation ordering and crystal structures in AGa_2X_4 compounds ($CoGa_2S_4$, $CdGa_2S_4$, $CdGa_2Se_4$, $HgGa_2Se_4$, $HgGa_2Te_4$), *Solid State Commun.*, **55**(7), 605–607 (1985).

Georgobiani A.N., Gruzintsev A.N., Ilyuhina Z.P., Myzina V.A., Radautsan S.I., Tiginianu I.M. Phase transitions in the $CdIn_2S_4$ and $CdGa_2S_4$ single crystals at the annealing in Cd melt [in Russian], *Izv. AN SSSR. Neorgan. materialy*, **21**(9), 1457–1461 (1985).

Gleize A., Cabane-Brouty F. Solubilité et diffusion de l'argent dans le sulfure de cadmium, *J. Solid State Chem.*, **14**(1), 1–7 (1975).

Gocke E., Schramm W., Dolscheid P., Schöllhorn R. Molybdenum cluster chalcogenides Mo_6X_8: Electrochemical intercalation of closed shell ions Zn^{2+}, Cd^{2+}, and Na^+, *J. Solid State Chem.*, **70**(1), 71–81 (1987).

Golovey M.I., Olekseyuk I.D., Voroshilov Yu.V. Chemical bond and solubility in the systems $A^{II}_3B^V_2$–$A^{II}C^{VI}$ [in Russian], in: *Khim. svyaz' v poluprovodn. i polumetallah.* Minsk, Nauka i Tekhnika Publish., 233–239 (1972).

Golovey M.I., Olekseyuk I.D., Voroshilov Yu.V. Obtaining and properties of CdP_2S_4 [in Russian], *Izv. AN SSSR. Neorgan. materialy*, **9**(8), 1346–1348 (1973).

Golovey M.I., Rigan M.Yu. Preparation of the Cd_3As_2–$2CdTe(Se,S)$ solid solution single crystals [in Russian], in: *Khimiya i physica khalcogenidov.* K.: Naukova dumka, 38–40 (1977).

Golovey M.I., Rigan M.Yu., Voroshilov Yu.V. The $(Cd_3As_2)_{1-x}$–$(2CdS)_x$ system [in Russian], *Izv. AN SSSR. Neorgan. materialy*, **10** (11), 1942— 1945 (1974).

Gordeev I.V., Karelin V.V. Equilibrium in the zinc sulfide–iodine system [in Russian], *Izv. AN SSSR. Neorgan. materialy*, **5**(7), 1190–1193 (1969).

Goriayev V.M., Pechkovski V.V., Pinaev G.F. Interaction of gaseous SeO_2 with CdS [in Russian], *Izv. AN SSSR. Neorgan. materialy*, **9**(7), 1126–1131 (1973).

Goriunova N.A. Complex diamond-like semiconductors [in Russian], *M.: Sov. Radio Publish.*, 268 p. (1968).

Grieshaber E., Nitsche R., Bubenzer A. New compounds with icosahedral structures of the type $Cd_{14-2x}P_4S_{24-x}I_{2x}$, *Mater. Res. Bull.*, **11**(9), 1169–1177 (1976).

Guseinov G.D. Searching and physical investigation of new semiconductors-analogues [in Russian], *Avtoref. dis. … doct. fiz.-mat. nauk.—Baku*, 82 p. (1969).

Guseinov F.H., Babanly M.B., Kuliev A.A. Phase equilibria in the TlX–Cd(Zn)X (X—S, Se, Te) systems [in Russian], *Izv. AN SSSR. Neorgan. materialy*, **18**(5), 759–763 (1982).

Guseinov G.D., Guseinov G.G., Ismailov M.Z., Godzhaev E.M. About structure and physical properties of new $CdTlS_2$ (Se_2, Te_2) semiconductor compounds [in Russian], *Izv. AN SSSR. Neorgan. materialy*, **5**(1), 33–39 (1969).

Guseinov G.G., Kerimova T.G., Nani R.U. Refinement of the $CdGa_2S_4$. crystal structure [in Russian], *Izv. AN AzSSR. Ser. fiz.-tehm. i mat. n.*, (4), 59–61 (1980).

Guseinov F.H., Kuliev A.A., Babanly M.B., Kuliev R.A. Investigation of the Tl_4S_3–Zn(Cd)S systems [in Russian], *Azerb. khim. zhurn.*, (4), 108–110 (1985).

Guseinov D.T., Mamedov Z.G. About phase transition in $CdIn_2S_4$ [in Russian], *Izv. AN SSSR. Neorgan. materialy*, **30**(12), 1597 (1994).

Hahn H., Frank G., Klingler W., Störger A.D., Störger G. Über ternäre Chalkogenide des Aluminiums, Galliums und Indiums mit Zink, Cadmium und Quecksilber, *Z. anorg. und allg. Chem.*, **279**(5/6), 241–270 (1955).

Hahn H., Klingler W. Über die Kristallstruktur einiger ternärer Sulfide, die sich vom Indium (III) Sulfid ableiten, *Z. anorg. und allg. Chem.*, **263**(4), 177–190 (1950).

Hansevarov R.Yu., Ryvkin S.M., Ageeva I.N. About composition dependence of energy gap of the CdS–CdSe solid solutions [in Russian], *Zhurn. tehn. fiziki*, **28**(3), 480–483 (1958).

Hiye Ya.V., Altosaar M.E., Mellikov E.Ya., Kerm K.V. Physico-chemical foundations of CdS doping by the cadmium chloride [in Russian], in: *Legir. poluprovodn.*, M.: Nauka, 59–62 (1982).

Hozhainov Yu.,M., Potapov I.I., Hozhainova T.I., Razvazhnoi E.M. Investigation of CdS–P_2S_5 system [in Russian], *Deposited in VINITI*, № 1243–76Dep. (1976).

Iglesias I.E., Pachali K.E., Steinfink H. Structural chemistry of Ba_2CdS_3, Ba_2CdSe_3, $BaCdS_2$, $BaCu_2S_2$ and $BaCu_2Se_2$, *J. Solid State Chem.*, **9**(1), 6–14 (1974).

Jones E.D., Mykura H. Diffusion of gallium into cadmium sulphide, *J. Phys. Chem. Solids*, **41**(11), 1261–1265 (1980).

Julien-Pouzol M., Jaulmes S. Redetermination de la structure du sulfure de cadmium et de germanium Cd_4GeS_6, *Acta Crystallogr. C*, **51**(10), 1966–1968 (1995).

Kaldis E., Krausbauer L., Widmer R. Cd_4SiS_6 and Cd_4SiSe_6, new ternary compounds, *J. Electrochem. Soc.*, **114** (10), 1074–1076 (1967).

Kaldis E., Widmer R. Nucleation and growth of single crystals by chemical transport. I. Cadmium–germanium sulphide, *J. Phys. Chem. Sol.*, **26**(12), 1697–1700 (1965).

Karasev N.M., Korenev Yu.M., Timoshin I.A., Sidorov L.N., Novoselova A.V. Mass-spectrum thermodynamic investigation of binary systems, forming by the chalcogenides of second group. II. The CdS–CdSe system, [in Russian], *Zhurn. fiz. khimii*, **46**(7), 1718–1721 (1972).

Kas'k R., Koppel H. About In_2S_3–CdS system [in Russian], *Izv. AN ESSR. Ser. Khimia, Geologia*, **22**(1), 42–45 (1973).

Kas'k R., Koppel H. About some chemical equilibria in the Cd–In–S ternary system [in Russian], *Izv. AN ESSR. Ser. Khimia, Geologia*, **23**(3), 214–217 (1974).

Kas'k R., Koppel H. Thermographic investigations of InS with $CdIn_2S_4$, CdS and Cd alloys [in Russian], *Izv. AN ESSR. Ser. Khimia, Geologia*, **24**(3), 210–214 (1975).

Kas'k R., Koppel H. About Cd-In-S ternary system [in Russian], *Izv. AN ESSR. Ser. Khimia, Geologia*, **25**(1), 28–31 (1976).

Kimura S. Phase equilibria in the systems Zn+S+Sn, Zn+Se+Sn and Cd+S+Sn, *J. Chem. Thermodyn.*, **3**(1), 7–17 (1971).

Kimura S., Panish M. B. Phase equilibria in the systems Zn+S+Bi, Zn+Se+Bi and Cd+S+Bi, *J. Chem. Thermodyn.*, **2**(1), 77–86 (1970).

Kislinskaya G.E. Investigation of coprecipitation of cadmium, indium and gallium with sulfides of some metals [in Russian], *Avtoref. dis. ... kand. khim. nauk, Kiev*, 27 p. (1974).

Klinger W., Eulenberger G., Hahn H. Über Hexachalkogeno-hypodiphosphate vom Typ $M_2P_2X_6$, *Naturwissenschaften*, **57**(2), 88 (1970).

Klinger W., Ott R., Hahn H. Über die Darstellung und Eigenschaften von Hexathio- und Hexaselenohypodiphosphaten, *Z. anorg. und allg. Chem.*, **396**(3), 271–278 (1973).

Kobayashi T., Susa K., Taniguchi S. High-pressure phase equilibrium study of the solid solution systems $Cd_{1-x}M_xS$ (M = Mg, Ca, Sr) with the rock salt structure, *High Temp.—High Pressures*, **8**(6), 652 (1976).

Kobayashi T., Susa K., Tanigushi S. Preparation and semiconductive properties of rock salt type solid solution systems $Cd_{1-x}M_xS$ (M = Sr, Ca, Mg, Pb, Sn), *J. Phys. Chem. Solids*, **40**(10), 781–785 (1979).

Kobayashi T., Susa K., Tanigushi S. P-T-x phase equilibrium study of new solid solution systems $Cd_{1-x}M_xS$ (M = Mg, Ca, Sr), *J. Solid State Chem.*, **33**(2), 203–207 (1980).

Koelmans H., Grimmeiss H. G. The photoconductivity of $CdIn_2S_4$, activated with Cu or Au, *Physica*, **25**(12), 1287–1288 (1959).

Kovach A.P., Potoriy M.V., Voroshilov Yu.V., Tovt V.V. Physico-chemical investigation of the $Cd(Zn)S(Se)–P_2S_4(Se_4)$ systems [in Russian], *Ukr. khim. zhurn.*, **63**(2), 92–96 (1997).

Kozer V.R., Fedorchuk A., Olekseyuk I.D., Parasyuk O.V. Phase equilibria in the quasi-ternary system $Ag_2S–In_2S_3–CdS$ at 870 K, *J. Alloys Compd.*, **480**(2), 360–364 (2009).

Krebs von B., Mandt J. Struktur und Eigenschaften von Cd_4SiS_6. Zur Kenntnis von Cd_4SiSe_6, *Z. anorg. und allg. Chem.*, **388**(3), 193–206 (1972).

Kremheller A., Levine A. K., Gashurov G. Hydrothermal preparation of two-component solid solutions from II–VI compounds, *J. Electrochem. Soc.*, **107**(1), 12–15 (1960).

Kshirsagar S.T., Thompson H.B., Edwards J.G. Vaporization chemistry and thermo-dynamics of solid solutions in the cadmium-indium-sulfide, indium-sulfide system, *J. Electrochem. Soc.*, **129**(8), 1835–1840 (1982).

Kulakov M.P. The melting temperature and vapor pressure of HgS [in Russian], *Izv. AN SSSR. Neorgan. materialy*, **11**(3), 553–554 (1975).

Kulakov M.P. Investigation of mutual solubility in the HgS–CdS system [in Russian], *Zhurn. neorgan. khimii*, **21**(2), 513–517 (1976).

Kulakov M.P., Sokolovskaya Zh.D., Thermal analysis of the $CdS–CdCl_2$ system [in Russian], *Izv. AN SSSR. Neorgan. materialy*, **7**(8), 1444–1446 (1971).

Kulkarni V.G., Garn P.D. Study of the formation of cadmium sulfoselenide, *Thermochim. Acta*, **99**, 33–36 (1986).

Lam S.T., Munir Z.A. Thermodynamics and kinetics of the evaporation of single crys-talline CdS–CdSe solid solutions, *High Temp. Sci.*, **12**(4), 249–258 (1980).

Larionkina L.S., Nani R.H. About some physical properties of $CdIn_2S_4$ [in Russian], in: *Troinyie poluprovodn. i ih primenienie*, Tez. dokl. Kishinev: Shtiinza, 154–156 (1976).

Loginova M.V., Andreev Yu.V. Melting diagrams of the $CdCl_2–CdX$ ($X=O$, S, Se, SO_4, SeO_3) systems [in Russian], *Izv. AN SSSR. Neorgan. materialy*, **6**(10), 1885–1886 (1970).

Lopatin G.S., Malkova A.S., Shumilin V.P. Investigation of vapor pressure in the CdS–CdSe system [in Russian], *Izv. AN SSSR. Neorgan. materialy*, **11**(7), 1322–1323 (1975).

Ma K.-J., Giriat W. Dependence of energy gap on temperature and composition in $Mn_xCd_{1-x}S$, *Phys. Status Solidi (a)*, **95**(2), K135-K138 (1986).

Mamedov K.K., Aliev M.M., Kerimov I.G., Nani R.Kh. Heat capacity of $A^{II}B^{III}_2C^{VI}_4$-type ternary semiconducting compounds at low temperatures, *Phys. status solidi (a)*, **9**(2), K149-K152 (1972).

Masaharu A., Masaoko W., Hiroshi N. Growth of Zn and Cd chalkogenides from solutions using Te as a solvent, *J. Fac. Eng. Univ. Tokyo*, **A**(18), 38–39 (1980).

Masako W., Ken-ichi S., Masaharu A. Solution growth of ZnS, ZnSe, CdS and their mixed compounds using tellurium as a solvent, *Jap. J. Appl. Phys.*, **18**(5), 869–872 (1979).

Maskayeva L.N., Markov V.F., Gusev A.I. Decomposition temperature interval and degradation of supersaturated substitutional solid solutions $Cd_xPb_{1-x}S$ [in Russian], *Dokl. RAN*, **390**(5), 639–643 (2003).

Meenakshi S., Vijayakumar V., Eifler A., Hochheimer H.D. Pressure-induced phase transition in defect chalcopyrites $HgAl_2Se_4$ and $CdAl_2S_4$, *J. Phys. Chem. Solids*, **71**(5), 832–835 (2010).

Mellikov E.Ya., Hiye Ya.V., Krunks M.I. The melted salts as the agents for cadmium sulfide recrystallization [in Russian], *Probl. sol'vatatsii i kompleksoobraz.*, Ivanovo, 158–162 (1980).

Mihalev A.A., Kalashnikova L.D., Levantsova N.S., Glagoleva A.A., Boiev E.I. The temperature dependence of copper sulfide solubility in the zinc and cadmium sulfides [in Russian], *Sb. nauch. tr./VNII liuminoforov i osobo chistyh veshchestv*, (3), 85–89 (1970).

Miller R.O., Dachille F., Ray R. High-pressure phase equilibrium studies of CdS and MnS by static and dynamic methods, *J. Appl. Phys.*, **37**(13), 4913–4918 (1966).

Mizetskaya I.B., Oleinik G.S., Trishchuk L.I. The Cu_2S–CdS phase diagram [in Russian], *Izv. AN SSSR. Neorgan. materialy*, **18**(4), 684–685 (1982).

Moldovyan N.A. Growth and investigation of photoconductivity of $ZnAl_2S_4$ and $CdAl_2S_4$ single crystals [in Russian], *Izv. AN SSSR. Neorgan. materialy*, **27**(9), 1969–1971 (1991).

Morozova N.K., Zimogorski V.S., Morozov A.V. About solubility of oxygen in CdS [in Russian], *Neorgan. materialy*, **29**(7), 1014–1016 (1993).

Motria S.F., Svitlinets V.P., Semrad E.E., Dovgoshey N.I. Thermogravimetric and tensimetric investigation of Cd_4GeS_6 и Cd_4GeSe_6 [in Russian], *Izv. AN SSSR. Neorgan. materialy*, **23**(9), 1543–1546 (1987).

Motria S.F., Tkachenko V.I., Chereshnia V.M., Kikineshi A.A., Semrad E.E. Physico-chemical and photoelectric properties of the crystals of cadmium hexathio-, hexaselenogermanate and solid solutions based on it [in Russian], *Izv. AN SSSR. Neorgan. materialy*, **22**(10), 1705–1708 (1986).

Movsum-zade A.A., Aliev Sh.B., Allazov M.R. The Cd–Ge–S phase diagram [in Russian], *Zhurn. neorgan. khimii*, **32**(4), 1025–1029 (1987).

Nebauer E. Vapour-pressure investigations of impurity diffusion and solubility in A^{II}–B^{VI} compounds demonstrated for the system $CdS:Au:S_2$, *Phys. status solidi*, **29**(1), 269–281 (1968).

Nitsche R., Wild P. Crystal growth of metal-phosphorus-sulfur compounds by vapor transport, *Mater. Res. Bull.*, **5**(6), 419–424 (1970).

Novoselova A.V., Sher A.A., Odin I.N. Investigation of interaction in the CdB^{VL}–$A_2^VB_3^{VI}$ (B^{VI}—S, Se; A^V—Sb, Bi) systems [in Russian], *Zhurn. neorgan. khimii*, **23**(12), 3321–3325 (1978).

Nunoue S.-Ya., Hemmi T., Kato E. Mass spectrometric study of the phase boundaries of the CdS–CdTe system, *J. Electrochem. Soc.*, **137** (4), 1248–1251 (1990).

Odin I.N. Physico-chemical analysis of ternary and ternary mutual systems, containing cadmium, zinc, silicon, bismuth chalcogenides and properties of ingots in these systems [in Russian], *Zhurn. neorgan. khimii*, **41**(6), 941–953 (1996).

Odin I.N. *T-x-y* diagrams for reciprocal systems $PbX + CdI_2 = CdX + PbI_2$ (X = S, Se, Te) [in Russian], *Zhurn. neorgan. khimii*, **46**(10), 1733–1738 (2001).

Odin I.N., Chukichev M.V. Physico-chemical analysis of the Cd–Sb(Bi)–S systems and properties of photosensitivity solid solutions based on cadmium sulfide compounds [in Russian], *Zhurn. neorgan. khimii*, **45**(2), 255–260 (2000).

Odin I.N., Chukichev M.V., Galiulin E.A. Preparation and luminescence properties of the CdX<Ge>, CdX<Sn> (X = S, Se) single crystals [in Russian], *Neorgan. materialy*, **35**(11), 1304–1306 (1999a).

Odin I.N., Chukichev M.V., Grin'ko V.V. Luminescence properties of the cadmium sulfide and selenide crystals doped by boron [in Russian], *Neorgan. materialy*, **35**(11), 1302–1303 (1999b).

Odin I.N., Galiulin E.A. Novoselova A.V. Investigation of cadmium sulfide with germanium sulfides interaction [in Russian], *Zhurn. neorgan. khimii*, **28**(9), 2362–2365 (1983).

Odin I.N., Grin'ko V.V. Tensimeter determination of the limits of Cd_4GeS_6 homogeneity region [in Russian], *Zhurn. neorgan. khimii*, **36**(5), 1332–1338 (1991a).

Odin I.N., Grin'ko V.V. Pressure and vapor composition over the ingots of the CdS–GeS_2 system [in Russian], *Zhurn. neorgan. khimii*, **36**(4), 1056–1061 (1991b).

Odin I.N., Grin'ko V.V., Petrovski A.Yu. The CdS–B_2S_3 and CdSe–B_2Se_3 phase diagrams and properties of compounds in these systems [in Russian], *Zhurn. neorgan. khimii*, **44**(12), 2061–2062 (1999c).

Odin I.N., Grin'ko V.V., Safronov E.V., Kozlovskiy V.F. Investigation of the CdX–B_2X_3–X (X = S, Se), CdTe–B–Te systems [in Russian], *Zhurn. neorgan. khimii*, **46**(7), 1210–1214 (2001).

Odin I.N., Ivanov V.A., Novoselova A.V. The CdS–SiS_2, CdSe–$SiSe_2$ systems [in Russian], *Zhurn. neorgan. khimii*, **30**(1), 207–211 (1985).

Odin I.N., Rubina M.E., Demidova E.D. Phase relations in the $Cd_3S_3 + In_2Te_3 = Cd_3Te_3 + In_2S_3$ system [in Russian], *Zhurn. neorgan. khimii*, **49**(5), 848–851 (2004).

Odin I.N., Rubina M.E., Gapanovich M.V., Demidova E.D. *T-x-y* phase diagrams of the GaS + CdTe = CdS + GaTe, CdS–CdTe–InTe systems [in Russian], *Zhurn. neorgan. khimii*, **50**(4), 714–716 (2005).

Ohata K., Saraie S., Tanaka T. Phase diagram of the CdS–CdTe pseudobinary system, *Jap. J. Appl. Phys.*, **12**(8), 1198–1204 (1973).

Oleinik G.S., Mizetski P.A., Nizkova A.I., Polivtsev L.A., Riadnina I.A. The PbS–CdS phase diagram [in Russian], *Izv. AN SSSR. Neorgan. materialy*, **19**(11), 1799–1801 (1983).

Oleinik G.S., Mizetski P.A., Nuzhnaya T.P. Liquidus surface of $CdCl_2$–CdS–CdTe ternary system [in Russian], *Izv. AN SSSR. Neorgan. materialy*, **22**(1), 164–165 (1986).

Olekseyuk I.D., Bogdanova A.V., Tovtin N.A., Sopko F.V., Chepur D.V. Investigation of interaction and glass region in the Cd–As–S ternary system [in Russian], *Zhurn. neorgan. khimii*, **21** (12), 3382—3387 (1976).

Olekseyuk I.D., Parasyuk O.V., Halka V.O., Piskach L.V., Pankevych V.Z., Romanyuk Ya.E. Phase equilibria in the quasi-ternary system Ag_2S–CdS–Ga_2S_3, *J. Alloys Compd.*, **325**(1–2), 167–179 (2001).

Olekseyuk I.D., Tsitrovskiy V.V., Turyanitsa I.D., Stoyka I.M., Chukhno T.A. Obtaining and properties of modulation and nonlinear materials based on some chalcogenides [in Russian], *Kvantovaya elektronika*, (13), 93–96 (1977).

Ouvrard G., Brec R., Rouxel J. Structural determination of some MPS_3 layered phases (M = Mn, Fe, Co, Ni and Cd), *Mater. Res. Bull.*, **20**(10), 1181–1189 (1985).

Paič M. Formation of cadmium sulphide–manganese (2) sulphide solid solutions by coprecipitation from aqueous solutions of corresponding sulphate by ammonium sulphide, *Croat. chem. acta*, **43**(3), 169–174 (1971).

Paič M., Despotovič Z. Thermogravimetric analysis of cadmium sulphide–manganese sulphide systems obtained by coprecipitation, *Croat. chem. acta*, **43**(3), 175–178 (1971).

Pajaczkowska A. Physicochemical properties and crystal growth of $A^{II}B^{VI}$–MnB^{VI} systems, *Progr. Crystal Growth Caract.*, **1**(3), 289–326 (1978).

Parasyuk O.V., Piskach L.V., Romanyuk Y.E., Olekseyuk I.D., Zaremba V.I., Pekhnyo V.I. Phase relations in the quasi-binary Cu_2GeS_3–ZnS and quasi-ternary Cu_2S–Zn(Cd)S–GeS_2 systems and crystal structure of Cu_2ZnGeS_4, *J. Alloys Compd.*, **397**(1–2), 85–94 (2005).

Pardo M.-P., Flahaut J. Étude des phases metastables. I. Cas du système Ga_2S_3–CdS, *Mater. Res. Bull.*, **20**(4), 399–405 (1985).

Pardo M.-P., Flahaut J., Gastaldi L. Système Ga_2S_3–CdS. Étude cristallographique— diagramme de phase, *Mater. Res. Bull.*, **19**(6), 735–743 (1984).

Pinch H.L., Berger S.B. The effects of nonstoichiometry on the magnetic properties of cadmium chromium chalcogenide spinels, *J. Phys. Chem. Sol.*, **29**(12), 2091–2099 (1968).

Piskach L.V., Parasyuk O.V., Olekseyuk I.D. The phase equilibria in the quasi-ternary Cu_2S–CdS–SnS_2 system, *J. Alloys Compd.*, **279**(2), 142–152 (1998).

Pleshchev V.G., Konev V.N., Gerasimov A.F. Crystallochemical investigation of $CdCr_2S_4$ and $CdCr_2Se_4$ and solid solutions on their base with different chalcogen contents [in Russian], *Izv. AN SSSR. Neorgan. materialy*, **14**(2), 223–227 (1978).

Pokrzywnicki S. Analysis of the magnetic susceptibility of $CdYb_2S_4$ spinel by means of the crystal field method, *Phys. Status Solidi B*, **71**(1), K111-K115 (1975).

Pokrzywnicki S., Czopnik A. Magnetic susceptibility of $CdTm_2S_4$ spinel, *Phys. status solidi (b)*, **70**(2), K85-K87 (1975).

Polyviannyi I.R., Lata V.A., Ivakina L.P., Antoniuk V.I. The Na_2S–CdS system [in Russian], *Zhurn. neorgan. khimii*, **26**(4), 1038–1042 (1981).

Polyvyanny I.R., Lata V.A. Thermal analysis of certain sulphide-sodium systems, *Thermochim. Acta*, **92**, 747–749 (1985).

Potoriy M.V., Voroshilov Yu.V., Tovt Yu.V. Physico–chemical investigation of the Cd–P–S(Se) systems [in Russian], *Neorgan. materialy*, **35**(11), 1297–1301 (1999).

Radautsan S.I., Derid Yu.O., Zhitar' V.F., Derid O.P., Trotsenko N.K., Tyuliupa A.G. The CdS–Ga_2S_3 phase diagram [in Russian], *Dokl. AN SSSR*, **267**(3), 673–675 (1982).

Radautsan S.I., Zhitar' V.F., Tezlevan V.E., Donika F.G. Investigation of structure and some properties of alloys based on In_2S_3 and zinc and cadmium sulfides [in Russian], in: *Khimicheskaya sviaz' v cristallah*, Minsk: Nauka i Tehnika Publish., 423–427 (1969).

Range K.-J., Becker W., Weiss A. Über Hochdruckphasen des $CdAl_2S_4$, $HgAl_2S_4$, $ZnAl_2Se_4$, $CdAl_2Se_4$ und $HgAl_2Se_4$ mit Spinellstruktur, *Z. Naturforsch.*, **23B**(7), 1009 (1968).

Rodic D., Spasojevic V., Bajorek A., Onnerud P. Similarity of structure properties of $Hg_{1-x}Mn_xS$ and $Cd_{1-x}Mn_xS$ (structure properties of HgMnS and CdMnS), *J. Magn. Magn. Mater.*, **152**(1–2), 159–164 (1996).

Rubenstein M. Solubilities of some II–VI compounds in bismuth, *J. Electrochem. Soc.*, **113**(6), 623–624 (1966).

Rubenstein M. Solution growth of some II–VI compounds using tin as a solvent, *J. Cryst. Growth*, **3–4**, 309–312 (1968).

Rudniev N.A., Dzhumayev R.M. Coprecipitation in the Hg^{2+}, Cd^{2+}, H^+ | | S^{2-}system [in Russian], *Zhurn. neorgan. khimii*, **11**(5), 1084–1090 (1966).

Ruut A., Koppel H. The CdS–Cd_3As_2 and CdS–$CdAs_2$ systems [in Russian], *Izv. AN ESSR. Ser. Khimia, Geologia*, **22**(2), 137–142 (1973).

Samal' G.I., Sadovnikova N.N. About solid phase interaction of cadmium and chromium sulfides [in Russian], *Vestn. Belorus. un-ta, Ser. 2*, (2) 17–20 (1974).

Sanitarov V.A., Kalinkin I.P., Vityuk V.Ya., Taratynov S.I. Phase diagram of the $CdS_{1-x}Te_x$ solid solution [in Russian], *Izv. AN SSSR. Neorgan. materialy*, **16**(3), 398–401 (1980).

Saraie J., Kato H., Yamada N., Kaida Sh., Tanaka T. Preparation and photoconductive properties of sintered films of CdS–CdTe mixed crystals, *Phys. Status Solidi A*, **39**(1), 331–336 (1977).

Schwer H., Krämer V. The crystal structures of $CdAl_2S_4$, $HgAl_2S_4$ and $HgGa_2S_4$, *Z. Kristallogr.*, **190**(1–2), 103–110 (1990).

Semenov V.N., Ostapenko O.V., Klyuev V.G. Films preparation and properties in the $CdS–Cr_2S_3$ system [in Russian], *Poverhnost': rentgen., sinhrotron. i neytron. issled.*, (4), 37–40 (2000b).

Semenov V.N., Ostapenko O.V., Lukin A.N., Zavalishin E.I., Zavrazhnov A.Yu. Solid phase interaction in the thin films of the $CdS–Bi_2S_3$ system [in Russian], *Neorgan. materialy*, **36**(12), 1424–1427 (2000a).

Serment J., Perez G., Hagenmuller P. Les systèmes SiS_2–MS et GeS_2–MS (M = Cd, Hg) entre 800 et 1000°C, *Bull. soc. chim. France*, (2), 561–566 (1968).

Sobolev V.V. Zone structure of $CdIn_2S_4$-type crystals [in Russian], *Izv. AN MSSR. Ser. fiz.-tehn. i mat. nauk*, (2), 60–63 (1976).

Soklakov A.I., Nechaeva V.V. New Cu, Fe, Ni, Zn and Cd thiophosphates [in Russian], *Izv. AN SSSR. Neorgan. materialy*, **6** (5), 998–999 (1970).

Sood A. K., Wu K., Zemel J. N. Metastable $Pb_{1-x}Cd_xS$ epitaxial films. I. Growth and physical properties, *Thin Solid Films*, **48**(1), 73–86 (1978).

Springford M. The luminescence of some ternary chalcogenides and mixed, *Proc. Phys. Soc.*, **82**(6), 1029–1037 (1963).

Stetiu P. Sur le diagramme d'équilibre du système PbS–CdS, *Phys. status solidi (a)*, **15**(1), K19-K22 (1973).

Suchow L., Stemple N. R. Fluorescent rare earths in semiconducting thiospinels, *J. Electrochem. Soc.*, **111**(2), 191–195 (1964).

Susa K., Kobayashi T., Tanigushi S. High-pressures synthesis of rock-salt type CdS using metal sulfide additives, *J. Solid State Chem.*, **33**(2), 197–202 (1980).

Susa K., Steinfink H. $GeCd_4S_6$, a new defect tetrahedral structure type, *Inorg. Chem*, **10**(8), 1754–1756 (1971a).

Susa K., Steinfink H. Ternary sulfide compounds AB_2S_4: The crystal structures of $GePb_2S_4$ and $SnBa_2S_4$, *J. Solid State Chem.*, **3**(1), 75–82 (1971b).

Susa K., Yadzima M., Taniguti T. Obtaining of the solid solutions cadmium sulfide—ytterbium sulfide, Pat. 47–47184 (Japan) (1975).

Sysoev L.A., Obuhova N.F. Obtaining and investigation of some properties of ZnS–MgS single crystals [in Russian], *Monokristaly i Tehnika*, 2(9), 54–57 (1973).

Sysoev L.A., Raiskin E.K., Gur'ev V.R. Measuring of melting temperatures for zinc and cadmium sulfides, selenides and tellurides [in Russian], *Izv. AN SSSR. Neorgan. materialy*, **3**(2), 390–391 (1967).

Tai H., Nakashima S., Hori S. Optical properties of $CdTe_{1-x}CdSe_x$ and $CdTe_{1-x}CdS_x$ systems, *Phys. status solidi (a)*, **30**(2), K115-K119 (1975).

Tai H., Nakashima S., Hori S. Band gap energies of solid solutions in $(CdTe)_{1-x}$–$(CdSe)_x$ and $(CdTe)_{1-x}$–$(CdS)_x$ systems [in Japanese]. *J. Jap. Inst. Metals*, **40**(5), 474–479 (1976).

Tarascon J.M., Hull G.W. On several new ternary molybdenum sulfide phases $M_{3.4}Mo_{15}S_{19}$ (M = vacancy, Li, Na, K, Zn, Cd, Sn and Tl), *Mater. Res. Bull.*, **21**(7), 859–869 (1986).

Tezlevan V.E., Zhitar' V.F. Physico-chemical properties of solid solutions in the $(CdS)_{3x}$— $(In_2S_3)_{1-x}$ system [in Russian], in: *Poluprovodn. soied. i ih tverdyie rastvory*, Kishinev: RIO AN MSSR, 207–210 (1970).

Tomas A., Guittard M., Flahaut J., Guymont M., Portier R., Gratias D. Structural study of CdY_2S_4, *Acta Crystallogr.*, **B42**(4), 364–371 (1986).

Tomas A., Shilo I., Guittard M. Structure cristalline du spinelle $CdEr_2S_4$, *Mater. Res. Bull.*, **13**(8), 857–859 (1978).

Tomas A., Tien Vovan, Guittard M., Flahaut J., Guymont M. Structures des composés $CdTm_2S_4$ et $CdYb_2S_4$, *Mater. Res. Bull.*, **20**(9), 1027–1030 (1985).

Tomashik V.N., Oleinik G.S., Mizetskaya I.B. Investigation of interaction in the $CdSe + ZnS \Leftrightarrow CdS + ZnSe$ ternary mutual system [in Russian], *Izv. AN SSSR. Neorgan. materialy*, **15**(2), 202–204 (1979).

Tovt V.V., Lukach P.M., Guranich P.P., Shusta V.C., Gerzanich O.I., Potoriy M.V., Voroshilov Yu.V. Growth and physical properties of $Cd_2P_2S_6$ and $Cd_2P_2Se_6$ single crystals [in Ukrainian], *Nauk. visnyk Uzhgorod. un-tu. Ser. Khimia*, (3), 52–54 (1998).

Tret'yakov Yu.D., Gordeev I.V., Alferov V.A., Saksonov Yu.G. Deviation from stoichiometry of chalcogenide chromites with the spinel structure [in Russian], *Izv. AN SSSR. Neorgan. materialy*, **8**(12), 2215–2216 (1972).

Tripathi B., Singh F., Avasthi D.K., Bhati A.K., Das D., Vijay Y.K. Structural, optical, electrical and positron annihilation studies of CdS:Fe system, *J. Alloys Compd.*, **454**(1–2), 97–101 (2008).

Trishchuk L.I., Oleinik G.S., Mizetskaya I.B. Thermal analysis determination of $A^{II}B^{VI}$ solid solubility in $A^I_2B^{VI}$. *Thermochim. Acta*, **92**, 611–613 (1985).

Tulva L., Koppel H. Physico-chemical investigation of Ag–Cd–S ternary system [in Russian], *Izv. AN ESSR. Ser. Khimia, Geologia*, **25**(1), 32–37 (1976).

Tulva L., Lepp A., Koppel H. Physico-chemical investigation of Ag–CdS and Ag_2S–CdS systems [in Russian], *Izv. AN ESSR. Ser. Khimia, Geologia*, **22**(1), 46–50 (1973).

Val'kovskaya M.I., Radautsan S.I., Tezlevan V.E. Investigation of anisotropy of mechanical properties for the $(CdS)_{3x}$–$(In_2S_3)_{1-x}$ system [in Russian], in: *Issled. slozhn. poluprovodn.* Kishinev: RIO AN MSSR, 35–44 (1970).

Val'kovskaya M.I., Tezlevan V.E., Donika F.G. Crystal chemistry peculiarities of solid solutions with spinel structure in the $(CdS)_{3x}$–$(In_2S_3)_{1-x}$ system [in Russian], in: *Slozhn. poluprovodn. i ih fiz. svoistva*, Kishinev: Shtiintsa, 82–87 (1971).

Valov Yu.A., Payonchkovska A. About synthesis of $A^{II}B_2^{III}C_4^{VI}$ ternary chalcogenides using chemical transport reactions [in Russian], *Izv. AN SSSR. Neorgan. materialy*, **6**(2), 241–246 (1970).

Vishniakov A.V., Bebiakin A.V., Zubkovskaya V.N. Phase equilibria in the region $Zn(Cd)$–$Zn(Cd)X$–Cu_2X (X = S, Se, Te) of Zn–Cu–X system [in Russian], *Zhurn. neorgan. khimii*, **27**(7), 1820–1826 (1982).

Vishniakov A.V., Dvoretskov G.A., Zubkovskaya V.N., Tyurin O.A., Kovtunenko P.V. Phase equilibria in the systems formed by the II-VI compounds and the elements of I and III groups of the periodic table [in Russian], *Tr. Mosk. Khim.-Tekhnol. In-t*, (120), 87–103 (1981).

Vishniakov A.V., Iofis B.G. Solubility of $Ag_2S(Se)$ in ZnS, CdS and CdSe [in Russian], *Izv. AN SSSR. Neorgan. materialy*, **10**(7), 1184–1186 (1974).

Vishniakov A.V., Kukleva T.V., Al'tah O.L., Zubkovskaya V.N., Kovtunenko P.V. The solid state solubility of cadmium chalcogenides in copper(I) chalcogenides [in Russian], *Zhurn. neorgan. khimii*, **25**(5), 1358–1361 (1980).

Vitrihovski N.I., Mizetskaya I.B. Obtaining of CdS CdSe mixte single crystals and some their characteristics [in Russian], *Fizika tverdogo tela*, **1**(3), 397–402 (1959a).

Vitrihovski N.I., Mizetskaya I.B. Obtaining of CdS CdTe mixte single crystals and some their characteristics [in Russian], *Fizika tverdogo tela*, **1**(6), 996–999 (1959b).

Vityuk V.Ya., Sanitarov V.A., Zavleshko N.N., Kalinkin I.P. Disintegration of the solid solutions in the films of the CdS–CdSe system [in Russian], *Izv. AN SSSR. Neorgan. materialy*, **18**(9), 1514–1517 (1982b).

Vityuk V.Ya., Zapleshko N.I., Sanitarov V.A., Kalinkin I.P. Condensation diagram of and phase changes in thin films solid solutions of the CdS_xTe_{1-x} system, *Thin Solid Films*, **91**(3), 183–190 (1982a).

Voroshilov Yu.V., Gurzan M.I., Pan'ko V.V., Peresh E.Yu., Rigan M.Yu., Koperles B.M. Preparation and some properties of the $CdGa_2S_4$ single crystals [in Russian], *Dokl. AN USSR. Ser. B*, (3), 163–165 (1979).

Vydyanath H.R., Kröger F.A. The defect structure of silver-doped CdS, *J. Phys. Chem. Solids*, **36**(6), 509–520 (1975).

Walker P.J., Ward R.C.C. The preparation of some ternary sulphides MR_2S_4 (M = Ca, Cd; R = La, Sm, Er) and the melt growth of $CaLa_2S_4$, *Mater. Res. Bull.*, **19**(6), 717–725 (1984).

Wiedemeier H., Kahn A. Phase studies in the system manganese sulfide–cadmium sulfide, *Trans. Metallurg. Soc. AIME*, **242**(9), 1969–1972 (1968).

Wu P., He X.-C., Dwight K., Wold A. Growth and characterization of zinc and cadmium thiogallate, *Mater. Res. Bull.*, **23**(11), 1605–1609 (1988).

Yim W.M., Fan A.K., Stofko E.J. Preparation and properties of II–Ln_2–S_4 ternary sulfides, *J. Electrochem. Soc.*, **120** (3), 441–446 (1973).

Yoshizawa M. Some crystal habits of lithium doped CdS single crystals, *J. Phys. Soc. Jap.*, **25**(2), 637 (1968).

Zargarova M.I., Alieva Sh.B., Allazov M.R., Sadyhova S.A., Movsum-zade A.A. View of liquidus surface of Cd–Sn–S system [in Russian], *Zhurn. neorgan. khimii*, **30**(5), 1279–1284 (1985).

Zhitar' V.F., Donu V.S., Val'kovskaya M.I., Markus M.M. Preparation and some physical properties of the $CdGa_2S(Se)_4$ [in Russian], in: *Fiz. i khim. slozhnyh poluprovodn.* Kishinev: Shtiintsa, 50–60 (1975).

Zuyev A.P., Kulakov M.P., Fadeev A.V., Kireiko V.V. Oxygen contents in the ZnS, ZnSe and CdS crystals, obtained from the melt [in Russian], *Izv. AN SSSR. Neorgan. materialy*, **17**(7), 1159–1161 (1981).

5

System Based on CdSe

5.1 Cadmium–Sodium–Selenium

CdSe–Na$_2$Se: The phase diagram is not constructed. Na$_4$Cd$_3$Se$_5$ ternary compound is formed in this system (Ansari and Ibers 1993). It crystallizes in an orthorhombic structure with lattice parameters a = 1402.6 ± 0.1, b = 424.4 ± 0.2, and c = 2010.5 ± 0.1 pm at 110 K and calculated density 4.582 g·cm^{-3}.

This compound has been prepared by the reaction of CdSe with Na$_2$Se$_4$ at 300°C–350°C (Ansari and Ibers 1993). Under similar conditions, the reaction of stoichiometric quantities of the elements produced only the mixtures of binary phases.

5.2 Cadmium–Potassium–Selenium

CdSe–K$_2$Se: The phase diagram is not constructed. K$_2$Cd$_3$Se$_4$ ternary compound is formed in this system (Axtell et al. 1996). It melts congruently at 784°C and crystallizes in an orthorhombic structure with lattice parameters a = 1432 ± 3, b = 1059 ± 2, and c = 679 ± 1 pm.

K$_2$Cd$_3$Se$_4$ was obtained by the interaction of K$_2$Se, Cd, and Se in the evacuated quartz tube (Axtell et al. 1996). The mixture was heated at 650°C for 2 days, followed by 5°C h^{-1} cooling to 400°C and quenching to room temperature. This compound is insoluble in water and common organic solvents but is sensitive to long exposure to air.

5.3 Cadmium–Rubidium–Selenium

CdSe–Rb$_2$Se: The phase diagram is not constructed. Rb$_2$Cd$_3$Se$_4$ ternary compound is formed in this system (Axtell et al. 1996). It melts congruently at 908°C and crystallizes in an orthorhombic structure with lattice parameters

$a = 1474.9 \pm 0.7$, $b = 1101.2 \pm 0.4$, and $c = 680.2 \pm 0.3$ pm and calculation density 4.95 g cm^{-3} and energy gap 2.37 eV.

$Rb_2Cd_3Se_4$ was obtained by the interaction of Rb_2Se, Cd, and Se in the evacuated quartz tube (Axtell et al. 1996). The mixture was heated at 650°C for 2 days, followed by 5°C h^{-1} cooling to 200°C and finally quenching to room temperature. This compound is insoluble in water and common organic solvents but is sensitive to prolonged exposure to air.

5.4 Cadmium–Copper–Selenium

The scheme of the phase equilibria in the region $CdSe–Cd–Cu_2Se–Cu$ of the Cd–Cu–Se ternary system is shown in Figure 5.1 (Vishniakov et al. 1982).

CdSe–Cu: The phase diagram is not constructed. Using static method, the equilibrium in the reaction $CdSe + 1.96Cu = Cu_{1.96}Se + Cd$ at 800°C was studied (Harif and Vishniakov 1976). The composition of equilibrium metal phase that was the solid solution of Cd in Cu (the solubility of Se was not higher than 0.05 at. %) was determined using the mass changing. It was constant within the interval of ratio Cu/CdSe = 2–10 and constituted 0.55 ± 0.07 at. % Cd. This indicates that there is monovariant equilibrium

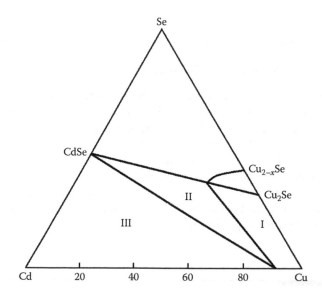

FIGURE 5.1
Scheme of the phase equilibria in the region $CdSe–Cd–Cu_2Se–Cu$ of the Cd–Cu–Se ternary system: 1, equilibrium of the phase based on Cu_2Se with solid solutions Cu_xCd_{1-x}; 2, equilibrium of the solid solutions based on Cu_2Se and CdSe with solid solutions Cu_xCd_{1-x}; and 3, equilibrium of the phase based on CdSe with solid solutions Cu_xCd_{1-x}. (From Vishniakov, A.V. et al., *Zhurn. neorgan. khimii*, 27(7), 1820, 1982.)

between solid solution of Cd in Cu and the phases based on CdSe and $Cu_{1.96}Se$ in this system. The solubility of Cu in CdSe at 800°C is not higher than 4 at. %.

According to the data of Epik and Varvas (1982), the predominant form of copper existence in the cadmium selenide in the wide region of selenium pressure and at lower cadmium pressure is Cu_2Se. The concentration of Cu increases with the increase of cadmium pressure. At 900°C, the distribution coefficient of Cu between CdSe and Cu_xCd_{1-x} alloys is equal to $5.1 \cdot 10^{-3}$.

CdSe–CuSe: The phase diagram is not constructed. The solubility of CuSe in CdSe at 550°C and 900°C determined using x-ray diffraction (XRD) is not higher than 1 mol. % (Reisman and Berkenblit 1962).

CdSe–Cu$_2$Se: The phase diagram belongs to the eutectic type (Figure 5.2) (Trishchuk et al. 1984). The eutectic composition and temperature are 53 mol. % CdSe and 910°C, correspondingly. At the eutectic temperature, solid solutions based on $Cu_{2-x}Se$ exist within the interval of 0–50 mol. % CdSe (Trishchuk et al. 1984, 1985). The solubility of CdSe in $Cu_{2-x}Se$ decreases sharply with the decrease in temperature and is equal to 2.0 mol. % at 500°C (Trishchuk et al. 1985). The solubility of $Cu_{2-x}Se$ in CdSe at 600°C, 700°C, 800°C, and 850°C constitutes 0.40, 0.95, 1.74, and 2.71 mol. %, respectively (Vishniakov

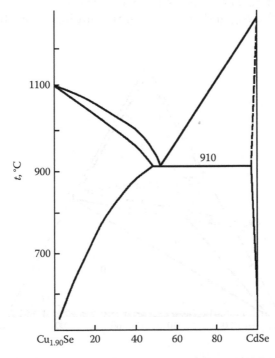

FIGURE 5.2
CdSe–$Cu_{1.90}$Se phase diagram. (From Vishniakov, A.V. and Harif, Ya.L., *Izv. AN SSSR. Neorgan. materialy*, 8(2), 217, 1972; Vishniakov, A.V. et al., *Zhurn. neorgan. khimii*, 25(5), 1358, 1980; Trishchuk, L.I. et al., *Izv. AN SSSR. Neorgan. materialy*, 20(4), 679, 1984.)

and Harif 1972). Determined by Trishchuk et al. (1984), degradation temperatures of solid solutions based on $Cu_{2-x}Se$ are on the average 50°C higher than
according to the data of Vishniakov et al. (1980).

According to the data of Harif and Vishniakov (1976), the solubility of CdSe in
Cu_2Se at 800°C reaches 42.2 ± 6.4 mol. %. It was determined that the formation
of solid solutions based both CdSe and Cu_2Se does not lead to the changes of
lattice parameters at the measurement accuracy of ± 1 pm (Datsenko et al. 1984).

This system was investigated using differential thermal analysis (DTA)
and metallography (Trishchuk et al. 1984, 1985). The solubility of CdSe in
$Cu_{2-x}Se$ was determined using DTA (Vishniakov et al. 1980, Trishchuk et al.
1984) and the solubility of $Cu_{2-x}Se$ in CdSe by the diffusion saturation of thin
layers (Vishniakov and Harif 1972).

5.5 Cadmium–Silver–Selenium

Isothermal section of the Cd–Ag–Se ternary system at 616°C–890°C is
shown in Figure 5.3 (Tiurin et al. 1982). At temperatures below 616°C, the
configuration of Se-rich side is simplified. The regions 2, 3, 7, and 9 dis-
appeared and chalcogenide phases are in equilibria with the melt based
on Se.

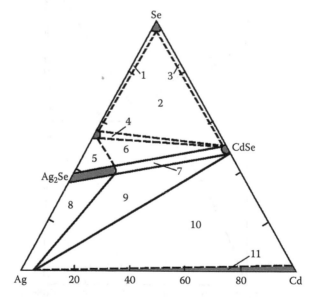

FIGURE 5.3

Isothermal section of the Cd–Ag–Se ternary system at 616°C–890°C: 1, $L_1 + L_2$; 2, $CdSe + L_1 + L_2$; 3,
$CdSe + L_2$; 4, $CdSe + L_1$; 5, $Ag_2Se + L_1$; 6, $CdSe + Ag_2Se + L_1$; 7, $Ag_2Se + CdSe$; 8, $Ag_2Se + Ag_{1-x}Cd_x$; 9,
$CdSe + Ag_2Se + Ag_xCd_{1-x}$; 10, $CdSe + Ag_xCd_{1-x}$; and 11, Ag_xCd_{1-x}. (From Tiurin, O.A. et al., *Zhurn.
neorgan. khimii*, 27(7), 1827, 1982.)

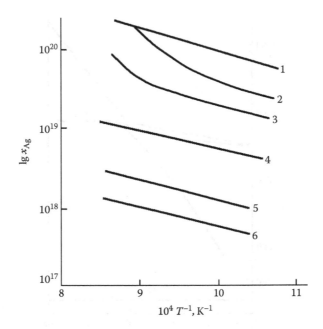

FIGURE 5.4
Temperature dependence of Ag solubility in CdSe at the different Ag contents in the Ag_xCd_{1-x} alloys: 1, Ag_2Se–CdSe–Ag; 2, the alloy $Ag_{0.9}Cd_{0.1}$; 3, the alloy $Ag_{0.8}Cd_{0.2}$; 4, the alloy $Ag_{0.5}Cd_{0.5}$; 5, the alloy $Ag_{0.2}Cd_{0.8}$; and 6, the alloy $Ag_{0.1}Cd_{0.9}$. (From Vishniakov, A.V. et al., *Zhurn. neorgan. khimii*, 28(5), 1274, 1983.)

CdSe–Ag: The phase diagram is not constructed. Temperature dependence of Ag solubility in CdSe at different Ag contents in the Ag_xCd_{1-x} alloys is shown in Figure 5.4 (Vishniakov et al. 1983). At 850°C–960°C, solubility changes weakly at the temperature changing. The solubility of Ag in CdSe decreases appreciably with the decrease of Ag concentration in the alloys and changes proportionally with Ag contents at the Ag concentration in the alloys less than 50 at. %.

CdSe–Ag_2Se: The phase diagram is a peritectic type (Figure 5.5) (Tiurin et al. 1982, Trishchuk et al. 1982). The peritectic composition and temperature are 62 mol. % CdSe [65 mol. % CdSe (Vishniakov et al. 1981)] and 970°C [960°C (Tiurin et al. 1982, Vishniakov et al. 1983], respectively (Trishchuk et al. 1984). At the peritectic temperature, solid solutions based on Ag_2Se exist within the interval of 0–68 mol. % CdSe (Trishchuk et al. 1982, 1985). The solubility of CdSe in Ag_2Se decreases sharply with the decrease in temperature and is equal to 10 mol. % at 550°C and 2.5 mol. % at 415°C. The solubility of Ag_2Se in CdSe at 790°C is equal to 0.94 mol. % (Vishniakov et al. 1974).

At 140°C [130 ± 5°C (Tiurin et al. 1982)], there are thermal effects that correspond to the polymorphous transformation of Ag_2Se (Trishchuk et al. 1982).

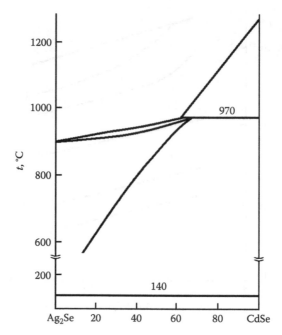

FIGURE 5.5
CdSe–Ag₂Se phase diagram. (From Tiurin, O.A. et al., *Zhurn. neorgan. khimii*, 27(7), 1827, 1982; Trishchuk, L.I. et al., *Izv. AN SSSR. Neorgan. materialy*, 18(11), 1795, 1982.)

This system was investigated using DTA, metallography, and XRD (Tiurin et al. 1982, Trishchuk et al. 1982, 1985). The solubility of Ag_2Se in CdSe was determined by the diffusion saturation of thin layers (Vishniakov and Iofis 1974).

5.6 Cadmium–Magnesium–Selenium

CdSe–MgSe: The phase diagram is not constructed. Shin Dong et al. (1997) reported the results on the growth of the $Cd_{1-x}Mg_xSe$ single crystals in the range of $0 \leq x \leq 0.46$ using chemical transport reactions technique and on the investigation of their structure and optical properties using XRD and optical absorption measurements. Iodine was used as a transporting agent. It was established that the solid solutions based on CdSe contain 46 mol. % MgSe. Atroshchenko et al. (1977) obtained $Cd_{1-x}Mg_xSe$ single crystals, using oriented crystallization under an argon pressure of 2–3 MPa (30–50 atm), containing only 10 mol. % MgSe. Atroshchenko et al. (1986) reported that CdSe dissolves 43 mol. % MgSe.

The forming solid solutions have a wurtzite-type structure (Sysoev and Obuhova 1973, Obuhova et al. 1978, Atroshchenko et al. 1979, 1986, Shin Dong et al. 1997). The lattice constants a and c decrease linearly from 429.38 and 700.03 pm to correspondingly 424.55 and 690.87 pm with the increase of MgSe content from 0 to 46 mol. % (Shin Dong et al. 1997) [according to Obuhova et al. (1978) and Atroshchenko et al. (1979), the lattice constants a and c decrease linearly at the MgSe content changing from 0 to 41 at. %]. The c/a ratio decreases from 1.630 for CdSe to 1.626 for the solid solution containing 20.5 at. % Mg (Obuhova et al. 1978, Atroshchenko et al. 1979).

The composition dependence of the optical energy gap shows that the optical energy gap increases linearly as the composition x increases from 0 to 0.46 (Shin Dong et al. 1997). The relationship is represented by the best-fit line given by the expression E_g (eV) $= 1.742 + 1.813x$, where $0 \leq x \leq 0.46$. The microhardness of $Cd_{1-x}Mg_xSe$ single crystals on the basic and lateral faces almost does not change and has the significances that are typical for CdSe (Sysoev and Obuhova 1973).

According to the data of Firszt et al. (1998), $Cd_{1-x}Mg_xSe$ mixed crystals for $0 < x < 0.55$ have been grown using the high-pressure Bridgman method under an Ar overpressure. They crystallize in a hexagonal structure with lattice constant decreasing with increasing x value.

5.7 Cadmium–Barium–Selenium

CdSe–BaSe: The phase diagram is not constructed. According to the data of Iglesias et al. (1974), Ba_2CdSe_3 compound was synthesized in this system. It crystallizes in an orthorhombic structure with lattice parameters $a = 922.47$, $b = 448.23$, and $c = 1787.06$ pm. $BaCdSe_2$ compound was not obtained in the CdSe–BaSe system.

Ba_2CdSe_3 was synthesized by heating of mixtures containing CdSe and BaSe in the stoichiometric ratio at 700°C–900°C.

5.8 Cadmium–Mercury–Selenium

CdSe–HgSe: The phase diagram belongs to the peritectic type (Figure 5.6) (Nelson et al. 1977). The peritectic composition and temperature are 57 mol. % CdSe and 947.5 ± 4°C, respectively. This phase diagram can be calculated using the model of regular associated solutions (Nelson et al. 1977, Gavaleshko et al. 1983). According to the data of Cruceanu (1965) and Kalb and Leute (1971), two-phase region exists at 500°C within the interval of (77 ± 1)–(81 ± 1) mol. % CdSe. In the hexagonal region, the lattice parameters change linearly with composition. In the cubic region, such

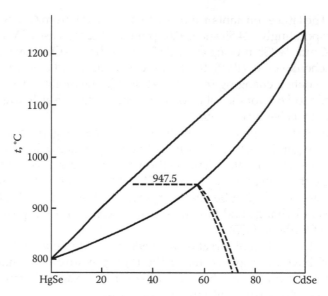

FIGURE 5.6
CdSe–HgSe phase diagram. (From Nelson, D.A. et al., *J. Electron. Mater.*, 6(5), 507, 1977.)

dependence is not linear: relatively sharp increase of lattice parameter takes place on the border of CdSe solubility (Słodowy and Giriat 1971).

Energy gap changes at room temperature with composition linearly (Słodowy and Giriat 1971) [or monotonically but nonlinearly (Kot et al. 1962, 1964)].

This system was investigated using DTA (Nelson et al. 1977, Gavaleshko et al. 1983). The ingots were annealed at the temperature 30°C below the solidus temperatures (Nelson et al. 1977). Pressed ingots were annealed for some days under the pressure 13.3 kPa for obtaining homogeneous mixtures (Kalb and Leute 1971).

5.9 Cadmium–Boron–Selenium

Liquidus surface of the CdSe–B_2Se_3–Se subsystem of the Cd–B–Se ternary system is shown in Figure 5.7 (Odin et al. 2001). Practically all liquidus surface is occupied by the field of CdSe primary crystallization, and the fields of B_2Se_3 and $Cd_2B_2Se_5$ primary crystallization occupy narrow bands near the B_2Se_3–Se subsystem. Ternary eutectic E (L $\Leftrightarrow$ B_2Se_3 + $Cd_2B_2Se_5$ + Se) and ternary peritectic U (L + CdSe $\Leftrightarrow$ $Cd_2B_2Se_5$ + Se) crystallize at 212°C and 213°C, respectively. The immiscibility region occupies the biggest part of the CdS primary crystallization field.

CdSe–B: According to the data of the atomic-absorption analysis, the doped CdSe contains 1.04–1.73 at. % B (0.06–0.10 mass. % B) (Odin et al. 1999a).

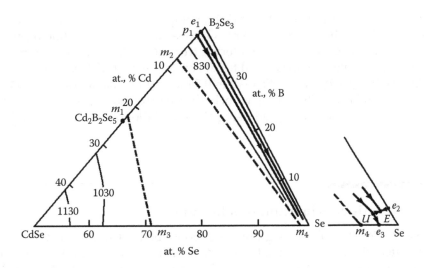

FIGURE 5.7
Liquidus surface of the CdSe–B_2Se_3–Se subsystem. (From Odin, I.N. et al., *Zhurn. neorgan. khimii*, 46(7), 1210, 2001.)

CdSe–B_2Se_3: The phase diagram belongs to the eutectic type (Figure 5.8) (Odin et al. 1999b). The eutectic composition and temperature are 5.0 ± 0.5 mol. % CdSe and 437°C, respectively. Immiscibility region exists in this system at 990°C within the interval from 15 to 55 mol. % CdSe. $Cd_2B_2Se_5$ ternary compound is formed at the interaction of CdSe and B_2Se_3 (Odin et al. 1999a,b).

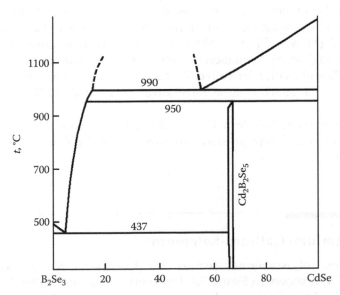

FIGURE 5.8
CdSe–B_2Se_3 phase diagram. (From Odin, I.N. et al., *Zhurn. neorgan. khimii*, 44(12), 2061, 1999c.)

It melts incongruently at 950°C and the composition of 14 ± 1 mol. % CdSe corresponds to the peritectic point (Odin et al. 1999b). The homogeneity region of $Cd_2B_2Se_5$ is within the interval from 66.2 ± 0.4 to 68.3 ± 0.3 mol. % CdSe, and the energy gap is equal to 2.02 eV (Odin et al. 1999b, 2001).

The solubility of B_2Se_3 in CdSe reaches 0.5 mol. % (0.4 mol. % Odin et al. 2001a) and the solubility of CdSe in B_2Se_3 is not higher than 0.4 mol. % (Odin et al. 1999b).

This system was investigated using DTA, metallography, and XRD (Odin et al. 1999b).

5.10 Cadmium–Aluminum–Selenium

$CdSe–Al_2Se_3$: The phase diagram is not constructed. $CdAl_2Se_4$ is formed in this system. It crystallizes in a tetragonal structure of chalcopyrite type with lattice parameters $a = 574.74$ and $c = 1075.51$ pm (Ouarani et al. 2010) [$a = 573.5$ and $c = 1065$ pm (Hahn et al. 1955); $a = 575$ and $c = 1068$ pm (Range et al. 1968); according to the calculations, $a = 576$ and $c = 1149.7$ pm (Mishra and Ganguli 2011)] and calculation and experimental density 4.542 [4.54 (Range et al. 1968)] and 4.50 g cm^{-3} (Hahn et al. 1955). This compound is a direct band-gap semiconductor with calculation energy band gap of 2.46 eV [2.927 eV (Ouarani et al. 2010)] and bulk modulus of 44.89 GPa (Mishra and Ganguli 2011). At high temperature and pressure (4.5 GPa or 45 kbar and 400°C), this compound can exist in a cubic structure of spinel type with lattice parameter $a = 1077$ pm and calculation density 5.13 g cm^{-3} (Range et al. 1968).

At 80 K, the ambient pressure phase of $CdAl_2Se_4$ is stable up to a pressure of 9.1 GPa above which a phase transition to a disordered rock-salt phase is observed (Meenakshi et al. 2006). A fit of the volume pressure data to a Birch–Murnaghan-type equation of state yields a bulk modulus of 52.1 GPa [59,854 GPa and its first derivative B' = 4.2444 (Ouarani et al. 2010)]. The relative volume change at the phase transition at ~9 GPa is about 10% (Meenakshi et al. 2006).

$CdAl_2Se_4$ was obtained from CdSe and powderlike Al and Se in equimolar ratio. The mixtures were pressed and annealed at 800°C for 12–24 h (Hahn et al. 1955).

5.11 Cadmium–Gallium–Selenium

The scheme of isothermal section for Cd–Ga–Se ternary system at 700°C–800°C is shown in Figure 5.9 (Dvoretskov et al. 1979a, 1982).

CdSe–Ga: According to the data of Dvoretskov et al. (1979a), this section is a nonquasibinary section of the Cd–Ga–Se ternary system.

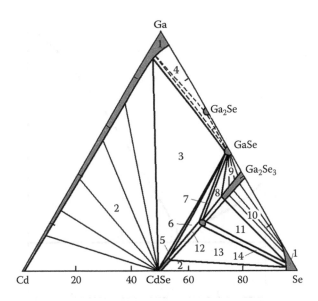

Ga

Ga₂Se

GaSe

Ga₂Se₃

Cd   20   40 CdSe 60   80   Se

FIGURE 5.9
Scheme of isothermal section for Cd–Ga–Se ternary system at 700°C–800°C: 1, L; 2, CdSe + L; 3, CdSe + GaSe + L; 4, GaSe + L; 5, CdSe + GaSe; 6, CdSe + CdGa₂Se₄ + GaSe; 7, CdGa₂Se₄ + GaSe; 8, CdGa₂Se₄ + GaSe + Ga₂Se₃; 9, GaSe + Ga₂Se₃; 10, Ga₂Se₃ + L; 11, CdGa₂Se₄ + Ga₂Se₃ + L; 12, CdGa₂Se₄ + CdSe; 13, CdSe + CdGa₂Se₄ + L; and 14, CdGa₂Se₄ + L. (From Dvoretskov, G.A. et al., *Zhurn. neorgan. khimii*, 24(9), 2505, 1979; Dvoretskov, G.A. et al., Solubility isotherm of gallium in cadmium and zinc selenides [in Russian], in *Legir. poluprovodn.*, M., Nauka, pp. 67–71, 1982.)

CdSe–GaSe: The phase diagram is a eutectic type (Figure 5.10) (Dvoretskov et al. 1979a). The eutectic composition and temperature are 36 ± 1 mol. % CdSe and 810 ± 5°C, respectively. The solubility of GaSe in CdSe at 790°C is equal to 2 mol. %.

This system was investigated using DTA and XRD. The ingots were annealed at 790°C for 290 h (Dvoretskov et al. 1979a).

CdSe–Ga₂Se₃: The phase diagram is shown in Figure 5.11 (Sosovska et al. 2008). The eutectic point coordinates are equal to 39 and 64 mol. % Ga₂Se₃ at 935°C and 933°C, respectively [38 and 59 mol. % Ga₂Se₃ and 940 ± 5°C and 920 ± 5°C, respectively (Tyrziu et al. 1970); 45 and 55 mol. % Ga₂Se₃ at 942°C and 930°C, respectively (Loireau-Lozac'h et al. 1985]. The invariant horizontal lines at 817°C and 790°C correspond to the phase transformation of CdGa₂Se₄ (Sosovska et al. 2008). The temperature difference is due to the presence of small solid-solution regions based on both modifications of the ternary compound; they extend up to ~2 mol. % and are shifted to the Ga₂Se₃ side. The solid immiscibility region reported in Loireau-Lozac'h et al. (1985) has been confirmed. Solid solution, which exhibits the highest CdGa₂Se₄ content at 715°C, undergoes a monotectoid decomposition into two solid solutions (Sosovska et al. 2008) (according the data of Loireau-Lozac'h et al. 1985, the monotectoid temperature is 704°C).

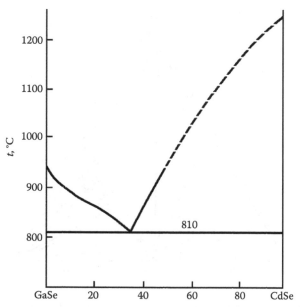

FIGURE 5.10
CdSe–GaSe phase diagram. (From Dvoretskov, G.A. et al., *Zhurn. neorgan. khimii*, 24(9), 2505, 1979.)

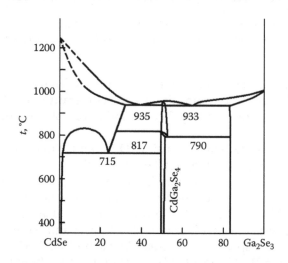

FIGURE 5.11
CdSe–Ga$_2$Se$_3$ phase diagram. (From Sosovska, S.M. et al., *J. Alloys Compd.*, 453(1–2), 115 2008.)

The CdSe solid solution extends up to 32 mol. % Ga$_2$Se$_3$ at the eutectic temperature and does not exceed a few mol. % at 600°C (Sosovska et al. 2008). The solid solubility in Ga$_2$Se$_3$ is 17 mol. % CdSe at 933°C, and it does not change with temperature decrease. Extraordinary peaks evidently connected with transfer "order–disorder" were observed on the curves of thermal analysis at 820°C and 720°C in Tyrziu et al. (1970).

The $CdGa_2Se_4$ compound is formed in this system. It characterizes by an order–disorder transition (Loireau-Lozac'h et al. 1985, Sosovska et al. 2008) and melts congruently at 953°C (Sosovska et al. 2008) [963°C (Loireau-Lozac'h et al. 1985) and 980 ± 5°C (Tyrziu et al. 1970, Zhitar' et al. 1975); according to the data of Agaev et al. 1970, it melts incongruently] and crystallizes in a tetragonal structure of thiogallate type with lattice parameters $a = 574.30 \pm 0.06$ and $c = 1075.2 \pm 0.2$ pm (Sosovska et al. 2008) [$a = 573$ and $c = 1072$ pm (Hahn et al. 1955, Kolosenko and Tyrziu 1970, Tyrziu et al. 1970, Tyrziu and Tyrziu 1971); $a = 573 \pm 1$ and $c = 1074 \pm 1$ pm (Lottici et al. 1989); $a = 574.4$ and $c = 1077$ pm (Loireau-Lozac'h et al. 1985); $a = 574.7$ and $c = 1075.8$ pm (Agaev et al. 1970); $a = 574.3$ and $c = 1075.6$ pm (Krämer et al. 1984); $a = 570$ and $c = 1080.5$ pm (Gastaldi et al. 1985); $a = 573$ and $c = 1075$ pm (Horinaka et al. 1990)] and experimental density 5.34 g/cm^{-3} (Tyrziu and Tyrziu 1971) and energy gap $E_g = 2.18$ eV (Tyrziu and Tyrziu 1971) [$E_g = 2.25$–2.43 eV (Beun et al. 1961)]. Homogeneity region of $CdGa_2Se_4$ is not higher than 5 mol. % (Dvoretskov et al. 1979b). According to the data of Dvoretskov et al. (1982), the lattice parameters of solid solutions based on CdSe do not change with cadmium pressure changing in a wide pressure range.

The high-pressure behavior of $CdGa_2Se_4$ with the defect-chalcopyrite structure is studied by in situ angle-dispersive synchrotron x-ray powder diffraction and optical reflectivity measurements in a diamond anvil cell at room temperature (Grzecznik et al. 2001). At 21 GPa, an order–disorder phase transition to the rock-salt structure occurs. Upon decompression, the metallic NaCl-type polymorph transforms into zinc blende at pressures of 7.5–4 GPa. The recovered metastable semiconducting material is of the zinc-blende type.

This system was investigated using DTA, metallography, and XRD (Kolosenko and Tyrziu 1970, Tyrziu et al. 1970, Loireau-Lozac'h et al. 1985, Sosovska et al. 2008). The alloys were annealed at 600°C for 250 h (Sosovska et al. 2008) [were annealed for 200–400 h (Kolosenko and Tyrziu 1970, Tyrziu et al. 1970)]. Single crystals of the $CdGa_2Se_4$ compound were grown using chemical transport reactions (Beun et al. 1961, Agaev et al. 1970, Tyrziu and Tyrziu 1971, Zhitar' et al. 1975, Mavlonov et al. 1976, Krämer et al. 1984, Gastaldi et al. 1985, Lottici et al. 1989) and Bridgman method (Zhitar' et al. 1975, Horinaka et al. 1990).

5.12 Cadmium–Indium–Selenium

The liquidus surface of the CdSe–Cd–In subsystem of the Cd–In–Se ternary system is shown in Figure 5.12 (Stanchu and Derid 1973). It was constructed using mathematical planning of the experiment and can be described by the following equation: t (°C) $= 156x_1 + 321x_2 + 1258x_3 - 194x_1x_2 + 1252x_1x_3 + 1342x_2x_3 - 157.6x_1x_2(x_1-x_2) + 1712x_1x_3(x_1-x_3) + 2061.6x_2x_3(x_2-x_3) - 653.6x_1x_2(x_1-x_2)^2 +$

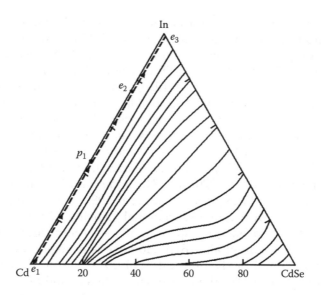

FIGURE 5.12

Liquidus surface of the CdSe–Cd–In ternary system. (From Stanchu, A.V. and Derid, O.P., Obtaining of the equation of the Cd–In–CdSe liquidus surface by the mathematical planning of the experiment [in Russian], in *Poluprovodn. pribory i materialy*, Kishinev, Shtiintsa, pp. 33–38, 1973.)

$$922.4x_1x_3(x_1-x_3)^2+1288x_2x_3(x_2-x_3)^2+11005.6x_1^2x_2x_3+7381.6x_1x_2^2x_3-4643.2x_1x_2x_3^2,$$

where x_1, x_2, and x_3 are the mole fractions of In, Cd, and CdSe, respectively.

CdSe–In: According to the data of Stanchu and Derid (1973), this section is a quasibinary section of the Cd–In–Se ternary system.

CdSe–InSe: The phase diagram is not constructed. According to the data of Guseinov (1969), the $CdInSe_2$ ternary compound is formed in this system. It crystallizes in a tetragonal structure with lattice parameters $a=1212.5$ and $c=714$ pm and experimental density 4.97 g cm^{-3} and energy gap $E_g=1.40$ eV.

CdSe–In$_2$Se$_3$: The phase diagram is shown in Figure 5.13 (Zmiy et al. 2004). Ternary compounds $CdIn_2Se_4$ and $CdIn_6Se_{10}$ are formed according to peritectic reactions at 912°C [910 ± 5°C (Derid et al. 1971)] and 850°C, respectively. Between the high-temperature modifications of In_2Se_3 and $CdIn_6Se_{10}$, eutectic interaction takes place with the coordinates of the eutectic point 842°C and 82.5 mol. % In_2Se_3. The solubility of In_2Se_3 in CdSe at 550°C is 7 mol. % In_2Se_3, and the solubility of CdSe in In_2Se_3 at this temperature is ~2 mol. % CdSe.

According to the data of Derid et al. (1971), the eutectic composition and temperature are 93.5 mol. % In_2Se_3 (the system is considered as $3CdSe–In_2Se_3$) and 825°C, respectively. Eutectoid transformations at 525°C and 210°C correspond to the polymorphous transformation of solid solutions based on In_2Se_3.

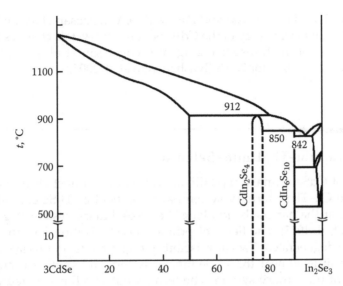

FIGURE 5.13
$3CdSe-In_2Se_3$ phase diagram. (From Zmiy, O.F. et al., *J. Alloys Compd.*, 367(1–2), 49, 2004.)

$CdIn_2Se_4$ crystallizes in a tetragonal pseudocubic (Hahn et al. 1955) or a cubic structure of the sphalerite type (Kolomiets and Mal'kova 1959), which weekly distinguishes one from the other (Sobolev 1976), with lattice parameters $a=c \cdot 587.1$ pm (Hahn et al. 1955) [$a=585$ pm (Vengatesan et al. 1987); $a=c=581.5 \pm 0.2$ pm (Koval et al. 1972, Przedmojski and Pałosz 1979, Lottici et al. 1989); $a=581.5$ and $c=1163$ pm (Koval et al. 1972, Przedmojski and Pałosz 1979)]. According to the data of Schwer and Krämer (1988), this compound has two modifications, which crystallize in a tetragonal structure of chalcopyrite type with close lattice parameters: $a=581.7$ and $c=2323.3$ pm and $a=581$ and $c=2326$ pm. The homogeneity region of $CdIn_2Se_4$ is within the interval of 73–77 mol. % In_2Se_3. Energy gap of this compound is equal to 1.61–1.72 eV (Beun et al. 1961).

This compound is characterized by three polytypes (1T, 2T, and 4T), which crystallize in a tetragonal structure with the same a lattice parameter and its parameters c are in ratio 1: 2: 4 (Manolikas et al. 1980, Ivashchenko et al. 2004). The lattice parameters for $4T-CdIn_2Se_4$ are equal to $a=580.6 \pm 0.2$ and $c=2325.2 \pm 0.6$ pm (Ivashchenko et al. 2004).

$CdIn_6Se_{10}$ crystallizes in a hexagonal structure with lattice parameters $a=407$ and $c=1983$ pm (Derid et al. 1971). The homogeneity region of $CdIn_6Se_{10}$ is insignificant. The solubility of $CdIn_6Se_{10}$ in high-temperature modification δ-In_2Se_3 is equal to 2–3 mol. % $3CdSe$.

This system was investigated using DTA, metallography, and XRD (Derid et al. 1971, Zmiy et al. 2004). The ingots were annealed at 750°C–780°C for 500–700 h. $CdIn_2Se_4$ and $CdIn_6Se_{10}$ ternary compounds were obtained using chemical transport reactions (Valov and Payonchkovska 1970, Derid et al. 1971,

Koval et al. 1972, Przedmojski and Pałosz 1979, Vengatesan et al. 1987, Lottici et al. 1989). The morphology of the $CdIn_2Se_4$ grown crystals changes with the concentrations of the transporting agent (Vengatesan et al. 1987). $CdIn_2Se_4$ was annealed at 550°C for 300 h (Ivashchenko et al. 2004).

5.13 Cadmium–Thallium–Selenium

The field of CdSe primary crystallization occupies almost all liquidus surface of the Cd–Tl–Se ternary system, and the field of Tl_2Se crystallization is situated at the narrow band along the Tl–Se binary system (Figure 5.14) (Babanly et al. 1985). The fields of primary crystallization for other binary phases are degenerated. Two immiscibility regions exist in this system. One of them is situated near the Tl–Se binary system and another occupies the Se corner of the ternary system. The ternary eutectic E is situated near the binary eutectic in the Tl–Se system (e_1) and crystallizes at 292°C.

Isothermal section of the Cd–Tl–Se ternary system at room temperature is shown in Figure 5.15 (Babanly et al. 1985).

CdSe–Tl: This section is a nonquasibinary section of the Cd–Tl–Se ternary system (Figure 5.16) (Babanly et al. 1985). There is a large immiscibility region in this section. The mixtures of CdSe and Tl crystallize at temperatures below solidus temperatures. Thermal effects, corresponding

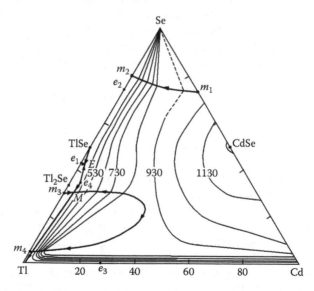

FIGURE 5.14
Liquidus surface of the Cd–Tl–Se ternary system. (From Babanly, M.B. et al., *Zhurn. neorgan. khimii*, 30(5), 1269, 1985.)

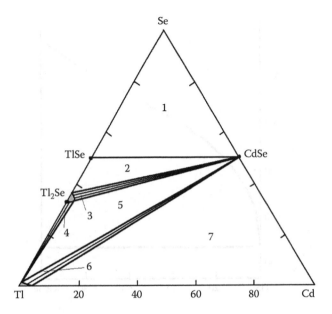

FIGURE 5.15
Isothermal section of the Cd–Tl–Se ternary system at room temperature: 1, CdSe + TlSe + Se; 2, CdSe + TlSe + Tl$_2$Se; 3, CdSe + Tl$_2$Se; 4, Tl$_2$Se + α-Tl; 5, CdSe + Tl$_2$Se + α-Tl; 6, CdSe + α-Tl; and 7, CdSe + Cd + α-Tl. (From Babanly, M.B. et al., *Zhurn. neorgan. khimii*, 30(5), 1269, 1985.)

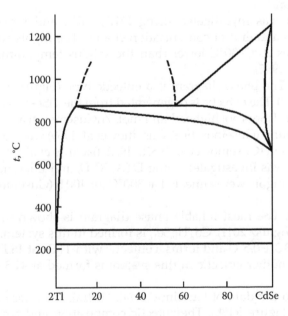

FIGURE 5.16
Phase relations in the CdSe–Tl system. (From Babanly, M.B. et al., *Zhurn. neorgan. khimii*, 30(5), 1269, 1985.)

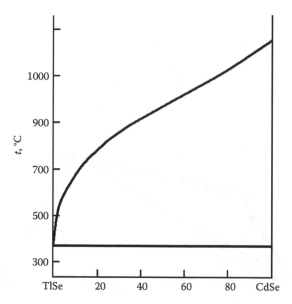

FIGURE 5.17
CdSe–TlSe phase diagram. (From Guseinov, F.H. et al., *Zhurn. neorgan. khimii*, 26(1), 215, 1981; Guseinov, F.H. et al., *Izv. AN SSSR. Neorgan. materialy*, 18(5), 759, 1982.)

to the polymorphous transformation of Tl, exist in the solid state of the CdSe–Tl section.

This system was investigated using DTA, XRD, and measurements of microhardness and emf of concentration chains. The ingots were annealed at temperatures 20°C–30°C lower than the solidus temperatures for 400 h (Babanly et al. 1985).

CdSe–TlSe: The phase diagram is a eutectic type (Figure 5.17) (Guseinov et al. 1981a, 1982). The eutectic is degenerated from the TlSe-rich side. The solubility of CdSe in TlSe is not higher than 1 mol. % (Guseinov et al. 1981a). $CdTlSe_2$ ternary compound (Guseinov 1969, Guseinov et al. 1969) was not found in the CdSe–TlSe system (Guseinov et al. 1981a, 1982, Babanly et al. 1985).

This system was investigated using DTA, XRD, and microhardness measurement. The ingots were annealed at 300°C for 400 h (Guseinov et al. 1981a, 1982).

CdSe–Tl₂Se: The most reliable phase diagram is shown in Figure 5.18a (Mucha and Wiglusz 2011). $Cd_3Tl_{16}Se_{11}$ is formed in this system. It melts congruently at 408.3 ± 0.5°C and forms a eutectic with Tl_2Se at 16.1 mol. % CdSe and 351.1°C. Another eutectic in this system is formed at 41.5 mol. % CdSe and 346°C.

According to the data of Guseinov et al. (1981a), the phase diagram is a eutectic type (Figure 5.18b). The eutectic composition and temperature are approximately 10 mol. % CdSe and 353°C, respectively. The solubility of CdSe in Tl_2Se reaches 3 mol. %.

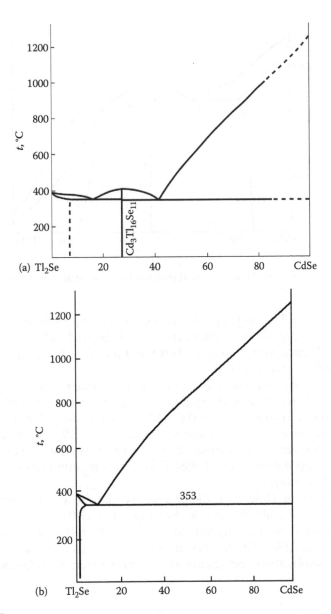

FIGURE 5.18
CdSe–Tl$_2$Se phase diagram: (a) (From Mucha, I. and Wiglusz, K., *Thermochim. Acta*, 526(1–2), 107, 2011.) and (b) (From Guseinov, F.H. et al., *Zhurn. neorgan. khimii*, 26(1), 215, 1981.)

This system was investigated using DTA, XRD, and microhardness measurement (Guseinov et al. 1981a, Mucha and Wiglusz 2011). The ingots were annealed at 300°C for 400 h (Guseinov et al. 1981a).

CdSe–Tl$_2$Se$_3$: The phase relations in this system are shown in Figure 5.19 (Karimov et al. 1981). The eutectic composition and temperature are 36 mol. %

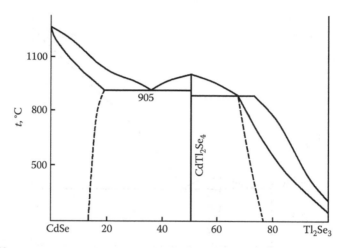

FIGURE 5.19
Phase relations in the CdSe–Tl$_2$Se$_3$ system. (From Karimov, S.K. et al., *Izv. AN SSSR. Neorgan. materialy*, 17(8), 1346, 1981.)

Tl$_2$Se$_3$ and 905 ± 3°C, respectively. Tl$_2$Se$_3$ compound melts incongruently; therefore, interaction in this system from the Tl$_2$Se$_3$-rich side is complicated. Additional thermal effects exist within the interval of 18–67 mol. % Tl$_2$Se$_3$, the nature of which is not clear.

CdTl$_2$Se$_4$ ternary compound is formed in this system, which crystallizes in a tetragonal structure of thiogallate type with lattice parameters $a = 420$ and $c = 670$ pm (Karimov et al. 1981) [$a = 428 \pm 2$ and $c = 667 \pm 2$ pm (Mavlonov et al. 1976, Sultonov et al. 1976, Sultonov and Karimov 1979)] and energy gap $E_g = 0.84$ eV (Sultonov and Karimov 1979). According to the data of Guseinov et al. (1981b) and Babanly et al. (1985), CdTl$_2$Se$_4$ compound does not exist in the CdSe–Tl$_2$Se$_3$ system.

The solubility of Tl$_2$Se$_3$ in CdSe reaches 14 mol. %, and the solubility of CdSe in Tl$_2$Se$_3$ is equal to 24 mol. % (Karimov et al. 1981).

This system was investigated using DTA, metallography, and XRD (Karimov et al. 1981). Single crystals of CdTl$_2$Se$_4$ were grown using the oriented crystallization and chemical transport reactions (Mavlonov et al. 1976).

5.14 Cadmium–Lanthanum–Selenium

CdSe–La$_2$Se$_3$: The phase diagram is shown in Figure 5.20 (Aliev et al. 1984). The eutectic compositions and temperatures are 18 and 80 mol. % La$_2$Se$_3$ and 1117°C and 1537°C, respectively. CdLa$_2$Se$_4$ ternary compound is formed in this system, which melts congruently at 1852°C (Aliev et al. 1984)

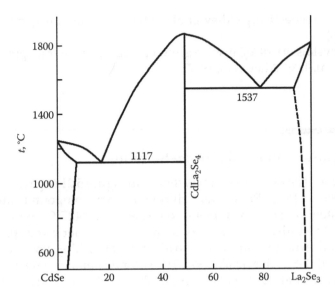

FIGURE 5.20
CdSe–La$_2$Se$_3$ phase diagram. (Aliev, O.M. et al., *Zhurn. neorgan. khimii*, 29(10), 2705, 1984.)

[1441°C (Agayev et al. 1989); 1442°C (Aliev et al. 1997)] and crystallizes in a cubic structure of Th$_3$P$_4$ type with lattice parameter $a = 916$ pm (Aliev et al. 1984) [$a = 894$ pm (Agayev et al. 1989, Aliev et al. 1997)] and calculation and experimental density 6.58 and 6.50 g cm^{-3}, respectively (Aliev et al. 1997) [6.60 and 6.63 g cm^{-3} (Agayev et al. 1989)].

The solubility of La$_2$Se$_3$ in CdSe is not higher than 1 mol. % and the solubility of CdSe in La$_2$Se$_3$ is equal to 2 mol. % (Aliev et al. 1984).

This system was investigated using DTA, metallography, XRD, and microhardness measurement. The ingots were annealed at 1000°C (up to 50 mol. % La$_2$Se$_3$) and 1200°C (within the interval of 50–100 mol. % La$_2$Se$_3$) for 240 h (Aliev et al. 1984). Single crystals of CdLa$_2$Se$_4$ were grown using chemical transport reactions (Aliev et al. 1984, 1997, Agayev et al. 1989).

5.15 Cadmium–Cerium–Selenium

CdSe–Ce$_2$Se$_3$: The phase diagram is not constructed. CdCe$_2$Se$_4$ ternary compound is formed in this system that melts congruently at 1447°C (Aliev et al. 1997) [1427°C (Agayev et al. 1989)] and crystallizes in a cubic structure of Th$_3$P$_4$ type with lattice parameter $a = 893.5$ pm (Agayev et al. 1989, Aliev et al. 1997) and calculation and experimental density 6.60 and

6.63 g cm^{-3}, respectively (Aliev et al. 1997) [6.65 and 6.67 g cm^{-3} (Agayev et al. 1989)].

Single crystals of CdCe$_2$Se$_4$ were grown using chemical transport reactions (Agayev et al. 1989, Aliev et al. 1997).

5.16 Cadmium–Praseodymium–Selenium

CdSe–Pr$_2$Se$_3$: The phase diagram is shown in Figure 5.21 (Agayev et al. 1984, Kulieva 1985). The CdPr$_2$Se$_4$ and CdPr$_4$Se$_7$ ternary compounds are formed in this system. First of them melts congruently at 1427°C and crystallizes in a cubic structure of Th$_3$P$_4$ type with lattice parameter $a = 892$ pm and calculation and experimental density 6.67 and 6.64–6.65 g cm^{-3}, respectively (Agayev et al. 1989, Aliev et al. 1997). The second compound melts incongruently (Agayev et al. 1984, Kulieva 1985). The solubility of Pr$_2$Se$_3$ in CdSe is equal to 3 mol. %.

This system was investigated using DTA, metallography, XRD, and microhardness, density, and electrical conductivity measurements. The ingots were annealed at 900°C for 200 h (Agayev et al. 1984, Kulieva 1985). Single crystals of CdPr$_2$Se$_4$ were grown using chemical transport reactions (Agayev et al. 1989, Aliev et al. 1997).

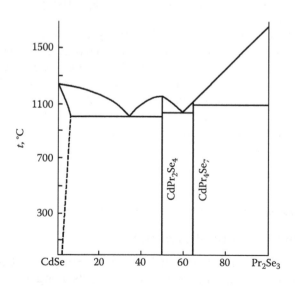

FIGURE 5.21
CdSe–Pr$_2$Se$_3$ phase diagram. (Agayev, A.B., *Azerb. khim. zhurn.*, (6), 95, 1984; Kulieva, S.A., Investigation of the electrical properties of the alloys in the CdSe–Pr$_2$Se$_3$ system [in Russian], in *Issled. v obl. sinteza i primenienie neorgan, soed.*, Baku, 78–82, 1985.)

5.17 Cadmium–Neodymium–Selenium

CdSe–Nd$_2$Se$_3$: The phase diagram is not constructed. CdNd$_2$Se$_4$ ternary compound is formed in this system that melts congruently at 1467°C and crystallizes in a cubic structure of Th$_3$P$_4$ type with lattice parameter $a = 890$ pm (Aliev et al. 1997) [$a = 893$ pm (Agayev et al. 1989)] and calculation and experimental density 6.67 and 6.68 g cm^{-3}, respectively (Agayev et al. 1989, Aliev et al. 1997).

Single crystals of CdNd$_2$Se$_4$ were grown using chemical transport reactions (Agayev et al. 1989, Aliev et al. 1997).

5.18 Cadmium–Samarium–Selenium

CdSe–Sm$_2$Se$_3$: The phase diagram is not constructed. CdSm$_2$Se$_4$ ternary compound is formed in this system that crystallizes in a cubic structure of Th$_3$P$_4$ type with lattice parameter $a = 887$ pm and calculation and experimental density 6.80 and 6.77 g cm^{-3}, respectively (Agayev et al. 1989, Aliev et al. 1997).

Single crystals of CdSm$_2$Se$_4$ were grown using chemical transport reactions (Agayev et al. 1989, Aliev et al. 1997).

5.19 Cadmium–Gadolinium–Selenium

CdSe–Gd$_2$Se$_3$: The phase diagram is shown in Figure 5.22 (Azadaliev and Alieva 1985). The CdGd$_2$Se$_4$ and CdGd$_4$Se$_7$ ternary compounds are formed in this system. First of them melts congruently at 1547°C [1537°C (Azadaliev and Alieva 1985)] and crystallizes in a cubic structure of Th$_3$P$_4$ type with lattice parameter $a = 886$ pm [$a = 587.2$ pm (Azadaliev and Alieva 1985)] and calculation and experimental density 7.10 and 6.94 [5.89 (Azadaliev and Alieva 1985)] g cm^{-3}, respectively (Agayev et al. 1989, Azadaliev 1990, Aliev et al. 1997). The second compound melts incongruently at 1277°C and crystallizes in a cubic structure with lattice parameter $a = 617.5$ pm and experimental density 6.02 g cm^{-3} (Azadaliev and Alieva 1985). The solubility of Gd$_2$Se$_3$ in CdSe reaches 5 mol. %.

This system was investigated using DTA, metallography, XRD, and microhardness and density measurements. The ingots were annealed at 830°C for 8–12 days (Azadaliev and Alieva 1985). Single crystals of CdGd$_2$Se$_4$ were grown using chemical transport reactions (Agayev et al. 1989, Azadaliev 1990, Aliev et al. 1997).

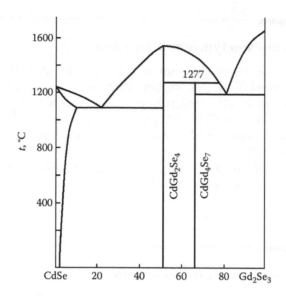

FIGURE 5.22

CdSe–Gd$_2$Se$_3$ phase diagram. (From Azadaliev, R.A. and Alieva, U.R. Physico-chemical investigation of the CdSe–Gd$_2$Se$_3$ system [in Azerbaijani], in *Issled. v obl. sinteza i primenienie neorgan. soed.*, Baku, pp. 93–96, 1985.)

5.20 Cadmium–Terbium–Selenium

CdSe–Tb$_2$Se$_3$: The phase diagram is not constructed. CdTb$_2$Se$_4$ ternary compound is formed in this system, which melts congruently at 1472°C and crystallizes in a cubic structure of Th$_3$P$_4$ type with lattice parameter $a = 880$ pm and calculation and experimental density 6.94 and 6.90 g cm^{-3}, respectively (Aliev et al. 1997).

Single crystals of CdPr$_2$Se$_4$ were grown using the chemical transport reactions (Aliev et al. 1997).

5.21 Cadmium–Dysprosium–Selenium

CdSe–Dy$_2$Se$_3$: The phase diagram is not constructed. CdDy$_2$Se$_4$ and CdDy$_4$Se$_7$ ternary compounds are formed in this system. CdDy$_2$Se$_4$ melts congruently at 1417°C and crystallizes in a cubic structure of spinel type with lattice parameter $a = 1169$ pm and calculation and experimental density 6.28 and 6.18 g cm^{-3}, respectively (Aliev et al. 1997).

CdDy$_4$Se$_7$ crystallizes in a monoclinic structure with lattice parameters $a = 1334$, $b = 412$, $c = 1228$ pm, and $\beta = 106.30°$ (Aliev et al. 1986).

Single crystals of $CdDy_2Se_4$ were grown using chemical transport reactions (Aliev et al. 1997). $CdDy_4Se_7$ was obtained from the binary components in quartz ampoules (Aliev et al. 1986).

5.22 Cadmium–Holmium–Selenium

$CdSe-Ho_2Se_3$: The phase diagram is not constructed. $CdHo_2Se_4$ and $CdHo_4Se_7$ ternary compounds are formed in this system. $CdHo_2Se_4$ melts congruently at 1397°C and crystallizes in a cubic structure of spinel type with lattice parameter $a = 1164$ pm (Aliev et al. 1997) [$a = 1163.8$ pm (Fujii Horonobu et al. 1972), $a = 1573.4$ pm (Range and Eglmeier 1991)] and calculation and experimental density 6.41 [6.40 (Range and Eglmeier 1991)] and 6.32 g cm^{-3}, respectively (Aliev et al. 1997).

$CdHo_4Se_7$ crystallizes in a monoclinic structure with lattice parameters $a = 1330$, $b = 410$, $c = 1225$ pm, and $\beta = 106.00°$ (Aliev et al. 1986).

$CdHo_2Se_4$ was obtained by the heating of mixtures from chemical elements in stoichiometric ratio at 800°C–900°C for 1 week (Fujii Horonobu et al. 1972). Its single crystals were grown using chemical transport reactions (Aliev et al. 1997). $CdHo_4Se_7$ was produced from the binary components in quartz ampoules (Aliev et al. 1986).

5.23 Cadmium–Erbium–Selenium

$CdSe-Er_2Se_3$: The phase diagram is not constructed. $CdEr_2Se_4$ ternary compound is formed in this system that crystallizes in a cubic structure of spinel type with lattice parameter $a = 1160$ pm [$a = 1160.3$ pm (Fujii Horonobu et al. 1972)] and calculation and experimental density 6.51 and 6.45 g cm^{-3}, respectively (Aliev et al. 1997).

$CdEr_2Se_4$ compound was obtained by the heating of mixtures from chemical elements in stoichiometric ratio at 800°C–900°C for 1 week (Fujii Horonobu et al. 1972). Its single crystals were grown using chemical transport reactions (Aliev et al. 1997).

5.24 Cadmium–Thulium–Selenium

$CdSe-Tm_2Se_3$: The phase diagram is not constructed. $CdTm_2Se_4$ and $CdTm_4Se_7$ ternary compounds are formed in this system. $CdTm_2Se_4$ crystallizes in a cubic structure of spinel type with lattice parameter $a = 1157$ pm

[$a = 1544.8$ pm (Range and Eglmeier 1991)] and calculation and experimental density 6.50 [6.59 (Range and Eglmeier 1991)] and 6.54 g cm^{-3}, respectively (Aliev et al. 1997).

$CdTm_4Se_7$ crystallizes in a monoclinic structure with lattice parameters $a = 1324$, $b = 400$, $c = 1213$ pm, and $\beta = 105.75°$ (Aliev et al. 1986).

Single crystals of $CdTm_2Se_4$ were grown using chemical transport reactions (Aliev et al. 1997). $CdTm_4Se_7$ was produced from the binary components in quartz ampoules (Aliev et al. 1986).

5.25 Cadmium–Ytterbium–Selenium

$CdSe–Yb_2Se_3$: The phase diagram is not constructed. $CdYb_2Se_4$ ternary compound is formed in this system, which crystallizes in a cubic structure of spinel type with lattice parameter $a = 1154$ pm and calculation and experimental density 6.71 and 6.60 g cm^{-3}, respectively (Aliev et al. 1997).

Single crystals of $CdTm_2Se_4$ were grown using chemical transport reactions (Aliev et al. 1997).

5.26 Cadmium–Lutetium–Selenium

$CdSe–Lu_2Se_3$: The phase diagram is not constructed. $CdLu_2Se_4$ ternary compound is formed in this system, which crystallizes in a cubic structure of spinel type with lattice parameter $a = 1150$ pm and calculation and experimental density 6.76 and 6.70 g cm^{-3}, respectively (Aliev et al. 1997).

Single crystals of $CdTm_2Se_4$ were grown using chemical transport reactions (Aliev et al. 1997).

5.27 Cadmium–Silicon–Selenium

The field of Si primary crystallization occupies the biggest part of the liquidus surface of the Cd–Si–Se ternary system (Figure 5.23) (Odin and Ivanov 1991). It adjoins with the fields of primary crystallization for SiSe, α- and β-Cd_4SiSe_6, CdSe, and Cd phases. The field of β-Cd_4SiSe_6 crystallization is tapered with the increase of Se contents. There is also the field of $SiSe_2$ crystallization and degenerated fields of Cd and Se crystallization on the liquidus surface. Immiscibility region exists in the field of CdSe primary crystallization. The data about three- and four-phase equilibria in the Cd–Si–Se ternary system are given in Table 5.1.

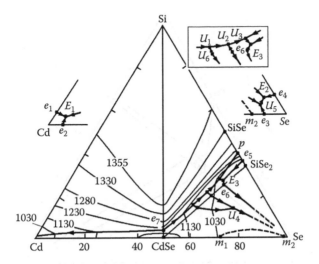

FIGURE 5.23

Liquidus surface of the Cd–Si–Se ternary system. (From Odin, I.N. and Ivanov, V.A., *Zhurn. neorgan. khimii*, 36(11), 2937, 1991.)

TABLE 5.1

Three- and Four-Phase Equilibria
with Participation of Liquid in the Cd–Si–Se
Ternary System

Symbol	Reaction	$t, °C$
e_1	$L \Leftrightarrow Cd + Si$	320
e_2	$L \Leftrightarrow Cd + CdSe$	319
e_3	$L_1 \Leftrightarrow CdSe + Se^*$	219
e_4	$L_1 \Leftrightarrow SiSe_2 + Se^*$	216
e_5	$L \Leftrightarrow SiSe_2 + SiSe$	946
M_1M_2	$L \Leftrightarrow CdSe + L_1^*$	997
P	$L + Si \Leftrightarrow SiSe$	1190
E_1	$L \Leftrightarrow Cd + Si + CdSe$	317
E_2	$L_1 \Leftrightarrow \alpha\text{-}Cd_4SiSe_6 + SiSe_2 + Se^*$	213
E_3	$L \Leftrightarrow \alpha\text{-}Cd_4SiSe_6 + SiSe_2 + SiSe$	905
U_1	$L + CdSe \Leftrightarrow \beta\text{-}Cd_4SiSe_6 + Si$	1097
U_2	$L + \beta\text{-}Cd_4SiSe_6 \Leftrightarrow SiSe + Si$	959
U_3	$L + \alpha\text{-}Cd_4SiSe_6 \Leftrightarrow SiSe + Si$	950
U_4	$L + \beta\text{-}Cd_4SiSe_6 \Leftrightarrow \alpha\text{-}Cd_4SiSe_6 + CdSe$	940
U_5	$L_1 + CdSe \Leftrightarrow \alpha\text{-}Cd_4SiSe_6 + Se^*$	217

Source: Odin, I.N. and Ivanov, V.A., *Zhurn. neorgan. khimii*, 36(11), 2937, 1991.

* Se-rich liquid.

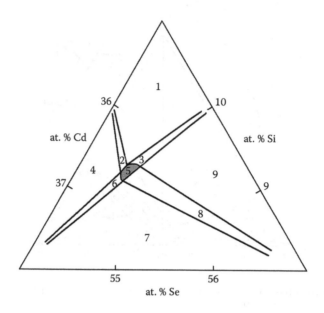

FIGURE 5.24
Part of isothermal section of the Cd–Si–Se ternary system at 730°C: 1, α-Cd$_4$SiSe$_6$+Si + SiSe$_2$; 2, α-Cd$_4$SiSe$_6$+Si; 3, α-Cd$_4$SiSe$_6$+SiSe$_2$; 4, α-Cd$_4$SiSe$_6$+Si+CdSe; 5, α-Cd$_4$SiSe$_6$; 6, α-Cd$_4$SiSe$_6$+CdSe; 7, α-Cd$_4$SiSe$_6$+CdSe+L; 8, α-Cd$_4$SiSe$_6$+L; and 9, α-Cd$_4$SiSe$_6$+SiSe$_2$+L. (From Odin, I.N. and Ivanov, V.A., *Zhurn. neorgan. khimii*, 36(11), 2937, 1991.)

CdSe, SiSe$_2$, α-Cd$_4$SiSe$_6$, Si, and liquid phases involve in equilibrium at 730°C (Figure 5.24) (Odin and Ivanov 1991). The homogeneity region of α-Cd$_4$SiSe$_6$ is elongated along the CdSe–SiSe$_2$ quasibinary system.

CdSe–Si: The phase diagram is a eutectic type (Figure 5.25) (Odin and Ivanov 1991). The eutectic composition and temperature are 6 ± 1 at. % Si and 1210°C, respectively. The solubility of Si in CdSe is not higher than 0.4 at. %.

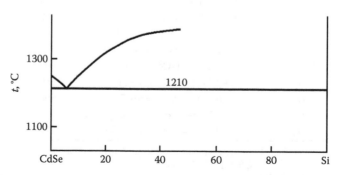

FIGURE 5.25
CdSe–Si phase diagram. (From Odin, I.N. and Ivanov, V.A., *Zhurn. neorgan. khimii*, 36(11), 2937, 1991.)

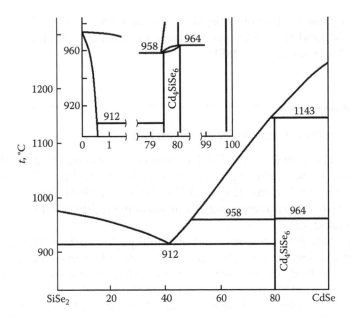

FIGURE 5.26
CdSe–SiSe₂ phase diagram. (From Odin, I.N. and Ivanov, V.A., *Zhurn. neorgan. khimii*, 36(11), 2937, 1991.)

This system was investigated using DTA, metallography, XRD, and micro-hardness and density measurements. The ingots were annealed at 730°C for 720 h (Odin and Ivanov 1991).

CdSe–SiSe₂: The phase diagram is shown in Figure 5.26 (Odin and Ivanov 1991). Cd₄SiSe₆ compound is formed in this system, which melts incongruently at 1143°C and has polymorphous transformation at 958°C–964°C (within the homogeneity region) (Odin et al. 1985, Odin and Ivanov 1991). The eutectic composition and temperature are 41 mol. % CdSe and 912°C, respectively. Homogeneity region of α-Cd₄SiSe₆ along the CdSe–SiSe₂ system is within the interval of (79.5 ± 0.25)–(80.1 ± 0.15) mol. % CdSe. The solubility of SiSe₂ in CdSe is equal to 0.4 mol. % and the solubility of CdSe in SiSe₂ is not higher than 0.6 mol. % (Odin and Ivanov 1991).

α-Cd₄SiSe₆ crystallizes in a monoclinic structure with lattice parameters $a = 1282.66 \pm 0.05$, $b = 735.91 \pm 0.04$, $c = 1281.97 \pm 0.05$ pm, and $\beta = 110.052 \pm 0.001°$ and calculation density 5.5591 ± 0.0007 g cm⁻³ (Parasyuk et al. 2003) [$a = 1281$, $b = 735.2$, $c = 1280$ pm, and $\beta = 110.00°$ and calculation and experimental density 5.58 and 5.56 g cm⁻³, respectively (Odin and Ivanov 1991); $a = 1281.6 \pm 0.6$, $b = 735.5 \pm 0.4$, $c = 1281.1 \pm 0.6$ pm, and $\beta = 110.06 \pm 0.05°$ and calculation and experimental density 5.571 and 5.52 ± 0.01 g cm⁻³, respectively (Krebs von and Mandt 1972)].

This system was investigated using DTA, metallography, XRD, and microhardness and density measurements. The ingots were annealed at 730°C for 720 h (Odin et al. 1985, Odin and Ivanov 1991). Cd₄SiSe₆ has been prepared by the reaction of CdSe, Si, and Se at 800°C–1000°C (Krebs von and Mandt 1972), and its single crystals were grown using chemical transport reactions (Odin and Ivanov 1991).

5.28 Cadmium–Germanium–Selenium

On the liquidus surface of the Cd–Ge–Se ternary system, there are the fields of Cd_4GeSe_6, CdSe 9GeSe (γ-phase), CdSe, $GeSe_2$, α- and β-GeSe, Ge, Se, and Cd primary crystallization (Figure 5.27) (Odin et al. 1985). Phase relations in this system are determined using 19 nonvariant equilibria with participation of liquid (Table 5.2).

Three immiscibility regions exist on the liquidus surface. One of them is situated in the field of Ge crystallization, the second covers a part of the CdSe crystallization field, and the third starts in the CdSe–Se subsystem and occupies a small part of the ternary system from the Se-rich side.

At 600°C, solid phases Cd_4GeSe_6, γ-phase ($CdGe_9Se_{10}$), CdSe, $GeSe_2$, α-GeSe, and Ge and liquid are in equilibria (Figure 5.28) (Odin et al. 1985). Liquid regions are situated from Se- and Cd-rich sides and near the eutectics e_3 and E_2.

Homogeneity region of Cd_4GeSe_6 at 544°C in the Cd–Ge–Se ternary system is shown in Figure 5.29 (Odin and Grin'ko 1991b). It is elongated along the CdSe–$GeSe_2$ quasibinary system. $CdGe_7Se_8$ compound exists at 547°C–552°C and does not appear at 544°C.

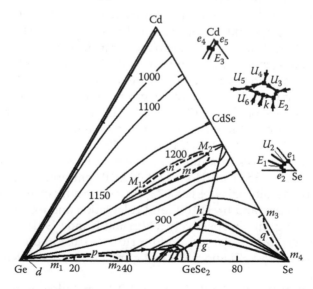

FIGURE 5.27
Liquidus surface of the Cd–Ge–Se ternary system. (From Odin, I.N. et al., *Zhurn. neorgan. khimii*, 30(1), 201, 1985.)

TABLE 5.2

Nonvariant Equilibria with Participation of Liquid
in the Cd–Ge–Se Ternary System

Symbol	Reaction	$t, °C$
e_1	$L \Leftrightarrow CdSe + Se$	214
e_2	$L \Leftrightarrow GeSe_2 + Se$	213
e_3	$L \Leftrightarrow GeSe_2 + \alpha\text{-}GeSe$	590
e_4	$L \Leftrightarrow Cd + Ge$	316
e_5	$L \Leftrightarrow CdSe + Cd$	318
p_1	$L + Ge \Leftrightarrow \beta\text{-}GeSe$	675
g	$\beta\text{-}GeSe \Leftrightarrow \alpha\text{-}GeSe$ (in the presence of L)	664
d	$L \Leftrightarrow CdSe + Ge$	929
f	$L \Leftrightarrow Cd_4GeSe_6 + GeSe_2$	715
h	$L + CdSe \Leftrightarrow Cd_4GeSe_6$	863
E_1	$L \Leftrightarrow Se + CdSe + Cd_4GeSe_6$	205
E_2	$L \Leftrightarrow \alpha\text{-}GeSe + \gamma + GeSe_2$	570
E_3	$L \Leftrightarrow CdSe + Cd + Ge$	311
U_2	$L + CdSe \Leftrightarrow Cd_4GeSe_6 + Se$	212
U_3	$L + Cd_4GeSe_6 \Leftrightarrow \gamma + GeSe_2$	590
U_4	$L + CdSe \Leftrightarrow Cd_4GeSe_6 + \gamma$	664
U_5	$L + CdSe + Ge \Leftrightarrow \gamma$	668
U_6	$L + Ge \Leftrightarrow \beta\text{-}GeSe + \gamma$	660
k	$\alpha\text{-}GeSe \Leftrightarrow \beta\text{-}GeSe$ (in the presence of L and γ)	650
m_3qm_4	$L \Leftrightarrow L_1 + CdSe$	992
m_1pm_2	$L \Leftrightarrow L_1 + Ge$	905
M_1nM_2m	$L \Leftrightarrow L_1 + CdSe$	1211

Source: Odin, I.N. et al., *Zhurn. neorgan. khimii*, 30(1), 201, 1985.
γ—$CdSe \cdot 9GeSe$ ($CdGe_9Se_{10}$).

CdSe–Ge: The phase diagram is a eutectic type (Figure 5.30) (Galiulin et al. 1983). The eutectic composition and temperature are 1.5 mol. % CdSe and 929 ± 6°C, respectively. The immiscibility region exists within the interval of 33–90 mol. % CdSe with monotectic temperature 1211°C. Mutual solubility of CdSe and Ge is not higher than 0.1 mol. %.

The Ge content in the CdSe-doped single crystals reaches 0.65 at. % (0.25 mass. %) (Odin et al. 1999c) (GeSe was used as a dopant).

This system was investigated using DTA, metallography, and XRD. The ingots were annealed at 550°C for 250 h (Galiulin et al. 1983).

CdSe–GeSe: This section is a nonquasibinary section of the Cd–Ge–Se ternary system that intersects the fields of CdSe and Ge primary crystallization (Figure 5.31) (Odin et al. 1985). Secondary crystallization of CdSe and Ge takes place at 683°C, and at 668°C, there is twice peritectic equilibrium CdSe + Ge + L $\Leftrightarrow$ CdSe 9GeSe (γ). Peritectic reaction L + Ge $\Leftrightarrow$ γ + β-GeSe takes place at 660°C. Thermal effects at 625°C–627°C are determined by the

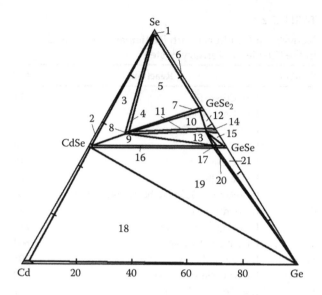

FIGURE 5.28

Isothermal section of the Cd–Ge–Se ternary system at 600°C: 1, L_1; 2, $L_1 + CdSe$; 3, $L_1 + CdSe + Cd_4GeSe_6$; 4, $L_1 + Cd_4GeSe_6$; 5, $L_1 + GeSe_2 + Cd_4GeSe_6$; 6, $L_1 + GeSe_2$; 7, $GeSe_2 + Cd_4GeSe_6$; 8, Cd_4GeSe_6; 9, $CdGe_9Se_{10} + Cd_4GeSe_6 + CdSe$; 10, $L_2 + GeSe_2 + Cd_4GeSe_6$; 11, $L_2 + Cd_4GeSe_6$; 12, $L_2 + GeSe_2$; 13, $L_2 + Cd_4GeSe_6 + CdGe_9Se_{10}$; 14, L_2; 15, $L_2 + \alpha$-$GeSe$; 16, $CdGe_9Se_{10} + CdSe$; 17, $CdGe_9Se_{10}$; 18, $L + CdSe + Ge$; 19, $CdGe_9Se_{10} + CdSe + Ge$; 20, $CdGe_9Se_{10} + Ge$; and 21, $CdGe_9Se_{10} + \alpha$-$GeSe + Ge$. (From Odin, I.N. et al., *Zhurn. neorgan. khimii*, 30(1), 201, 1985.)

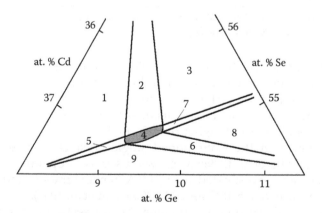

FIGURE 5.29

Homogeneity region of Cd_4GeSe_6 at 544°C in the Cd–Ge–Se ternary system: 1, $L + CdSe + Cd_4GeSe_6$; 2, $L + Cd_4GeSe_6$; 3, $L + Cd_4GeSe_6 + GeSe_2$; 4, Cd_4GeSe_6; 5, $CdSe + Cd_4GeSe_6$; 6, $GeSe + Cd_4GeSe_6$; 7, $GeSe_2 + Cd_4GeSe_6$; 8, $GeSe + GeSe_2 + Cd_4GeSe_6$; and 9, $CdSe + GeSe + Cd_4GeSe_6$. (From Odin, I.N. and Grin'ko, V.V., *Zhurn. neorgan. khimii*, 36(9), 2387, 1991.)

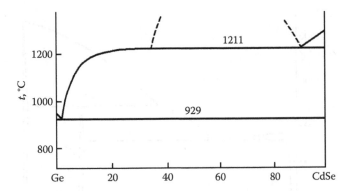

FIGURE 5.30
CdSe–Ge phase diagram. (From Galiulin, E.A. et al., *Zhurn. neorgan. khimii*, 28(5), 1281, 1983.)

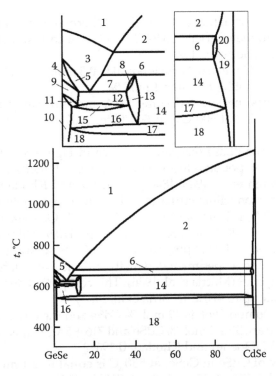

FIGURE 5.31
Phase relations in the CdSe–GeSe system: 1, L; 2, L+CdSe; 3, L+Ge; 4, L+β-GeSe; 5, L+Ge+ β-GeSe; 6, L+CdSe+Ge; 7, L+γ+Ge; 8, γ+Ge; 9, β-GeSe; 10, α-GeSe; 11, β-GeSe+α-GeSe; 12, γ+β-GeSe; 13, γ; 14, γ+CdSe; 15, γ+β-GeSe+α-GeSe; 16, α-GeSe+γ; 17, α-GeSe+γ+CdSe; 18, α-GeSe+CdSe; 19, CdSe+Ge; and 20, CdSe. (From Odin, I.N. et al., *Zhurn. neorgan. khimii*, 30(1), 201, 1985.)

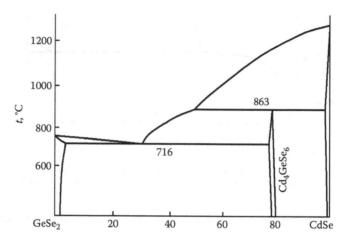

FIGURE 5.32
CdSe–GeSe$_2$ phase diagram. (From Galiulin, E.A. et al., *Zhurn. neorgan. khimii*, 28(5) 1281, 1983.)

polymorphic transformation of GeSe. Small thermal effects are on the heating curves within the interval of 540°C–560°C that correspond to the formation of γ from α-GeSe and CdSe.

The solubility of CdSe in α-GeSe is equal to 0.5 mol. % and the solubility of α-GeSe in CdSe is not higher than 1 mol. %.

This system was investigated using DTA, metallography, and XRD. The ingots were annealed at 500°C and 600°C for 1100 and 1000 h, respectively (Odin et al. 1985).

CdSe–GeSe$_2$: The phase diagram is shown in Figure 5.32 (Galiulin et al. 1983). Cd$_4$GeSe$_6$ compound is formed in this system. It melts incongruently at 863 ± 6°C (Galiulin et al. 1983) [840°C (Quenez and Khodadad 1969)] and crystallizes in a monoclinic structure with lattice parameters $a = 1281 \pm 2$, $b = 738 \pm 1$, $c = 1279 \pm 2$ pm, and $\beta = 109°34' \pm 0.1'$ (Quenez and Khodadad 1969) [$a = 1293.7$, $b = 740.2$, $c = 1285.2$ pm, and $\beta = 110.9°$ (Motria et al. 1986); $a = 1282.3$, $b = 740.9$, $c = 1280.2$ pm, and $\beta = 109.60°$ (Susa and Steinfink 1971)] and calculation and experimental density 5.75 [5.444 (Hahn and Lorent 1958)] and 5.57 g cm^{-3} (Motria et al. 1986). This compound decomposes at the heating on the air up to 470°C (Motria et al. 1987).

The peritectic composition is 49 mol. % CdSe and the eutectic composition and temperature are 30 ± 2 mol. % CdSe and 716 ± 5°C, respectively (Galiulin et al. 1983) [708°C (Quenez and Khodadad 1969)].

The solubility of CdSe in GeSe$_2$ at 550°C is equal to 2.1 mol. % (Galiulin et al. 1983) and 1.57 ± 0.05 mol. % at 710°C (Odin and Grin'ko 1991b). The solubility of GeSe$_2$ in CdSe at 550°C is not higher than 0.7 mol. % (Galiulin et al. 1983) and 0.44 ± 0.06 mol. % at 785°C (Odin and Grin'ko 1991b). Homogeneity region of Cd$_4$GeSe$_6$ is within the interval of 78.5–80.3 mol. % CdSe (Galiulin et al. 1983) [78.9–80.1 mol. % CdSe (Odin and Grin'ko 1991b)].

Ternary compounds CdGeSe$_3$ (Radautsan and Ivanova 1961) and Cd$_2$GeSe$_4$ (Hahn and Lorent 1958) were not found in this system (Quenez and Khodadad 1969, Galiulin et al. 1983). Galiulin et al. (1983) did not confirm the existence of thermal effects at 450°C near the GeSe$_2$-rich side that characterize devitrification of ingots (Quenez and Khodadad 1969). According to the data of Zhao et al. (2005), no glass was obtained in the CdSe–GeSe$_2$ system.

Vapor composition in the CdSe–GeSe$_2$ system corresponds to GeSe$_2$ compound, that is, it lies on this section (Odin et al. 1986, Motria et al. 1987, Odin and Grin'ko 1991a,b). The vapor above Cd$_4$GeSe$_6$ consists mainly from GeSe and Se$_2$ in ratio of 2 to 1, and the Cd concentration in vapor is 0.01 mass. % (Odin and Grin'ko 1991b). The limit of Cd$_4$GeSe$_6$ homogeneity region from the CdSe side does not change with temperature and from the GeSe$_2$-side changes at temperatures above a eutectic temperature (Figure 5.33) (Odin and Grin'ko 1991b). $(T–x)_p$ sections of the CdSe–GeSe$_2$ system at $p=133$ hPa are shown in Figure 5.34 (Odin and Grin'ko 1991b).

On $p–T$ projection of CdSe–GeSe$_2$ phase diagram (Figure 5.35) (Odin et al. 1986), the lines represent two-phase equilibria [AB, S(GeSe$_2$)/V; BC, L(GeSe$_2$)/V; KG, S(CdSe)/V; GF, L(CdSe)/V] and three-phase equilibria [BD, S(GeSe$_2$)/L/V; MD, V/S(GeSe$_2$)/S(Cd$_4$GeSe$_6$); DE, S(Cd$_4$GeSe$_6$)/L/V]. The lines of three-phase equilibria intersect in the nonvariant points: D, V/S(GeSe$_2$)/L/S(Cd$_4$GeSe$_6$), and E, V/S(Cd$_4$GeSe$_6$)/L/S(CdSe).

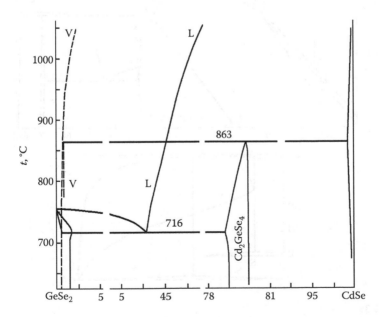

FIGURE 5.33
$T–x$ projection of CdSe–GeSe$_2$ phase diagram close to Cd$_4$GeSe$_6$, CdSe, and GeSe$_2$. (From Odin, I.N. and Grin'ko, V.V., *Zhurn. neorgan. khimii*, 36(9), 2387, 1991.)

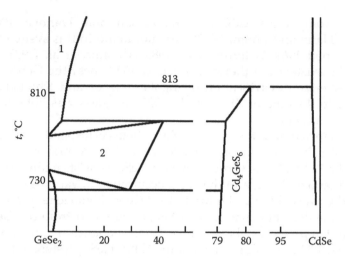

FIGURE 5.34
$(T–x)_p$ sections of CdSe–GeSe$_2$ phase diagram at 133 hPa. (From Odin, I.N. and Grin'ko, V.V., *Zhurn. neorgan. khimii*, 36(9), 2387, 1991.)

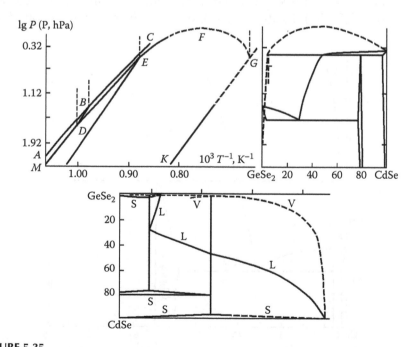

FIGURE 5.35
$P_{total}–T–x$ phase diagram of the CdSe–GeSe$_2$ system: 1, GeSe$_2$; 2, 7.0; 3, 60.0; 4, 80.5; and 5, 85.0 mol. % CdSe. (From Odin, I.N. et al., *Zhurn. neorgan. khimii*, 31(5), 1274, 1986.)

The coordinates of D and E points are 716°C and 4213 Pa and 863°C and 71.3 kPa, respectively. The lines, representing equilibria of condensed phases [L(GeSe$_2$)/S(GeSe$_2$), S(GeSe$_2$)/L/S(Cd$_4$GeSe$_6$), S(Cd$_4$GeSe$_6$)/L/ S(CdSe), and S(CdSe)/L(CdSe)], are practically vertical at small pressures. The region of 1000°C–1250°C is not investigated, but it can be confirmed that the line of three-phase equilibrium V/L/S(CdSe) passes trough maximum.

This system was investigated using DTA, metallography, XRD and chemical analysis, microhardness measurement, and tensimeter (Quenez and Khodadad 1969, Galiulin et al. 1983, Odin et al. 1986, Odin and Grin'ko 1991a,b). The ingots were annealed at 550°C for 750 h (Galiulin et al. 1983) [at 650°C for 720 h (Odin and Grin'ko 1991b)]. Single crystals of Cd$_4$GeSe$_6$ were grown using chemical transport reaction (Motria et al. 1986, 1987).

5.29 Cadmium–Tin–Selenium

There are seven fields of primary crystallization on the liquidus surface of the Cd–Sn–Se ternary system (Figure 5.36) (Odin et al. 1983). The field of CdSe primary crystallization occupies almost all liquidus surface. Three immiscibility regions exist in this system. One of them begins in the Cd–Se binary system and occupies a part of CdSe crystallization

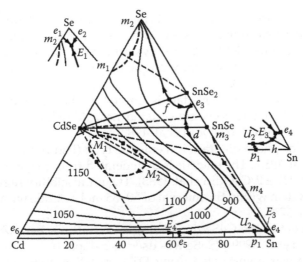

FIGURE 5.36
Liquidus surface of the Cd–Sn–Se ternary system. (From Odin, I.N. et al., *Zhurn. neorgan. khimii*, 28(3), 764, 1983.)

TABLE 5.3

Nonvariant Equilibria in the Cd–Sn–Se
Ternary System at the Participation
of Liquid

Symbol	Reaction	t, °C
e_1	L $\Leftrightarrow$ CdSe + Se	214
e_2	L $\Leftrightarrow$ SnSe$_2$ + Se	219
e_3	L $\Leftrightarrow$ SnSe$_2$ + SnSe	627
e_4	L $\Leftrightarrow$ SnSe + Sn	231
e_5	L $\Leftrightarrow$ Cd + β*	196
e_6	L $\Leftrightarrow$ CdSe + Cd	318
p_1	L + Sn $\Leftrightarrow$ β*	223
d	L $\Leftrightarrow$ CdSe + SnSe	822
f	L $\Leftrightarrow$ CdSe + SnSe$_2$	622
h	L $\Leftrightarrow$ CdSe + Sn	231
E_1	L $\Leftrightarrow$ CdSe + SnSe$_2$ + Se	212
E_2	L $\Leftrightarrow$ CdSe + SnSe + SnSe$_2$	605
E_3	L $\Leftrightarrow$ CdSe + SnSe + Sn	230
E_4	L $\Leftrightarrow$ CdSe + Cd + β*	190
U	L + Sn $\Leftrightarrow$ CdSe + β*	220

Source: Odin, I.N. et al., *Zhurn. neorgan. khimii,*
　　　　　28(3), 764, 1983.
* β-phase from the Cd–Sn system.

field, the second occupies a part of SnSe crystallization field and is situated near the Sn–Se binary system and the third is situated inside the ternary system in the field of CdSe crystallization. Nonvariant equilibria in the Cd–Sn–Se ternary system with participation of liquid are given in Table 5.3 (Odin et al. 1983).

At 550°C, the region of solid state is limited by the triangle CdSe–SnSe–SnSe$_2$, and the liquid regions are situated near Se-rich corner and Cd–Sn binary system (Figure 5.37) (Odin et al. 1983).

CdSe–Sn: The phase diagram is a eutectic type (Figure 5.38) (Rubenstein 1968, Odin et al. 1983). The eutectic is degenerated from the Sn-rich side and crystallizes at 231 ± 2°C (Odin et al. 1983). The immiscibility region exists in the CdS–Sn system within the interval of 50–85 mol. % CdSe with monotectic temperature 1157°C.

The Sn content in the CdSe-doped single crystals reaches 0.76 at. % (0.45 mass. %) (Odin et al. 1999c) (SnSe was used as a dopant).

This system was investigated using DTA, metallography, and XRD. The ingots were annealed at 550°C for 250 h (Odin et al. 1983). The solubility of CdSe in the liquid Sn was determined using high-temperature filtration (Rubenstein 1968).

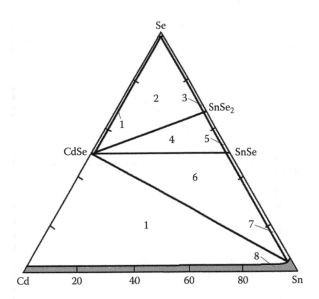

FIGURE 5.37
Isothermal section of the Cd–Sn–Se ternary system at 550°C: 1, L+CdSe; 2, L+CdSe+SnSe₂;
3, L+SnSe₂; 4, CdSe+SnSe+SnSe₂; 5, SnSe+SnSe₂; 6, L+CdSe+SnSe; 7, L+SnSe; and 8, L.
(From Odin, I.N. et al., *Zhurn. neorgan. khimii*, 28(3), 764, 1983.)

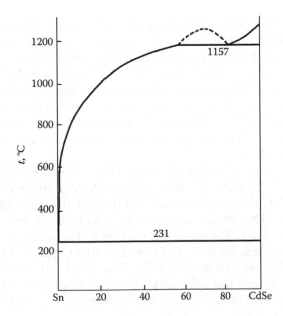

FIGURE 5.38
CdSe–Sn phase diagram. (From Odin, I.N. et al., *Zhurn. neorgan. khimii*, 28(3), 764, 1983.)

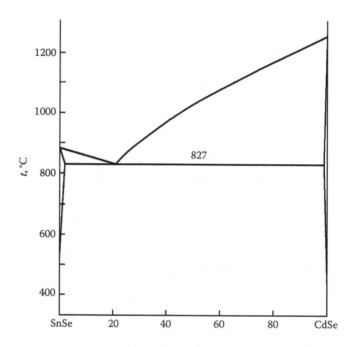

FIGURE 5.39

CdSe–SnSe phase diagram. (From Leute, V. and Menge, D., Z. *Phys. Chem. (Munchen)*, 176(1), 47, 1992.)

CdSe–SnSe: The phase diagram is a eutectic type (Figure 5.39) (Leute and Menge 1992). The eutectic composition and temperature are 21 ± 1 mol. % CdSe and 827°C (Leute and Menge 1992) [19 mol. % CdSe and 822°C (Galiulin et al. 1981)]. The solubility of SnSe in CdSe could not be detected (Leute and Menge 1992). According to the data of Galiulin et al. (1981) and Odin et al. (1983), the solubility of SnSe in CdSe at the eutectic temperature reaches 15 mol. % and decreases to 7 mol. % at 650°C, and the solubility of CdSe in SnSe at the eutectic temperature is equal to 2 mol. % and 1 mol. % at 550°C.

This system was investigated using DTA, metallography, and XRD (Galiulin et al. 1981, Odin et al. 1983, Leute and Menge 1992). The ingots were annealed at 650°C [550°C (Odin et al. 1983)] for 250 h (Galiulin et al. 1981).

CdSe–SnSe$_2$: The phase diagram is a eutectic type (Figure 5.40) (Galiulin et al. 1982, Parasyuk et al. 1999). The eutectic composition and temperature are 23 mol. % CdSe and 618°C [22 ± 3 mol. % CdSe and 620 ± 5°C (Galiulin et al. 1982)]. The solubility of SnSe$_2$ in CdSe reaches 1 mol. % and the solubility of CdSe in SnSe$_2$ is not higher than 0.5 mol. % (Galiulin et al. 1982). CdSnSe$_3$ ternary compound (Radautsan

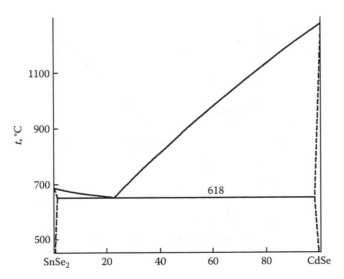

FIGURE 5.40
CdSe–SnSe$_2$ phase diagram. (From Parasyuk, O.V. et al., *Zhurn. neorgan. khimii*, 44(8), 1363, 1999.)

and Ivanova 1961) is not found in the CdSe–SnSe$_2$ system (Galiulin et al. 1982, Parasyuk et al. 1999).

This system was investigated using DTA, metallography, and XRD (Galiulin et al. 1982, Parasyuk et al. 1999). The ingots were annealed at 400°C (Parasyuk et al. 1999) [550°C (Galiulin et al. 1982)] for 250 h.

5.30 Cadmium–Lead–Selenium

CdSe–PbSe: The phase diagram is a eutectic type (Figure 5.41) (Tomashik et al. 1980). The eutectic composition and temperature are 46 mol. % CdSe and 995°C. The data about nonquasibinarity of this system (Wald and Rosenberg 1965) were not confirmed (Tomashik et al. 1980). The solubility of CdSe in PbSe at the eutectic temperature is equal to 30 mol. % (Tomashik et al. 1980) and at 940°C, 800°C, and 600°C, respectively, 26, 12, and 6 mol. % (Wald and Rosenberg 1965, Sealy and Crocker 1973). The solubility of PbSe in CdSe at the eutectic temperature is not higher than 1 mol. % (Tomashik et al. 1980).

This system was investigated using DTA, metallography, and microhardness measurement (Tomashik et al. 1980).

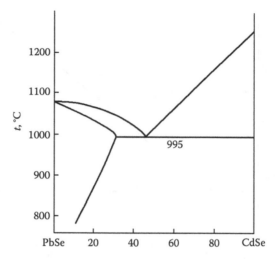

FIGURE 5.41
CdSe–PbSe phase diagram. (From Tomashik, Z.F. et al., *Izv. AN SSSR. Neorgan. materialy*, 16(2), 261, 1980.)

5.31 Cadmium–Phosphorus–Selenium

Isothermal section of the Cd–P–Se ternary system at 500°C is shown in Figure 5.42 (Potoriy et al. 1999). CdPSe$_4$ ternary compound was not obtained in this system.

CdSe–Cd$_3$P$_2$: The phase diagram is a peritectic type (Figure 5.43) (Olekseyuk et al. 1972, Golovey and Rigan 1974). The peritectic interaction takes place at 825°C (Golovey and Rigan 1974). The solubility of 2CdSe in Cd$_3$P$_2$ decreases from 30 mol. % at 825°C to 0.5 mol. % at room temperature, and the solubility of Cd$_3$P$_2$ in 2CdSe decreases at the same temperatures from 15 to 12 mol. %.

This system was investigated using DTA, metallography, XRD, and microhardness measurement (Olekseyuk et al. 1972, Golovey and Rigan 1974).

CdSe–"P$_2$Se$_4$": The phase diagram is a eutectic type (Figure 5.44) (Kovach et al. 1997, Potoriy et al. 1999). The eutectic composition and temperature are 12 mol. % "P$_2$S$_4$" and 710 ± 5°C, respectively. Cd$_2$P$_2$Se$_6$ (CdPSe$_3$) was obtained in this system (Klinger et al. 1970, 1973, Brusilovets and Fedoruk 1976, Brec et al. 1980, Kovach et al. 1997, Potoriy et al. 1999), which melts incongruently at 755 ± 5°C and crystallizes in a rhombohedral structure with lattice parameters $a = 660.1 \pm 0.3$ and $c = 2007.9 \pm 0.8$ pm (Kovach et al. 1997, Potoriy et al. 1999) [$a = 649$ and $c = 1997$ pm (Klinger et al. 1970, 1973), $a = 651.2 \pm 0.1$ and $c = 2006.4 \pm 0.4$ (Brec et al. 1980)], calculation and experimental density 5.20 and 5.12 [5.03 (Tovt et al. 1998)] g cm^{-3}, respectively (Klinger et al. 1973), and energy gap $E_g = 2.28$ eV (Tovt et al. 1998). The heating of Cd$_2$P$_2$Se$_6$

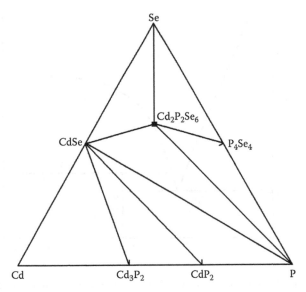

FIGURE 5.42
Isothermal section of the Cd–P–Se ternary system at 500°C. (From Potoriy, M.V. et al., *Neorgan. materialy*, 35(11), 1297, 1999.)

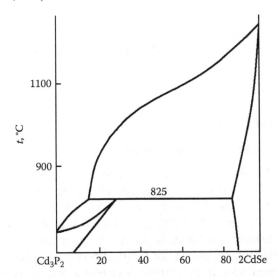

FIGURE 5.43
CdSe–Cd$_3$P$_2$ phase diagram. (From Olekseyuk, I.D. et al., *Izv. AN SSSR. Neorgan. materialy*, 8(4), 696, 1972; Golovey, M.I. and Rigan, M.Yu, *Izv. AN SSSR. Neorgan. materialy*, 10(5), 919, 1974.)

at 450°C–500°C in the ampoule with cold zone leads to its decomposition into CdSe and volatile phosphorus sulfides.

The Cd$_2$P$_2$Se$_6$ single crystals have been grown using chemical transport reactions with iodine in closed silica ampoules (Tovt et al. 1998, Potoriy et al. 1999).

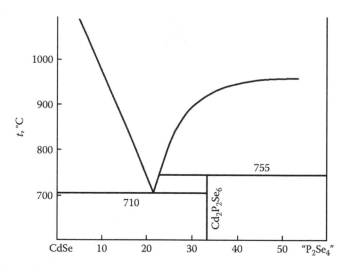

FIGURE 5.44

CdSe–"P$_2$Se$_4$" phase diagram. (From Kovach, A.P. et al., *Ukr. khim. zhurn.*, 63(2), 92, 1997; Potoriy, M.V. et al., *Neorgan. materialy*, 35(11), 1297, 1999.)

The ingots were annealed at 500°C–530°C for 2 weeks and this system was investigated using DTA and XRD (Kovach et al. 1997, Potoriy et al. 1999).

5.32 Cadmium–Arsenic–Selenium

CdSe–Cd$_3$As$_2$: The phase diagram is a peritectic type (Figure 5.45) (Golovey et al. 1972, 1973). The peritectic interaction takes place at 735 ± 5°C (Golovey and Shpyrko 1975). The solubility of Cd$_3$As$_2$ in 2CdSe decreases from 10 mol. % at 735°C to 2 mol. % at 200°C, and the solubility of 2CdSe in Cd$_3$As$_2$ decreases from 20 mol. % at 600°C to 12 mol. % at temperatures below 200°C (Golovey et al. 1972, 1973, Golovey and Shpyrko 1975). β-solid solutions based on Cd$_3$As$_2$ has peritectic transformation at 600 ± 5°C (Golovey et al. 1973, Golovey and Shpyrko 1975).

Single crystals of the solid solutions based on CdSe were grown using chemical transport reaction (Golovey and Rigan 1977). This system was investigated using DTA, metallography, XRD, and microhardness and density measurements (Golovey et al. 1973).

CdSe–As$_2$Se$_3$: The phase diagram is a eutectic type (Figure 5.46) (Aliev et al. 1987). The eutectic temperature is 355°C. The solubility of CdSe in As$_2$Se$_3$ reaches 2.5 mol. %, and the solubility of As$_2$Se$_3$ in CdSe is equal to

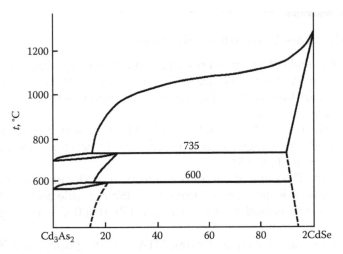

FIGURE 5.45
2CdSe–Cd$_3$As$_2$ phase diagram. (From Golovey, M.I. et al., Chemical bond and solubility in the systems A$^{II}_3$B$^V_2$–A^{II}C^{VI} [in Russian], in *Khim. svyaz' v poluprovodn. i polumetallah*. Minsk, Nauka i Tekhnika Publish., pp. 233–239, 1972; Golovey, M.I. et al., *Izv. AN SSSR. Neorgan. materialy*, 9(9), 1520, 1973.)

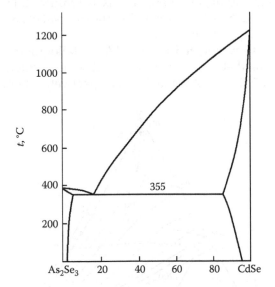

FIGURE 5.46
CdSe–As$_2$Se$_3$ phase diagram. (From Aliev, I.I. et al., *Izv. AN SSSR. Neorgan. materialy*, 23(12), 1965, 1987.)

5 mol. %. Glass region exists in this system at the CdSe contents up to 10 mol. % that reaches 15 mol. % CdSe at the quenching of ingots.

This system was investigated using DTA, metallography, XRD, and microhardness and density measurements. The ingots were annealed at 230°C for 600 h (Aliev et al. 1987).

5.33 Cadmium–Antimony–Selenium

There are six fields of primary crystallization on the liquidus surface of the Cd–Sb–Se ternary system (Figure 5.47) (Sher et al. 1981). Two immiscibility regions exist in this system. Nonvariant equilibria with participation of liquid are given in Table 5.4.

At 400°C, the region of solid state is limited by the subsystem CdSe–CdSb–Sb–Sb$_2$Se$_3$, and liquid regions are situated near Se-rich and Cd-rich corners (Figure 5.48) (Sher et al. 1981).

CdSe–CdSb: The phase diagram is a eutectic type (Figure 5.49) (Sher et al. 1981). The eutectic is degenerated from the CdSb-rich side and crystallizes at 455 ± 3°C. Mutual solubility of CdSe and CdSb at 400°C is not higher than 0.5 mol. %.

This system was investigated using DTA, metallography, and XRD. The ingots were annealed at 400°C for 1000 h, and some of them were annealed also at 550°C and 700°C for 720 and 360 h, respectively (Sher et al. 1981).

CdSe–Sb: The phase diagram is a eutectic type (Figure 5.50) (Sher et al. 1981). The eutectic is degenerated from the Sb-rich side and crystallizes at 628 ± 4°C. Mutual solubility of CdSe and Sb is insignificant. The immiscibility region exists in the CdSe–Sb system within the interval of 28–82 mol. % CdSe with monotectic temperature 1205°C.

This system was investigated using DTA, metallography, and XRD. The ingots were annealed at 400°C for 1000 h, and some of them were annealed also at 550°C and 700°C for 720 and 360 h, respectively (Sher et al. 1981).

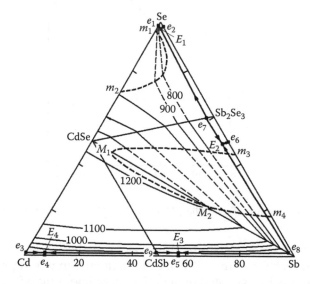

FIGURE 5.47
Liquidus surface of the Cd–Sb–Se ternary system. (From Sher, A.A. et al., *Zhurn. neorgan. khimii*, 26(4), 1052, 1981.)

TABLE 5.4

Nonvariant Equilibria in the Cd–Sn–Se
Ternary System at the Participation of Liquid

Symbol	Reaction	t, °C
e_1	$L \Leftrightarrow CdSe + Se$	213
e_2	$L \Leftrightarrow Sb_2Se_3 + Se$	220
e_3	$L \Leftrightarrow CdSe + Cd$	317
e_4	$L \Leftrightarrow CdSb + Cd$	290
e_5	$L \Leftrightarrow CdSb + Sb$	445
e_6	$L \Leftrightarrow Sb_2Se_3 + Sb$	540
e_7	$L \Leftrightarrow CdSe + Sb_2Se_3$	592
e_8	$L \Leftrightarrow CdSe + Sb$	628
e_9	$L \Leftrightarrow CdSe + CdSb$	455
m_1m_2	$L \Leftrightarrow CdSe + L_1$	991
m_3m_4	$L \Leftrightarrow Sb + L_1$	580
M_1M_2	$L \Leftrightarrow CdSe + L_1$	1205
E_1	$L \Leftrightarrow CdSe + Sb_2Se_3 + Se$	210
E_2	$L \Leftrightarrow CdSe + Sb_2Se_3 + Sb$	512
E_3	$L \Leftrightarrow CdSb + Sb_2Se_3 + Sb$	431
E_4	$L \Leftrightarrow CdSe + CdSb + Cd$	275

Source: Sher, A.A. et al., *Zhurn. neorgan. khimii,*
26(4), 1052, 1981.

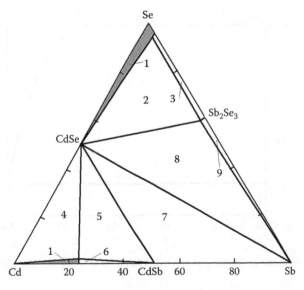

FIGURE 5.48
Isothermal section of the Cd–Sb–Se ternary system at 400°C: 1, L; 2, L+CdSe+Sb$_2$Se$_3$; 3, L+
Sb$_2$Se$_3$; 4, L+CdSe; 5, L+CdSe+CdSb; 6, L+CdSb; 7, CdSe+CdSb+Sb; 8, CdSe+Sb+Sb$_2$Se$_3$; and
9, Sb+Sb$_2$Se$_3$. (From Sher, A.A. et al., *Zhurn. neorgan. khimii,* 26(4), 1052, 1981.)

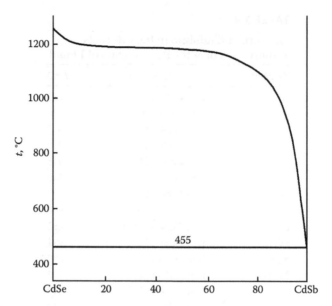

FIGURE 5.49
CdSe–CdSb phase diagram. (From Sher, A.A. et al., *Zhurn. neorgan. khimii*, 26(4), 1052, 1981.)

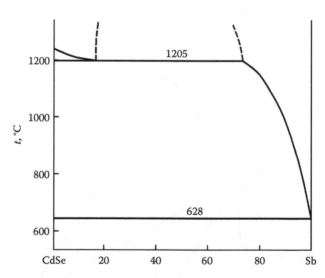

FIGURE 5.50
CdSe–Sb phase diagram. (From Sher, A.A. et al., *Zhurn. neorgan. khimii*, 26(4), 1052, 1981.)

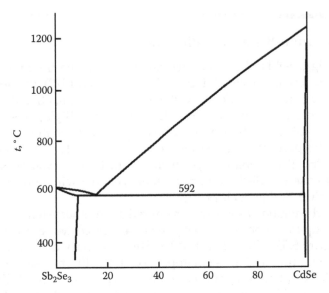

FIGURE 5.51
CdSe–Sb₂Se₃ phase diagram. (From Novoselova, A.V. et al., *Zhurn. neorgan. khimii*, 23(12), 3321, 1978.)

CdSe–Sb₂Se₃: The most reliable phase diagram is a eutectic type (Figure 5.51) (Novoselova et al. 1978). The eutectic composition and temperature are 15 ± 2 mol. % CdSe and 592 ± 3°C, respectively (Novoselova et al. 1978) [26 mol. % CdSe and 485°C (Safarov 1988)]. The solubility of CdSe in Sb₂Se₃ at 550°C and 450°C is equal to 6 mol. %, and the solubility of Sb₂Se₃ in CdSe at 700°C and 450°C constitutes 2 and 1.5 mol. % (Novoselova et al. 1978) [1 mol. % at 550°C (Sher et al. 1981), 0.75 ± 0.10 mol. % at 530°C (Odin et al. 2000)]. According to the data of Safarov (1988), solid solutions based on Sb₂Se₃ and CdSe contain 8.5 mol. % CdSe and 16 mol. % Sb₂Se₃, respectively.

Thermal effects at 610±3°C within the interval of 20–95 mol. % CdSe that can correspond to the metastable transformations were determined (Novoselova et al. 1978). However, these effects can be dictated by the formation of CdSb₂Se₄ ternary compounds that melt incongruently at 593°C (Safarov 1988).

According to the data of Safarov (1988), solid solutions based on CdSe have polymorphous transformation; however, CdSe has no any transformation in investigated interval of temperatures.

This system was investigated using DTA, metallography, XRD, and microhardness and density measurements (Novoselova et al. 1978, Safarov 1988, Odin et al. 2000). The ingots were annealed at 450°C for 1000 h (Novoselova et al. 1978) [at 350°C for 200 h or at 400°C for 300 h (Safarov 1988), at 530°C for 640 h (Odin et al. 2000)].

5.34 Cadmium–Bismuth–Selenium

There are eight fields of primary crystallization on the liquidus surface of the Cd–Bi–Se ternary system (Figure 5.52) (Sher et al. 1979): the degenerated fields of Cd, Bi, and Se; the field of CdSe primary crystallization that occupies almost all liquidus surface; and the fields of $CdBi_2Se_4$, Bi_2Se_3, BiSe, and Bi_3Se_2 primary crystallization. Two immiscibility regions exist in this system: one of them occupies parts of the CdSe and Bi_2Se_3 fields, and the second is situated along the CdSe–Bi quasibinary system inside the ternary system. Nonvariant equilibria in the Cd–Bi–Se ternary system with participation of liquid are given in Table 5.5 (Sher et al. 1979).

At 450°C, the region of solid state is limited by the subsystem CdSe–Bi_2Se_3–Bi_3Se_2, and liquid regions are situated near Se-rich corner and Cd–Bi binary system (Figure 5.53) (Sher et al. 1979). $CdBi_2Se_4$ ternary compound exists at higher temperatures and does not appear on the isothermal section at 450°C.

CdSe–Bi: The phase diagram is a eutectic type (Figure 5.54) (Rubenstein 1966, Sher et al. 1979). The eutectic is degenerated from the Bi-rich side. The immiscibility region exists in the CdSe–Bi system within the interval of 25–85 mol. % CdSe with monotectic temperature 1210°C.

This system was investigated using DTA, metallography, and XRD. The ingots were annealed at 230°C for 1000 h (Sher et al. 1979). Temperature

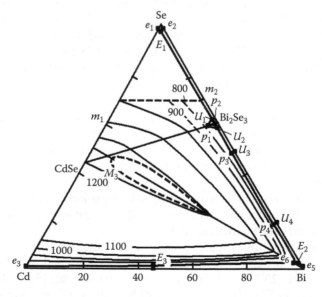

FIGURE 5.52

Liquidus surface of the Cd–Bi–Se ternary system. (From Sher, A.A. et al., *Zhurn. neorgan. khimii*, 24(9), 2509, 1979.)

TABLE 5.5

Nonvariant Equilibria in the Cd–Bi–Se
Ternary System with Participation
of Liquid

Symbol	Reaction	t, °C
e_1	$L \Leftrightarrow CdSe + Se$	213
e_2	$L \Leftrightarrow Bi_2Se_3 + Se$	217
e_3	$L \Leftrightarrow CdSe + Cd$	317
e_4	$L \Leftrightarrow Cd + Bi$	146
e_5	$L \Leftrightarrow CdSe + Bi$	265
e_6	$L \Leftrightarrow Bi_2Se_3 + Bi$	270
p_1	$L + CdSe \Leftrightarrow CdBi_2Se_4$	736
p_2	$L + CdBi_2Se_4 \Leftrightarrow Bi_2Se_3$	712
p_3	$L + Bi_2Se_3 \Leftrightarrow BiSe$	607
p_4	$L + BiSe \Leftrightarrow Bi_3Se_2$	468
m_2	$L_1 \Leftrightarrow L + Bi_2Se_3$	618
m_1	$L_1 \Leftrightarrow L + CdSe$	991
M_3	$L_1 \Leftrightarrow L + CdSe$	1210
E_1	$L \Leftrightarrow CdSe + Bi_2Se_3 + Se$	210
E_2	$L \Leftrightarrow CdSe + Bi_2Se_3 + Bi$	253
E_3	$L \Leftrightarrow CdSe + Bi + Cd$	128
U_1	$L + CdBi_2Se_4 \Leftrightarrow CdSe + Bi_2Se_3$	≈700
U_2	$L + CdBi_2Se_4 \Leftrightarrow CdSe + Bi_2Se_3$	≈700
U_3	$L + Bi_2Se_3 \Leftrightarrow CdSe + BiSe$	576
U_4	$L + BiSe \Leftrightarrow CdSe + Bi_3Se_2$	460

Source: Sher, A.A. et al., *Zhurn. neorgan. khimii,*
24(9), 2509, 1979.

dependence of CdSe solubility in the liquid bismuth was determined using high-temperature filtration (Rubenstein 1966).

CdSe–BiSe: This section is a nonquasibinary section of the Cd–Bi–Se ternary system since BiSe melts incongruently (Figure 5.55) (Sher et al. 1979). Solid solutions based on BiSe contain 2 mol. % CdSe, and CdSe dissolves less than 1 mol. % BiSe. *AF* line is the line of CdSe and Bi_2Se_3 secondary crystallization. The crystallization ends in the ternary peritectic at 576°C. Mixtures of CdSe and BiSe exist in the solid state of the CdSe–BiSe system.

This system was investigated using DTA, metallography, and XRD. The ingots were annealed at 450°C for 1000 h (Sher et al. 1979).

CdSe–Bi$_2$Se$_3$: The phase diagram is a peritectic type (Figure 5.56) (Novoselova et al. 1978). $CdBi_2Se_4$ compound is formed in this system. It melts incongruently at 736°C and decomposes at a temperature below 604°C forming CdSe and Bi_2Se_3. The solubility of Bi_2Se_3 in CdSe is equal to 0.25 mol. % at 450°C and approximately 1 mol. % at 700°C [less than

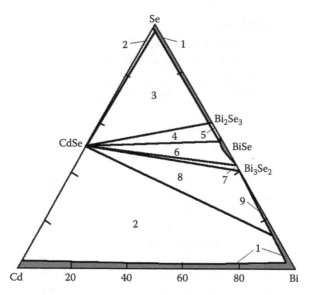

FIGURE 5.53
Isothermal section of the Cd–Bi–Se ternary system at 450°C: 1, L; 2, L+CdSe; 3, L+CdSe+ Bi$_2$Se$_3$; 4, CdSe+Bi$_2$Se$_3$+BiSe; 5, Bi$_2$Se$_3$+BiSe; 6, CdSe+BiSe; 7, CdSe+BiSe+Bi$_3$Se$_2$; 8, L+CdSe+Bi$_2$Se$_2$; and 9, L+Bi$_3$Se$_2$. (From Sher, A.A. et al., *Zhurn. neorgan. khimii*, 24(9), 2509, 1979.)

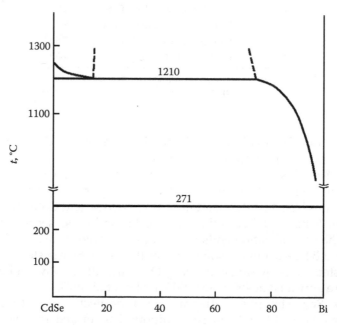

FIGURE 5.54
CdSe–Bi phase diagram. (From Sher, A.A. et al., *Zhurn. neorgan. khimii*, 24(9), 2509, 1979.)

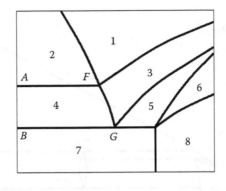

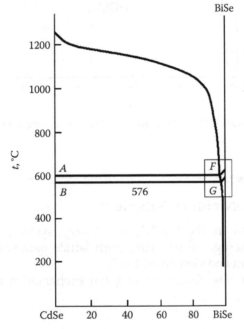

FIGURE 5.55
Phase relations in the CdSe–BiSe system: 1, L; 2, L+CdSe; 3, L+Bi$_2$Se$_3$; 4, L+CdSe+Bi$_2$Se$_3$; 5, L+Bi$_2$Se$_3$+BiSe; 6, L+BiSe; 7, CdSe+BiSe; and 8, BiSe. (From Sher, A.A. et al., *Zhurn. neorgan. khimii*, 24(9), 2509, 1979.)

0.15 mol. % at 530°C (Odin et al. 2000)], and the solubility of CdSe in Bi$_2$Se$_3$ at 450°C is not higher than 3 mol. %.

This system was investigated using DTA, metallography, and XRD (Novoselova et al. 1978, Odin et al. 2000). The ingots were annealed at 450°C for 1000 h (Novoselova et al. 1978) or at 530°C for 640 h (Odin et al. 2000).

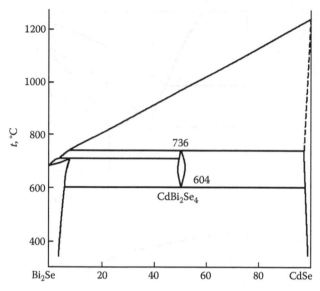

FIGURE 5.56

CdSe–Bi$_2$Se$_3$ phase diagram. (From Novoselova, A.V. et al., *Zhurn. neorgan. khimii*, 23(12), 3321, 1978.)

5.35 Cadmium–Niobium–Selenium

Cd$_{0.50}$Nb$_6$Se$_8$ forms in the Cd–Nb–Se ternary system. This compound crystallizes in a hexagonal structure with lattice parameters $a = 999.8$ and $c = 346.51$ pm (Huan and Greenblatt 1987).

This compound was obtained using ion exchange in melts containing Tl$_x$Nb$_6$Se$_8$.

5.36 Cadmium–Oxygen–Selenium

The phase diagram is not constructed. Isothermal section of the Cd–O–Se ternary system at room temperature, constructing according to thermodynamic calculations, is shown in Figure 5.57 (Medvedev and Berchenko 1993).

CdSe–SeO$_2$: The phase diagram is not constructed. Cadmium selenide interacts with selenium dioxide forming CdSeO$_3$ (Goriayev et al. 1973): 2CdSe + 3SeO$_2$ = 2CdSeO$_3$ + 3Se. The reaction rate increases with an increasing temperature.

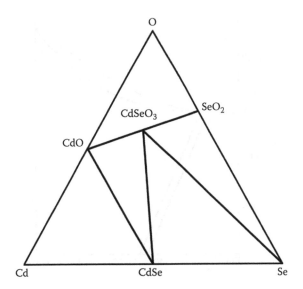

FIGURE 5.57
Isothermal section of the Cd–O–Se ternary system at room temperature. (From Medvedev, Yu.V. and Berchenko, N.N., *Zhurn. neorgan. khimii*, 38(12), 1940, 1993.)

5.37 Cadmium–Tellurium–Selenium

Thermodynamic calculations of the Cd–Te–Se ternary system were performed using a regular associated liquid solution model (Laugier 1973). The result of computing of the chalcogen-rich part of this ternary system is given in Figure 5.58. Immiscibility region exists in this system that penetrates deep in the ternary system from the Cd–Se binary system.

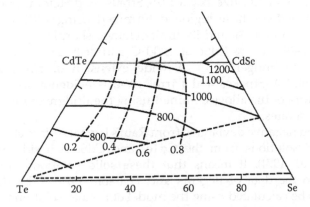

FIGURE 5.58
Calculated chalcogen-rich part of the Cd–Te–Se ternary system: solid lines, liquidus isotherms; dashed lines, solid isoconcentrated curves. (From Laugier, A., *Rev. Phys. Appl.*, 8(3), 259, 1973.)

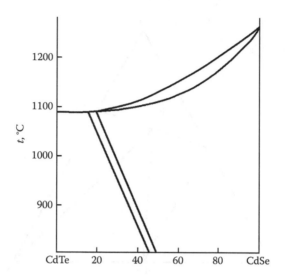

FIGURE 5.59
CdSe–CdTe phase diagram. (From Strauss, A.J. and Steininger, J., *J. Electrochem. Soc.*, 117(11), 1420, 1970. With permission.)

CdSe–CdTe: The phase diagram is shown in Figure 5.59 (Oleinik and Mizetskaya 1975, Strauss and Steininger 1970). The composition and temperature of azeotropic point are approximately 15 mol. % CdSe and 1089 ± 1°C (Oleinik and Mizetskaya 1975) [1091 ± 1°C (Strauss and Steininger 1970)], respectively. Solid solutions forming in this system crystallize depending from their compositions both in a hexagonal and in a cubic structure (Savickas and Baršauskas 1960, Litwin 1964, Oleinik and Mizetskaya 1975, Strauss and Steininger 1970). Lattice parameters of these solid solutions increase with increasing CdTe contents (Vitrihovskiy et al. 1971, Shevchenko et al. 1974, Oleinik and Mizetskaya 1975, Strauss and Steininger 1970).

The width of two-phase region determined using XRD is within the interval of 2.5–3 mol. %, and its limits change sharply with temperature changing: 26–29 mol. % CdSe at 1000°C and 46.5–49.5 mol. % at 800°C (Strauss and Steininger 1970). A wide extension of two-phase region (up to 20 mol. %) (Stuckes and Farrel 1964) is evidently associated with nonequilibrium of the alloys obtained using the Bridgman method at high crystallization rates.

Positive and negative deviations from Raul's law take place at 522°C–822°C for the solid solutions from the CdTe- and CdSe-rich side, respectively (Lopatin et al. 1974). It means that these solid solutions are not ideal. According to the data of Ilegems and Pearson (1975), CdSe–CdTe phase diagram can be calculated using the model of regular solutions.

Dependence of energy gap from composition has a minimum at 40 mol. % CdSe that is equal to 1.32 eV (Prytkina et al. 1968, Vitrihovskiy et al. 1971, Tai Hideo et al. 1976).

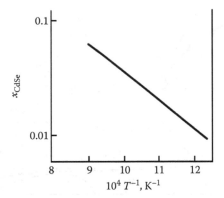

FIGURE 5.60
Temperature dependence of CdSe solubility in liquid Te. (From Masaharu, A. et al., *J. Fac. Eng. Univ. Tokyo*, A(18), 38, 1980.)

This system was investigated using DTA and XRD (Oleinik and Mizetskaya 1975, Strauss and Steininger 1970). The ingots were annealed at 950°C for 21 days (Strauss and Steininger 1970).

CdSe–Te: The phase diagram is not constructed. Temperature dependence of CdSe solubility in liquid Te was determined using the saturation method within the interval of 550°C–800°C (Figure 5.60) (Masaharu et al. 1980).

CdTe–Se: The phase diagram is not constructed. Selenium impurity in CdTe is characterized by uniform distribution at the initial Se concentrations from 10^{17} to 10^{19} cm^{-3} (Fochuk et al. 1985), which can be explained by isovalency of Se relative to Te replaced atom.

5.38 Cadmium–Chromium–Selenium

The fields of CdSe and Cr_2Se_3 primary crystallization occupy almost all liquidus surface of the $CdSe–Cr_2Se_3–Se$ subsystem in the Cd–Cr–Se ternary system (Figure 5.61) (Shabunina et al. 1981). The field of $CdCr_2Se_4$ crystallization is small and situated near the Se-rich corner, and the field of Se crystallization is degenerated.

The homogeneity region of $CdCr_2Se_4$ in the Cd–Cr–Se ternary system at 800°C is shown in Figure 5.62 (Bel'skiy et al. 1984, Bel'ski et al. 1988). The composition of this compound changes within the interval of 1.5–2 at. % for all three chemical elements (Bel'skiy and Ochertyanova 1980, Bel'skiy et al. 1984, Bel'ski et al. 1988). The stoichiometric composition of $CdCr_2Se_4$ is situated inside of the homogeneity region.

This system was investigated using DTA, XRD, and chemical analysis and using atomic-absorption spectroscopy (Shabunina et al. 1981, Bel'skiy et al. 1984, Bel'ski et al. 1988). The ingots were annealed at 800°C for 24 days (Bel'skiy et al. 1984, Bel'ski et al. 1988).

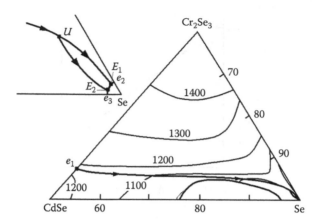

FIGURE 5.61
Liquidus surface of the CdSe–Cr$_2$Se$_3$–Se subsystem in the Cd–Cr–Se ternary system.
(Shabunina, G.G. et al., *Zhurn. neorgan. khimii*, 26(2), 476, 1981.)

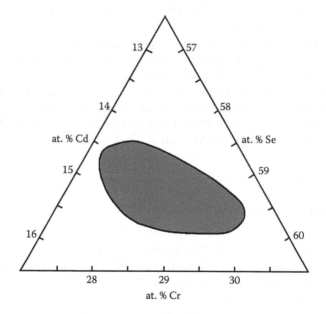

FIGURE 5.62
Homogeneity region of CdCr$_2$Se$_4$ in the Cd–Cr–Se ternary system at 800°C. (From Bel'skiy, N.K.
et al., *Izv. AN SSSR. Neorgan. materialy*, 20(5), 762, 1984; Bel'ski, N.K. et al., *Izv. AN SSSR. Neorgan.*
materialy, 24(7), 1089, 1988.)

CdSe–Cr$_2$Se$_3$: The phase diagram is shown in Figure 5.63 (Barraclough
and Meyer 1973). The eutectic composition and temperature are 7 ± 1 mol. %
(12 ± 1 mass. %) Cr$_2$Se$_3$ and 1185 ± 5°C. CdCr$_2$Se$_4$ ternary compound is
formed in this system. It melts incongruently at 885 ± 10°C (Barraclough
and Meyer 1973) [810 ± 5°C (Chebotaru and Derid 1973)] and crystallizes in

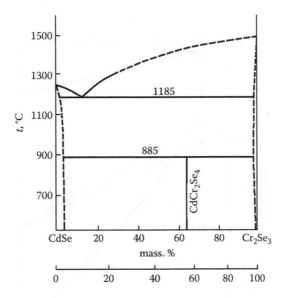

FIGURE 5.63
CdSe–Cr$_2$Se$_3$ phase diagram. (From Barraclough, K.G. and Meyer, A., *J. Cryst. Growth*, 20(3), 212, 1973.)

a cubic structure of spinel type with lattice parameter $a = 1070.0$–1074.5 pm (Pinch and Berger 1968) [$a = 1074.1$ pm (Czerwonko and Węglowski 1976); $a = 1075.5$ pm (Marunia et al. 1975); $a = 1074.0$ pm (Raccah et al. 1966)] and calculation and experimental density 5.702 and 5.51 g cm^{-3}, respectively (Hahn and Schröder 1952). According to the data of Tret'yakov et al. (1972), CdCr$_2$Se$_{4-x}$ has a narrow gap of homogeneity with $0 \leq x \leq 0.08$, and the lattice parameter does not change at the changing x. Pleshchev et al. (1978) indicate that the lattice parameter of this compound increases when the partial chalcogen pressure increases at the annealing. Deviation from stoichiometry of the CdCr$_2$Se$_4$ compound takes place relative to all three elements (Bel'skiy and Ochertyanova 1980).

The selenospinel, CdCr$_2$Se$_4$, transforms under high pressure and temperature to a new structure with monoclinic symmetry related to a defect-NiAs structure and with the lattice parameters: $a = 14.62$ Å, $b = 6.90$ Å, $c = 11.45$ Å, and $\beta = 91.0°$ (Banus and Lavine 1969). The high-pressure phase is retained indefinitely at atmospheric pressure and room temperature but retransforms at 125°C. The pressure–temperature boundary between the phases has a slope of $-15.2°$C kbar^{-1} over the temperature range of 400°C–750°C. Under pressure, the structural change is accompanied by a change from semiconducting to metallic behavior.

Review of Kalinnikov et al. (2003) is devoted to the phase diagram, homogeneity region, and vapor pressure above CdCr$_2$Se$_4$, growth and doping of its crystals, as well as to the problem of its structure imperfection.

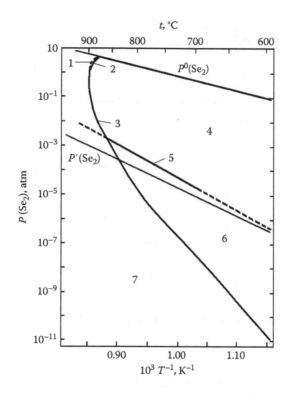

FIGURE 5.64

p–T phase diagram for $CdCr_2Se_4$ at 600°C–964°C: 1, $CdSe+CdCr_2Se_4+L+V$; 2, $CdCr_2Se_4+Cr_2Se_3+L+V$; 3, $CdSe+CdCr_2Se_4+Cr_2Se_3+V$; 4, $CdCr_2Se_4+V$ *or* $CdSe+CdCr_2Se_4+V$ *or* $CdCr_2Se_4+Cr_2Se_3+V$; 5, $Cr_2Se_3+Cr_3Se_4+Se_2(V)$ $(CdCr_2Se_4+Cr_2Se_3+Cr_3Se_4+V)$; 6, $CdCr_2Se_4+V$ *or* $CdSe+CdCr_2Se_4+V$ *or* $CdCr_2Se_4+Cr_3Se_4+V$; and 7, $CdSe+CdCr_2Se_4+Cr_3Se_4+V$. (From Kiyosawa, T. and Masumoto, K., *J. Phys. Chem. Solids*, 38(6), 609, 1977.)

Maximum solubility of Cr_2Se_3 in CdSe is equal to 2.3 mol. % (4 mass. %) and the solubility of CdSe in Cr_2Se_3 is insignificant (Barraclough and Meyer 1973).

Undoping single crystals of $CdCr_2Se_4$ has p type of electrical conductivity (Lehmann and Harbeke 1967).

p–T-phase diagram for $CdCr_2Se_4$ at 600°C–964°C is shown in Figure 5.64 (Kiyosawa and Masumoto 1977). It can be seen that this compound is stable up to approximately 900°C, which coincides with the data of Barraclough and Meyer (1973). $CdCr_2Se_4$ dissolves in Se and is in equilibrium with Se-rich melts. The vapor pressure above $CdCr_2Se_4$ is determined using the pressure of Se_2 (Zhegalina et al. 1985). Within the temperature interval 740°C–900°C, the dependence of the Se_2 vapor pressure above $CdCr_2Se_4$ can be expressed by the next equation: log p(Pa) $= -(6550 \pm 450)/T + (9.20 \pm 0.41)$. The enthalpy and the entropy of Se_2 evaporation from $CdCr_2Se_4$ are equal to $\Delta H_T = 125.57 \pm 8.46$ kJ mol^{-1} and $\Delta S_T = 80.51 \pm 7.53$ J (mol·K)$^{-1}$, respectively.

According to the data of Bel'ski et al. (1986), cadmium vapor pressure above $CdCr_2Se_4$ crystals annealing at the CdSe presence is 1.5–2 times lower than the one above the unannealing crystals. At the hot pressing, $CdCr_2Se_4$ decomposes into CdSe and Cr_2Se_3 (Filatov et al. 1990a).

At heating in the air, $CdCr_2Se_4$ decomposes according to a three-stage scheme (Zhukov et al. 1997). Selenium detachment takes place at temperatures below 520°C. At 520°C, $CdCr_2O_4$ intermediate oxide is formed, which decomposes into CdO and Cr_2O_3 at the next heating. Selenium deficiency in the $CdCr_2Se_{4-y}$ spinel depends from the temperature and can be described by the next equation: $y = (0.25 \pm 0.01)T - (0.080 \pm 0.005)$.

According to the data of Chebotaru and Derid (1973), the formation of peritectic phase with approximate composition $Cd_3Cr_2Se_6$ takes place in the $CdSe–Cr_2Se_3$ system.

This system was investigated using DTA, metallography, XRD, and microhardness measurement (Barraclough and Meyer 1973). The ingots were annealed at 800°C for 20 days (Bel'ski et al. 1986). Influence of high pressures on the $CdCr_2Se_4$ phase composition was investigated at 500°C–1000°C and 0.5–3 GPa (Filatov et al. 1990a).

$CdCr_2Se_4$ can be prepared by the direct combination of the elements or by the reaction of the cadmium oxide spinel with hydrogen sulfide at elevated temperatures (Raccah et al. 1966). Its single crystals were grown using chemical transport reactions (Kolowos et al. 1974) or crystallization from the solutions in the melt (Pinch and Berger 1968, Shabunina et al. 1989). The $CdCr_2Se_4$-pressed ingots with desired geometry and properties can be obtained at 0.5 GPa and 600°C–770°C or 1.5–2 GPa and 500°C (Filatov et al. 1990b).

5.39 Cadmium–Molybdenum–Selenium

$Cd_{2.5}Mo_{15}Se_{19}$ and $Cd_4Mo_{15}Se_{19}$ ternary compounds (Tarascon et al. 1985) and Chevrel phases $CdMo_6Se_8$ and $Cd_2Mo_6Se_8$ (Gocke et al. 1987, Janssen et al. 1998) are formed in the Cd–Mo–Se ternary system. $Cd_{2.5}Mo_{15}Se_{19}$ and $Cd_4Mo_{15}Se_{19}$ crystallize in a hexagonal structure with lattice parameters $a = 997.0$ and $c = 5670$ pm and $a = 1003$ and $c = 5689$ pm, respectively, and decompose at temperatures above 800°C (Tarascon et al. 1985).

The Chevrel phases $CdMo_6Se_8$ and $Cd_2Mo_6Se_8$ crystallize in a rhombohedral structure with lattice parameters $a = 675.7$ pm and $\alpha = 92.53°$ and $a = 683.2$ pm and $\alpha = 95.25°$, respectively (Gocke et al. 1987, Janssen et al. 1998) [in a hexagonal structure with the lattice parameters $a = 976.5$ and $c = 1117.8$ pm and $a = 1025.0$ and $c = 1084.0$ pm, respectively (Gocke et al. 1987)]. $CdMo_6Se_8$ undergoes a phase transition near 130 K, but the exact crystal structure of the low-temperature modification is not known, although the analogy with several other Chevrel phases suggests that a triclinic phase is formed (Janssen et al. 1998). $Cd_2Mo_6Se_8$ is a semiconductor and decomposes at elevated temperatures (Gocke et al. 1987, Janssen et al. 1998).

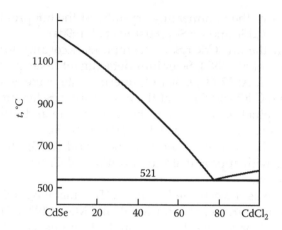

FIGURE 5.65
CdSe–CdCl$_2$ phase diagram. (From Reisman, A. and Berkenblit, M., *J. Electrochem. Soc.*, 109(11), 1111, 1962. With permission.)

The Chevrel phases were obtained using an intercalation of cadmium via cathodic reduction in a galvanic cell Cd/CdSO$_4$(2 M), H$_2$O/Mo$_6$Se$_8$ (Gocke et al. 1987, Janssen et al. 1998). Their structures have been studied using several complementary ^{111}Cd NMR spectroscopic techniques.

5.40 Cadmium–Chlorine–Selenium

CdSe–CdCl$_2$: The phase diagram is a eutectic type (Figure 5.65) (Reisman and Berkenblit 1962). The eutectic composition and temperature are 77.5 mol. % CdCl$_2$ and 521°C, respectively (Reisman and Berkenblit 1962) [77.8 mol. % CdCl$_2$ and 510°C (Loginova and Andreev 1970)]. Maximum solubility of CdCl$_2$ in CdSe is equal to 0.5 mol. % (Reisman and Berkenblit 1962).

This system was investigated using DTA and XRD (Reisman and Berkenblit 1962, Loginova and Andreev 1970).

5.41 Cadmium–Iodine–Selenium

CdSe–CdI$_2$: The phase diagram is a eutectic type (Figure 5.66) (Odin 2001). The eutectic composition and temperature are 18 ± 1 mol. % CdS and 364 ± 2°C, respectively. The solubility of CdI$_2$ in CdSe reaches 0.2 mol. %.

This system was investigated using DTA, metallography, and XRD (Odin 2001).

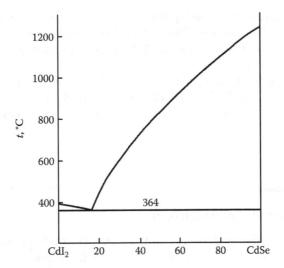

FIGURE 5.66
CdSe–CdI$_2$ phase diagram. (From Odin, I.N., *Zhurn. neorgan. khimii*, 46(10), 1733, 2001.)

5.42 Cadmium–Manganese–Selenium

CdSe–MnSe: The phase equilibria in this system have been discussed in Pajaczkowska (1978) based on papers that have appeared up to 1977. The phase diagram is a eutectic type (Figure 5.67) (Cooke 1968). Distance between solidus and liquidus lines from the CdSe-rich side is within the interval of 5°C.

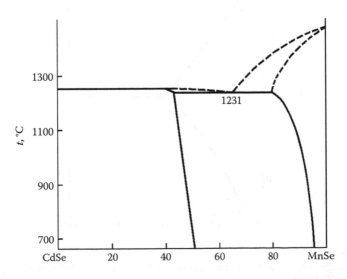

FIGURE 5.67
CdSe–MnSe phase diagram. (From Cooke, W.R., *J. Amer. Ceram. Soc.*, 51(9), 518, 1968.)

Minimum temperature on the liquidus is equal to 1231°C. Maximum solubility of MnSe in CdSe reaches 50 mol. % at 700°C and decreases to 44 mol. % at 1175°C. The solubility of CdSe in MnSe increases from 5 mol. % at 700°C to 19 mol. % at 1265°C.

This system was investigated using DTA and XRD. The ingots were annealed at 1140°C–1175°C for 18–24 h (Cooke 1968). The crystal growth of the $Cd_{1-x}Mn_xSe$ solid solutions is given by Pajaczkowska (1978).

5.43 Cadmium–Iron–Selenium

CdSe–FeSe$_{1.24}$: CdSe and $FeSe_{1.24}$ form substitution-type solid solutions with Fe atoms randomly replacing Cd atom in hexagonal lattice of the Cd-rich region of the phase diagram (Figure 5.68) (Wiedemeier and Huang Xuejun 1993). The solid solubility of $FeSe_{1.24}$ in CdSe increases with temperature to a maximum of about 19 mol. % at about 925°C. Upon further increase in temperature, the solubility decreases accompanied by an apparent partial melting of the sample. According to the data of Smith et al. (1988), the limit of substitution of Fe for Cd in CdSe was determined

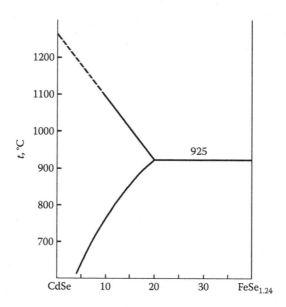

FIGURE 5.68
Part of the CdSe–FeSe$_{1.24}$ phase diagram. (From Wiedemeier, H. and Huang X., *J. Electron. Mater.*, 22(6), 695, 1993.)

to be 11.5 ± 1 at. % [a single-phase region up to 13.2 at. % Fe in $Cd_{1-x}Fe_xSe$ solid solutions was observed by Dynowska et al. (1990)]. At 850°C or below, the solid solubility of CdSe in $FeSe_{1.24}$ is less than 4 mol. % (Wiedemeier and Huang Xuejun 1993).

This system was investigated using metallography, XRD, and atomic-absorption spectrometry (Smith et al. 1988, Dynowska et al. 1990, Wiedemeier and Huang Xuejun 1993). Single crystals of $Cd_{1-x}Fe_xSe$ were grown using chemical transport reaction and a modified Bridgman method (Smith et al. 1988). The iron concentration of Bridgman-grown crystals was homogeneous in a radial direction but shows 0.5–0.8 mol. % change in Fe concentration along the length of the boule.

References

Agaev V.G., Antonov V.B., Nani R.Kh., Salaev E.Yu. Synthesis and growth of the $CdGa_2Se_4$ single crystals [in Russian], *Dokl. An. AzSSR*, **26**(10–11), 8–12 (1970).

Agayev A.B., Kulieva S.A., Kulieva U.A. Investigation of the $CdSe-Pr_2Se_3$ system [in Russian], *Azerb. khim. zhurn.*, (6), 95–97 (1984).

Agayev A.B., Rustamov P.G., Aliev O.M., Azadaliev R.A. Synthesis and properties of cadmium–lanthanide chalcogenides [in Russian], *Izv. AN SSSR. Neorgan. materialy*, **25**(2), 294–298 (1989).

Aliev I.I., Rustamov P.G., Il'yasov T.M., Maksudova T.F. Phase diagram of the As_2Se_3–CdSe system [in Russian], *Izv. AN SSSR. Neorgan. materialy*, **23**(12), 1965–1967 (1987).

Aliev O.M., Agayev A.B., Azadaliev R.A. Synthesis and properties of $CdLn_2S_4(Se_4,Te_4)$ and $CdLn_4S_7$ [in Russian], *Neorgan. materialy*, **33**(11), 1327–1332 (1997).

Aliev O.M., Agayev A.B., Azadaliev R.A., Abdullaieva M.I. The $CdSe-La_2Se_3$ system [in Russian], *Zhurn. neorgan. khimii*, **29**(10), 2705–2798 (1984).

Aliev O.M., Khasaev G.G., Kurbanov T.Kh. Synthesis and physico-chemical study of the $Me^{2+}Ln_4S_7(Se_7)$ type compounds, *Bull. Soc. chim. France*, (1), 26–28 (1986).

Ansari M.A., Ibers J.A. Synthesis and structure of the new layered ternary selenide $Na_4Cd_3Se_5$, *J. Solid State Chem.*, **103**(2), 293–297 (1993).

Atroshchenko L.V., Kukol V.V., Lakin E.E., Nosachev B.G., Sysoev, L.A. Growth from the melt and investigation of the crystal structure, defect and phase transformation of pure, doping and mixed $A^{II}B^{VI}$ single crystals [in Russian], in: *Rost i legirovanie poluprovodn. kristallov i plenok*, Pt.2, 298–301 (1977).

Atroshchenko L.V., Obuhova N.F., Pisarevski Y.V., Sil'vestrova I.M., Sysoev, L.A. The regularity of properties change for ZnS–MgS, CdS–MgS and CdSe–MgSe solid solution single crystals [in Russian], in: *Fiz. slozhn. poluprovodn. soedin.* Kishinev: Shtiinza, 189–198 (1979).

Atroshchenko L.V., Obuhova N.F., Sysoev L.A. Solid solutions with wurtzite type structure in the systems based on $A^{II}B^{VI}$ [in Russian], *Izv. AN SSSR. Neorgan. materialy*, **22**(1), 162–163 (1986).

Axtell E.A. (III), Liao J.-H., Pikramenou Z., Kanatzidis M.G. Dimensional reduction in II-VI materials: $A_2Cd_3Q_4$ (A = K, Q = S, Se, Te; A = Rb, Q = S, Se), novel ternary low-dimensional cadmium chalcogenides produced by incorporation of A_2Q in CdQ, *Chem. Eur. J.*, **2**(6), 656–666 (1996).

Azadaliev R.A. Synthesis and properties of the $CdGd_2Se_4$ and $CdGd_2Te_4$ compounds [in Russian], in: *Neorgan. soed. – sintez i svoystva*, Baku, 26–27 (1990).

Azadaliev R.A., Alieva U.R. Physico-chemical investigation of the $CdSe–Gd_2Se_3$ system [in Azerbaijani], in: *Issled. v obl. sinteza i primenienie neorgan. soed.*, Baku, 93–96 (1985).

Babanly M.B., Guseinov F.H., Kuliev A.A. The Tl–Cd–Se system [in Russian], *Zhurn. neorgan. khimii*, **30**(5), 1269–1273 (1985).

Banus M.D., Lavine M.C. Polymorphism in selenospinels. A high pressure phase of $CdCr_2Se_4$, *J. Solid State Chem.*, **1**(1), 109–116 (1969).

Barraclough K. G., Meyer A. The system $CdSe–Cr_2Se_3$ and formation of the ferromagnetical semiconductor with structure of the spinel $CdCr_2Se_4$, *J. Cryst. Growth*, **20**(3), 212–216 (1973).

Bel'ski N.K., Ochertianova L.I., Zhegalina V.A. Determination of cadmium vapor pressure above dichromium-cadmium tetraselenide by the atomic-absorption method [in Russian], *Izv. AN SSSR. Neorgan. materialy*, **22**(2), 211–216 (1986).

Bel'skiy N.K., Ochertianova L.I., Zhegalina V.A., Kalinnikov V.T. Homogeneity region of the $CdCr_2Se_4$ ferromagnetic semiconductor [in Russian], *Izv. AN SSSR. Neorgan. materialy*, **20**(5), 762–764 (1984).

Bel'ski N.K., Ochertianova L.I., Zhegalina V.A., Koroleva L.I. Nonstoichiometry and physical properties of the $CdCr_2Se_4$ ferromagnetic semiconductor [in Russian], *Izv. AN SSSR. Neorgan. materialy*, **24**(7), 1089–1093 (1988).

Bel'skiy N.K., Ochertyanova L.I. Determination of the $CdCr_2Se_4$ crystals deviation from stoichiometric composition by the atomic absorption analysis [in Russian], *Zhurn. analit. khimii*, **35**(3), 604–606 (1980).

Beun J.A., Nitsche R., Lichensteiger M. Optical and electrical properties of ternary chalcogenides, *Physica*, **27**(5) 448–452 (1961).

Brec R., Ouvrard G., Louisy A., Rouxel J. Proprietés structurales de phases $M^{2+}PX_3$ (X = S, Se), *Ann. chim. (France)*, **5**(6), 499–512 (1980).

Brusilovets A.I., Fedoruk T.I. Investigation of interaction in the $CdSe–P_2Se_5$, $CdS–P_2S_5$ systems [in Russian], *Zhurn. neorgan. khimii*, **21**(6), 1648–1651 (1976).

Chebotaru V.Z., Derid O.P. Investigation of phase transformations in the $CdSe–Cr_2Se_3$ system [in Russian], in: *Materialy dokl. na IX nauchno-tehnich. konf. Kishin. polytehn. in-ta*, Kishinev: Shtiintsa, 104p. (1973).

Cooke W. R. The CdS–MnS and CdSe–MnSe phase diagrams, *J. Amer. Ceram. Soc.*, **51**(9), 518–519 (1968).

Cruceanu E., Niculescu D. Solutions solides dans les systèmes CdS–CdTe et HgSe–CdSe, *C. r. Acad. sci.*, **261**(4), 935–938 (1965).

Czerwonko A., Węglowski S. Badania krystalograficzne i magnetyczne faz powstalych w procesie monokrystalizacji selenku kadmu domieszkowanego selenkiem chromu [in Polish], *Pr. nauk. Inst. chem. nieorgan. i metallurg. perwiast. rzadkich PWr.*, (28), 73–81 (1976).

Datsenko A.M., Kukleva T.V., Vishniakov A.V., Kovtunenko P.V. Investigation of the solid solutions in the $CdSe–Cu_2Se$ system by high temperature roentgenography [in Russian], *Zhurn. neorgan. khimii*, **29**(1), 263–265 (1984).

Derid O.P., Markus M.M., Radautsan S.I., Stanchu A.V. Phase diagram of the Cd_3Se_3–In_2Se_3 pseudobinary system [in Russian], *Izv. AN SSSR. Neorgan. materialy*, **7**(8), 1328–1330 (1971).

Dvoretskov G.A., Vishniakov A.V., Al'tah O.L., Kovtunenko P.V. The Cd–Ga–Se system [in Russian], *Zhurn. neorgan. khimii*, **24**(9), 2505–2508 (1979a).

Dvoretskov G.A., Vishniakov A.V., Kovtunenko P.V. Equilibria solid-vapor in the $CdSe$–Ga_2Se_3 system, *Deposited in VINITI*, № 2457–79 Dep (1979b).

Dvoretskov G.A., Vishniakov A.V., Kovtunenko P.V. Solubility isotherm of gallium in cadmium and zinc selenides [in Russian], in: *Legir. poluprovodn.*, M.: Nauka, 67–71 (1982).

Dynowska E., Sarem A., Orłowski A., Mycielski A. Structural and chemical studies of the $Cd_{1-x}Fe_x$ semimagnetic semiconductors, *Mater. Res. Bull.*, **25**(9), 1109–1113 (1990).

Epik A.I., Varvas Yu.A. Solubility of copper in CdSe [in Russian], *Izv. AN SSSR. Neorgan. materialy*, **18**(6), 904–907 (1982).

Filatov A.V., Davydov V.A., Volkov E.A., Baneeva M.I., Novotortsev V.M. Obtaining of the $CdCr_2Se_4$ ingots with the spinel structure by the hot pressing and their properties [in Russian], *Izv. AN SSSR. Neorgan. materialy*, **26**(8), 1618–1621 (1990).

Filatov A.V., Volkov E.A., Davydov V.A., Baneeva M.I., Novotortsev V.M. Phase transformations in $CdCr_2Se_4$ at high pressures and temperatures [in Russian], *Zhurn. neorgan. khimii*, **35**(5), 1137–1141 (1990).

Firszt F., Łęgowski S., Męczyńska H., Szatkowski J., Paszkowicz W., Marczak M. Growth and characterization of $Cd_{1-x}Mg_xSe$ mixed crystal, *J. Cryst. Growth*, **184/185,** 1053–1056 (1998).

Fochuk P.M., Belotski D.P., Panchuk O.E., Boychuk R.F. Distribution of Se in CdTe [in Russian], Deposited in UkrNIINTI 6.09.85, N 2091-Uk (1985).

Fujii Horonobu, Okamoto Tetsuhiko, Kamigaichi Takahiko. Crystallographic and magnetic studies of rare earth chalcogenide spinels, *J. Phys. Soc. Jap.*, **32**(5), 1432 (1972).

Galiulin E.A., Odin I.N., Astaf'ev S.A., Novoselova A.V. Physico-chemical investigation of the $CdSe$–$GeSe_2$, CdSe–Ge systems [in Russian], *Zhurn. neorgan. khimii*, **28**(5) 1281–1283 (1983).

Galiulin E.A., Odin I.N., Novoselova A.V. Physico-chemical investigation of the CdS–SnS and CdSe–SnSe systems [in Russian], *Zhurn. neorgan. khimii*, **26**(7), 1881–1883 (1981).

Galiulin E.A., Odin I.N., Novoselova A.V. The $ZnSe$–SnSe, $ZnSe$–$SnSe_2$, $CdSe$–$SnSe_2$ systems [in Russian], *Zhurn. neorgan. khimii*, **27**(1), 266–268 (1982).

Gastaldi L., Simeone L.G., Viticoli S. Cation ordering and crystal structures in AGa_2X_4 compounds ($CoGa_2S_4$, $CdGa_2S_4$, $CdGa_2Se_4$, $HgGa_2Se_4$, $HgGa_2Te_4$), *Solid State Commun.*, **55**(7), 605–607 (1985).

Gavaleshko N.P., Gorley P.N., Paranchich S.Yu., Frasuniak V.M., Homiak V.V. Phase diagrams of CdSe–HgSe, ZnSe–HgSe, MgTe–HgTe quasibinary systems [in Russian], *Izv. AN SSSR. Neorgan. materialy*, **10**(2), 327–329 (1983).

Gocke E., Schramm W., Dolscheid P., Schöllhorn R. Molybdenum cluster chalcogenides Mo_6X_8: Electrochemical intercalation of closed shell ions Zn^{2+}, Cd^{2+}, and Na^+, *J. Solid State Chem.*, **70**(1), 71–81 (1987).

Golovey M.I., Olekseyuk I.D., Voroshilov Yu.V. Chemical bond and solubility in the systems $A^{II}_3B^V_2$–$A^{II}C^{VI}$ [in Russian], in: *Khim. svyaz' v poluprovodn. i polumetallah*. Minsk, Nauka i Tekhnika Publish., 233–239 (1972).

Golovey M.I., Rigan M.Yu. Phase diagram of the $(Cd_3P_2)_{1-x}-(2CdSe)_x$ system [in Russian], *Izv. AN SSSR. Neorgan. materialy*, **10**(5), 919 (1974).

Golovey M.I., Rigan M.Yu. Preparation of the $Cd_3As_2-2CdTe(Se,S)$ solid solution single crystals [in Russian], in: *Khimiya i physika khalkogenidov*. K.: Naukova dumka, 38–40 (1977).

Golovey M.I., Rigan M.Yu., Olekseyuk I.D., Voroshilov Yu.V. The $(Cd_3As_2)_{1-x}-(2CdSe)_x$ system [in Russian], *Izv. AN SSSR. Neorgan. materialy*, **9**(9), 1520–1523 (1973).

Golovey M.I., Shpyrko G.N. Enthalpy of formation and heat capacity of solid solutions in the $Cd_3As_2-2CdS(Se,Te)$ systems [in Russian], in: *Termodinamicheskie svoistva metallicheskih splavov*, Baku: Elm, 321–323 (1975).

Goriayev V.M., Pechkovski V.V., Pinaev G.F. Interaction of gaseous SeO_2 with CdS [in Russian], *Izv. AN SSSR. Neorgan. materialy*, **9**(7), 1126–1131 (1973).

Grzecznik A., Ursaki V.V., Syassen K., Loa I., Tiginyanu I.M., Hanfland M. Pressure-induced phase transition in cadmium thiogallate $CdGa_2Se_4$, *J. Solid State Chem.*, **160**(1), 205–211 (2001).

Guseinov G.D. Searching and physical investigation of new semiconductors-analogues [in Russian], *Avtoref. dis. ... doct. fiz.-mat. nauk.—Baku*, 82 p. (1969).

Guseinov F.H., Babanly M.B., Kuliev A.A. Phase equilibria and intermolecular interaction in the $TlSe(Tl_2Se)-CdSe$ systems [in Russian], *Zhurn. neorgan. khimii*, **26**(1), 215–217 (1981a).

Guseinov F.H., Babanly M.B., Kuliev A.A. Phase equilibria in the $Tl_2Te_3-Zn(Cd)Te$ systems [in Russian], *Izv. vuzov. Khimia i khim. tehnologia*, **24**(10), 1245–1248 (1981b).

Guseinov F.H., Babanly M.B., Kuliev A.A. Phase equilibria in the $TlX-Cd(Zn)X$ $(X - S, Se, Te)$ systems [in Russian], *Izv. AN SSSR. Neorgan. materialy*, **18**(5), 759–763 (1982).

Guseinov G.D., Guseinov G.G., Ismailov M.Z., Godzhaev E.M. About structure and physical properties of new $CdTlS_2$ (Se_2, Te_2) semiconductor compounds [in Russian], *Izv. AN SSSR. Neorgan. materialy*, **5**(1), 33–39 (1969).

Hahn H., Frank G., Klingler W., Störger A.D., Störger G. Über ternäre Chalkogenide des Aluminiums, Galliums und Indiums mit Zink, Cadmium und Quecksilber, *Z. anorg. und allg. Chem.*, **279**(5/6), 241–270 (1955).

Hahn H., Lorent C. Untersuchungen über ternäre Chalkogenide. Über ternäre Sulfide und Selenide des Germaniums mit Zink, Cadmium und Quecksilber, *Naturwissenschaften*, **45**(24), 621–622 (1958).

Hahn H., Schröder K. F. Über die Structur des $ZnCr_2Se_4$ und $CdCr_2Se_4$, *Z. anorg. und allg. Chem.*, **269**(3), 135–140 (1952).

Harif Ya.L., Vishniakov A.V. Phase equilibria in the Cd–Cu–Se system [in Russian], *Izv. AN SSSR. Neorgan. materialy*, **12**(8), 1361–1364 (1976).

Horinaka H., Uemura A., Yamamoto N. Growth and electric properties of $CdGa_2Se_4$, *J. Cryst. Growth*, **99**(1–4), Pt. 2, 785–789 (1990).

Huan G., Greenblatt M. New $A_xNb_6Se_8$ (A = Na, K, Rb, Cu Ag, Zn, Cd, Pb) phases with the Nb_3Te_4 structure, *Mater. Res. Bull.*, **22**(4), 505–512 (1987).

Iglesias I.E., Pachali K.E., Steinfink H. Structural chemistry of Ba_2CdS_3, Ba_2CdSe_3, $BaCdS_2$, $BaCu_2S_2$ and $BaCu_2Se_2$, *J. Solid State Chem.*, **9**(1), 6–14 (1974).

Ilegems M., Pearson G.L. Phase studies in III–V, II–VI and IV–VI compound semiconductor alloy systems, *Ann. Rev. Mater. Sci.*, **5**, 345–371 (1975).

Ivashchenko I.A., Aksel'rud L.G., Olekseyuk I.D., Zmiy O.F. Polytypes of the $CdIn_2Se_4$ compound [in Ukrainian], *Ukr. khim. zhurn.*, **70**(1–2), 67–71 (2004).

Janssen M., Eckert H., Müller-Warmuth W., Stege U., Schöllhorn R. Solid-state NMR structural studies of the ternary molybdenum cluster chalcogenides $Cd_xMo_6Se_8$ ($x = 1, 2$), *Chem. Mater.*, **10**(11), 3459–3466 (1998).

Kalb A., Leute V. The miscibility gap of the system CdSe–HgSe, *Phys. Status Solidi* (a), **5**(3), K199–K201 (1971).

Kalinnikov V.T., Aminov T.G., Novotortsev V.M. Physical chemistry of the $CdCr_2Se_4$ magnetic semiconductor [in Russian], *Neorgan. materialy*, **39**(10), 1159–1176 (2003).

Karimov S.K., Gafarov S., Sultonov S. Investigation of phase equilibria in the $Cd–Tl–C^{VI}$ (C^{VI} – Se, Te) systems [in Russian], *Izv. AN SSSR. Neorgan. materialy*, **17**(8), 1346–1349 (1981).

Kiyosawa T., Masumoto K. P-T phase diagram for ferromagnetic semiconductor $CdCr_2Se_4$, *J. Phys. Chem. Solids*, **38**(6), 609–616 (1977).

Klinger W., Eulenberger G., Hahn H. Über Hexachalkogeno-hypodiphosphate vom Typ $M_2P_2X_6$, *Naturwissenschaften*, **57**(2), 88 (1970).

Klinger W., Ott R., Hahn H. Über die Darstellung und Eigenschaften von Hexathio- und Hexaselenohypodiphosphaten, *Z. anorg. und allg. Chem.*, **396**(3), 271–278 (1973).

Kolomiets B.T., Mal'kova A.A. Properties and structure of ternary semiconductor systems. V. Limits of ordered solid solution in the $CdSe–In_2Se_3$ system and electrical properties of $CdIn_2Se_4$ [in Russian], in: *Fizika tverdogo tela*, Moscow; Leningrad: Izd-vo AN SSSR, vol. 2, 32–38 (1959).

Kolosenko S.M., Tyrziu M.P. Investigation of some physico-chemical properties of $CdSe–Ga_2Se_3$ alloys [in Russian], in: *Issled. slozhn. poluprovodn.*, Kishinev: RIO AN MSSR, 112–117 (1970).

Kolowos I., Knobloch A., Sieler J. Darstellung von $CdCr_2Se_4$ – Einkristallen durch chemische Transportreaktionen im System $CdCr_2Se_4/CrCl_3/Se$, *Krist. und Techn.*, **9**(2), 157–159 (1974).

Kot M.V., Mshenski V.A. Structure and electrophysical properties of the CdSe–HgSe system [in Russian], *Izv. AN SSSR. Ser. fiz.*, **28**(6), 1069–1072 (1964).

Kot M.V., Tyrziu V.G., Simashkevich A.V., Maronchuk Yu.E., Mshenski V.A. Dependence of energy activation in thin films from molar composition for some $A^{II}B^{VI}–A^{II}B^{VI}$ systems [in Russian], *Fizika tverdogo tela*, **4**(6), 1535–1541 (1962).

Kovach A.P., Potoriy M.V., Voroshilov Yu.V., Tovt V.V. Physico-chemical investigation of the $Cd(Zn)S(Se)–P_2S_4(Se_4)$ systems [in Russian], *Ukr. khim. zhurn.*, **63**(2), 92–96 (1997).

Koval L.S., Markus M.M., Radautsan S.I., Sobolev V.V., Stanchu A.V. Optical properties of the two modifications of $CdIn_2Se_4$, *Phys. Status Solidi (a)*, 9(1), K69-K72 (1972).

Krämer V., Siebert D., Febbraro S. Structure refinement of cadmium gallium selenide $CdGa_2Se_4$, *Z. Kristallogr.*, **169**(1–4), 283–287 (1984).

Krebs von B., Mandt J. Struktur und Eigenschaften von Cd_4SiS_6. Zur Kenntnis von Cd_4SiSe_6, *Z. anorg. und allg. Chem.*, **388**(3), 193–206 (1972).

Kulieva S.A. Investigation of the electrical properties of the alloys in the $CdSe–Pr_2Se_3$ system [in Russian], in: *Issled. v obl. sinteza i primenienie neorgan. soed.*, Baku, 78–82 (1985).

Laugier A. Thermodynamics and phase diagram calculations in II–VI and IV–VI ternary systems using an associated solution model, *Rev. Phys. Appl.*, **8**(3), 259–270 (1973).

Lehmann H. W., Harbeke G. Semiconducting and optical properties of ferromagnetic $CdCr_2S_4$ and $CdCr_2Se_4$, *J. Appl. Phys.*, **38**(3), 946 (1967).

Leute V., Menge D. Thermodynamic investigations of the quasibinary systems $(Cd_kSn_{1-k})Te$, $(Cd_kSn_{1-k})Se$ and $Sn(Se_lTe_{1-l})$, *Z. Phys. Chem.* (Munchen), **176**(1), 47–64 (1992).

Litwin J. X-ray examination of the binary system CdTe–CdSe, *Phys. Status Solidi*, **5**(3), 551–553 (1964).

Loginova M.V., Andreev Yu.V. Melting diagrams of the $CdCl_2$–CdX (X = O, S, Se, SO_4, SeO_3) systems [in Russian], *Izv. AN SSSR. Neorgan. materialy*, **6**(10), 1885–1886 (1970).

Loireau-Lozac'h A.M., Guittard M., Flahaut J. Diagramme de phase du system Ga_2Se_3–CdSe, *Mater. Res. Bull.*, **20**(4), 443–451 (1985).

Lopatin G.S., Malkova A.S., Shumilin V.P. Investigation of vapor pressure in the CdSe–CdTe system [in Russian], *Zhurn. fiz. khimii*, **48**(5), 1283–1285 (1974).

Lottici P.P., Antonioli G., Razzetti C. Ordered-vacancy compound semiconductors: an EXAFS study of the structure of α-$CdIn_2Se_4$, *J. Phys. Chem. Solids*, **50**(9), 967–973 (1989).

Manolikas C., Bartzokas D., van Tendeloo G., van Landuyt J., Amelinckx S. High resolution electron microscopic study of polytypism in $CdIn_2Se_4$, *Phys. Status Solidi (a)*, **59**(2), 425–436 (1980).

Marunia M.S., Boruhovich A.S., Bamburov V.G., Lobachevskaya N.I., Rokeah O.P., Gel'd P.V. Heat capacity and phase transition in the $CdCr_2Se_4$ spinel [in Russian], *Fizika tverdogo tela*, **17**(4), 1202–1204 (1975).

Masaharu A., Masaoko W., Hiroshi N. Growth of Zn and Cd chalcogenides from solutions using Te as a solvent, *J. Fac. Eng. Univ. Tokyo*, **A**(18), 38–39 (1980).

Mavlonov Sh., Sultonov S., Kurbanov H.M., Ahmedov C., Husainov B., Karimov S. Obtaining, physico-chemical and electrical properties of some $A^{II}B_2^{III}C_4^{VI}$ compounds [in Russian], in: *Troinyie poluprovodn. i ih primenienie: Tez. dokl.*, Kishinev: Shtiinza, 143 (1976).

Medvedev Yu.V., Berchenko N.N. Peculiarities of the interface between solid solutions based on mercury chalcogenides and native oxides [in Russian], *Zhurn. neorgan. khimii*, **38**(12), 1940–1945 (1993).

Meenakshi S., Vijyakumar V., Godwal B.K., Eifler A., Orgzall I., Tkachev S., Hochheimer H.D. High pressure X-ray diffraction study of $CdAl_2Se_4$ and Raman study of AAl_2Se_4 (A = Hg, Zn) and $CdAl_2X_4$ (X = Se, S), *J. Phys. Chem. Solids*, **67**(8), 1660–1667 (2006)

Mishra S., Ganguli B. Electronic and structural properties of AAl_2Se_4 (A = Ag, Cu, Cd, Zn) chalcopyrite semiconductors, *J. Solid State Chem.*, **184**(7), 1614–1621 (2011).

Motria S.F., Svitlinets V.P., Semrad E.E., Dovgoshey N.I. Thermogravimetric and tensiometric investigation of Cd_4GeS_6 and Cd_4GeSe_6 [in Russian], *Izv. AN SSSR. Neorgan. materialy*, **23**(9), 1543–1546 (1987).

Motria S.F., Tkachenko V.I., Chereshnia V.M., Kikineshi A.A., Semrad E.E. Physico-chemical and photoelectric properties of the crystals of cadmium hexathio-, hexaselenogermanate and solid solutions based on it [in Russian], *Izv. AN SSSR. Neorgan. materialy*, **22**(10), 1705–1708 (1986).

Mucha I., Wiglusz K. Phase studies on the quasi-binary thallium(I) selenide–cadmium selenide system, *Thermochim. Acta*, **526**(1–2), 107–110 (2011).

Nelson D.A., Summers C.J., Whitsett C.R. Phase diagram and crystal growth of pseudobinary HgSe–CdSe alloys, *J. Electron. Mater.*, **6**(5), 507–529 (1977).

Novoselova A.V., Sher A.A., Odin I.N. Investigation of interaction in the CdB^{VI}–A$_2^V$B$_3^{VI}$ (B^{VI} – S, Se; A^V – Sb, Bi) systems [in Russian], *Zhurn. neorgan. khimii*, **23**(12), 3321–3325 (1978).

Obuhova N.F., Atroshchenko L.V., Putiatin V.D. Microhardness anisotropy and structure of Cd$_x$Mg$_{1-x}$Se single crystals, growing from the melt [in Russian], *Izv. AN SSSR. Neorgan. materialy*, **14**(5), 852–854 (1978).

Odin I.N. *T-x-y* diagrams for reciprocal systems PbX + CdI$_2$ = CdX + PbI$_2$ (X = S, Se, Te) [in Russian], *Zhurn. neorgan. khimii*, **46**(10), 1733–1738 (2001).

Odin I.N., Chukichev M.V., Galiulin E.A. Preparation and luminescence properties of the CdX<Ge>, CdX<Sn> (X = S, Se) single crystals [in Russian], *Neorgan. materialy*, **35**(11), 1304–1306 (1999a).

Odin I.N., Chukichev M.V., Grin'ko V.V. Luminescence properties of the cadmium sulfide and selenide crystals doped by boron [in Russian], *Neorgan. materialy*, **35**(11), 1302–1303 (1999b).

Odin I.N., Chukichev M.V., Rubina M.E. Luminescence properties of the cadmium selenide single crystals doped by antimony (bismuth) [in Russian], *Neorgan. materialy*, **36**(3), 298–301 (2000).

Odin I.N., Galiulin E.A., Chudakova L.S., Novoselova A.V. The Cd–Sn–Se system [in Russian], *Zhurn. neorgan. khimii*, **28**(3), 764–769 (1983).

Odin I.N., Galiulin E.A., Novoselova A.V. The Cd–Ge–Se system [in Russian], *Zhurn. neorgan. khimii*, **30**(1), 201–206 (1985).

Odin I.N., Grin'ko V.V. Pressure and vapor composition over the ingots of the CdSe–GeSe$_2$ system [in Russian], *Zhurn. neorgan. khimii*, **36**(7), 1860–1864 (1991a).

Odin I.N., Grin'ko V.V. Tensimetric determination of Cd$_4$GeSe$_6$ homogeneity range along CdSe–GeSe$_2$ section of the Cd–Ge–Se ternary system [in Russian], *Zhurn. neorgan. khimii*, **36**(9), 2387–2392 (1991b).

Odin I.N., Grin'ko V.V., Novoselova A.V. P-T-x phase diagram of the CdSe–GeSe$_2$ system [in Russian], *Zhurn. neorgan. khimii*, **31**(5), 1274–1377 (1986).

Odin I.N., Grin'ko V.V., Petrovski A.Yu. The CdS–B$_2$S$_3$ and CdSe–B$_2$Se$_3$ phase diagrams and properties of compounds in these systems [in Russian], *Zhurn. neorgan. khimii*, **44**(12), 2061–2062 (1999c).

Odin I.N., Grin'ko V.V., Safronov E.V.,Kozlovskiy V.F. Investigation of the CdX–B$_2$X$_3$–X (X = S, Se), CdTe–B–Te systems [in Russian], *Zhurn. neorgan. khimii*, **46**(7), 1210–1214 (2001).

Odin I.N., Ivanov V.A. Phase equilibria in the Cd–Si–Se system and properties of Cd$_4$SiSe$_6$ compound [in Russian], *Zhurn. neorgan. khimii*, **36**(11), 2937–2943 (1991).

Odin I.N., Ivanov V.A., Novoselova A.V. The CdS–SiS$_2$, CdSe–SiSe$_2$ systems [in Russian], *Zhurn. neorgan. khimii*, **30**(1), 207–211 (1985).

Oleinik G.S., Mizetskaya I.B. Phase transformations in the CdSe–CdTe pseudobinary system and obtaining of solid solution single crystals [in Russian], in: *Protsessy rosta i sinteza poluprovodn. kristallov i plenok*, Novosibirsk: Nauka Publish., P. 2, 22–25 (1975).

Olekseyuk I.D., Golovey M.I., Rigan M.Yu., Voroshilov Yu.V., Gurzan M.I. Investigation of the Cd$_3$As$_2$–2CdTe and Cd$_3$P$_2$–2CdSe systems [in Russian], *Izv. AN SSSR. Neorgan. materialy*, **8**(4), 696–700 (1972).

Ouarani T., Reshak A.H., Khenata R., Amrani B., Mebrouki M., Otero-de-la-Roza A., Luaña V. *Ab-initio* study of the structural, linear and nonlinear optical properties of CdAl$_2$Se$_4$ defect-chalcopyrite, *J. Solid State Chem.*, **183**(1), 46–51 (2010).

Pajaczkowska A. Physicochemical properties and crystal growth of $A^{II}B^{VI}$–MnB^{VI} systems, *Progr. Crystal Growth Caract.*, **1**(3), 289–326 (1978).

Parasyuk O.V., Olekseyuk I.D., Gulay L.D., Piskach L.V. Phase diagram of the Ag$_2$Se–Zn(Cd)Se–SiSe$_2$ systems and crystal structure of the Cd$_4$SiSe$_6$ compound, *J. Alloys Compd.*, **354**(1–2), 138–142 (2003).

Parasyuk O.V., Piskach L.V., Olekseyuk I.D. The Cu$_2$Se–CdSe–SnSe$_2$ system [in Russian], *Zhurn. neorgan. khimii*, **44**(8), 1363–1367 (1999).

Pinch H. L., Berger S. B. The effects of nonstoichiometry on the magnetic properties of cadmium chromium chalcogenide spinels, *J. Phys. Chem. Sol.*, **29**(12), 2091–2099 (1968).

Pleshchev V.G., Konev V.N., Gerasimov A.F. Crystallochemical investigation of CdCr$_2$S$_4$ and CdCr$_2$Se$_4$ and solid solutions based on them with different chalcogen contents [in Russian], *Izv. AN SSSR. Neorgan. materialy*, **14**(2), 223–227 (1978).

Potoriy M.V., Voroshilov Yu.V., Tovt Yu.V. Physico–chemical investigation of the Cd–P–S(Se) systems [in Russian], *Neorgan. materialy*, **35**(11), 1297–1301 (1999).

Prytkina L.V., Volkov V.V., Mentser A.N., Vaniukov A.V., Kireev P.S. Dependence of energy gap of CdSe$_x$CdTe$_{1-x}$ alloys from composition [in Russian], *Fizika i tehnika poluprovodn.*, **2**(4), 611–612 (1968).

Przedmojski J., Pałosz B. X-ray investigation of needle-like crystal of CdIn$_2$Se$_4$, *Phys. Status Solidi (a)*, **51**(1), K1-K3 (1979).

Quenez P., Khodadad P. Étude du système GeSe$_2$–CdSe. Identification du composé Cd$_4$GeSe$_6$, *C. r. Acad. Sci.*, **C268**(26), 2294–2297 (1969).

Raccah P.M., Bouchard R.J., Wold A. Crystallographic study of chromium spinels, *J. Appl. Phys.*, **37**(3), 1436–1437 (1966).

Radautsan S.I., Ivanova R.A. Formation of the solid solutions based on $A^{II}B^{IV}C_3^{VI}$ compounds [in Russian], *Izv. AN MSSR*, № 10 (88), 64–70 (1961).

Range K.-J., Becker W., Weiss A. Über Hochdruckphasen des CdAl$_2$S$_4$, HgAl$_2$S$_4$, ZnAl$_2$Se$_4$, CdAl$_2$Se$_4$ und HgAl$_2$Se$_4$ mit Spinellstruktur, *Z. Naturforsch.*, **23B**(7), 1009 (1968).

Range K.-J., Eglmeier Ch. Structure refinement of two cadmium rare earth selenospinels, CdTm$_2$Se$_4$ and CdHo$_2$Se$_4$, *J. Alloys Compd.*, **176**(2), L13-L16 (1991).

Reisman A., Berkenblit M. Impurity incorporation into CdSe and equilibria in the system CdSe–CdCl$_2$, *J. Electrochem. Soc.*, **109**(11), 1111–1113 (1962).

Rubenstein M. Solubilities of some II–VI compounds in bismuth, *J. Electrochem. Soc.*, **113**(6), 623–624 (1966).

Rubenstein M. Solution growth of some II–VI compounds using tin as a solvent, *J. Cryst. Growth*, **3–4**, 309–312 (1968).

Safarov M.G. The Sb$_2$Se$_3$–CdSe system [in Russian], *Zhurn. neorgan. khimii*, **33**(6), 1536–1540 (1980).

Savickas R., Baršauskas K. ZnSe–CdSe ir CdSe–CdTe trejinių sistemų struktūros klausimu [in Lithuanian], *Tr. Kaunas. polytehn. in-ta*, **14**(1), 15–18 (1960).

Schwer H., Krämer V. Neue Überstrukturen von AB$_2$Z$_4$-Defekt-Tetraheder-Verbindungen, *Z. Kristallogr.*, **182**(1–4), 245–246 (1988).

Sealy B. J., Crocker A. J. Some physical properties of the system Pb$_{1-x}$Mg$_x$Se and Pb$_{1-x}$Cd$_x$Se, *J. Mater. Sci.*, **8**(9), 1247–1252 (1973).

Shabunina G.G., Luzhnaya N.P., Kalinnikov V.T. The CdSe–Cr$_2$Se$_3$–Se system [in Russian], *Zhurn. neorgan. khimii*, **26**(2), 476–478 (1981).

Shabunina G.G., Shibanova N.M., Aminov T.G., Kalinnikov V.T. Growth of $CdCr_2Se_4$ single crystals from the melts of the $CdCl_2$ with CdJ_2 and $PbCl_2$ mixtures [in Russian], *Zhurn. neorgan. khimii*, **34**(6), 1586–1588 (1989).

Sher A.A., Odin I.N., Galiulin E.A., Novoselova A.V. Heterogeneous equilibria and electrophysical properties of alloys in the Cd–Sb–Se system [in Russian], *Zhurn. neorgan. khimii*, **26**(4), 1052–1057 (1981).

Sher A.A., Odin I.N., Novoselova A.V. Physico-chemical investigation of interactions in the Cd–Bi–Se system [in Russian], *Zhurn. neorgan. khimii*, **24**(9), 2509–2515 (1979).

Shevchenko I.B., Nikol'ski Yu.V., Smirnova E.M., Darashkevich V.R., Sutyrin Yu.E. Solid solutions in the CdSe–CdTe system [in Russian], *Izv. AN SSSR. Neorgan. materialy*, **10**(2), 217–220 (1974).

Shin Dong-Ho, Kim Chang-Dae, Jang Hyang-Hee, Choe Sung-Hyu, Kim Duck-Tae, Yoon Chang-Sun, Kim Wha-Tek. Growth and characterization of $Mg_xCd_{1-x}Se$ single crystals, *J. Cryst. Growth*, **177**(1/2), 167–170 (1997).

Słodowy P.A., Giriat W. The dependence of the energy gap on the composition in the mixed crystals $Cd_xHg_{1-x}Se$, *Phys. Status Solidi* (b), **48**(2), 463–466 (1971).

Smith K., Marsella J., Kershaw R., Dwight K., Wold A. Preparation and characterization of $Cd_{1-x}Fe_xSe$ single crystals, *Mater. Res. Bull.*, **23**(10), 1423–1428 (1988).

Sobolev V.V. Zone structure of $CdIn_2S_4$-type crystals [in Russian], *Izv. AN MSSR. Ser. fiz.-tehn. i mat. nauk*, (2), 60–63 (1976).

Sosovska S.M., Olekseyuk I.D., Parasyuk O.V. The $CdSe–Ga_2Se_3–PbSe$ system, *J. Alloys Compd.*, **453**(1–2), 115–120 (2008).

Stanchu A.V., Derid O.P. Obtaining of the equation of the Cd–In–CdSe liquidus surface by the mathematical planning of the experiment [in Russian], in: *Poluprovodn. pribory i materialy*, Kishinev: Shtiintsa, 33–38 (1973).

Strauss A. J., Steininger J. Phase diagram of the CdTe–CdSe pseudobinary system, *J. Electrochem. Soc.*, **117**(11), 1420–1426 (1970).

Stuckes A. D., Farrel G. Electrical and thermal properties of alloys of CdTe and CdSe, *J. Phys. Chem. Sol.*, **25**(5), 477–482 (1964).

Sultonov C., Karimov S.K. Some electrophysical properties of $CdTl_2Se_4$ and $CdTl_2Te_4$ ternary chalcogenides [in Russian], *Zhurn. neorgan. khimii*, **15**(7), 1191–1193 (1979).

Sultonov S., Mavlonov Sh., Karimov S., Kurbanov H.M., Madazimov A. Obtaining and electrical conductivity of $CdTl_2Se_4$ single crystals [in Russian], *Izv. AN SSSR. Neorgan. materialy*, **12**(1), 115–116 (1976).

Susa K., Steinfink H. $GeCd_4S_6$, a new defect tetrahedral structure type, *Inorg. Chem.*, **10**(8), 1754–1756 (1971).

Sysoev L.A., Obuhova N.F. Obtaining and investigation of some properties of ZnS–MgS single crystals [in Russian], *Monokristaly i tehnika*, [2(9)], 54–57 (1973).

Tai Hideo, Nakashima Shinichi, Hori Shigenori. Bandgap energies of solid solutions in $(CdTe)_{1-x}–(CdSe)_x$ and $(CdTe)_{1-x}–(CdS)_x$ systems [in Japanese]. *J. Jap. Inst. Metals*, **40**(5), 474–479 (1976).

Tarascon J.M., Hull G.W., Waszczak J.V. Synthesis, structural and physical properties of the binary molybdenum chalcogenide phase $Mo_{15}Se_{19}$ and of the related compounds $M_2Mo_{15}Se_{19}$ and $M_3Mo_{15}Se_{19}$ (M = group IA metal; Sn; Pb; Cd), *Mater. Res. Bull.*, **20**(8), 935–946 (1985).

Tiurin O.A., Dvoretskov G.A., Vishniakov A.V., Kovtunenko P.V. The Ag-Cd-Se system [in Russian], *Zhurn. neorgan. khimii*, **27** (7), 1827–1831 (1982).

Tomashik Z.F., Oleinik G.S., Tomashik V.N. Phase diagram of the PbSe–CdSe system [in Russian], *Izv. AN SSSR. Neorgan. materialy*, **16**(2), 261–263 (1980).

Tovt V.V., Lukach P.M., Guranich P.P., Shusta V.C., Gerzanich O.I., Potoriy M.V., Voroshilov Yu.V. Growth and physical properties of $Cd_2P_2S_6$ and $Cd_2P_2Se_6$ single crystals [in Ukrainian], *Nauk. visnyk Uzhgorod. un-tu. Ser. Khimia*, (3), 52–54 (1998).

Tret'yakov Yu.D., Gordeev I.V., Alferov V.A., Saksonov Yu.G. Deviation from stoichiometry of chalcogenide chromites with the spinel structure [in Russian], *Izv. AN SSSR. Neorgan. materialy*, **8**(12), 2215–2216 (1972).

Trishchuk L.I., Oleinik G.S., Mizetskaya I.B. Phase equilibria in the Ag_2Se–ZnSe and Ag_2Se–CdSe systems, *Izv. AN SSSR. Neorgan. materialy*, **18**(11), 1795–1797 (1982).

Trishchuk L.I., Oleinik G.S., Mizetskaya I.B. The $Cu_{2-x}Se$–CdSe phase diagram [in Russian], *Izv. AN SSSR. Neorgan. materialy*, **20** (4), 679–681 (1984).

Trishchuk L.I., Oleinik G.S., Mizetskaya I.B. Thermal analysis determination of $A^{II}B^{VI}$ solid solubility in $A^I_2B^{VI}$. *Thermochim. Acta*, **92**, 611–613 (1985).

Tyrziu M. P., Radautsan S. I., Markus M. M., Kolosenko S. M. State diagram of CdSe–Ga_2Se_3, *Phys. Status Solidi* (a), 3(4), K293–K296 (1970).

Tyrziu M.P., Tyrziu V.G. Physical properties of the $CdGa_2Se_4$ compound [in Russian], *Izv. AN SSSR. Neorgan. materialy*, **7**(10), 1855–1856 (1971).

Valov Yu.A., Payonchkovska A. About synthesis of $A^{II}B_2^{III}C_4^{VI}$ ternary chalcogenides using chemical transport reactions [in Russian], *Izv. AN SSSR. Neorgan. materialy*, **6**(2), 241–246 (1970).

Vengatesan B., Kanniah N., Gobinathan R., Ramasamy P. Growth of cadmium indium selenide crystals by closed tube chemical vapor transport, *Indian J. Phys.*, **A61**(5), 393–396 (1987).

Vishniakov A.V., Bebiakin A.V., Zubkovskaya V.N. Phase equilibria in the region $Zn(Cd)$–$Zn(Cd)X$–Cu_2X (X = S, Se, Te) of Zn–Cu–X system [in Russian], *Zhurn. neorgan. khimii*, **27**(7), 1820–1826 (1982).

Vishniakov A.V., Dvoretskov G.A., Zubkovskaya V.N., Tyurin O.A., Kovtunenko P.V. Phase equilibria in the systems formed by the II-VI compounds and the elements of I and III groups of the periodic table [in Russian], *Tr. Mosk. khim.-tekhnol. in-t*, (120), 87–103 (1981).

Vishniakov A.V., Harif Ya.L. Solubility of Cu_2Se in ZnSe and CdSe [in Russian], *Izv. AN SSSR. Neorgan. materialy*, **8**(2), 217–219 (1972).

Vishniakov A.V., Iofis B.G. Solubility of Ag_2S (Se) in ZnS, CdS and CdSe [in Russian], *Izv. AN SSSR. Neorgan. materialy*, **10** (7), 1184–1186 (1974).

Vishniakov A.V., Kukleva T.V., Al'tah O.L., Zubkovskaya V.N., Kovtunenko P.V. The solid state solubility of cadmium chalcogenides in copper(I) chalcogenides [in Russian], *Zhurn. neorgan. khimii*, **25**(5), 1358–1361 (1980).

Vishniakov A.V., Tiurin O.A., Kovtunenko P.V. Solubility of Ag in the cadmium chalcogenides [in Russian], *Zhurn. neorgan. khimii*, **28** (5), 1274–1280 (1983).

Vitrihovskiy N.I., Mizetskaya I.B., Oleinik G.S. Properties of alloys in the CdSe–CdTe system [in Russian], *Izv. AN SSSR. Neorgan. materialy*, **7**(5), 757–760 (1971).

Wald P., Rosenberg A. J. Solid solutions of CdSe and InSe in PbSe, *J. Phys. Chem. Solids*, **26**(7), 1087–1091 (1965).

Wiedemeier H., Huang Xuejun. Phase studies of the Cd–Fe–Se system in the Cd-rich region, *J. Electron. Mater.*, **22**(6), 695–699 (1993).

Zhao D., Xia F., Chen G., Zhang X., Ma H., Adam J.L. Formation and properties of chalcogenide glasses in the $GeSe_2$–As_2Se_3–CdSe system, *J. Amer. Ceram. Soc.*, **88**(11), 3143–3146 (2005).

Zhegalina V.A., Arakelyan Z.S., Aminov T.G. Vapor pressure above $CdCr_2Se_4$ [in Russian], *Izv. AN SSSR. Neorgan. materialy*, **21**(2), 221–224 (1985).

Zhitar' V.F., Donu V.S., Val'kovskaya M.I., Markus M.M. Preparation and some physical properties of the $CdGa_2S(Se)_4$ [in Russian], in: *Fiz. i khim. slozhnyh poluprovodn*. Kishinev: Shtiintsa, 50–60 (1975).

Zhukov E.G., Poluliak E.S., Varnakova E.S., Fedorov V.A. Investigation of chalcogenide spinel stability at the heating in the air [in Russian], *Neorgan. materialy*, **33**(8), 939–941 (1997).

Zmiy O.F., Mishchenko I.A., Olekseyuk I.D. Phase equilibria in the quasi-ternary system Cu_2Se–CdSe–In_2Se_3, *J. Alloys Compd.*, **367**(1–2), 49–57 (2004).

6

System Based on CdTe

6.1 Cadmium–Potassium–Tellurium

CdTe–K$_2$Te: The phase diagram is not constructed. K$_2$Cd$_3$Te$_4$ ternary compound is formed in this system (Axtell et al. 1996). It melts incongruently at 690°C and crystallizes in an orthorhombic structure with lattice parameters $a = 1513.7 \pm 0.8$, $b = 1130.7 \pm 0.4$, and $c = 729.2 \pm 0.7$ pm at 193 K and calculation density 4.93 g cm^{-3} and energy gap 2.26 eV.

K$_2$Cd$_3$Te$_4$ was obtained by the interaction of CdTe and K$_2$Te in the evacuated quartz tube (Axtell et al. 1996). The mixture was heated at 800°C for 2 days, followed by 10°C h^{-1} cooling to 400°C and finally at 25°C h^{-1} to room temperature. This compound is insoluble in water and common organic solvents but is sensitive to exposure to air.

6.2 Cadmium–Copper–Tellurium

The Cd–Cu–Te isothermal section at 400°C is given in Figure 6.1 (Cordes and Schmid-Fetzer 1995a). The experimentally observed phase diagram that essentially confirms the approximate calculation is in fact dominated by the CdTe–Cu tie-line. However, the solubility of Cd in Cu leads to a shift of the line to the Cd side. Hence, the connection from CdTe to Cu crosses the three-phase field CdTe–Cu–Cu$_2$Te, and the formation of Cu$_2$Te is expected in CdTe/Cu diffusion couples. In samples that were annealed at 500°C or 550°C, the precipitation of Cu$_2$Te was observed in parts of the interface, which grow in the direction of CdTe. At higher annealing temperatures, the whole CdTe wafer had reacted and Cu$_2$Te is formed.

At 700°C alloy of Cd with Cu, phases based on Cu$_{2-x}$Te and Cu$_5$Te$_3$ and melt based on Te are in equilibrium with CdTe (Figure 6.2) (Vishniakov et al. 1989).

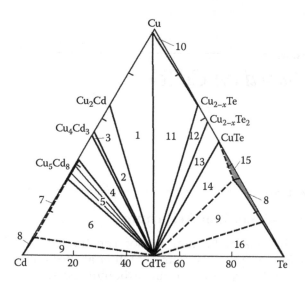

FIGURE 6.1
Isothermal section of the Cd–Cu–Te ternary system at 400°C (the ternary equilibria with the liquid phase are estimated): 1, $CdTe + Cu + Cu_2Cd$; 2, $CdTe + Cu_2Cd + Cu_4Cd_3$; 3, $CdTe + Cu_4Cd_3$; 4, $CdTe + Cu_4Cd_3 + Cu_5Cd_8$; 5, $CdTe + Cu_5Cd_8$; 6, $L + CdTe + Cu_5Cd_8$; 7, $L + Cu_5Cd_8$; 8, L; 9, $L + CdTe$; 10, $Cu + Cu_{2-x}Te$; 11, $CdTe + Cu + Cu_{2-x}Te$; 12, $CdTe + Cu_{2-x}Te + Cu_{3-x}Te_2$; 13, $CdTe + Cu_{3-x}Te_2 + CuTe$; 14, $L + CdTe + CuTe$; 15, $CuTe$; and 16, $L + CdTe + Te$. (From Cordes, H. and Schmid-Fetzer, R., *Semicond. Sci. Technol.*, 10(1), 77, 1995.)

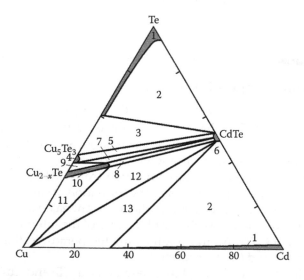

FIGURE 6.2
Scheme of isothermal section of Cd–Cu–Te ternary system at 700°C: 1, L; 2, $L + CdTe$; 3, $CdTe + Cu_5Te_3 + L$; 4, Cu_5Te_3; 5, $CdTe + Cu_5Te_3$; 6, $CdTe$; 7, $CdTe + Cu_{2-x}Te + Cu_5Te_3$; 8, $CdTe + Cu_{2-x}Te$; 9, $Cu_{2-x}Te + Cu_5Te_3$; 10, $Cu_{2-x}Te$; 11, $Cu_{2-x}Te + Cu$; 12, $CdTe + Cu_{2-x}Te + Cu$; and 13, $L + CdTe + Cu$ (the region of phases based on CdTe, $Cu_{2-x}Te$, and Cu_5Te_3 is enlarged). (From Vishniakov, A.V. et al., *Izv. AN SSSR. Neorgan. materialy*, 25(4), 578, 1989.)

Single-crystal $Cu_{1.70}Cd_{0.05}Te$ that corresponds to the $CdTe–Cu_{1.8}Te$ section was grown using the Bridgman method by Asadov et al. (1996). It is shown that isomorphous substitution of a part of copper atoms by cadmium atoms stabilizes the high-temperature cubic phase of $Cu_{1.70}Cd_{0.05}Te$ at room temperature. Low-temperature hexagonal phase with lattice parameters $a = 833.5$ and $c = 2165.4$ at 300°C changes into another hexagonal phase with lattice parameters $a = 828.0$ and $c = 725.6$ pm, reproduces at 400°C, and transforms at 500°C into new hexagonal phase with lattice parameters $a = 837.8$ and $c = 1092.3$ pm, which changes into a primitive cubic phase with lattice parameter $a = 609.2$ pm at 528°C (Asadov et al. 1996).

The ingots were annealing at 400°C for 120–480 h and the reaction products were analyzed using x-ray diffraction (XRD), metallography, and scanning electron microscopy, including energy-dispersive x-ray microanalysis (Cordes and Schmid-Fetzer 1995a).

CdTe–Cu: This section is a nonquasibinary section of the Cd–Cu–Te ternary system (Panchuk et al. 1973a, 1974a). XRD detects CdTe and Cu in all ingots and Cu_2Te and Cu_5Cd_8 were obtained also at 80–95 at. % Cu. An immiscibility region with monotectic temperature 825°C ± 5°C exists in this system within the interval of 50–95 at. % Cu.

Temperature dependence of Cu solubility in CdTe under cadmium vapor pressure close to its maximum is retrograde with the maximum of 10^{20} cm^{-3} at 700°C ± 10°C (Figure 6.3) (Grytsiv 1975). Above 1000°C, the solubility of Cu was calculated. Within the interval of 200°C–700°C, the solubility of Cu (cm^{-3}) can be described by the equation: $x_{Cu} = 2.40 \cdot 10^{22} \exp(E/kT)$, where $E = -0.48$ eV.

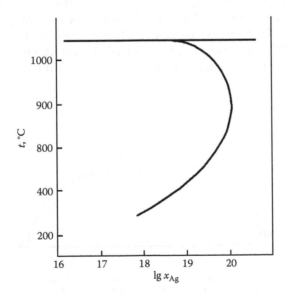

FIGURE 6.3
Temperature dependence of Cu solubility in CdTe. (From Grytsiv, V.I., *Kand. dis.*, Chernovtsy, 141 p., 1975.)

According to the data of Vishniakov et al. (1989), temperature dependence of Cu solubility in CdTe can be described by the equation: $x_{Cu} = 3.73 \cdot 10^{24}$ exp $(-7900/T)$.

Jimenéz-Sandoval et al. (1999) noted that a novel semiconducting alloy of the type $Cu_xCd_{1-x}Te$ was successfully prepared for Cu concentrations lower than 7 at. %. It was determined that the crystalline structure and the optical band gap of CdTe were nearly unaffected by the incorporation of Cu into the lattice. Moreover, the best structural properties were found for samples in which $x \le 0.02$.

The interaction of CdTe with Cu was investigated using differential thermal analysis (DTA), metallography, XRD, Auger spectroscopy, energy-dispersive x-ray analysis, and microhardness measurement (Panchuk et al. 1973a, 1974a, Grytsiv 1975, Jimenéz-Sandoval et al. 1999). The solubility of Cu in CdTe was determined using radioactive isotopes (the time of ingots annealing was increased from 3 h at 1000°C to 60 h at 450°C) (Grytsiv 1975).

CdTe–Cu₅Cd₈: The phase diagram is a eutectic type (Figure 6.4) (Grytsiv 1975). The eutectic composition and temperature are 3.9 mol. % CdTe and 556°C, respectively. The solubility of Cu_5Cd_8 in CdTe at 1000°C is not higher than 1 mol. %.

This system was investigated using DTA, metallography, and XRD. The ingots were annealed at 520°C for 600 h (Grytsiv 1975).

CdTe–Cu₂₋ₓTe: The phase diagram is a eutectic type (Figure 6.5) (Vishniakov et al. 1980, Parasyuk et al. 1981, 1997, Trishchuk et al. 1984, 2008). The eutectic composition and temperature are 53 mol. % CdTe and 842°C, respectively (Trishchuk et al. 1984, 2008) [52 mol. % CdTe and 840°C (Parasyuk et al. 1997), 65 ± 5 mol. % CdTe and 820°C ± 10°C (Vishniakov et al. 1981)]. The solubility of CdTe in $Cu_{2-x}Te$ decreases from 47 mol. % at the eutectic temperature to 6 mol. % at 600°C and 2 mol. % at 535°C (Trishchuk et al. 1984, 1985, Parasyuk et al. 1997), and the solubility of $Cu_{2-x}Te$ in CdTe at 850°C is not higher than

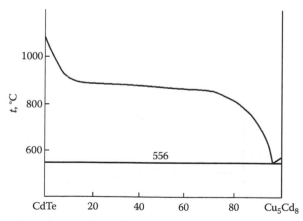

FIGURE 6.4

CdTe–Cu₅Cd₈ phase diagram. (From Grytsiv, V.I., *Kand. dis.*, Chernovtsy, 141 p., 1975.)

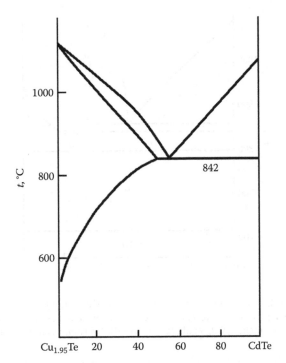

FIGURE 6.5
CdTe–Cu$_{1.95}$Te phase diagram. (From Trishchuk, L.I. et al., *Izv. AN SSSR. Neorgan. materialy*, 20(9), 1486, 1984.)

3 mol. % (Trishchuk et al. 1984) [less than 2 mol. % (Parasyuk et al. 1997)]. In the subsolidus region, this phase diagram is complicated by the polymorphic transformations of Cu$_{2-x}$Te (Parasyuk et al. 1997).

This system was investigated using DTA, metallography, and XRD (Trishchuk et al. 1984, Parasyuk et al. 1997). The ingots were annealed at 400°C for 250 h (Parasyuk et al. 1997).

6.3 Cadmium–Silver–Tellurium

CdTe–Ag: This section is a nonquasibinary section of the Cd–Ag–Te ternary system (Figure 6.6) (Panchuk et al. 1974b). The immiscibility region exists in this system within the interval of 50–70 at. % Ag. At the alloying of CdTe and Ag, the next interaction takes place: CdTe + 2Ag = Ag$_2$Te + Cd. Cadmium that is formed at the interaction dissolves in the silver.

The data of Panchuk et al. (1974b) contradict the thermochemical calculations of Shuh and Williams (1988), who predicted that CdTe–Ag system is quasibinary and their experimental results confirmed such prediction.

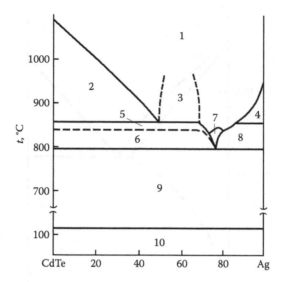

FIGURE 6.6
Phase relations in the CdTe–Ag system: 1, L; 2, L_1+CdTe; 3, L_1+L_2; 4, L_2+Ag; 5, L_2+CdTe; 6, L_2+CdTe+β-Ag_2Te; 7, L_2+β-Ag_2Te; 8, L_2+β-Ag_2Te+Ag; 9, CdTe+β-Ag_2Te+Ag; and 10, CdTe+α-Ag_2Te+Ag. (From Panchuk, O.E. et al., *Izv. AN SSSR. Neorgan. materialy*, 10(6), 980, 1974b.)

Temperature dependence of Ag solubility in CdTe of *n* type under cadmium vapor pressure closed to the maximum is retrograde with maximum of $3.4 \cdot 10^{19}$ cm^{-3} at 793°C ± 5°C (Figure 6.7) (Panchuk et al. 1974c). At 1000°C–1092°C, this curve is calculated using equilibrium distribution coefficient for Ag in cadmium telluride value equal to $2.14 \cdot 10^{-3}$. A part of solidus curve with positive temperature coefficient can be described by the following equation: x_{Ag} $(cm^{-3}) = 1.29 \cdot 10^{27}$ exp (E/RT), where $E = -148.5$ kJ mol^{-1} (-35.5 kcal mol^{-1}).

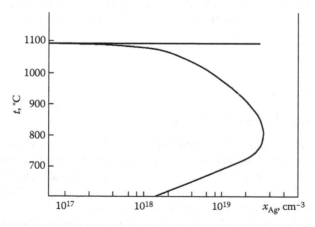

FIGURE 6.7
Temperature dependence of Ag solubility in CdTe. (From Panchuk, O.E. et al., *Izv. AN SSSR. Neorgan. materialy*, 10(4), 581, 1974c.)

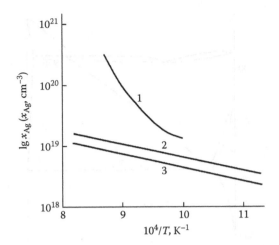

FIGURE 6.8
Solubility of Ag in CdTe at 600°C–950°C in the dependence of the conjugated phase composition: 1, $Ag_{0.5}Cd_{0.5}$; 2, $Ag_{0.2}Cd_{0.8}$; and 3, $Ag_{0.1}Cd_{0.9}$. (From Vishniakov, A.V. et al., *Zhurn. neorgan. khimii*, 28(5), 1274, 1983.)

The solubility of Ag in CdTe at 600°C–950°C depends on Ag contents in the conjugated Ag_xCd_{1-x} alloys as shown in Figure 6.8 (Vishniakov et al. 1983).

This system was investigated using DTA, metallography, and XRD (Panchuk et al. 1974b, Shuh and Williams 1988). The ingots were annealed at 750°C for 500 h (Panchuk et al. 1974b). The solubility of Ag in CdTe was determined using radioactive isotopes and the crystals with different deviations from stoichiometry.

CdTe–Ag₂Te: The most reliable phase diagram is a peritectic type (Figure 6.9) (Trishchuk et al. 1986, 2008, Trishchuk 2009). The peritectic composition and temperature are 24 mol. % Ag_2Te and 955°C, and the composition and temperature of azeotropic point are 32 mol. % CdTe and 928°C, respectively. The limit solubility of CdTe in γ-Ag_2Te is equal to 83 mol. % at the peritectic temperature. γ-Solid solution undergoes the eutectoid transformation. The eutectoid composition and temperature are 13.5 mol. % CdTe and 532°C, respectively. The maximum solubility of CdTe in β-Ag_2Te is equal to 12 mol. % and decreases sharply with the decreasing temperature.

According to the data of Boncheva-Mladenova et al. (1985), the phase diagram of the CdTe–Ag_2Te belongs to the eutectic type. The eutectic reaction at 900°C ± 10°C with the eutectic point at 50 mol. % CdTe and the two eutectoid decompositions at 500°C ± 10°C and 140°C ± 10°C with eutectoid points at about 20 and 15 mol. % CdTe, respectively, were determined.

The solubility of CdTe in Ag_2Te reaches 83 mol. % at the peritectic temperature and decreases up to 1.0 mol. % at 345°C (Trishchuk et al. 1985, 1986, 2008, Trishchuk 2009). The solubility of Ag_2Te in CdTe is equal to 2.5 mol. % and 0.5 mol. % at 700°C. Thermal effect at 145°C corresponds to the polymorphous transformation of Ag_2Te (Trishchuk et al. 1986, 2008, Trishchuk 2009). Much higher region of solid solutions based on CdTe and low-temperature Ag_2Te

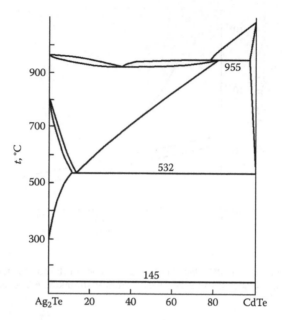

FIGURE 6.9

CdTe–Ag₂Te phase diagram. (From Trishchuk, L.I. et al., *Ukr. khim. zhurn.*, 52(8), 799, 1986; Trishchuk, L.I. et al., *Chem. Met. Alloys*, 1(1), 58, 2008; Trishchuk, L.I., *Optoelektronika i polupro-vodn. tekhnika*, (44), 99, 2009.)

have been determined by Boncheva-Mladenova et al. (1985). No glasses were obtained in the CdTe–Ag₂Te system (Vassilev et al. 2008).

This system was investigated using DTA, XRD, metallography, and microhardness measurement (Boncheva-Mladenova et al. 1985, 1986, 2008, Trishchuk 2009). The ingots were annealed at 870°C for 20 h (Trishchuk et al. 1986, 2008, Trishchuk 2009).

6.4 Cadmium–Gold–Tellurium

CdTe–Au: The phase diagram is a eutectic type (Figure 6.10) (Grytsiv et al. 1978). The eutectic composition and temperature are 75 ± 3 at. % Au and 810°C, respectively.

The thermochemical calculations of Shuh and Williams (1988) predicted that CdTe–Au system is quasibinary and experimental results confirmed such prediction.

At 600°C–1000°C, the solubility of Au (cm⁻³) in CdTe can be described by the next equation: $x_{Au} = 6.2 \cdot 10^{20} \exp (E/kT)$, where $E = -0.59$ eV (Grytsiv et al. 1978). According to the data of Akutagava et al. (1975), the measured Au solubility at 800°C and 900°C varied from ~10¹⁸ cm⁻³ to ~10²⁰ cm⁻³ and was found

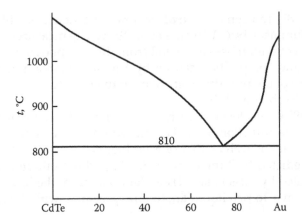

FIGURE 6.10
CdTe–Au phase diagram. (From Grytsiv, V.I. et al., *Izv. AN SSSR. Neorgan. materialy*, 14(7), 1348, 1978.)

to increase with an increasing temperature and with a decreasing Cd partial pressure roughly as $p_{Cd}^{-1/3}$.

This system was investigated using DTA, metallography, and XRD. The ingots were annealed at 800°C for 600 h (Grytsiv et al. 1978, Shuh and Williams 1988). CdTe samples were doped with Au either in the melt or by diffusion into undoped crystals (Akutagava et al. 1975).

6.5 Cadmium–Magnesium–Tellurium

CdTe–MgTe: The phase diagram is not constructed. The solubility of MgTe in CdTe reaches 30 mol. % (Rusnak et al. 1990). The concentration dependence of energy gap of the $Cd_{1-x}Mg_xTe$ solid solution ($0 \le x \le 0.3$) at 77 K can be described by the next equation: $E_g(x) = 1.595 + 0.78x + 5.22x^2$.

6.6 Cadmium–Mercury–Tellurium

Liquidus isotherms of the Cd–Hg–Te ternary system between 470°C and 600°C in the Te corner were calculated in Dub et al. (1983) using ideal solution model. Vanyukov et al. (1973, 1977) determined liquidus isotherms in the Hg-rich corners of the ternary system using the technique of quenching and subsequent analysis and dew-point method. The liquidus temperatures have been measured for some $(Cd_xHg_{1-x})_yTe_{1-y}$ compositions using a visual examination and DTA (Ueda et al. 1972, Harman 1980, Mroczkowski and

Vydyanath 1981, Tomson et al. 1981, Szofran and Lehoczky 1983, Meschter et al. 1985, Gumiński 1986, Wermke et al. 1992). The measured values are in good agreement with those calculated from the thermodynamic model. The change in liquid composition caused by Hg vaporization into free volume of the DTA ampoule introduces an error in the liquidus temperature of less than 1.5°C (Meschter et al. 1985).

Herning (1984) developed a vertical "infinite-melt" reflux technique that allowed the determination of the Hg-rich liquidus from solubility measurements. Their data as well as that of Meschter et al. (1985) fall on calculated isotherms predicted by Tung et al. (1982). Liquidus isotherms at 460°C and 480°C calculated by Lusson and Triboulet (1987) from the kinetics of epitaxial growth from a Te-rich liquid are in fair agreement with the experimental results.

The liquidus isotherms and solid solutions isoconcentration lines of the Cd–Hg–Te ternary system according to Tung et al. (1982) and Kalisher et al. (1994) are presented in Figure 6.11 for the whole diagram and in Figure 6.12 for the Hg-rich corner. Harman (1980) and Sanz-Maudes et al. (1990) investigated the Te-rich corner, and the results are shown in Figure 6.13. The numbers on the isoconcentration lines give the values of x in the solid solution $Hg_{1-x}Cd_xTe$.

Some isothermal sections of the Cd–Hg–Te ternary system are given in Figures 6.14 through 6.17 (Yang et al. 1995, Voronin and Pentin 2005, Pentin et al. 2006). Isothermal section of the Cd–CdTe–HgTe–Te subsystem at 120°C was estimated based on known thermodynamic properties of the

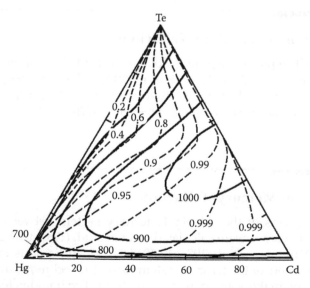

FIGURE 6.11
Liquidus isotherms (solid lines) and solidus isoconcentration (dashed lines) for the Cd–Hg–Te ternary system. (From Kalisher, M.H. et al., *Progr. Cryst. Growth Charact.*, 29, 41, 1994.)

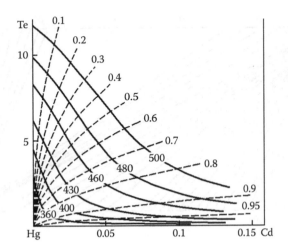

FIGURE 6.12
Liquidus isotherms (solid lines) and isoconcentration solidus line (dashed lines) in the Hg-rich corner of the Cd–Hg–Te ternary system. (From Tung Tse et al., *J. Vac. Sci. Technol.*, 21(1), 117, 1982.)

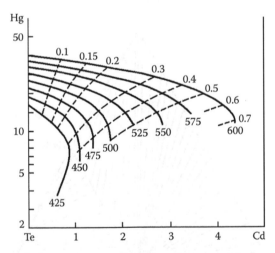

FIGURE 6.13
Liquidus isotherms (solid lines) and isoconcentration solidus line (dotted lines) in the Te-rich corner of the Cd–Hg–Te ternary system. (From Harman, T.C. et al., *J. Electron. Mater.*, 9(6), 945, 1980.)

binary system, and the obtained by Voronin and Pentin (2005) and Pentin et al. (2006) results for the CdTe–HgTe quasibinary system. New tie-lines in the CdTe–HgTe–Te triangle are experimentally determined by Liu and Brebrick (1995).

The most essential successes in the thermodynamic description of the Cd–Hg–Te ternary system were achieved with the help of the associated liquid solution model (Laugier 1973, Kikuchi 1982, Tung et al. 1982, Brebrick et al.

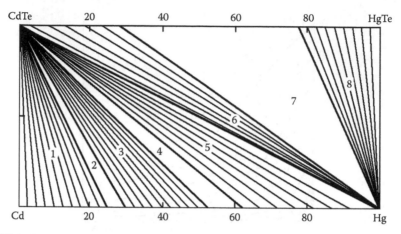

FIGURE 6.14
Isothermal section of the Cd–CdTe–HgTe–Te subsystem at 120°C: 1, $Cd_{1-x}Hg_x + CdTe$; 2, $Cd_{1-x}Hg_x + \omega + CdTe$; 3, $\omega + CdTe$; 4, $L + \omega + CdTe$; 5, $L + CdTe$; 6, $L + Cd_{1-x}Hg_xTe$; 7, $L + Cd_{1-x}Hg_xTe + Cd_xHg_{1-x}Te$; and 8, $L + Cd_xHg_{1-x}Te$. (From Voronin, G.F. and Pentin, I.V., *Zhurn, Fiz. Khimii*, 79(10), 1771, 2005; Pentin, I.V. et al., *Calphad*, 30(2), 191, 2006.)

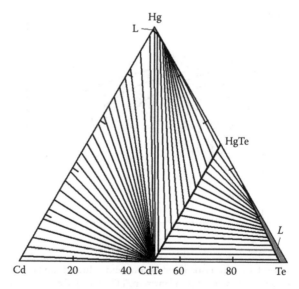

FIGURE 6.15
Isothermal section of the Cd–Hg–Te ternary system at 540°C. (From Yang, J. et al., *Calphad*, 19(3), 415, 1995.)

1983, Su 1986, Brebrick 1988). According to the data of Moskvin et al. (2008), the polyassociative solutions model satisfactorily describes the *p-T-x* equilibrium in this system, and obtained data can serve as the first approximation for the choice of the growth conditions of the $Cd_xHg_{1-x}Te$ solid solutions of the necessary composition.

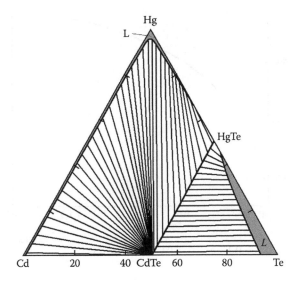

FIGURE 6.16
Isothermal section of the Cd–Hg–Te ternary system at 660°C. (From Yang, J. et al., *Calphad*, 19(3), 415, 1995.)

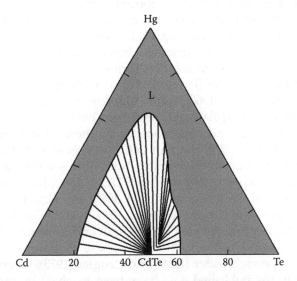

FIGURE 6.17
Isothermal section of the Cd–Hg–Te ternary system at 960°C. (From Yang, J. et al., *Calphad*, 19(3), 415, 1995.)

CdTe–Hg: The phase diagram is not constructed. Temperature dependence of CdTe solubility in Hg is given in Figure 6.18 (Vanyukov et al. 1977). Herning (1984) measured the solubility of CdTe in Hg in the temperature range 200°C–400°C with the result that confirmed the very low solubility.

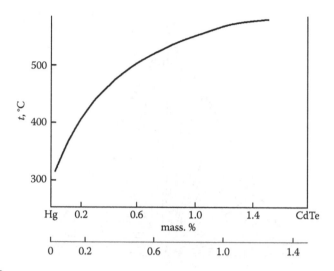

FIGURE 6.18

Temperature dependence of the CdTe solubility in Hg. (From Vanyukov, A.V. et al., *Inorg. Mater. (Engl. Trans.)*, 13(5), 667, 1977, transl. from *Izv. Akad. Nauk SSSR, Neorg. Mater.* 13(5), 815, 1977.)

CdTe–HgTe: The first information about the phase relations in the CdTe–HgTe system appeared in 1953 when the conclusion was made by Goriunova and Fedorova (1953) that CdTe and HgTe form a complete series of solid solutions with a zinc-blende structure. Some interesting semiconducting properties of $Cd_xHg_{1-x}Te$ solid solution series stimulated further investigations of the crystal structure (Lawson et al. 1959, Blair and Newnham 1960, Shneider and Gavrishchak 1960, Ivanov-Omskiy et al. 1964), phase relations in the CdTe–HgTe system (Blair and Newnham 1960, Woolley and Ray 1960a, Balagurova et al. 1974) and conditions of synthesis (Korovin et al. 1968, Dziuba 1969, Ueda et al. 1972, Harman 1980, Kalisher et al. 1994).

Experimental problems in the investigations of this system are due to the strong tendency to supercooling during DTA studies (Blair and Newnham 1960, Gromakov et al. 1964, Meschter et al. 1985) and also due to high components' vapor pressure (Gromakov et al. 1964, Ivanov-Omskiy et al. 1964, Lutsiv et al. 1968, Ray and Spencer 1967, Schmit and Speerschneider 1968).

The first assessment of the CdTe–HgTe phase diagram was performed by Schmit and Speerschneider (1968) and Laugier (1973), where significant discrepancies in the published data have been resolved by presenting p-T phase diagram for $Cd_{0.2}Hg_{0.8}Te$. Pressure–temperature phase diagrams for the $Cd_xHg_{1-x}Te$ melts and, partly, $Cd_{0.3}Hg_{0.7}Te$–Te melts were studied by Steininger (1976) and for the $Cd_xHg_{1-x}Te$ at $0 < x < 0.3$ by Farrar et al. (1977). Later, the assessed CdTe–HgTe phase diagrams were proposed by Brebrick et al. (1983), Marbeuf et al. (1985), Brice et al. (1986), Brebrick (1988), Patrick et al. (1988), Vasilyev et al. (1990b), Litvak and Charykov (1991), Nasar and Shamsuddin (1991), and Yu and Brebrick (1992), which demonstrated the special importance of such data for practical purposes.

The CdTe–HgTe section does not exhibit any invariants and continuous solid solutions exist between the two compounds (Capper 1982). The system presents a miscibility gap in the solid state with a critical point at $x_{CdTe} = 0.53$ and 182.2°C (Voronin et al. 2003, Voronin and Pentin 2005, Pentin et al. 2006) [$x_{CdTe} = 0.60 \pm 0.05$ and 221 ± 12°C (Chen and Wiedemeier 1995); $x_{CdTe} = 0.55$ and 172°C (Schmit 1983); $x_{CdTe} = 0.55$ and 182°C (Gambino et al. 1991); $x_{CdTe} = 0.40$ and 84 K (Wei et al. 1990)]. Similar conclusion was confirmed in Vasilyev et al. (1990a, 1990b), Nasar and Shamsuddin (1991), Sidorko and Goncharuk (1995), Goncharuk and Sidorko (1996). Based on the XRD, Wiedemeier and Chen (1994) showed that the phase separation occurs primarily in a thin surface layer at 140°C and is reversible after annealing at 530°C. The compositions of the two solid solutions along the tie-line at 140°C are $Cd_{0.78}Hg_{0.22}Te$ and $Cd_{0.37}Hg_{0.63}Te$. Small Hg clusters, from 0.5 to 2 nm in radius, were observed on the $Cd_xHg_{1-x}Te$ surfaces by x-ray photoemission study (Sporken et al. 1988). Besides, high pressure of Hg above the temperatures of the $Cd_xHg_{1-x}Te$ alloy synthesis causes an appearance of Te precipitates in the crystal (Gillham and Farrar 1977, Anderson et al. 1982).

The phase diagram of the CdTe–HgTe system, shown in Figure 6.19, is proposed by Yang et al. (1995), and the immiscibility gap in this system is taken from Voronin and Pentin (2005) and Pentin et al. (2006). According to the data of Marbeuf et al. (1992), contrary to other $A^{II}B^{VI}$ solid solutions, $Cd_xHg_{1-x}Te$ alloys are nearly ideal. This conclusion contradicts the thermodynamics that forbids the presence of a miscibility gap for an ideal solution. Observations of Krotov et al. (1979) and Chen and Wiedemeier (1995) based on mercury vapor tension measurements and data of Shamsuddin and Nasar (1995) and Goncharuk and Sidorko (1996) based on emf measurements showed that the solid solutions $Cd_xHg_{1-x}Te$ present a positive departure toward ideality,

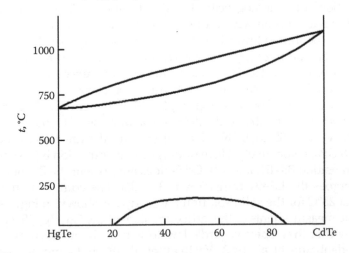

FIGURE 6.19
CdTe–HgTe phase diagram. (From Yang, J. et al., *Calphad*, 19(3), 415, 1995; Voronin, G.F. and Pentin, I.V., *Zhurn, fiz. khimii*, 79(10), 1771, 2005; Pentin, I.V. et al., *Calphad*, 30(2), 191, 2006.)

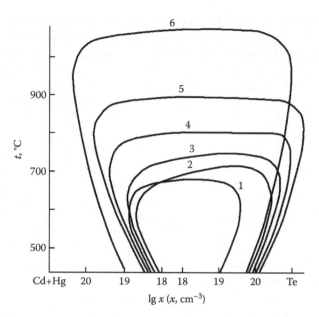

FIGURE 6.20

Homogeneity region of the $Cd_xHg_{1-x}Te$ solid solutions: 1, x = 0; 2, 0.2; 3, 0.4; 4, 0.6; 5, 0.8; and 6, 1.0. (From Glazov, V.M. and Pavlova, L.M., *Dokl. RAN*, 342(2), 189, 1995; Glazov, V.M. and Pavlova, L.M., *Zhurn. fiz. khimii*, 70(3), 479, 1996.)

which means repulsive interactions between Cd and Hg in the lattice. The differences between thermodynamic properties of $Cd_xHg_{1-x}Te$ alloys and that of the other $A^{II}B^{VI}$ solid solutions are confirmed in Barlow (2004) based on the phase diagram calculation in framework of a binary regular solutions model.

The position of the homogeneity domain of solid solutions along the CdTe–HgTe quasibinary section was studied in Glazov and Pavlova (1995, 1996) and was shown to be shifted toward the tellurium corner (Figure 6.20). Far-infrared lattice-vibration spectroscopy data led (Kozyrev et al. 1998) to the conclusion about the disorder in CdTe–HgTe alloys due to a difference in the CdTe (semiconductor) and the HgTe (semimetals) chemical bond type.

High-pressure studies of $Cd_{0.2}Hg_{0.8}Te$ alloy by energy-dispersive diffraction spectra in Quadri et al. (1986) show that the B3(sphalerite type) to B9(HgS type) and B9 to B1(NaCl type) structural transitions occurred at 2.2 and 6.8 GPa, respectively. HgTe undergoes the transition B3–B9 at 1.6 GPa and the transition B9–B1 at 8 GPa. CdTe does not present the B9 modification and undergoes the B3–B1 transition at 3.2 GPa. The composition–pressure diagram at 25°C for the solid solution $Hg_{1-x}Cd_xTe$ is shown in Figure 6.21.

Accurate measurements of the lattice constant of the $Cd_xHg_{1-x}Te$ solid solutions show that they obey Vegard's law to within the measurement uncertainty (Indenbaum et al. 1983, Vasilyev et al. 1990b, Gambino et al. 1991, Skauli and Colin 2001). Concentration dependence of the energy gap for these solid solutions is linear (Kot et al. 1962, Ivanov-Omskiy et al. 1964).

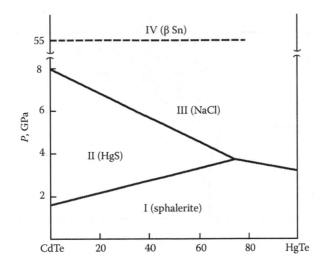

FIGURE 6.21
Pressure diagram for the solid solution $Hg_{1-x}Cd_xTe$ at 25°C. (From Quadri, S.B. et al., *J. Vac. Sci. Technol.*, A4(4), 1974, 1986.)

Temperature and concentration dependences of the energy gap of these solid solutions within the interval $0 \leq x \leq 0.6$ and $4.2 \leq T \leq 300$ K can be expressed by the next empiric equation (Hansen et al. 1982): $E_g = -0.302 + 1.93x + 5.35 \cdot 10^{-4} \cdot T(1-2x) - 0.810x^2 + 0.832x^3$.

The CdTe–HgTe system may be reasonably described by the regular solution approximation (Bublik and Zaitsev 1977). The mixing energy of this system is found to be 4.61 ± 0.84 kJ mol^{-1}. The solid solutions $Cd_xHg_{1-x}Te$ were found to exhibit a positive deviation from ideality (Nasar and Shamsuddin 1991, Sidorko and Goncharuk 1995, Goncharuk and Sidorko 1996), and the stability of the $Cd_xHg_{1-x}Te$ solid solutions at different compositions decreases with increasing temperature, except for $x = 0.1$ and 0.9 where it is constant (Nasar and Shamsuddin 1991).

Integral and excess thermodynamic functions for the formation of the $Cd_xHg_{1-x}Te$ solid solutions were given by Bublik and Zaitsev (1977) with a pseudopotential method and experimentally determined at room temperatures in Rugg (1995), at 370°C (Goncharuk and Sidorko 1996), and at 462°C–567°C (Nasar and Shamsuddin 1991). The bonding energy, measured at −1.10 eV for Cd–Te and at −0.815 eV for Hg–Te (Chen et al. 1983), remains constant within 0.01 eV through the solid solution $Cd_xHg_{1-x}Te$.

Assuming that the solution of HgTe in CdTe is quasi-regular, Basu (1979) and Tung et al. (1982) calculated the enthalpy of mixing of the solid solution. A cluster theory based on the quasichemical approximation has been applied by Patrick et al. (1988) to study the local correlation, bond-length distribution, and CdTe–HgTe phase diagram. Solid (liquid)–vapor equilibria for the $Cd_xHg_{1-x}Te$ are investigated in Bailly et al. (1975), Krotov et al. (1979), Galchenko et al. (1984), Jianrong (1993), Chen and Wiedemeier (1995), and Sang and Wu (1996).

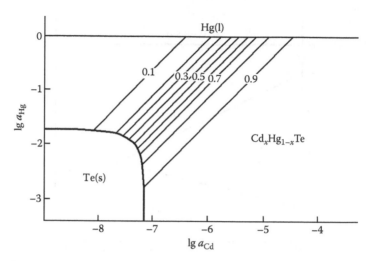

FIGURE 6.22
Predominance area of $Cd_xHg_{1-x}Te$ in equilibrium with the gas phase in the system Cd–Hg–Te at 430°C. (From Diehl, R. and Nolaeng, B.I., *J. Cryst. Growth*, 66, 91, 1984.)

The development of vapor-phase epitaxy process for $Cd_xHg_{1-x}Te$ is effectively guided by the evaluation of predominance areas in the O_2-free Cd–Hg–Te system. The pertinent phase relationships in the system are displayed in Figure 6.22 with $lg\ a_{Hg}$ and $lg\ a_{Cd}$ as coordinates at 430°C (Diehl and Nolaeng 1984).

The temperature dependence of the density in liquid $Cd_xHg_{1-x}Te$ for $x \le$ 20 mol. % CdTe has been investigated using γ-ray attenuation method, and the anomalous character of temperature–density dependence for the investigated melts has been noted (Glazov and Pavlova 1998). According to the hypothesis of the authors, the reasons of these density gradient and concentration occurring in the melts are the formation of complex polyanions on Hg_2Te_3 basis. An expression for the segregation coefficient of Cd into $Cd_xHg_{1-x}Te$ as a function of x was obtained in Lusson and Triboulet (1987): $k_{Cd} = x(1.080 - x)/0.080$.

$Cd_xHg_{1-x}Te$ single crystals could be obtained using the Bridgman method (Korovin et al. 1968) and using vertical (Dziuba 1969) or horizontal zone recrystallization (Lutsiv et al. 1968). A big value of the coefficient of the HgTe segregation in CdTe as well as a tellurium excess in the melt prevents high quality and homogeneity of growth of single crystals. To obtain more homogeneous single crystals, it is necessary to use ultrasound at their growth (Dubrovin et al. 1975).

HgTe–Cd: This section is a nonquasibinary section of the Cd–Hg–Te ternary system. Cadmium telluride and Hg are formed at the interaction of HgTe with Cd (Vengel' and Tomashik 1988).

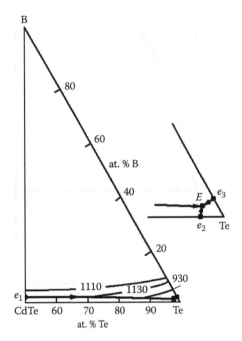

FIGURE 6.23
Liquidus surface of the CdTe–B–Te subsystem. (From Odin, I.N. et al., *Zhurn. neorgan. khimii*, 46(7), 1210, 2001.)

6.7 Cadmium–Boron–Tellurium

A part of the liquidus surface of the Cd–B–Te ternary system is shown in Figure 6.23 (Odin et al. 2001). The fields of the primary crystallization of the CdTe, B, and Te exist on this liquidus surface, and the existence of ternary compounds was not determined.

CdTe–B: The phase diagram belongs to the eutectic type (Figure 6.24) (Odin et al. 2001). The eutectic is degenerated from CdTe side and crystallizes at 1086°C.

This system was investigated using DTA, XRD, metallography, and vapor pressure measurement (Odin et al. 2001).

6.8 Cadmium–Aluminum–Tellurium

CdTe–Al$_2$Te$_3$: The phase diagram is not constructed. The CdAl$_2$Te$_4$ ternary compound is formed in this system, which crystallizes in a tetragonal structure with lattice parameters $a = 599.9$ and $c = 1219$ pm (609.5 pm) (Hahn et al. 1955)

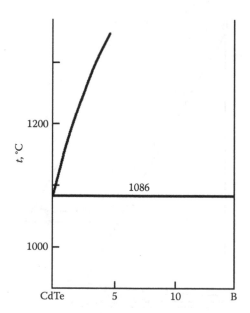

FIGURE 6.24

Part of the CdTe–B phase diagram. (From Odin, I.N. et al., *Zhurn. neorgan. khimii*, 46(7), 1210, 2001.)

[$a = 855.4$ and $c = 4898.0$ pm (Schwer and Krämer 1988)] and calculation and experimental density 5.099 and 5.10 g cm^{-3}, respectively (Hahn et al. 1955).

6.9 Cadmium–Gallium–Tellurium

Liquidus surface of the Cd–Ga–Te ternary system was constructed using mathematical planning of experiment (Figure 6.25) (Demina and Derid 1979). In the subsystem CdTe–Ga$_2$Te$_3$–Ga, the region of primary crystallization for all phases is constructed qualitatively.

The composition of liquid phase in the subsystem CdTe–GaTe–Ga$_2$Te$_3$ changes along the e_2U curve by the crystallization of CdTe + GaTe binary eutectic, and the composition of liquid phase at three-phase peritectic reaction L + CdTe $\Leftrightarrow$ β changes along the U_1U curve. U point determines the composition of liquid phase and temperature of four-phase peritectic reaction L + CdTe $\Leftrightarrow$ GaTe + β. The composition of liquid phase at the crystallization of binary eutectics with the participation of phases forming in the CdTe–Ga$_2$Te$_3$ system according to peritectic reactions changes along the broken line U_1UE.

In the subsystem CdTe–Ga–GaTe, the eutectic curve e_2U_4 separates the fields of CdTe and GaTe primary crystallization. U_4 point corresponds to the liquid-phase composition that participates in the four-phase peritectic equilibrium: L + GaTe $\Leftrightarrow$ CdTe + Ga$_3$Te$_2$. Below p_4 point, the eutectic curve

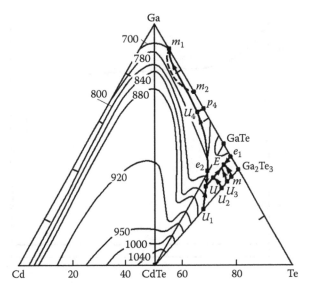

FIGURE 6.25
Liquidus surface of the Cd–Ga–Te ternary system. (From Demina, T.V. and Derid, O.P., Phase diagram of the cadmium–gallium–tellurium system [in Russian], in *Fizika Slozhn. Poluprovodn. Soied*, Kishinev, Shtiintsa, pp. 118–127, 1979.)

separates the fields of CdTe and Ga_3Te_2 primary crystallization. The dotted line indicates the immiscibility region.

Liquidus surface of the CdTe–Cd–Ga subsystem was constructed also by Lesyna et al. (2007) with the help of the mathematical planning of experiment. There are some discrepancies between the results of Demina and Derid (1979) and Lesyna et al. (2007) that must be resolved using experimental investigations or thermodynamic simulation of this ternary system.

CdTe–Ga: The phase diagram is a eutectic type (Figure 6.26) (Derid et al. 1974). The eutectic is degenerated from the Ga-rich side and crystallizes at 28°C. The solubility of CdTe in liquid gallium increases sharply at temperatures above 650°C (Andronik et al. 1975, Derid et al. 1974). The mutual solubility of CdTe and Ga is insignificant. According to the data of Fochuk et al. (1985) and Panchuk et al. (1985), the solubility of Ga in CdTe is retrograde and at the melting temperature of CdTe is not less than 10^{19} cm^{-3} (Figure 6.27).

This system was investigated using DTA, metallography, and XRD (Derid et al. 1974). The solubility of Ga in CdTe was determined using radioactive isotopes (Fochuk et al. 1985, Panchuk et al. 1985).

CdTe–GaTe: The phase diagram is a eutectic type (Figure 6.28) (Derid et al. 1974). The eutectic composition and temperature are 25 mol. % CdTe and 785°C, respectively. The solubility of GaTe in CdTe at the eutectic temperature is not higher than 4 mol. % (Derid et al. 1974) and 2.85 ± 0.01 mol. % at 737°C (Odin et al. 2011). The lattice parameter decreases linearly with the increase of the GaTe content (Odin et al. 2011).

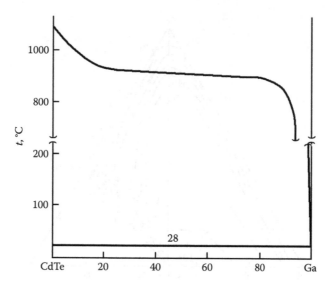

FIGURE 6.26

CdTe–Ga phase diagram. (From Derid, O.P. et al., *Izv. AN SSSR. Neorgan. materialy*, 10(1), 18, 1974.)

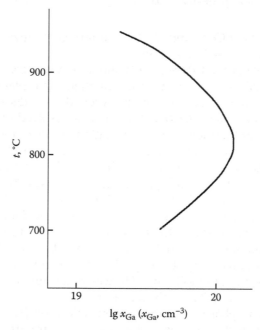

FIGURE 6.27

Solubility of Ga in CdTe at p_{Cd}^{max}. (From Fochuk, P.M. et al., *Deposited in UkrNIINTI*, 6.09.85, N 2090-Uk, 1985.)

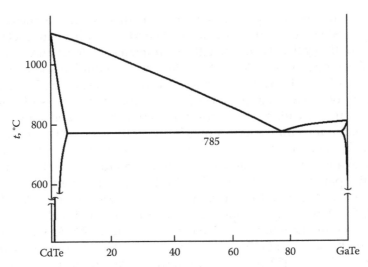

FIGURE 6.28
CdTe–GaTe phase diagram. (From Derid, O.P. et al., *Izv. AN SSSR. Neorgan. materialy*, 10(1), 18, 1974.)

This system was investigated using DTA, metallography, and XRD. The ingots were annealed at 790°C for 300 h (Derid et al. 1974) [at 737°C for 720 h with the next quenching in cold water (Odin et al 2011)].

CdTe–Ga₂Te₃: The calculated phase diagram of the $3CdTe–Ga_2Te_3$ system including the experimental data is shown in Figure 6.29 (Leute and Bolwin 2001). The calculated critical point for spinodal demixing is 20 mol. % Ga_2Te_3 and 617°C.

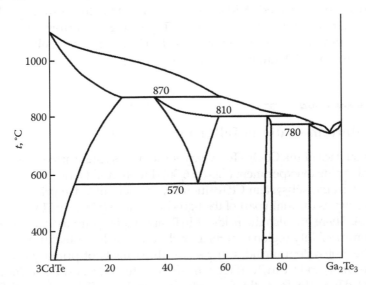

FIGURE 6.29
$3CdTe-Ga_2Te_3$ phase diagram. (From Leute, V. and Bolwin, H., *Solid State Ionics*, 141–142, 279, 2001.)

$CdGa_2Te_4$ ternary compound is formed in this system. It melts incongruently at 810°C ± 5°C (Derid et al. 1976) and crystallizes in a tetragonal structure of thiogallate type with lattice parameters $a = 608.1$ and $c = 1179$ pm [$a = 611.7$ and $c = 1180$ pm (Derid et al. 1976)] and calculation and experimental density 5.771 and 5.63 g · cm^{-3}, respectively (Hahn et al. 1955). According to the data of Bredol and Leute (1985), the lattice parameters of $CdGa_2Te_4$ depend from annealing temperature: $a = 610$, 608, 605 and $c = 1186$, 1188, 1201 pm at the annealing temperatures 500°C, 600°C, and 700°C, respectively. At gradual cooling of the alloys containing 73–77 mol. % Ga_2Te_3, this compound has a cubic body-centered tetragonal structure of chalcopyrite type with lattice parameters $a = 852.9$ and $c = 4844.4$ pm (Schwer and Krämer 1988) [$a = 856$ and $c = 4824$ pm (Demina and Radautsan 1984, Derid et al. 1976)].

At 870°C ± 10°C, a high-temperature β-phase is formed with a homogeneity region within the interval of 47–53 mol. % Ga_2Te_3 at 700°C. It decomposes by eutectoid reaction at 570°C ± 5°C (Derid et al. 1976). According to the data of Bredol and Leute (1985), $Cd_5Ga_2Te_6$ ternary compound is formed in the $CdTe–Ga_2Te_3$ system at 600°C.

The region of solid solutions based on Ga_2Te_3 reaches 10 mol. % 3CdTe (Bredol and Leute 1985, Derid et al. 1976). Limit solid solution is formed by the peritectic reaction at 780°C ± 5°C. The minimum on the liquidus corresponds to the composition of 96 mol. % Ga_2Te_3 (Derid et al. 1976). According to the data of Woolley and Ray (1960b), the solubility of Ga_2Te_3 in 3CdTe is equal to 42 mol. % at 765°C. Homogeneous solid solutions based on Ga_2Te_3 are formed at the same temperature within the intervals of 0–13 and 17–28 mol. % 3CdTe.

This system was investigated using DTA, metallography, electron probe microanalyzer (EPMA), and XRD (Derid et al. 1976, Leute and Bolwin 2001). The ingots were annealed at 500°C, 600°C, and 700°C for 19–76 days (Bredol and Leute 1985). Single crystals of $CdGa_2Te_4$ were grown from the melt containing 87 mol. % Ga_2Te_3 (Demina et al. 1976), using zone melting and the Bridgman method (Demina and Radautsan 1984).

6.10 Cadmium–Indium–Tellurium

Liquidus surface of the Cd–In–Te ternary system was constructed using mathematical planning of experiment (Figure 6.30) (Dyntu and Derid 1976, Radautsan. et al. 1982). In the subsystem $CdTe–In–InTe$, the field of CdTe primary crystallization occupies the main part of the liquidus surface (Figure 6.31a) (Radautsan et al. 1975). There are also the fields of InTe and In_9Te_7 primary crystallizations and the immiscibility region near the In-rich coin. The liquidus surface of CdTe–InTe–In_2Te_3 subsystem includes in general three fields of primary crystallization: CdTe, InTe, and $CdIn_2Te_4$ (Figure 6.31b) (Dyntu and Derid 1979). The phases based on CdTe, $CdIn_2Te_4$, $CdIn_8Te_{13}$, $CdIn_{30}Te_{46}$, In_2Te_3, In_3Te_5, In_2Te_5, and Te crystallize primarily in the subsystem $CdTe–In_2Te_3–Te$ (Figure 6.31c) (Dyntu 1977).

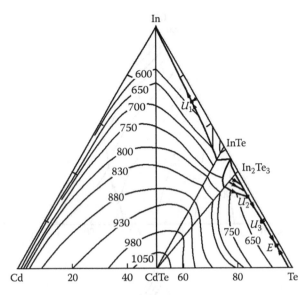

FIGURE 6.30
Liquidus surface of the Cd–In–Te ternary system. (From Dyntu, G.M. and Derid, O.P., Phase diagram of the Cd–In–Te system and conditions of solid solution formation based on CdIn$_2$Te$_4$ [in Russian], in *Troin. Poluprovodn. i ih Primenenie: Tez. Dokl*, Kishinev, Shtiintsa, pp. 139–141, 1976.)

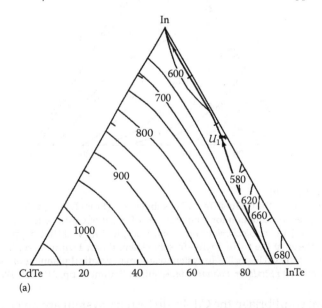

FIGURE 6.31
(a) Liquidus surfaces of the CdTe–In–InTe subsystem. (From Radautsan, S.I. et al., *Izv. AN MSSR. Ser. fiz.-tehn. i mat. nauk*, (3), 37, 1975.)

(continued)

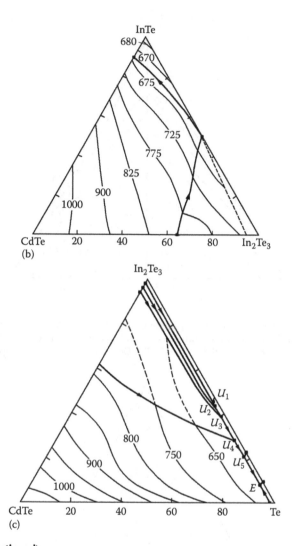

FIGURE 6.31 (continued)
(b) Liquidus surfaces of the CdTe–InTe–In$_2$Te$_3$ subsystem. (From Dyntu, G.M. and Derid, O.P., Obtaining of regression equation for temperatures of onset of Cd-In-Te ternary alloy crystallization [in Russian], in *Fiz. Slozhnyh Poluprovodn. Soied*, Kishinev, Shtiintsa, pp. 136–141, 1979.)
(c) Liquidus surfaces of the CdTe–In$_2$Te$_3$–Te subsystem. (From Dyntu, G.M., Investigation of liquidus surface of the CdTe–In$_2$Te$_3$–Te system using mathematical planning of experiment [in Russian], in *Kristallich. i Stekloobr. Poluprovodn*, Kishinev, Shtiintsa, pp. 215–221, 1977.)

Nonvariant equilibria in the Cd–In–Te ternary system are given in Table 6.1 (Dyntu 1977).

CdTe–In: The phase diagram is a eutectic type (Figure 6.32) (Yokozawa et al. 1966, Panchuk and Shcherbak 1979, Tai et al. 1980). The eutectic is degenerated from the In-rich side and crystallizes at 156°C [the eutectic composition and temperature are $4 \cdot 10^{-3}$ mol. % CdTe and 156.5°C, respectively

TABLE 6.1

Nonvariant Equilibria in the Cd–In–Te
Ternary System

Symbol	Reaction	t, °C
U_1	$L + \varepsilon \Leftrightarrow \delta + In_3Te_5$	600
U_2	$L + \delta \Leftrightarrow \gamma + In_3Te_5$	560
U_3	$L + \gamma \Leftrightarrow \beta + In_3Te_5$	455
U_4	$L + \beta \Leftrightarrow \alpha + In_3Te_5$	435
U_5	$L + In_3Te_5 \Leftrightarrow \alpha + In_2Te_5$	430
E	$L \Leftrightarrow \alpha + In_2Te_3 + Te$	425

Source: Dyntu, G.M., Investigation of liquidus surface of the CdTe–In2Te3–Te system using mathematical planning of experiment [in Russian], in *Kristallich. i stekloobr. poluprovodn,* Kishinev, Shtiintsa, pp. 215–221, 1977.

Note: α, (CdTe); β, CdIn$_2$Te$_4$; γ, CdIn$_8$Te$_{13}$; δ, CdIn$_{30}$Te$_{46}$; and ε, (In$_2$Te$_3$).

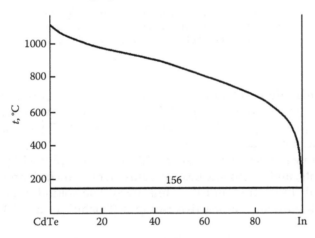

FIGURE 6.32

CdTe–In phase diagram. (From Panchuk, O.E. et al., *Izv. AN SSSR. Neorgan. materialy,* 9(8), 1437, 1979; Tai, H. et al., *J. Jap. Inst. Metals,* 44(3), 276, 1980.)

(Tai et al. 1980)]. The solubility of CdTe in the liquid indium increases sharply at temperatures above 500°C (Andronik et al. 1975, Panchuk and Shcherbak 1979). This solubility can be represented as (Tai et al. 1980) ln x_{CdTe} (mole fraction) $= -(6339/T) + 4.729$, $(x < 0.1)$.

The thermochemical calculations of Shuh and Williams (1988) predicted that CdTe–In system is quasibinary and experimental results confirmed such prediction.

The solubility of In in CdTe is retrograde with a maximum at 950°C (Figure 6.33) (Feichuk et al. 1982, Panchuk et al. 1983). The decrease of cadmium vapor pressure increases the In solubility. Temperature dependence

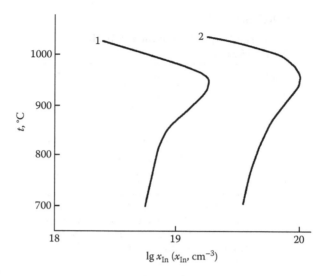

FIGURE 6.33
Temperature dependences of In maximum solubility in CdTe at $p_{Cd}{}^{max}$ (1) and $p_{Cd}{}^{min}$ (2). (From Feichuk, P.I. et al., Defect structure of CdTe doping by In [in Russian], in *Legir. Poluprovodn.*, Moscow, Russia, pp. 72–75, 1982; Panchuk, O.E. et al., *Izv. AN SSSR. Neorgan. materialy*, 19(3), 362, 1983.)

of maximum indium solubility at 700°C–950°C is described by the next equations: x_{In} (cm^{-3}) = 1.6 · 10^{22} exp[–(0.54 ± 0.04)]/[kT] at minimum cadmium vapor pressure and x_{In} (cm^{-3}) = 5.93 · 10^{21} exp [–(0.62 ± 0.07)]/kT] at maximum cadmium vapor pressure.

This system was investigated using DTA, metallography, XRD, dissolution measurements, and microhardness measurement (Panchuk and Shcherbak 1979, Tai et al. 1980, Shuh and Williams 1988). The solubility of In in CdTe was determined using radioactive isotopes (Feichuk et al. 1982, Panchuk et al. 1983).

CdTe–InTe: The phase diagram belongs to the eutectic type (Figure 6.34) (Radautsan et al. 1975). The eutectic composition and temperature are 10 mol.% CdTe and 655°C, respectively. CdInTe$_2$ ternary compound (Guseinov 1969) was not found in the CdTe–InTe system (Radautsan et al. 1975, Odin et al. 2009). The solubility of CdTe in InTe at the eutectic temperature is equal to 5–6 mol. %, and the maximum solubility of InTe in CdTe reaches 5 mol. % (Radautsan et al. 1975) [4.7 mol. % at 630°C (Odin et al. 2009)].

This system was investigated using DTA, metallography, XRD, and density measurement (Radautsan et al. 1975, Odin et al. 2009). The ingots were annealed at 350°C–500°C for 300–500 h (Radautsan et al. 1975) [at 630°C for 720 h (Odin et al. 2009)].

CdTe–In$_2$Te$_3$: The phase diagram is shown in Figure 6.35 (Weitze and Leute 1996). CdIn$_2$Te$_4$ melts incongruently at 785°C (Thomassen et al. 1963, Weitze and Leute 1996). The system 3CdTe–In$_2$Te$_3$ is characterized by a spinodal

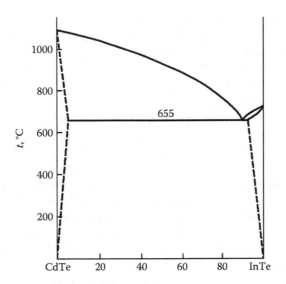

FIGURE 6.34
CdTe–InTe phase diagram. (From Radautsan, S.I. et al., *Izv. AN MSSR. Ser. fiz.-tehn. i mat. nauk,* (3), 37, 1975.)

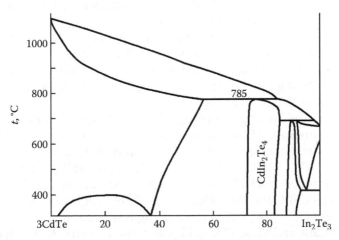

FIGURE 6.35
3CdTe–In$_2$Te$_3$ phase diagram. (From Weitze, D. and Leute, V., *J. Alloys Compd.,* 236(1–2), 229, 1996.)

miscibility gap with a critical point at 20 mol. In$_2$Te$_3$ and $T_c \approx 620°C$ (these data are taken from Figure 6.35) and by the extended regions of ordered structures in the In$_2$Te$_3$-rich part of the phase diagram (Weitze and Leute 1996, Leute and Bolwin 2001). The small existence region of the cubic phase between this spinodal miscibility gap and the structural gap (cubic + tetragonal) extends down to 330°C.

A new procedure for the calculation of the CdTe–In$_2$Te$_3$ phase diagram has been developed by Leute (1996). The diagram was calculated from excess

Gibbs energy function that is modulated in the solid state at the specified stoichiometric compositions by distribution functions of the Gauss type. The obtained results coincide with experimental data.

According to the data of Thomassen et al. (1963), in addition to $CdIn_2Te_4$, two compounds $CdIn_8Te_{13}$ and $CdIn_{30}Te_{46}$, which decompose at 702°C and 695°C, respectively, are formed in this system.

$CdIn_2Te_4$ ternary compound crystallizes in a tetragonal structure with lattice parameters $a = 619.2$ and $c = 1238$ pm; calculation and experimental density 5924 and 5.88 g cm^{-3}, respectively (Hahn et al. 1955); and energy gap $E_g = 0.9$ eV (Busch 1956). The value of a parameter increases nonlinearly with the temperature and, within the limits of experimental error, can be expressed at 27°C–377°C by the next equation: $a(T) = 619.41 - 4.0290 \cdot 10^{-4}T + 8.5193 \cdot 10^{-6}T^2 - 4.4941 \cdot 10^{-9}T^3$ (pm), where T is the temperature in K (Quintero et al. 1996). The thermal expansion coefficient of a parameter can be given by the next equation: $\alpha_a(T) = -6.4993 \cdot 10^{-7} + 2.7486 \cdot 10^{-8}T - 2.1749 \cdot 10^{-11}T^2$ (K^{-1}).

$CdIn_8Te_{13}$ ternary compound has a cubic structure of chalcopyrite type, the lattice parameter of which changes stepwise from 619.7 to 618.0 pm (Thomassen et al. 1963). $CdIn_{30}Te_{46}$ ternary compound also crystallizes in a cubic structure of chalcopyrite type. Its lattice parameters change stepwise from 618.0 to 617.2 pm. At 550°C–600°C, solid solution based on In_2Te_3 transforms into ε'-phase (Thomassen et al. 1963).

The solubility of In_2Te_3 in CdTe is equal to 6 mol. % at 630°C (Odin et al. 2009) [the solubility of In_2Te_3 in 3CdTe at 665°C reaches 50 mol. % and the solubility of 3CdTe in In_2Te_3 is equal to 28 mol. % (Woolley and Ray 1960c)]. According to the data of Thomassen et al. (1963), the solubility of $CdIn_2Te_4$ in CdTe is equal to 25 mol. % In_2Te_3.

This system was investigated using DTA, metallography, EPMA, XRD, and density measurement (Thomassen et al. 1963, Quintero et al. 1996, Weitze and Leute 1996, Leute and Bolwin 2001, Odin et al. 2009). Samples were annealed at 330°C, 430°C, 530°C, 630°C, 705°C, and 863°C with the annealing times extended from 21 h at 863°C up to 280 days at 330°C (Weitze and Leute 1996) [at 630°C for 720 h (Odin et al. 2009)]. Single crystals of $CdIn_2Te_4$, $CdIn_8Te_{13}$, and $CdIn_{30}Te_{46}$ were grown using floating-zone refining and zone leveling (O'Kane and Mason 1963). Liquid phase was enriched by the indium telluride for $CdIn_2Te_4$ extracting from the solution in the melt (Mason and O'Kane 1960).

CdTe–In$_3$Te$_2$: This section is a nonquasibinary section of the Cd–In–Te ternary system (Figure 6.36) (Radautsan et al. 1975). Binary eutectic crystallizes within the interval of 580°C–450°C, and at 450°C, the next peritectic reaction takes place: $L + InTe \Leftrightarrow \alpha(CdTe) + In_4Te_3(In_9Te_7)$.

At temperatures below 450°C, $CdTe + In_4Te_3(In_9Te_7)$ binary eutectic goes on crystallizing, and at 420°C, the next monotectic reaction takes place: $L_1 + CdTe \Leftrightarrow L_2 + In_4Te_3(In_9Te_7)$.

This section was investigated using DTA, metallography, and XRD (Radautsan et al. 1975). The ingots were annealed at 350°C–500°C for 300–500 h.

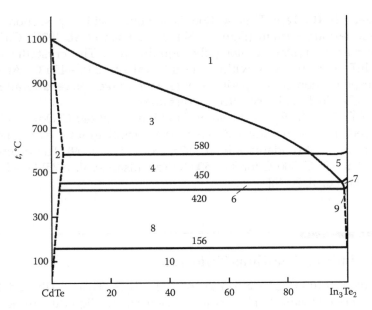

FIGURE 6.36
Phase relations in the CdTe–In$_3$Te$_2$ system: 1, L; 2, CdTe; 3, L + CdTe; 4, L + CdTe + InTe; 5, L + InTe; 6, L + CdTe + In$_4$Te$_3$(In$_9$Te$_7$); 7, L + In$_4$Te$_3$(In$_9$Te$_7$); 8, L$_2$ + CdTe + In$_4$Te$_3$(In$_9$Te$_7$); 9, L$_2$ + In$_4$Te$_3$(In$_9$Te$_7$); and 10, CdTe + In$_4$Te$_3$(In$_9$Te$_7$) + In. (From Radautsan, S.I. et al., *Izv. AN MSSR. Ser. fiz.-tehn. i mat. nauk*, (3), 37, 1975.)

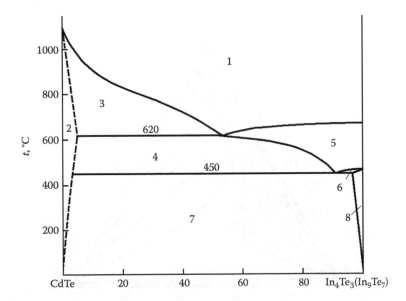

FIGURE 6.37
Phase relations in the CdTe–In$_4$Te$_3$(In$_9$Te$_7$) system: 1, L; 2, CdTe; 3, L + CdTe; 4, L + CdTe + InTe; 5, L + InTe; 6, L + In$_4$Te$_3$(In$_9$Te$_7$); 7, CdTe + In$_4$Te$_3$(In$_9$Te$_7$); and 8, In$_4$Te$_3$(In$_9$Te$_7$). (From Radautsan, S.I. et al., *Izv. AN MSSR. Ser. fiz.-tehn. i mat. nauk*, (3), 37, 1975.)

CdTe–In$_4$Te$_3$(In$_9$Te$_7$): This section is a nonquasibinary section of the Cd–In–Te ternary system (Figure 6.37) (Radautsan et al. 1975). CdTe and InTe primarily crystallize along the liquidus line. The crystallization of CdTe + InTe eutectic runs within the interval of 620°C–450°C. At 450°C, the four-phase nonvariant peritectic interaction takes place, and after that, CdTe + In$_4$Te$_3$(In$_9$Te$_7$) binary eutectic crystallizes.

The solubility of In$_4$Te$_3$ in CdTe is equal to 1.1 mol. % at 440°C (Odin et al. 2009).

This section was investigated using DTA, metallography, XRD, and density measurement (Radautsan et al. 1975, Odin et al. 2009). The ingots were annealed at 350°C–500°C for 300–500 h (Radautsan et al. 1975) [at 440°C for 1000 h (Odin et al. 2009)].

6.11 Cadmium–Thallium–Tellurium

The fields of primary crystallization of CdTe, Tl$_2$Te, Tl$_5$Te$_3$, TlTe, and Te and the degenerated fields of Tl$_2$Te$_3$, Tl, and Cd exist on the liquidus surface of the Cd–Tl–Te ternary system (Figure 6.38) (Babanly et al. 1984). The composition of U_1 ternary peritectic is 58 at. % Te, 1 at. % Cd, and 41 at. % Tl. The immiscibility region exists in this ternary system from the Tl–Te binary system.

Isothermal section of the Cd–Tl–Te ternary system at room temperature indicates the absence of ternary compounds in this system (Figure 6.39)

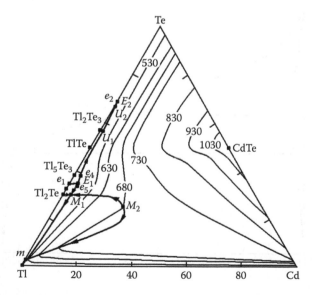

FIGURE 6.38
Liquidus surface of the Cd–Tl–Te ternary system. (From Babanly, M.B. et al., *Izv. AN SSSR. Neorgan. materialy,* 20(1), 46, 1984.)

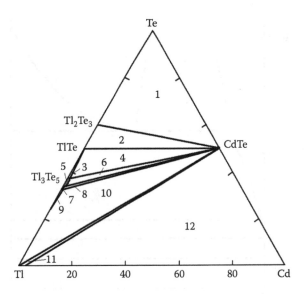

FIGURE 6.39
Isothermal section of the Cd–Tl–Te ternary system at 25°C: 1, $Tl_2Te_3 + CdTe + Te$; 2, $TlTe + Tl_2Te_3 + CdTe$; 3, $TlTe + Tl_5Te_3$; 4, $Tl_5Te_3 + TlTe + CdTe$; 5, Tl_5Te_3; 6, $Tl_5Te_3 + CdTe$; 7, Tl_2Te; 8, $Tl_2Te + Tl_5Te_3 + CdTe$; 9, $Tl_2Te + Tl$; 10, $Tl + Tl_2Te + CdTe$; 11, $Tl + CdTe$; and 12, $Tl + Cd + CdTe$. (From Asadov, M.M. et al., *Zhurn. fiz. khimii*, 57(11), 2865, 1983; Babanly, M.B. et al., *Izv. AN SSSR. Neorgan. materialy*, 20(1), 46, 1984.)

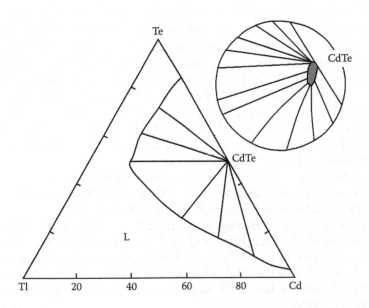

FIGURE 6.40
Isothermal section of the Cd–Tl–Te ternary system at 730°C. (Babanly, M.B. et al., *Izv. AN SSSR. Neorgan. materialy*, 20(1), 46, 1984.)

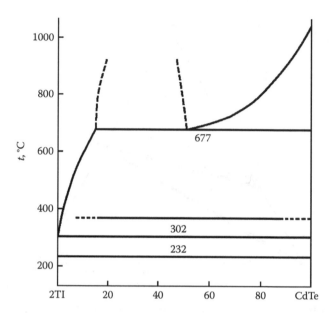

FIGURE 6.41
CdTe–2Tl phase diagram. (Babanly, M.B. et al., *Izv. AN SSSR. Neorgan. materialy*, 20(1), 46, 1984.)

(Asadov et al. 1983, Babanly et al. 1984) [according to the data of Gawel et al. (2001a,b), two compounds $CdTl_{12}Te_7$ and $Cd_3Tl_2Te_4$ are formed in this system on the section $CdTe–Tl_2Te$]. At 730°C, there is a narrow homogeneity region based on CdTe (Figure 6.40) (Babanly et al. 1984). In the authors' opinion, a position of the tie-lines indicates on nonquasibinarity of CdTe–Tl system.

CdTe–Tl: The phase diagram is a eutectic type (Figure 6.41) (Babanly et al. 1984, Feichuk et al. 1977). The eutectic composition and temperature are 0.5 mol. % CdTe and 302°C [300°C (Feichuk et al. 1977)], respectively (Babanly et al. 1984). The immiscibility region exists in this system with monotectic temperature 677°C (Babanly et al. 1984) [390°C (Feichuk et al. 1977)]. Thermal effect at 230°C [232°C (Babanly et al. 1984)] corresponds to the polymorphous transformation of Tl (Feichuk et al. 1977).

Temperature dependences of thallium solubility in CdTe at maximum and minimum cadmium vapor pressure are retrograde with a maximum solubility at 730°C in both cases (Figure 6.42) (Panchuk et al. 1986).

This system was investigated by using DTA, metallography, XRD, and measurements of microhardness and emf of concentrated chains (Babanly et al. 1984, Feichuk et al. 1977). The ingots were annealed at the temperatures 20°C–30°C below the solidus temperatures for 400–500 h (Babanly et al. 1984). The solubility of Tl in CdTe was determined using radioactive isotopes (Panchuk et al. 1986).

CdTe–TlTe: This section is a nonquasibinary section of Cd–Tl–Te ternary system (Figure 6.43) (Guseinov et al. 1982). Thermal effects at 350°C

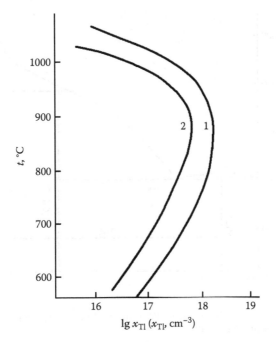

FIGURE 6.42
Temperature dependences of thallium solubility in CdTe at minimum (1) and maximum (2) cadmium vapor pressure. (From Panchuk, O.E. et al., *Izv. AN SSSR. Neorgan. materialy*, 22(10), 1642, 1986.)

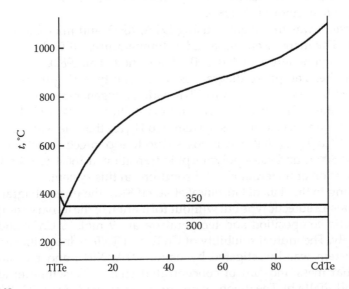

FIGURE 6.43
Phase relations in the CdTe–TlTe system. (From Guseinov, F.H. et al., *Izv. AN SSSR. Neorgan. materialy*, 18(5), 759, 1982.)

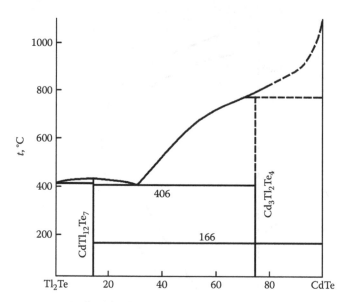

FIGURE 6.44

CdTe–Tl₂Te phase diagram. (From Gawel, W. et al., *J. Phase. Equil.*, 22(6), 656, 2001; Gawel, W. et al., *Pol. J. Chem.*, 75(10), 1553, 2001.)

correspond to secondary crystallization of Tl_5Te_3 and at 300°C to the four-phase equilibrium $L + Tl_5Te_3 \Leftrightarrow ZnTe + TlTe$. The mutual solubility of ZnTe and TlTe is less than 1 mol. %.

$CdTlTe_2$ compound (Guseinov 1969, Guseinov et al. 1969) was not found in this system (Guseinov et al. 1982).

This system was investigated using DTA, XRD, and microhardness measurement. The ingots were annealed at temperatures 20°C–30°C lower than the solidus temperature for 400–500 h (Guseinov et al. 1982).

CdTe–Tl₂Te: The phase diagram is a eutectic type (Figure 6.44) (Gawel et al. 2001a,b). The eutectic from Tl₂Te side is degenerated and the second eutectic contains 31.9 mol. % Tl₂Te and crystallizes at 406.3°C. Two ternary compounds are formed in this system: $CdTl_{12}Te_7$ that melts congruently at 431°C and $Cd_3Tl_2Te_4$ that decomposes due to a peritectic reaction between 730°C and 800°C and has a polymorphic transition at 166.4°C ± 4.5°C. There is an indication of no terminal solid solutions in this system.

According to the data of Guseinov et al. (1981a), the phase diagram of this system is also a eutectic type but without formation of the ternary compounds. The eutectic composition and temperature are 9 mol. % CdTe and 377°C, respectively. The mutual solubility of CdTe and Tl₂Te is insignificant.

This system was investigated by using DTA, XRD, and measurements of microhardness and emf of concentrated chains (Guseinov et al. 1981a, Gawel et al. 2001a,b). The ingots were annealed at 350°C for 400 h (Guseinov et al. 1981a).

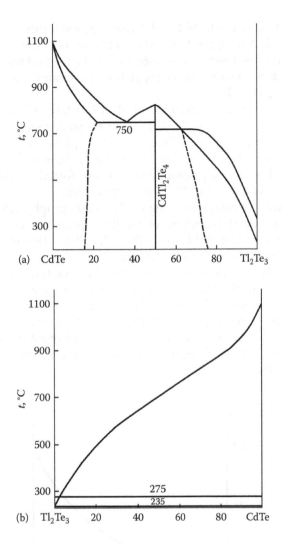

FIGURE 6.45
Phase relations in the CdTe–Tl$_2$Te$_3$ system. (a: From Karimov, S.K. et al., *Izv. AN SSSR. Neorgan. materialy*, 17(8), 1346, 1981; b: From Guseinov, F.H. et al., *Izv. vuzov. Khimia i Khim. Ttehnologia*, 24(10), 1245, 1981.)

CdTe–Tl$_2$Te$_3$: The results of this system investigation are contradictory. According to the data of Mavlonov et al. (1976), Sultonov and Karimov (1979), and Karimov et al. (1981), CdTl$_2$Te$_4$ ternary compound is formed in the CdTe–Tl$_2$Te$_3$ system (Figure 6.45a). It melts congruently at 822°C and crystallizes in a tetragonal structure of thiogallate type with lattice parameters $a = 561$ and $c = 1060$ pm and energy gap $E_g = 0.34$ eV. The eutectic composition and temperature are 36.7 mol. % Tl$_2$Te$_3$ and 750°C ± 3°C, respectively. The solubility of Tl_2Te_3 in CdTe is equal to 15.4 mol. % and the solubility of CdTe in Tl_2Te_3 reaches 24 mol. %. Additional thermal effects were observed at 780°C, 871°C, and 880°C within the interval of 19.5–66 mol. % Tl_2Te_3, and their nature is not clear. The equilibrium from the Tl_2Te_3-rich side is complicated by the incongruently melting of Tl_2Te_3.

According to the data of Guseinov et al. (1981b), the CdTe–Tl_2Te_3 section is nonquasibinary section of the Cd–Tl–Te ternary system (Figure 6.45b). Incongruent melting point of Tl_2Te_3 leads to four-phase peritectic reaction at 235°C: $L + TlTe \Leftrightarrow CdTe + Tl_2Te_3$. The mixture of CdTe and Tl_2Te_3 exists in the solid state. $CdTl_2Te_4$ ternary compound was not found in this system. The mutual solubility of CdTe and Tl_2Te_3 is insignificant.

This system was investigated using DTA, metallography, XRD, and microhardness measurement (Karimov et al. 1981, Guseinov et al. 1981b). The ingots were annealed at 700°C and 800°C for 350 h (Karimov et al. 1981) [at 230°C for 400 h (Guseinov et al. 1981b)].

CdTe–Tl_5Te_3: The phase diagram is a eutectic type (Figure 6.46) (Babanly et al. 1984). The eutectic composition and temperature are 10 mol. % CdTe and 397°C. The solubility of 4CdTe in Tl_5Te_3 is equal to 2 mol. %.

This system was investigated by using DTA, XRD, and measurements of microhardness and emf of concentrated chains. The ingots were annealed at the temperatures 20°C–30°C below the solidus temperatures for 400–500 h (Babanly et al. 1984).

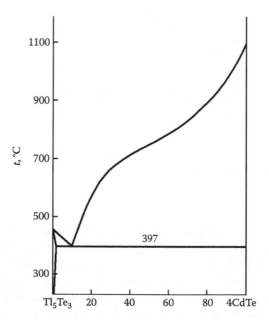

FIGURE 6.46
4CdTe–Tl_5Te_3 phase diagram. (Babanly, M.B. et al., *Izv. AN SSSR. Neorgan. materialy*, 20(1), 46, 1984.)

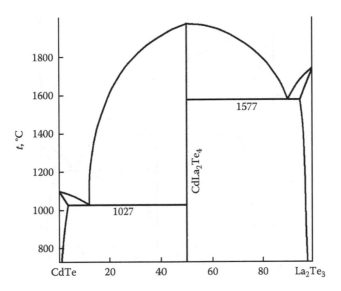

FIGURE 6.47
CdTe–La$_2$Te$_3$ phase diagram. (From Aliev, O.M. et al., *Zhurn. neorgan. khimii*, 30(4), 1041, 1985.)

6.12 Cadmium–Lanthanum–Tellurium

CdTe–La$_2$Te$_3$: The phase diagram is shown in Figure 6.47 (Aliev et al. 1985). The eutectic compositions and temperatures are 12 and 90 mol. % La$_2$Te$_3$ and 1027°C and 1577°C, respectively. CdLa$_2$Te$_4$ ternary compound is formed in the CdTe–La$_2$Te$_3$ system. It melts congruently at 1977°C and crystallizes in a cubic structure of Th$_3$P$_4$ type with lattice parameter $a = 890$ pm. The solubility of La$_2$Te$_3$ in CdTe is equal to 1 mol. %, and the solubility of CdTe in La$_2$Te$_3$ reaches 1.5 mol. %.

This system was investigated using DTA, metallography, XRD, and microhardness measurement. The ingots were annealed at 900°C for 40 h (Aliev et al. 1985).

6.13 Cadmium–Neodymium–Tellurium

CdTe–Nd$_2$Te$_3$: The phase diagram is shown in Figure 6.48 (Zul'fugarly et al. 1982). The eutectic temperature is 960°C. CdNd$_2$Te$_4$ ternary compound is formed in the CdTe–Nd$_2$Te$_3$ system. It melts incongruently at 1130°C and crystallizes in a rhombohedral structure with lattice parameters $a = 772$ pm and $\alpha = 33.16°$. The solubility of CdTe in Nd$_2$Te$_3$ decreases from 10 mol. % at 1100°C to 5 mol. % at 200°C.

This system was investigated using DTA, metallography, XRD, and microhardness measurement (Zul'fugarly et al. 1982).

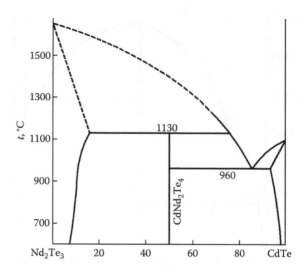

FIGURE 6.48
CdTe–Nd$_2$Te$_3$ phase diagram. (From Zul'fugarly et al., *Azerb. khim. zhurn.*, (3), 103, 1982.)

6.14 Cadmium–Gadolinium–Tellurium

CdTe–Gd$_2$Te$_3$: The phase diagram is shown in Figure 6.49 (Agaev et al. 1983). The eutectic temperature is 930°C. CdGd$_2$Te$_4$ ternary compound is formed in the CdTe–Gd$_2$Te$_3$ system. It melts incongruently at 1150°C [at 977°C (Azadaliev 1990)] and crystallizes in a cubic structure of Th$_3$P$_4$ type

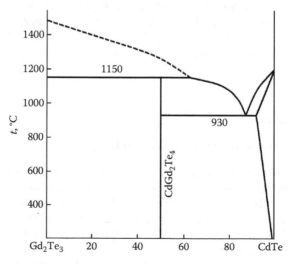

FIGURE 6.49
CdTe–Gd$_2$Te$_3$ phase diagram. (From Agaev, A.B. et al., *Zhurn. Neorgan. khimii*, 28(1), 256, 1983.)

with lattice parameter $a = 892$ pm [$a = 948$ pm (Azadaliev 1990)] and calculation and experimental density 6.79 and 6.70 g cm^{-3}, respectively (Agaev et al. 1983). The solubility of Gd$_2$Te$_3$ in CdTe at room temperature is equal to 1.5 mol. % (Zul'fugarly et al. 1984).

This system was investigated using DTA, metallography, XRD, and microhardness measurement. The ingots were annealed at 950°C–1000°C for 250 h (Agaev et al. 1983, Zul'fugarly et al. 1984, Azadaliev 1990).

6.15 Cadmium–Dysprosium–Tellurium

CdTe–DyTe: The phase diagram is a eutectic type (Figure 6.50) (Agaev et al. 1987). The eutectic composition and temperature are 21 mol. % DyTe and 877°C, respectively. The solubility of DyTe in CdTe reaches 5 mol. % and the solubility of CdTe in DyTe is equal to 3 mol. %.

This system was investigated using DTA, metallography, XRD, and microhardness measurement. The ingots were annealed at 730°C for 240 h (Agaev et al. 1987).

CdTe–Dy$_2$Te$_3$: The phase diagram is shown in Figure 6.51 (Agaev and Kulieva 1984, Agaev et al. 1987). The eutectic composition and temperature

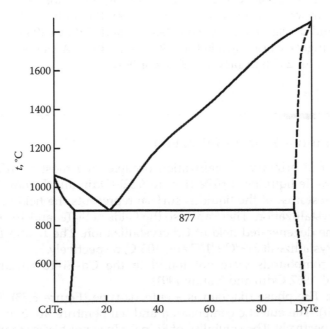

FIGURE 6.50
CdTe–DyTe phase diagram. (From Agaev, A.B. et al., *Zhurn. Neorgan. khimii*, 32(10), 2554, 1987.)

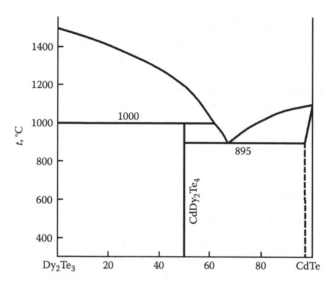

FIGURE 6.51

CdTe–Dy$_2$Te$_3$ phase diagram. (From Agaev, A.B. et al., *Zhurn. Neorgan. khimii*, 32(10), 2554, 1987.)

are 33 mol. % Dy$_2$Te$_3$ and 895°C, respectively. CdDy$_2$Te$_4$ ternary compound is formed in the CdTe–Dy$_2$Te$_3$ system. It melts incongruently at 1000°C and crystallizes in a cubic structure of spinel type with lattice parameter $a = 1138$ pm. The solubility of Dy$_2$Te$_3$ in CdTe reaches 3 mol. %.

This system was investigated using DTA, metallography, XRD, and microhardness and density measurements (Agaev and Kulieva 1984, Agaev et al. 1987). The ingots were annealed at 730°C for 240 h (Agaev et al. 1987) [at 850°C–900°C for 200 h (Agaev and Kulieva 1984)].

6.16 Cadmium–Silicon–Tellurium

The field of Si primary crystallization occupies the most significant part of the Cd–Si–Te liquidus surface (Figure 6.52) (Odin and Ivanov 1991). The second largest area of the liquidus surface represents the field of the CdTe primary crystallization. There are also the fields of Si$_2$Te$_3$ and Te crystallization and the degenerated field of Cd crystallization. The E_1 and E_2 ternary eutectics crystallize at 319°C ± 1°C and 405°C, respectively.

Ternary compounds were not found in the Cd–Si–Te ternary system (Kaldis et al. 1967, Odin and Ivanov 1991).

CdTe–Si: The phase diagram is a eutectic type (Figure 6.53) (Odin and Ivanov 1991). The eutectic composition and temperature are 5 at. % Si and 1083°C, respectively. The solubility of Si in CdTe is not higher than 0.5 at. % and the solubility of CdTe in Si is equal to 0.1 at. %.

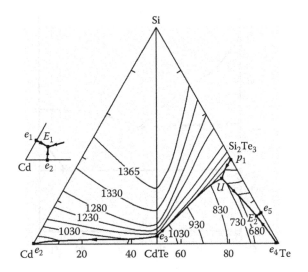

FIGURE 6.52
Liquidus surface of the Cd–Si–Te ternary system. (From Odin, I.N. and Ivanov, V.A., *Zhurn. Neorgan. khimii*, 36(7), 1837, 1991.)

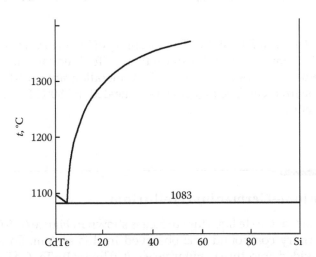

FIGURE 6.53
CdTe–Si phase diagram. (From Odin, I.N. and Ivanov, V.A., *Zhurn. Neorgan. khimii*, 36(7), 1837, 1991.)

This system was investigated using DTA, metallography, XRD, and microhardness measurement. The ingots were annealed at 750°C for 1040 h (Odin and Ivanov 1991).

CdTe–Si$_2$Te$_3$: This section is a nonquasibinary section of the Cd–Si–Te ternary system (Figure 6.54) (Odin and Ivanov 1991). Silicon primary crystallizes from the Si$_2$Te$_3$-rich side. Crystallization of all melts ends at 810°C by the peritectic reaction $L + Si \Leftrightarrow CdTe + \beta\text{-Si}_2Te_3$. Polymorphous transformation of

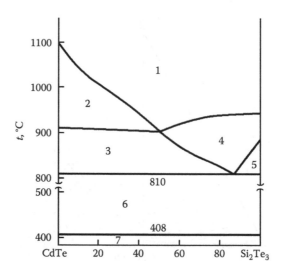

FIGURE 6.54

Phase relations in the CdTe–Si$_2$Te$_3$ system: 1, L; 2, L+CdTe; 3, L+CdTe+Si; 4, L+Si; 5, L+Si+β-Si$_2$Te$_3$; 6, CdTe+β-Si$_2$Te$_3$; and 7, CdTe+α-Si$_2$Te$_3$. (From Odin, I.N. and Ivanov, V.A., *Zhurn. Neorgan. khimii*, 36(7), 1837, 1991.)

Si$_2$Te$_3$ takes place at 407°C–409°C. The solubility of Si$_2$Te$_3$ in CdTe is not higher than 0.3 mol. % and the solubility of CdTe in Si$_2$Te$_3$ is equal to 0.2 mol. %.

This system was investigated using DTA, metallography, XRD, and vapor pressure measurement. The ingots were annealed at 750°C for 1040 h (Odin and Ivanov 1991).

6.17 Cadmium–Germanium–Tellurium

The scheme of Cd–Ge–Te liquidus surface is shown in Figure 6.55 (Dichi et al. 1995). No ternary compound was observed in this system. So the ternary system is divided into three subternaries (CdTe–GeTe–Te, CdTe–GeTe–Ge, and CdTe–Cd–Ge), each of which is characterized by a ternary invariant. Two ternary eutectics and one ternary peritectic have been found. A ternary metatectic reaction has been observed owing to the occurrence of the phase transition in the GeTe. The glass area covers largely the subternary CdTe–GeTe–Te system.

The nonvariant equilibria in the Cd–Ge–Te ternary system are given in Table 6.2 (Dichi et al. 1995).

CdTe–"CdGeTe$_3$": The phase diagram is not constructed. The mixture of two or more phases with different crystal lattices is formed in this system at the alloying of CdTe and "CdGeTe$_3$" (Radautsan and Ivanova 1961).

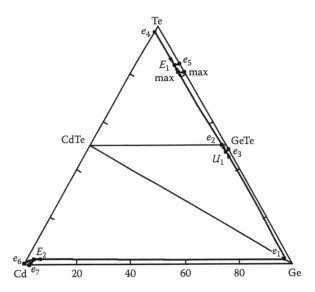

FIGURE 6.55

Scheme of Cd–Ge–Te liquidus surface. (From Dichi, E. et al., *J. Alloys Compd.*, 217(2), 193, 1995.)

TABLE 6.2

Nonvariant Equilibria in the Cd–Ge–Te
Ternary System

Symbol	Reaction	$t, °C$
e_1	$L \Leftrightarrow CdTe + Ge$	921
e_2	$L \Leftrightarrow CdTe + \beta\text{-}GeTe$	723
e_3	$L \Leftrightarrow \beta\text{-}GeTe + Te$	720
e_4	$L \Leftrightarrow CdTe + Te$	446
e_5	$L \Leftrightarrow \alpha\text{-}GeTe + Te$	383
e_6	$L \Leftrightarrow CdTe + Cd$	321
e_7	$L \Leftrightarrow Cd + Ge$	320
p	$\beta\text{-}GeTe + Ge \Leftrightarrow \alpha\text{-}GeTe$	420
m	$\beta\text{-}GeTe \Leftrightarrow \alpha\text{-}GeTe + L$	404
E_1	$L \Leftrightarrow CdTe + \alpha\text{-}GeTe + Te$	380
E_2	$L \Leftrightarrow CdTe + Ge + Cd$	320
U_1	$L + CdTe \Leftrightarrow \beta\text{-}GeTe + Ge$	721
U_2	$\beta\text{-}GeTe + Ge \Leftrightarrow \alpha\text{-}GeTe + CdTe$	400
M	$\beta\text{-}GeTe \Leftrightarrow \alpha\text{-}GeTe + CdTe + L$	397

Source: Dichi, E. et al., *J. Alloys Compd.*, 217(2), 193, 1995.

This system was investigated using metallography, XRD, and microhardness measurement (Radautsan and Ivanova 1961).

CdTe–Ge: The phase diagram belongs to the eutectic type (Figure 6.56) (Dichi et al. 1995, Panchuk et al. 1976, Shcherbak et al. 1997). The monotectic lies at 1074°C ± 3°C in the 4–83 at. % Ge composition range (Shcherbak et al.

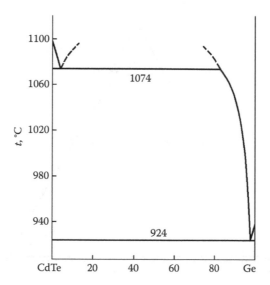

FIGURE 6.56
CdTe–Ge phase diagram. (From Shcherbak, L. et al., *Calphad*, 21(4), 463, 1997.)

1997) [the immiscibility region exists within the interval of 7–93 at. % Ge with monotectic temperature 1050°C (Panchuk et al. 1976)]. Dichi et al. (1995) did not observe the immiscibility region, but they did not use the metallography. The position of liquidus on CdTe–Ge phase diagram indicates the presence of immiscibility region. (Panchuk et al. 1976) found four phases (CdTe, GeTe, Cd, and Ge) in the CdTe–Ge system that can be explained by the nonequilibrium state of the alloys. However, the stability of monotectic temperature is an evidence of this system quasibinarity.

The eutectic composition and temperature are 2.5 mol. % CdTe and 924°C ± 1°C, respectively (Shcherbak et al. 1997) [3 mol. % CdTe and 908°C (Panchuk et al. 1976); according to the data of Dichi et al. (1995), the eutectic is degenerated and crystallizes at 921°C]. Thermal effects at 720°C from Panchuk et al. (1976) correspond to the melting of GeTe that is formed as a result of the nonequilibrium in this system.

The maximum mutual solid solubility of CdTe and Ge is less than 1 mol. % (Shcherbak et al. 1997). Temperature dependences of Ge solubility in CdTe at maximum and minimum cadmium vapor pressure are retrograde (Figure 6.57) (Panchuk and Shcherbak 1979). The maximum solubility of Ge at minimum and maximum cadmium vapor pressure ($4 \cdot 10^{19}$ and $4 \cdot 10^{18}$ cm^{-3}) corresponds to 680 and approximately 900°C, respectively. The solubility of Ge (cm^{-3}) in CdTe at p_{Cd}^{max} up to maximum x_{Ge} value can be described by the next equation: $x_{Ge} = 1.48 \cdot 10^{23} \exp(E/kT)$, where $E = -1.00$ eV.

This system was investigated using DTA, differential scanning calorimetry (DSC), metallography, XRD, and microhardness measurement

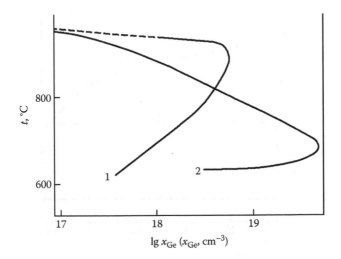

FIGURE 6.57
Temperature dependences of Ge solubility in CdTe: 1, p_{Cd}^{max}; 2, p_{Cd}^{min}. (From Panchuk, O.E. and Shcherbak, L.P., *Izv. AN SSSR. Neorgan. materialy*, 15(8), 1339, 1979.)

supplemented by optical and scanning electron microscopy combined with EPMA (Panchuk et al. 1976, Dichi et al. 1995, Shcherbak et al. 1997). The solubility of Ge in CdTe was determined using radioactive isotopes (Panchuk and Shcherbak 1979).

CdTe–GeTe: The results of this system investigation are contradictory. According to the data of Glazov et al. (1970, 1972, 1975) and Dichi et al. (1995) the phase diagram is a eutectic type (Figure 6.58a). The eutectic composition and temperature are 3.5 mol. % CdTe and 723°C, respectively (Dichi et al. 1995) [700°C (Glazov et al. 1970, 1972, 1975)]. The solubility of CdTe in GeTe at 230°C and 530°C is equal to 2 and 3 mol. %, respectively (Kutsia and Stavrianidis 1983). This section loses its quasibinary character in the vicinity of the GeTe solid solution (Dichi et al. 1995). The plateau at 400°C characterizes the phase transition of GeTe according to the reaction of type U_2 as follows: β-GeTe + Ge $\Leftrightarrow$ α-GeTe + CdTe.

According to the data of Quenez and Khodadad (1969), the phase diagram is a peritectic type with the peritectic temperature 724°C (Figure 6.58b). Two regions of α- and β-solid solutions, based on GeTe and limited by the CdGe$_9$Te$_{10}$ composition, were found in the CdTe–GeTe system. α-β-Phase transition for the CdGe$_9$Te$_{10}$ alloy takes place at 350°C.

This system was investigated using DTA, DSC, metallography, and XRD (Quenez and Khodadad 1969, Glazov et al. 1970, 1972, 1975, Kutsia and Stavrianidis 1983, Dichi et al. 1995). The ingots were annealed at 530°C, 430°C, 330°C, and 230°C for 200, 400, 600, and 800 h, respectively (Kutsia and Stavrianidis 1983).

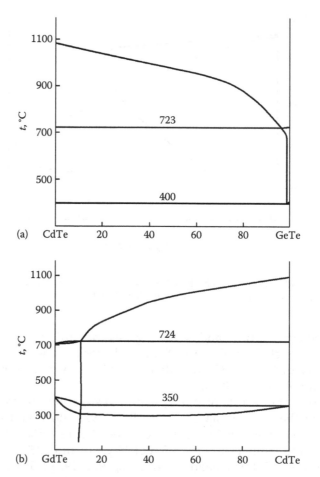

FIGURE 6.58
CdTe–GeTe phase diagram. (a: From Dichi, E. et al., *J. Alloys Compd.*, 217(2), 193, 1995; b: From Quenez, P. and Khodadad, P., *Bull. Soc. Chim. France*, (1), 3, 1969.)

6.18 Cadmium–Tin–Tellurium

In the CdTe–SnTe–Te subsystem of the Cd–Sn–Te ternary system, there is a ternary eutectic at 405°C that is degenerated and practically coincides with eutectic in the SnTe–Te system (Morgant et al. 1981).

CdTe–"CdSnTe₃:" The phase diagram is not constructed. The mixture of two or more phases with different crystal lattices is formed in this system at the alloying of CdTe and "CdSnTe₃" (Radautsan and Ivanova 1961).

This system was investigated using metallography, XRD, and microhardness measurement (Radautsan and Ivanova 1961).

CdTe–Sn: The phase diagram is a eutectic type (Figure 6.59) (Panchuk et al. 1973b, Tai and Hori 1974, Morgant et al. 1981). The eutectic is degenerated

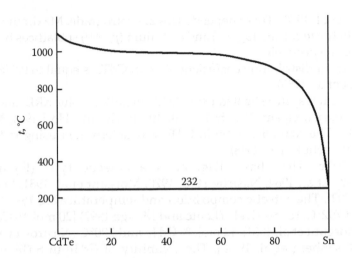

FIGURE 6.59
CdTe–Sn phase diagram. (From Panchuk, O.E. *Izv. AN SSSR. Neorgan. materialy,* 9(4), 572, 1973.)

from the Sn-rich side, contains 99.995 at. % Sn, and crystallizes at 232°C. The mutual solubility of CdTe and Sn is insignificant.

At 700°C–750°C, tin occupies mainly cadmium vacancies and at 850°C–925°C, tellurium vacancies. Amphoteric behavior of Sn in CdTe leads to the complicated temperature dependences of Sn dissolution in CdTe at maximum and minimum cadmium vapor pressure (Figure 6.60)

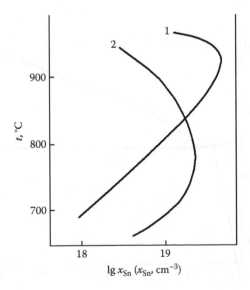

FIGURE 6.60
Temperature dependences of Sn solubility in CdTe: 1, p_{Cd}^{max}; 2, p_{Cd}^{min}. (From Panchuk, O.E. et al., *Izv. AN SSSR. Neorgan. materialy,* 14(1), 50, 1978.)

(Panchuk et al. 1978). These dependences are retrograde. Maximum solubility of Sn in the tellurium ($p_{Cd}{}^{max}$) and cadmium ($p_{Cd}{}^{min}$) sublattices is at 925°C and 775°C, respectively.

Equilibrium distribution coefficient of Sn in CdTe is equal to 0.066 ± 0.003 (Burachek et al. 1985).

This system was investigated using DTA, metallography, XRD, and microhardness measurement (Panchuk et al. 1973b, Tai and Hori 1974, Morgant et al. 1981). The solubility of Sn in CdTe was determined using radioactive isotopes (Panchuk et al. 1978).

CdTe–SnTe: The phase diagram is a eutectic type (Figure 6.61) (Rosenberg et al. 1964, Nasirov et al. 1970, Morgant et al. 1981, Leute and Menge 1992). The eutectic composition and temperature are 18 ± 1 mol. % CdTe and 792°C, respectively (Leute and Menge 1992) [20 mol. % CdTe and 792°C (Morgant et al. 1981), 12 mol. % CdTe and 780°C (Nasirov et al. 1970), 784°C (Rosenberg et al. 1964)]. The solubility of CdTe in SnTe at 530°C, 630°C, and 730°C is equal correspondingly to 4, 6.5, and 11 mol. % (Leute and Menge 1992) [according to the data of Tairov et al. (1969) and Morgant et al. (1981), the solubility of CdTe in SnTe at the eutectic temperature is equal to 10 mol. % and decreases to 8 mol. % at 760°C]. The solubility of SnTe in CdTe at the eutectic temperature reaches 4 mol. % (Morgant et al. 1981).

This system was investigated using DTA, metallography, XRD, and EPMA (Rosenberg et al. 1964, Nasirov et al. 1970, Morgant et al. 1981, Leute and Menge 1992).

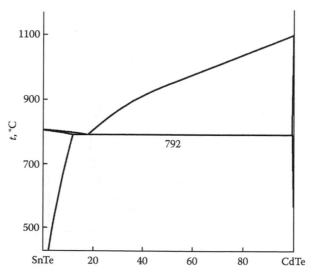

FIGURE 6.61
CdTe–SnTe phase diagram. (From Leute, V. and Menge, D., *Z. Phys. Chem. (Munchen)*, 176(1), 47, 1992. With permission.)

6.19 Cadmium–Lead–Tellurium

The regions of phase secondary crystallization in the subsystem CdTe–PbTe–Te of the Cd–Pb–Te ternary system have been determined by Lesyna et al. (1998). It was noted that ternary eutectic in this subsystem crystallizes at 355°C and contains 2 mol. % CdTe, 94 mol. % Te, and 4 mol. % PbTe.

CdTe–Pb: The phase diagram is a eutectic type (Figure 6.62) (Tai and Hori 1974). The eutectic is degenerated from Pb-rich side, contains 99.94 at. % Pb, and crystallizes at 327°C. The mutual solubility of CdTe and Pb is insignificant.

This system was investigated using DTA, metallography, and XRD (Tai and Hori 1974).

CdTe–PbTe: The phase diagram is a eutectic type (Figure 6.63) (Hirai and Kurata 1968, Morgant et al. 1980, Tomashik and Tomashik 1982). The eutectic composition and temperature are 35 mol. % CdTe and 870°C, respectively (Tomashik and Tomashik 1982) [36 mol. % CdTe and 884°C (Morgant et al. 1980), 40 mol. % CdTe and 840°C (Hirai and Kurata 1968), 866°C (Rosenberg et al. 1964)]. The solubility of CdTe in PbTe at 250°C, 630°C, 720°C, and 800°C is correspondingly equal to 3, 4.6, 10, and 17 mol. %

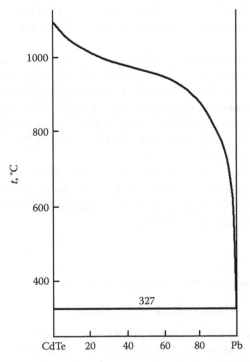

FIGURE 6.62
CdTe–Pb phase diagram. (From Tai, H. and Hori, S., *J. Jap. Inst. Metals*, 38(5), 451, 1974.)

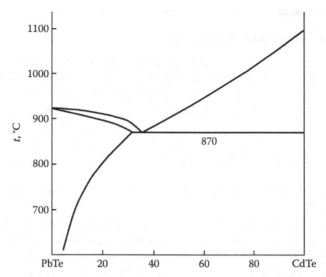

FIGURE 6.63
CdTe–PbTe phase diagram. (From Tomashik, Z.F. and Tomashik, V.N., *Izv. AN SSSR. Neorgan. materialy*, 18(12), 1994, 1982.)

(Rosenberg et al. 1964, Nikolič 1966, Crocker 1968, Hirai and Kurata 1968) [at 450°C, 550°C, and 650°C correspondingly 2, 3, and 4 mol. % (Rogacheva et al. 1988)].

Non-monotonic nature of concentration dependences of electrophysical properties was observed in the region of diluted solid solutions based on PbTe (Rogacheva et al. 1988). It can be explained by the transition of diluted solid solutions, where inter-impurity interaction can be ignored, to concentrated solid solutions, where the interaction between particles transfers the crystal in the qualitatively new state. In the CdTe–PbTe system, this transition takes place at 0.4 mol. % CdTe.

This system was investigated using DTA, metallography, XRD, and microhardness measurement (Hirai and Kurata 1968, Morgant et al. 1980, Tomashik and Tomashik 1982). The ingots were annealed at 450°C, 550°C, and 650°C for 400 and 100 h (Rogacheva et al. 1988).

6.20 Cadmium–Phosphorus–Tellurium

CdTe–Cd$_3$P$_2$: The phase diagram is a peritectic type (Figure 6.64) (Rigan et al. 1974). The results of determining of ingot phase composition by different methods distinguish from each other (Goriunova et al. 1970).

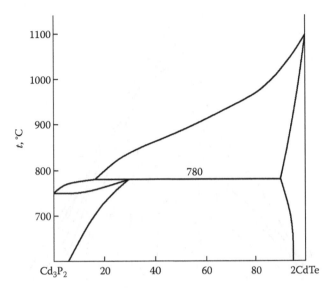

FIGURE 6.64

CdTe–Cd$_3$P$_2$ phase diagram. (From Rigan, M.Yu. et al., *Izv. vuzov. ser. khimia i khim. tehnologia*, 17(12), 1865, 1974.)

Using XRD, there was determined an existence of solid solutions within the interval of concentrations $0.7 < x < 1.0$ in the system $(2CdTe)_x–(Cd_3P_2)_{1-x}$. At room temperature, the region of solid solutions determined using DTA and metallography is within the interval of $0.95 < x < 1.0$. The solubility of 2CdTe in Cd$_3$P$_2$ at the same conditions is not higher than 1 mol. %. The increasing temperature up to 780°C (the peritectic temperature) leads to the increase of Cd$_3$P$_2$ solubility in 2CdTe up to 10 mol. % and 2CdTe in Cd$_3$P$_2$ up to 30 mol. %.

This system was investigated using DTA, metallography, and XRD (Goriunova et al. 1970, Rigan et al. 1974).

6.21 Cadmium–Arsenic–Tellurium

Seven fields of primary crystallization of CdTe, Cd, Cd$_3$As$_2$, CdAs$_2$, As, As$_2$Te$_3$, and Te exist on the liquidus surface of the Cd–As–Te ternary system (Figure 6.65) (Lakiza and Olekseyuk 1977). The field of CdTe primary crystallization occupies the most part of the liquidus surface, and CdTe determines the triangulation of this ternary system. There are five nonvariant equilibria in this system: four ternary eutectics and one ternary peritectic (Table 6.3).

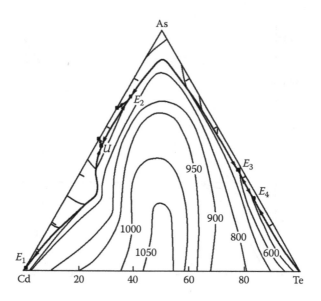

FIGURE 6.65
Liquidus surface of the Cd–As–Te ternary system. (From Lakiza, S.N. and Olekseyuk, I.D., *Zhurn. Neorgan. khimii*, 22(7), 1925, 1977.)

TABLE 6.3

Nonvariant Equilibria in the Cd–As–Te
Ternary System

Symbol	Reaction	t, °C
E_1	$L \Leftrightarrow CdTe + Cd_3As_2 + Cd$	315
E_2	$L \Leftrightarrow CdTe + CdAs_2 + As$	617
E_3	$L \Leftrightarrow CdTe + As + As_2Te_3$	372
E_4	$L \Leftrightarrow CdTe + As_2Te_3 + Te$	360
U	$L + CdTe \Leftrightarrow CdAs_2 + Cd_3As_2$	615

Source: Lakiza, S.N. and Olekseyuk, I.D., *Zhurn. Neorgan. khimii*, 22(7), 1925, 1977.

At 250°C, the fields of solid solutions based on α-Cd_3As_2 and CdTe exist in the Cd–As–Te ternary system (Figure 6.66) (Lakiza and Olekseyuk 1977). The extents of some one-phase regions based on binary compounds are insignificant. Two-phase regions are situated near the quasibinary and binary systems.

CdTe–CdAs₂: The phase diagram is a eutectic type (Figure 6.67) (Lakiza and Olekseyuk 1977). The eutectic composition and temperature are equal to 6 mol. % CdTe and 618°C, respectively.

This system was investigated using DTA, metallography, XRD, and microhardness measurement. The ingots were annealed at 600°C for 75 h with additional annealing at 250°C for 240 h (Lakiza and Olekseyuk 1977).

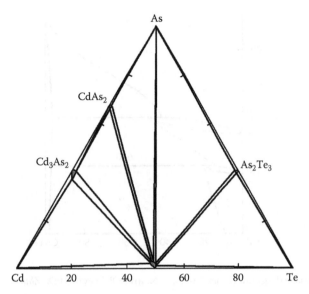

FIGURE 6.66
Isothermal section of the Cd–As–Te ternary system at 250°C. (From Lakiza, S.N. and Olekseyuk, I.D., *Zhurn. Neorgan. khimii*, 22(7), 1925, 1977.)

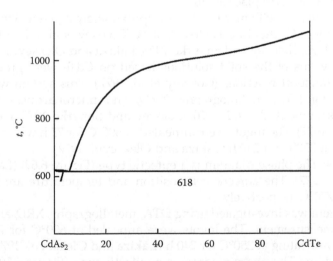

FIGURE 6.67
CdTe–CdAs₂ phase diagram. (From Lakiza, S.N. and Olekseyuk, I.D., *Zhurn. Neorgan. khimii*, 22(7), 1925, 1977.)

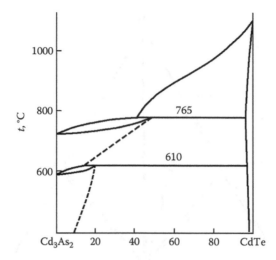

FIGURE 6.68

CdTe–Cd$_3$As$_2$ phase diagram. (From Lakiza, S.N. and Olekseyuk, I.D., *Zhurn. Neorgan. khimii*, 22(7), 1925, 1977.)

CdTe–Cd$_3$As$_2$: The phase diagram is a peritectic type (Figure 6.68) (Golovey et al. 1972, Lakiza and Olekseyuk 1977). Primary crystallization of γ-solid solution ends by the peritectic reaction L + γ ⇔ β at 765°C. The composition of the peritectic point is 40 mol. % CdTe. The peritectoid transformation of β-solid solutions takes place at 610°C.

The solubility of CdTe in α-Cd$_3$As$_2$ is approximately equal to 8 mol. % and the solubility of Cd$_3$As$_2$ in CdTe to 3 mol. % (Golovey et al. 1972, Olekseyuk et al. 1972, 1976, Golovey and Shpyrko 1975, Lakiza and Olekseyuk 1977).

Single crystals of the solid solutions based on CdTe were grown using chemical transport reactions (Golovey et al. 1977). This system was investigated using DTA, metallography, XRD, and microhardness measurement (Olekseyuk et al. 1972, 1976, Golovey and Shpyrko 1975, Lakiza and Olekseyuk 1977). The ingots were annealed at 600°C for 75 h with additional annealing at 250°C for 240 h (Lakiza and Olekseyuk 1977).

CdTe–As: The phase diagram is a eutectic type (Figure 6.69) (Lakiza and Olekseyuk 1977). The eutectic composition and temperature are 6 mol. % CdTe and 770°C, respectively.

This system was investigated using DTA, metallography, XRD, and micro-hardness measurement. The ingots were annealed at 600°C for 75 h with additional annealing at 250°C for 240 h (Lakiza and Olekseyuk 1977).

CdTe–As$_2$Te$_3$: The phase diagram is a eutectic type (Figure 6.70) (Lakiza and Olekseyuk 1977). The eutectic composition and temperature are 5 mol. % CdTe and 376°C ± 3°C, respectively.

This system was investigated using DTA, metallography, XRD, and micro-hardness measurement. The ingots were annealed at 350°C for 75 h (Lakiza and Olekseyuk 1977).

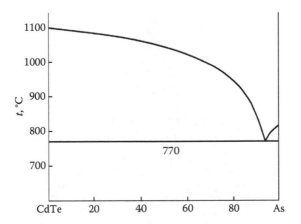

FIGURE 6.69
CdTe–As phase diagram. (From Lakiza, S.N. and Olekseyuk, I.D., *Zhurn. Neorgan. khimii*, 22(7), 1925, 1977.)

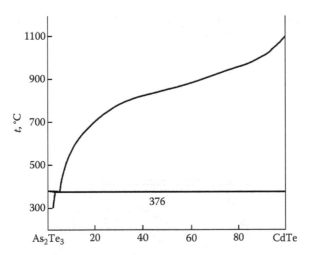

FIGURE 6.70
CdTe–As$_2$Te$_3$ phase diagram. (From Lakiza, S.N. and Olekseyuk, I.D., *Zhurn. Neorgan. khimii*, 22(7), 1925, 1977.)

6.22 Cadmium–Antimony–Tellurium

Liquidus surface of the CdTe–Cd–Sb subsystem of the Cd–Sb–Te ternary system has been constructed by Lesyna et al. (2004) using the experimental data of DTA and mathematical planning of experiment (Figure 6.71). Ternary eutectics E_1 and E_2 crystallize at 259°C and 393°C, respectively.

Isothermal section of the Cd–Sb–Te ternary system at 350°C is shown in Figure 6.72 (Mayer et al. 1978). In this system, no stable non-marginal phases have been found.

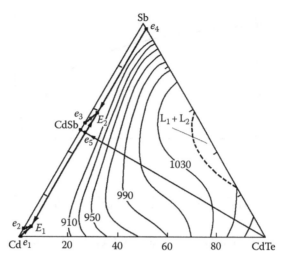

FIGURE 6.71
Liquidus surface of the CdTe–Cd–Sb subsystem. (From Lesyna, N.V. et al., *Nauk. visnyk Cherniv. un-tu. Fizyka. Elektronika*, (201), 72, 2004.)

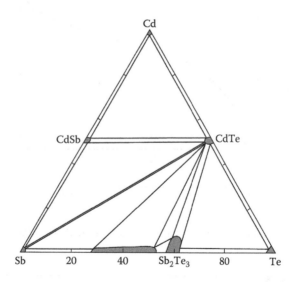

FIGURE 6.72
Isothermal section of the Cd–Sb–Te ternary system at 350°C. (From Mayer, H.W. et al., *J. Less-Common Metals*, 59(1), 43, 1978.)

CdTe–CdSb: The phase diagram is a eutectic type (Figure 6.73) (Lesyna et al. 1999). The eutectic composition and temperature are 3 mol. % CdTe and 418°C, respectively. The mutual solubility of CdTe and CdSb is insignificant (Belotski et al. 1969, 1970, Lesyna et al. 1999).

This system was investigated using DTA, metallography, and XRD and microhardness measurement (Belotski et al. 1969, 1970, Lesyna et al. 1999).

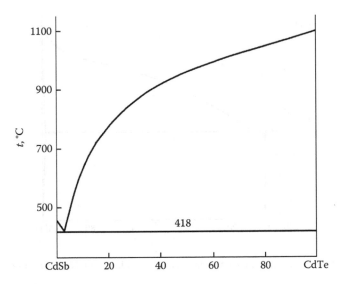

FIGURE 6.73
CdTe–CdSb phase diagram. (From Lesyna, N.V. et al., *Nauk. visnyk Cherniv. un-tu. Fizyka. Elektronika*, (66), 18, 1999.)

CdTe–Sb: The phase diagram is a eutectic type (Figure 6.74) (Tai and Hori 1970). The eutectic composition and temperature are 98.8 at. % Sb and 623°C. The immiscibility region within the interval of $(18 \pm 2)–(58 \pm 2)$ at. % Sb exists in the CdTe–Sb system with monotectic temperature 1049.5°C.

This system was investigated using DTA, metallography, and XRD (Tai and Hori 1970).

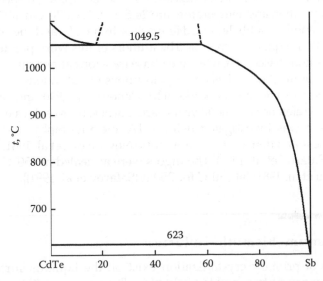

FIGURE 6.74
CdTe–Sb phase diagram. (From Tai, H. and Hori, S., *J. Jap. Inst. Metals*, 34(8), 843, 1970.)

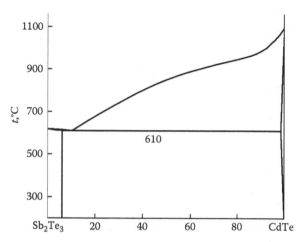

FIGURE 6.75
CdTe–Sb$_2$Te$_3$ phase diagram. (From Novoselova, A.V. et al., *Zhurn. Neorgan. khimii*, 26(4), 1048, 1981.)

CdTe–Sb$_2$Te$_3$: The most reliable phase diagram is a eutectic type (Figure 6.75) (Novoselova et al. 1981). The eutectic composition and temperature are 10 ± 2 mol. % CdTe and 610°C [40 mol. % CdTe and 606°C (Mavlonov et al. 1990)], respectively. The solubility of CdTe in Sb$_2$Te$_3$ at 450°C reaches 7 mol. % and the solubility of Sb$_2$Te$_3$ in CdTe at the same temperature is not higher than 0.5 mol. %.

According to the data of Safarov et al. (1991), CdSb$_2$Te$_4$ ternary compound is formed in this system, which melts incongruently at 590°C and crystallizes in an orthorhombic structure with lattice parameters $a = 1141$, $b = 402$, and $c = 1394$ pm and calculation and experimental density 4.67 and 4.68 g cm^{-3}, respectively. The eutectic composition and temperature are 24.5 mol. % CdTe and 465°C, respectively. The solubility of Sb$_2$Te$_3$ in CdTe reaches 16 mol. % and the solubility of CdTe in Sb$_2$Te$_3$ is equal to 7 mol. %. The authors of this work pointed out that the solid solutions based on CdTe have phase transformation that is impossible because cadmium telluride has no polymorphous modifications.

It is necessary to note that the works of Mavlonov et al. (1990) and Safarov et al. (1991) have no references of the 10 years earlier article (Novoselova et al. 1981).

This system was investigated using DTA, metallography, and XRD and microhardness and density measurements (Novoselova et al. 1981, Mavlonov et al. 1990, Safarov et al. 1991). The ingots were annealed at 450°C for 1000 h (Novoselova et al. 1981) [at 380°C for 250 h (Safarov et al. 1991)].

6.23 Cadmium–Bismuth–Tellurium

Eight fields of primary crystallization exist on the liquidus surface of the Cd–Bi–Te ternary system, and the field of CdTe primary crystallization occupies the most part of this liquidus surface (Figure 6.76) (Marugin et al. 1984).

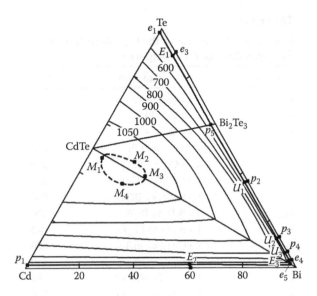

FIGURE 6.76
Liquidus surface of the Cd–Bi–Te ternary system. (From Marugin, V.V. et al., *Zhurn. Neorgan. khimii*, 29(6), 1566, 1984.)

The immiscibility region is situated within the concentrated triangle and is elongated along the CdTe–Bi quasibinary system. Nonvariant equilibria in the Cd–Bi–Te ternary system are given in Table 6.4.

At 500°C, the region of a solid state is limited by the CdTe–BiTe–Bi_2Te_3 subsystem (Figure 6.77). The liquid-state regions are adjoined to the Cd–Bi binary system and Te-rich corner (Marugin et al. 1984).

CdTe–Bi: The phase diagram is a eutectic type (Figure 6.78) (Rubenstein 1966, Tai and Hori 1970). The eutectic is degenerated from the Bi-rich side, contains 99.992 at. % Bi, and crystallizes at 270.5°C. The immiscibility region within the interval of (10 ± 2)–(39 ± 2) at. % Bi exists in the CdTe–Bi system with monotectic temperature 1061°C (Tai and Hori 1970). The solubility of Bi in CdTe is not higher than 0.5 at. % (Marugin et al. 1984).

This system was investigated using DTA, metallography, and XRD (Tai and Hori 1970).

CdTe–BiTe: This section is a nonquasibinary section of the Cd–Bi–Te ternary system (Marugin et al. 1984). It intersects the fields of CdTe and Bi_2Te_3 primary crystallization. The crystallization ends in the ternary peritectic point at 545°C. The solubility of BiTe in CdTe is not higher than 0.5 mol. % and the solubility of CdTe in BiTe is equal to 2 mol. %.

This system was investigated using DTA, metallography, and XRD. The ingots were annealed at a temperature 50°C below the temperatures of corresponding nonvariant equilibria for 700 h (Marugin et al. 1984).

CdTe–Bi_2Te: This section is a nonquasibinary section of the Cd–Bi–Te ternary system (Figure 6.79) (Marugin et al. 1984). It intersects the fields of CdTe

TABLE 6.4

Nonvariant Equilibria in the Cd–Bi–Te Ternary
System with Liquid Participation

Symbol	Reaction	t, °C
e_1	$L \Leftrightarrow CdTe + Te$	450
e_2	$L \Leftrightarrow Cd + Bi$	155
e_3	$L \Leftrightarrow Bi_2Te_3 + Te$	418
e_4	$L \Leftrightarrow Bi_{14}Te_6 + Bi$	266
e_5	$L \Leftrightarrow CdTe + Bi$	270
p_1	$L + CdTe \Leftrightarrow (Cd)$	324
p_2	$L + Bi_2Te_3 \Leftrightarrow BiTe$	550
p_3	$L + BiTe \Leftrightarrow Bi_2Te$	420
p_4	$L + Bi_2Te \Leftrightarrow Bi_{14}Te_6$	312
p_5	$L + CdTe \Leftrightarrow Bi_2Te_3$	600
E_1	$L \Leftrightarrow CdTe + Te + Bi_2Te_3$	410
E_2	$L \Leftrightarrow CdTe + Cd + Bi$	154
E_3	$L \Leftrightarrow Bi_{14}Te_6 + Bi + CdTe$	260
U_1	$L + Bi_2Te_3 \Leftrightarrow CdTe + BiTe$	545
U_2	$L + BiTe \Leftrightarrow CdTe + Bi_2Te$	415
U_3	$L + Bi_2Te \Leftrightarrow CdTe + Bi_{14}Te_6$	310
$M_1M_2M_3M_4$	$L \Leftrightarrow CdTe + L_1$	1061

Source: Marugin, V.V. et al., *Zhurn. Neorgan. khimii*,
29(6), 1566, 1984.

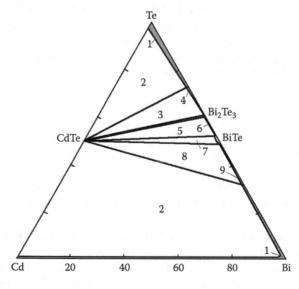

FIGURE 6.77

Isothermal section of the Cd–Bi–Te ternary system at 500°C: 1, L; 2, L + CdTe; 3, L + CdTe + Bi_2Te_3; 4, L + Bi_2Te_3; 5, CdTe + Bi_2Te_3 + BiTe; 6, Bi_2Te_3 + BiTe; 7, CdTe + BiTe; 8, L + CdTe + BiTe; and 9, L + BiTe. (From Marugin, V.V. et al., *Zhurn. Neorgan. khimii*, 29(6), 1566, 1984.)

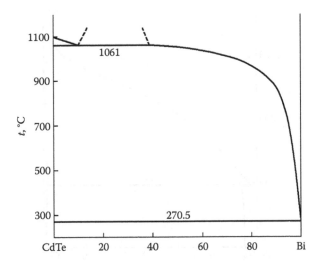

FIGURE 6.78
CdTe–Bi phase diagram. (From Rubenstein, M., *J. Electrochem. Soc.*, 113(6), 623, 1966; Tai, H. and Hori S., *J. Jap. Inst. Metals*, 34(8), 843, 1970. With permission.)

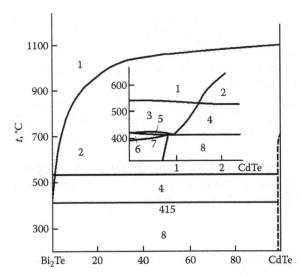

FIGURE 6.79
Phase relations in the CdTe–Bi$_2$Te system: 1, L; 2, L+CdTe; 3, L+BiTe; 4, L+CdTe+BiTe; 5, L+BiTe+Bi$_2$Te; 6, L+Bi$_2$Te; 7, Bi$_2$Te; and 8, Bi$_2$Te+CdTe. (From Marugin, V.V. et al., *Zhurn. Neorgan. khimii*, 29(6), 1566, 1984.)

and BiTe primary crystallization. The crystallization ends in the ternary peritectic point at 415°C. The solubility of Bi$_2$Te in CdTe is not higher than 0.5 mol. % and the solubility of CdTe in Bi$_2$Te is equal to 1 mol. %.

This system was investigated using DTA, metallography, and XRD. The ingots were annealed at a temperature 50°C below the temperatures of corresponding nonvariant equilibria for 700 h (Marugin et al. 1984).

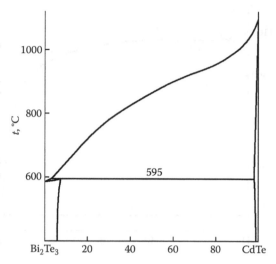

FIGURE 6.80

CdTe–Bi$_2$Te$_3$ phase diagram. (From Novoselova, A.V. et al., *Zhurn. Neorgan. khimii*, 26(4), 1048, 1981.)

CdTe–Bi$_2$Te$_3$: The most reliable phase diagram is a peritectic type (Figure 6.80) (Novoselova et al. 1981). Thermal effects at 595°C correspond to the peritectic reaction of solid solutions based on Bi$_2$Te$_3$ formation. The solubility of CdTe in Bi$_2$Te$_3$ at 450°C reaches 6 mol. % and the solubility of Bi$_2$Te$_3$ is not higher than 0.5 mol. %.

According the data of Datsenko et al. (1981), the phase diagram of this system is a eutectic type. The eutectic is degenerated from the Bi$_2$Te$_3$-rich side and crystallizes at 580°C ± 3°C. The mutual solubility of CdTe and Bi$_2$Te$_3$ is insignificant.

This system was investigated using DTA, metallography, and XRD (Datsenko et al. 1981, Novoselova et al. 1981). The ingots were annealed at 450°C for 1000 h (Novoselova et al. 1981).

6.24 Cadmium–Tantalum–Tellurium

No mutual solubility was found in the Cd–Ta–Te ternary system (Cordes and Schmid-Fetzer 1995b). The phase equilibria at 500°C according to the experimental data are in complete agreement with the approximate calculation of the solid-state equilibria. Tantalum and CdTe are in thermodynamic equilibrium, and this ternary system can be divided in four subsystems, CdTe–Cd–Ta, CdTe–Ta–TaTe$_2$, CdTe–TaTe$_2$–TaTe$_4$, and CdTe–TaTe$_4$–Te.

This system was investigated using XRD. The ingots were annealed at 500°C from 5 to 10 days (Cordes and Schmid-Fetzer 1995b).

6.25 Cadmium–Oxygen–Tellurium

The predominance area diagram of the phase relationships in the system Cd–O–Te at 430°C is shown in Figure 6.81 (Diehl and Nolaeng 1984). In the upper left part of the diagram in which the Cd activity is sufficiently high and oxygen activity low and CdTe coexists with the gas phase, the CdTe predominance area is bound by horizontal lines representing unity activity for a_{Te} and a_{Cd}. There is a forbidden area for the values of a_{Cd} above unity.

At higher values of a_{Cd}, CdTe reacts with the gas phase to form CdO, and at lower higher values of a_{Cd}, the reaction yields $CdTeO_3$. If, in case of lower but constant values of the cadmium activity, $CdTeO_3$ is allowed to react further with the gas phase under an increasing oxygen activity, CdO will form at the expense of monotellurite. It follows from Figure 6.81 that CdO is the stable oxide phase for higher values of both the cadmium and the oxygen activities. With a decreasing a_{Cd} and for the medium of the oxygen activities, for example, between 10^{-10} and 10^{-15}, the composition of the oxide phase changes from CdO via the tellurites $CdTeO_3$ and $CdTe_2O_5$ to TeO_2. For high values of the oxygen activities, it changes from CdO via the tellurates Cd_3TeO_6 and $CdTeO_4$ to TeO_3. It is important to note that CdTe cannot coexist with $CdTe_2O_5$ or TeO_2 under equilibrium conditions (Diehl and Nolaeng 1984).

According to the data of (Brandt and Moritz 1985) and (Weil 2004), ternary compounds $CdTe_2O_6$, $Cd_2Te_2O_7$, and $Cd_2Te_3O_9$ are formed in the Cd–O–Te ternary system. $Cd_2Te_2O_7$ crystallizes in a triclinic structure with the lattice

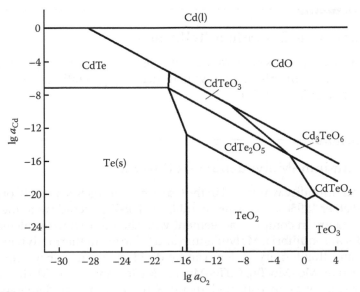

FIGURE 6.81
Predominance area of the condensed phases in equilibrium with the gas phase in the system Cd–O–Te at 430°C. (From Diehl, R. and Nolaeng, B.I., *J. Cryst. Growth*, 66, 91, 1984.)

parameters $a = 743.28 \pm 0.07$, $b = 833.46 \pm 0.06$, $c = 998.98 \pm 0.08$ pm, $\alpha = 87.005° \pm$ 0.006°, $\beta = 78.843° \pm 0.008°$, and $\gamma = 77.210° \pm 0.008°$. $Cd_2Te_3O_9$ crystallizes in a monoclinic structure with the lattice parameters $a = 930.39 \pm 0.07$, $b = 731.96 \pm$ 0.05, $c = 1324.79 \pm 0.07$ pm, and $\beta = 122.914° \pm 0.004°$ (Weil 2004). Both compounds are composed of $[CdO_x]$ polyhedra ($x = 7$ for $Cd_2Te_3O_9$, $x = 6$–8 for $Ca_2Te_2O_7$) and $Te^{IV}O_3$ and $Te^{VI}O_6$ groups, respectively.

Isothermal section of the Cd–O–Te ternary system at room temperature was constructed by (Medvedev 1993) according to the thermodynamic calculations with the account of the experimental data (it is the same as for the Cd–O–Se ternary system; Figure 5.57).

CdO–TeO$_2$: The phase diagram is not constructed. $CdTeO_3$ and $CdTe_2O_5$ ternary compounds are formed in this system (Robertson et al. 1978, Rhiger and Kvaas 1983, Diehl and Nolaeng 1984). $CdTeO_3$ crystallizes in a cubic structure with lattice parameter $a = 534 \pm 1$ pm and experimental density 5.99 ± 0.01 g cm^{-3}, transforms in a hexagonal structure at 497°C, and melts at 790°C (Markovskii and Pron' 1968). It is stable at the heating up to 1000°C. The heat of formation of the amorphous and crystalline $CdTeO_3$ is equal to 618 ± 0.8 and 645.6 ± 0.8 kJ mol^{-1} correspondingly (Pron' and Markovskii 1969). This compound has been obtained by the precipitation from water solutions at the mixing of Na_2TeO_3 and $CdSO_4$ (Markovskii and Pron' 1968).

CdO–TeO$_3$: The phase diagram is not constructed. $CdTeO_4$, Cd_3TeO_6, and $Cd_3Te_2O_9$ ternary compounds are formed in this system (Diehl and Nolaeng 1984, Brandt and Moritz 1985).

6.26 Cadmium–Chromium–Tellurium

CdTe–Cr$_2$Te$_3$: The phase diagram is not constructed. $CdCr_2Te_4$ ternary compound is formed in this system, which melts congruently at 985°C $\pm$ 10°C (Derid and Kosnichan 1973).

6.27 Cadmium–Molybdenum–Tellurium

No mutual solubility was found in the Cd–Mo–Te ternary system (Cordes and Schmid-Fetzer 1995b). The phase equilibria at 550°C according to the experimental data are in complete agreement with the approximate calculation of the solid-state equilibria. Molybdenum and CdTe are in thermodynamic equilibrium, and this ternary system can be divided in four subsystems CdTe–Cd–Mo, CdTe–Mo–Mo$_3$Te$_4$, CdTe–Mo$_3$Te$_4$–MoTe$_2$, and CdTe–MoTe$_2$–Te.

This system was investigated using XRD and energy-dispersive x-ray analysis. The ingots were annealed at 500°C for 5–10 days (Cordes and Schmid-Fetzer 1995b).

6.28 Cadmium–Tungsten–Tellurium

No mutual solubility was found in the Cd–W–Te ternary system (Cordes and Schmid-Fetzer 1995b). The phase equilibria at 550°C according to the experimental data are in complete agreement with the approximate calculation of the solid-state equilibria. Tungsten and CdTe are in thermodynamic equilibrium, and this ternary system can be divided in three subsystems, CdTe–Cd–W, CdTe–W–WTe$_2$, and CdTe–WTe$_2$–Te.

This system was investigated using XRD, energy-dispersive x-ray analysis, and scanning electron microscope. The ingots were annealed at 550°C at least 10 days (Cordes and Schmid-Fetzer 1995b).

6.29 Cadmium–Chlorine–Tellurium

CdTe–CdCl$_2$: The phase diagram is a eutectic type (Figure 6.82) (Tai and Hori 1976, Saraie et al. 1978). The eutectic composition and temperature are 74 mol. % CdCl$_2$ and 505°C [490°C ± 5°C (Andronik et al. 1976, Vishniakov et al. 1978)], respectively (Tai and Hori 1976, Saraie et al. 1978). The mutual solubility of CdTe and CdCl$_2$ in the solid state is insignificant.

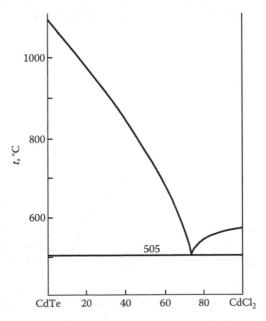

FIGURE 6.82
CdTe–CdCl$_2$ phase diagram. (From Tai, H. and Hori, S., *J. Jap. Inst. Metals*, 40(7), 722, 1976.)

This system was investigated using DTA, XRD, and determination of CdTe and $CdCl_2$ dissolution in the liquid state (Andronik et al. 1976, Tai and Hori 1976, Saraie et al. 1978, Vishniakov et al. 1978).

6.30 Cadmium–Bromine–Tellurium

CdTe–CdBr$_2$: The phase diagram is a eutectic type (Figure 6.83) (Tai and Hori 1976). The eutectic composition and temperature are 73 mol. % $CdBr_2$ and 513°C, respectively. The mutual solubility of CdTe and $CdBr_2$ in the solid state is insignificant.

This system was investigated using DTA, XRD, and determination of CdTe and $CdBr_2$ dissolution in the liquid state (Tai and Hori 1976).

6.31 Cadmium–Iodine–Tellurium

CdTe–CdI$_2$: The phase diagram is shown in Figure 6.84 (Odin 2001). The eutectic composition and temperature are 14 ± 1 mol. % CdS and 374°C ± 2°C, respectively. The solubility of CdI_2 in CdTe at 500°C is equal to 0.25 mol. %.

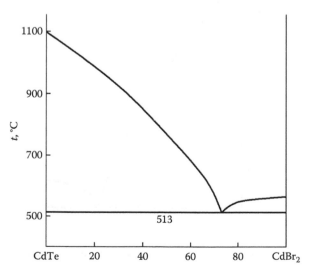

FIGURE 6.83
CdTe–CdBr$_2$ phase diagram. (From Tai, H. and Hori, S., *J. Jap. Inst. Metals*, 40(7), 722, 1976.)

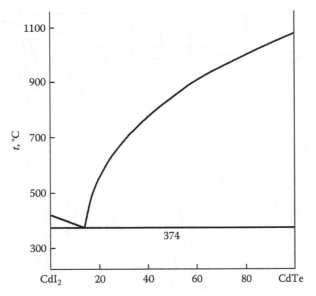

FIGURE 6.84
CdTe–CdI$_2$ phase diagram. (From Odin, I.N., *Zhurn. Neorgan. khimii*, 46(10), 1733, 2001.)

According to the data of Rodionov et al. (1972), Cd$_2$TeI$_2$ ternary compound is formed in this system, which is stable in a narrow temperature interval. The solubility of CdTe in the liquid CdI$_2$ within the interval of 500°C–700°C was determined using filtration method by Polistanski et al. (1978). The saturation time of the CdI$_2$ melt by cadmium telluride in the investigated temperature interval is not higher than 3 h.

At temperatures ranged from 20°C to 600°C, CdI$_2$ can be used as a diffusion source for the CdTe doping by iodine (Malzbender et al. 1994). During the diffusion annealing, the concentration of iodine in the surface of the CdTe slice was in the range of 10^{21}–10^{22} cm^{-3}.

This system was investigated using DTA, metallography, and XRD (Odin 2001).

CdTe–I$_2$: During diffusion anneals of CdTe with elemental iodine at temperatures between 200°C and 250°C, a chemical reaction occurs forming CdI$_2$ (Malzbender et al. 1994).

The system was investigated by SEM using EDAX attachment.

6.32 Cadmium–Manganese–Tellurium

The liquidus temperatures were determined using DTA method and verified by the growth experiments (Becla et al. 1985). An excess of tellurium in Cd$_{1-x}$Mn$_x$Te is accumulated in the liquid phase and forms MnTe$_2$, resulting in a reduction of the α-phase region with sphalerite-type structure.

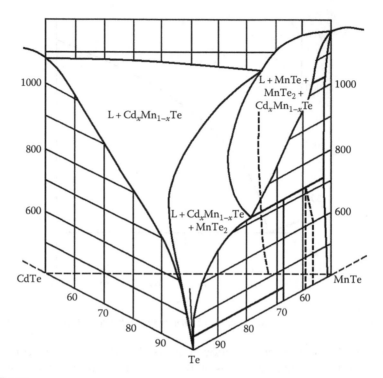

FIGURE 6.85
Phase-separation region in the Te-rich $Cd_xMn_{1-x}Te$ solid solutions. (From Becla, P. *J. Vac. Sci. Technol.*, A3(1), 116, 1985.)

The phase-separation region in the Te rich of $Cd_{1-x}Mn_xTe$ solutions is presented in Figure 6.85. As can be seen from this figure, the α-phase of $Cd_{1-x}Mn_xTe$ exists over the $0 < x < 0.75$ in stoichiometric melts. When excess Te is added, the α-phase region shrinks and disappears at 95 at. % Te.

Figure 6.86 presents experimentally determined liquidus isotherms for the different Te contents and variable Cd/(Cd+Mn) ratios (Becla et al. 1985). The dashed lines in Figure 6.86 denote the phase-separation region.

The existence of six phase regions was determined in the CdTe–MnTe–Te subsystem of the Cd–Mn–Te ternary system at 370°C (Figure 6.87) (Mamontov and Sevastianova 1996).

This system was investigated using XRD. The ingots were annealed at 370°C for 630–1630 h (Mamontov and Sevastianova 1996).

CdTe–Mn: The phase diagram is a eutectic type (Figure 6.88) (Fesh et al. 1978). The eutectic composition is 0.5 at. % Mn. The immiscibility region exists in this system at the Mn contents more than 1.5 at. % with monotectic temperature 1080°C. At 900°C, the solubility of Mn in CdTe is not higher than 0.25 at. %.

This system was investigated using DTA, metallography, and microhardness measurement. The ingots were annealed at 900°C for 200 h (Fesh et al. 1978).

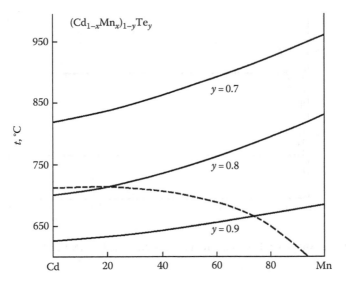

FIGURE 6.86
Liquidus surface at constant Te content for variable Cd/(Cd+Mn) ratios: the dashed curve denotes the separation between the α-phase and the sphalerite structure. (From Becla, P. *J. Vac. Sci. Technol.*, A3(1), 116, 1985.)

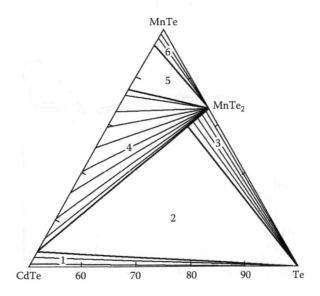

FIGURE 6.87
Isothermal section of the CdTe–MnTe–Te subsystem at 370°C: 1, $Cd_xMn_{1-x}Te + Te$; 2, $Cd_{0.94}Mn_{0.06}Te + Te + (CdTe)_{0.14}(MnTe_2)_{0.86}$; 3, $(CdTe)_x(MnTe_2)_{1-x} + Te$; 4, $Cd_xMn_{1-x}Te + (CdTe)_x(MnTe_2)_{1-x}$; 5, $Cd_{0.25}Mn_{0.75}Te + MnTe_2 + Cd_xMn_{1-x}Te$; and 6, $MnTe_2 + Cd_xMn_{1-x}Te$. (From Mamontov, M.N. and Sevastianova, L.G., *Neorgan. materialy*, 32(7), 810, 1996.)

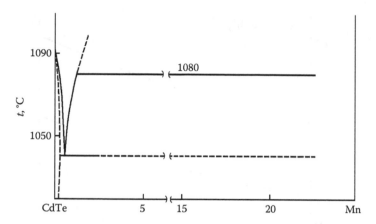

FIGURE 6.88
Part of the CdTe–Mn phase diagram. (From Fesh, R.N. et al., *Izv. AN SSSR. Neorgan. materialy*, 14(1), 170, 1978.)

CdTe–MnTe: The phase equilibria in this system have been discussed in Pajaczkowska (1978) based on papers that have appeared up to 1977. This section is a nonquasibinary section of the Cd–Mn–Te ternary system as MnTe melts incongruently (Figure 6.89) (Triboulet and Didier 1980, 1981, Odin et al. 2003). The $Cd_{1-x}Mn_xTe$ solid solutions at 800°C extend up to $x = 0.714$ (Odin et al. 2003) [$x = 0.77$ (Triboulet and Didier 1981)]. The lattice constant

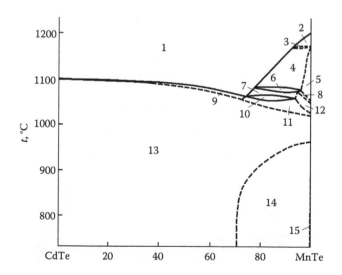

FIGURE 6.89
CdTe–MnTe phase diagram: 1, L; 2, L + γ-Mn; 3, L + γ-Mn + δ-MnTe; 4, L + δ-MnTe; 5, δ-MnTe; 6, L + δ-MnTe + γ-MnTe; 7, L + γ-MnTe; 8, δ-MnTe + γ-MnTe; 9, L + β-$Cd_{1-x}Mn_x$Te; 10, L + β-$Cd_{1-x}Mn_x$Te + γ-MnTe; 11, β-$Cd_{1-x}Mn_x$Te + γ-MnTe; 12, γ-MnTe; 13, β-$Cd_{1-x}Mn_x$Te; 14, β-$Cd_{1-x}Mn_x$Te + α-MnTe; and 15, α-MnTe. (From Odin, I.N. et al., *Neorgan. materialy*, 39(4), 425, 2003.)

follows Vegard's law in the range $0 \leq x \leq 0.4$ (Abrahams et al. 1989). Solid MnTe dissolves at this temperature 0.3 mol. % CdTe. There is a narrow two-phase region at the MnTe contents up to 80 mol. %, and at higher MnTe concentrations, the phase diagram becomes complicated.

Within the interval from 75 to 100 mol. % MnTe, solid solutions with a sphalerite structure transform into solid solutions with a wurtzite structure (Odin et al. 2003). The solubility of CdTe in δ-MnTe is not large and the content of CdTe in the solid solutions based on γ-MnTe can reach 7 mol. %.

At 370°C, the solubility of CdTe in MnTe is equal to 14 ± 2 mol. % (Mamontov and Sevastianova 1996). The tellurium-saturated limit of homogeneity region for $Cd_{1-x}Mn_x$Te solid solutions at $0.53 < x < 0.73$°C and $321 < t < 425$°C is situated at the tellurium mole fraction 0.51 (Mamontov 1997).

The ingots were annealing at 800°C and 1000°C for 600 and 300 h, respectively (Odin et al. 2003). This system was investigated by using DTA, metallography, XRD, EPMA (Triboulet and Didier 1981, Mamontov and Sevastianova 1996, Odin et al. 2003), and measurement of emf of concentration chains (Mamontov 1997). Solid-solution single crystals were grown using the Bridgman method (Triboulet and Didier 1980, 1981, Abrahams et al. 1989, Odin et al. 2003). The crystal growth of the $Cd_{1-x}Mn_x$Te solid solutions are also given by Pajaczkowska (1978).

6.33 Cadmium–Iron–Tellurium

CdTe–Fe: The phase diagram is a eutectic type (Figure 6.90) (Fesh et al. 1977). The eutectic concentration and temperature are 97.5 mol. % CdTe and 1055°C, respectively. The immiscibility region exists in this system at the Fe contents more than 5 at. % with monotectic temperature 1070°C.

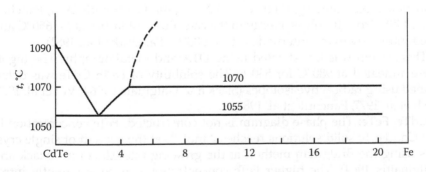

FIGURE 6.90
Part of the CdTe–Fe phase diagram. (From Fesh, R.N., *Izv. AN SSSR. Neorgan. materialy*, 13(1), 166, 1977.)

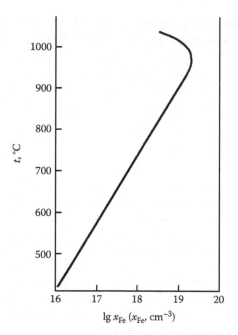

FIGURE 6.91
Temperature dependence of Fe solubility in CdTe of *n* type. (From Fesh, R.N., *Izv. AN SSSR. Neorgan. materialy*, 13(1), 166, 1977.)

The solubility of Fe in CdTe of *n* type is retrograde (Figure 6.91) and reaches $2 \cdot 10^{19}$ cm^{-3} at 980°C (Fesh et al. 1977) [$4 \cdot 10^{19}$ cm^{-3} at 950°C (Vul et al. 1972); 10^{20} cm^{-3} at the temperature close to the CdTe melting point (Slack and Galginaitis 1964)]. Excessive solubility from Slack and Galginaitis (1964) can be explained by the impurity capture of growing crystal or by the clusters formation.

The solubility of Fe in CdTe depends on cadmium vapor pressure (deviation from stoichiometry) (Figure 6.92) (Slack and Galginaitis 1964, Panchuk et al. 1981). Iron dissolves mainly in the cadmium sublattice at *t* < 650°C and predominantly in the internodes at *t* ≥ 900°C (Panchuk et al. 1981).

This system was investigated using DTA and metallography. The ingots were annealed at 980°C for 530 h. The solubility of Fe in CdTe was determined using radioactive isotopes (Slack and Galginaitis 1964, Vul et al. 1972, Fesh et al. 1977, Panchuk et al. 1981).

CdTe–FeTe: The phase diagram is not constructed. FeTe concentration in the $Cd_{1-x}Fe_xTe$ solid solutions reaches 2 mol. % at the growth of single crystals using the Bridgman method at the growing rate 0.5 cm h^{-1} (Slack and Galginaitis 1964). The higher FeTe concentration can be apparently introduced in the solid solutions at the smaller rate of growing.

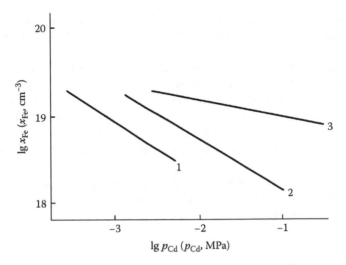

FIGURE 6.92
Dependences of Fe solubility in CdTe from cadmium vapor pressure at 680°C (1), 810°C (2), and 900°C (3). (From Panchuk, O.E. et al., *Izv. AN SSSR. Neorgan. materialy*, 17(8), 1354, 1981.)

6.34 Cadmium–Cobalt–Tellurium

CdTe–Co: The phase diagram is a eutectic type (Figure 6.93) (Fesh et al. 1978). The eutectic composition and temperature are 1 at. % Co and 1050°C, respectively. The immiscibility region exists in this system at the Co contents more than 3 at. % with monotectic temperature 1072°C. The solubility of Co in CdTe at 900°C is not higher than 0.25 at. %.

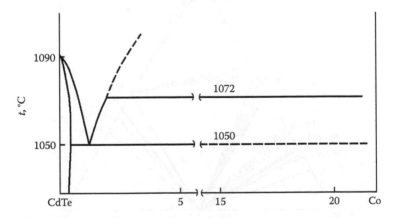

FIGURE 6.93
Part of the CdTe–Co phase diagram. (From Fesh, R.N. et al., *Izv. AN SSSR. Neorgan. materialy*, 14(1), 170, 1978.)

This system was investigated using DTA, metallography, and microhardness measurement. The ingots were annealed at 900°C for 200 h (Fesh et al. 1978).

CdTe–CoTe: The phase diagram is not constructed. The lattice parameter of the $Cd_{1-x}Co_xTe$ solid solutions decreased linearly with increasing x following Vegard's law, in the entire composition range studied ($0.001 \leq x \leq 0.009$) confirming the substitutional nature of the samples (Dwarakanadha et al. 2009). The EPR spectra indicated the existence of Co in Co^{2+} state in the present samples confirming the substitution of Co in place of Cd.

This system was investigated using XRD. $Cd_{1-x}Co_xTe$ crystals (x = 0.001, 0.003, 0.005, 0.007, and 0.009) were grown by vapor-phase technique (Dwarakanadha et al. 2009).

6.35 Cadmium–Platinum–Tellurium

The experimentally determined isothermal section of the Cd–Pt–Te ternary system at 500°C is given in Figure 6.94 (Cordes and Schmid-Fetzer 1994). Liquid phases are existent only at almost pure Te in the CdTe–Te–PtTe$_2$ subsystem or at almost pure Cd in the Cd–CdTe–Cd$_5$Pt triangle. These liquid phases vanish in the ternary eutectics below approximately 446°C or 314°C, respectively.

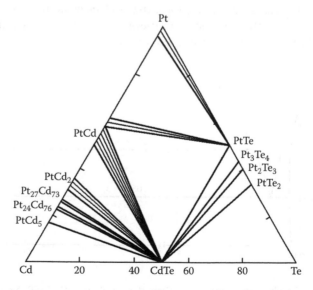

FIGURE 6.94

Isothermal section of the Cd–Pt–Te ternary system at 500°C. (From Cordes, H. and Schmid-Feltzer, R., *Semicond. Sci. Technol.*, 9(11), 2085, 1994.)

CdTe–Pt: Platinum is strongly reactive on CdTe surfaces (Cordes and Schmid-Fetzer 1994). In the early stages of interface reaction in the CdTe–Pt system, an approximately planar PtTe layer forms in direct contact with CdTe, followed by a CdPt layer. Upon further reaction, a cellular mixture of CdPt and PtTe is formed growing toward the CdTe side.

This system was investigated by XRD and SEM with EDX. The ingots were annealed at 500°C for 12 days and quenched (Cordes and Schmid-Fetzer 1994).

References

Abrahams S.C., Marsh P., Bridenbaugh P.M. Atomic substitution in $Cd_{1-x}Mn_xTe$ for $0.1 \leq x \leq 0.4$, *Acta Crystallogr.*, **C45**(4), 545–548 (1989).

Agaev A.B., Kulieva S.A. Investigation of $CdTe–Dy_2Te_3$ system [in Russian], *Azerb. khim. zhurn.*, (5), 61–63 (1984).

Agaev A.B., Kulieva U.A., Rustamov P.G. Interaction of CdTe with DyTe and Dy_2Te_3 [in Russian], *Zhurn. neorgan. khimii*, **32**(10), 2554–2557 (1987).

Agaev A.B., Zul'fugarly Dzh.I., Aliev O.M., Azadaliev R.A. The $CdTe–Gd_2Te_3$ system [in Russian], *Zhurn. neorgan. khimii*, **28**(1), 256–258 (1983).

Akutagava W., Turnbull D., Chu W.K., Mayer J.W. Solubility and lattice location of Au in CdTe by backscattering techniques, *J. Phys. Chem. Solids*, **36**(6), 521–528 (1975).

Aliev O.M., Agaev A.B., Azadaliev R.A. The $CdTe–La_2Te_3$ system [in Russian], *Zhurn. neorgan. khimii*, **30**(4), 1041–1042 (1985).

Anderson P.L., Schaake H.F., Tregilgas J.H. Precipitation and phase stability of (Hg,Cd)Te, *J. Vac. Sci. Technol.*, **21**(1), 125–128 (1982).

Andronik I.K., Kuleva Z.P., Sushkevich K.D. About solubility of cadmium telluride in some metals [in Russian], in: *Protsesy rosta i sinteza poluprovodn. kristallov i plenok.* Novosibirsk, Russia: Nauka, pp. 2, 239–240 (1975).

Andronik I.K., Kuleva Z.P., Sushkevich K.D. Investigation of possibility of CdTe crystals obtaining by the floating-zone refining from the solution in the $CdCl_2$ melt [in Russian], *Izv. AN SSSR. Neorgan. materialy*, **12**(4), 759–760 (1976).

Asadov M.M., Babanly M.B., Guseinov F.H., Kuliev A.A. Application of electromoving forces for the determination of physico-chemical properties of Zn(Cd)–Tl–Te ternary system [in Russian], *Zhurn. fiz. khimii*, **57**(11), 2865–2867 (1983).

Asadov Yu.G., Dzhafarov K.M., Asadov F.Yu., Gamidova S.S. Mechanism of structure transformation in the $Cu_{1.70}Cd_{0.05}Te$ crystals [in Russian], *Kristallografiya*, **41**(2), 230–234 (1996).

Axtell E.A. (III), Liao J.-H., Pikramenou Z., Kanatzidis M.G. Dimensional reduction in II-VI Materials: $A_2Cd_3Q_4$ (A=K, Q=S, Se, Te; A=Rb, Q=S, Se), novel ternary low-dimensional cadmium chalcogenides produced by incorporation of A_2Q in CdQ, *Chem. Eur. J.*, **2**(6), 656–666 (1996).

Azadaliev R.A. Synthesis and properties of the $CdGd_2Se_4$ and $CdGd_2Te_4$ compounds [in Russian], in: *Neorgan. soed.—sintez i svoystva*, Baku, Azerbaijan, pp. 26–27 (1990).

Babanly M.B., Guseinov F.H., Kuliev A.A. Phase equilibria in the Cd–Tl–Te system [in Russian], *Izv. AN SSSR. Neorgan. materialy*, **20**(1), 46–51 (1984).

Bailly F., Svob L., Cohen-Solal G., Triboulet R. Mercury pressure over HgTe and HgCdTe in a closed isothermal system, *J. Appl. Phys.*, **46**(10), 4244–4250 (1975).

Balagurova E.A., Vladimirov N.V., Ryazantsev A.A., Tovpentsev Yu.K., Khabarov E.N. Phase diagram of the HgTe–CdTe system [in Russian], *Izv. AN SSSR. Neorgan. materialy*, **10**(6), 1135–1136 (1974).

Barlow D.A. Calculation of pseudobinary phase diagrams for the infrared detector materials (Cd,Zn)Te, (Hg,Cd)Te and (Hg,Zn)Te, *Infrared Phys. Technol.*, **45**(1), 25–30 (2004).

Basu R., Computer estimation of the Cd-Hg-Te phase diagram, *Calphad*, **3**(2), 85–89 (1979).

Becla P., Wolff P.A., Aggarwal P.L., Yuen S.Y. LPE growth conditions for $Cd_{1-x}Mn_xTe$ and $Hg_{1-x}Mn_xTe$ epitaxial layers, *J. Vac. Sci. Technol.*, **A3**(1), 116–118 (1985).

Belotski D.P., Panchuk O.E., Panchuk I.E., Rybailo O.I. Investigation of the CdSb–CdTe system [in Russian], *Izv. AN SSSR. Neorgan. materialy*, **5**(10), 1703–1706 (1969).

Belotski D.P., Panchuk O.E., Panchuk I.E., Slyn'ko E.I. Properties of the ingots of the CdSb–CdTe quasibinary system [in Russian], *Izv. AN SSSR. Neorgan. materialy*, **6**(5), 992–993 (1970).

Blair J., Newnham R. Preparation and physical properties of crystals in the HgTe–CdTe, *Metallurgy Elemental Comp. Semiconduct.*, **12**, 393–402 (1960).

Boncheva-Mladenova Z., Vassilev V., Milenov T., Aleksandrova S. Investigation on phase-diagram of the Ag_2Te–CdTe system, *Thermochim. Acta*, **92**(Pt.A), 591–594 (1985).

Brandt G., Moritz R. Compound formation in the mercury-cadmium-tellurium-oxygen system, *Mater. Res. Bull.*, **20**(1), 49–56 (1985).

Brebrick R.F. Thermodynamic modeling of the Hg–Cd–Te and Hg–Zn–Te systems, *J. Cryst. Growth*, **86**(1–4), 39–48 (1988).

Brebrick R.F., Su C.-H., Liao P.-K. Associated solution model for Ga–In–Sb and Hg–Cd–Te, *Semicond. Semimetal.*, **19**, 171–253 (1983).

Bredol M., Leute V. Solid-state reactions in the quasibinary system Ga_2Te_3/CdTe, *J. Solid State Chem.*, **60**(1), 29–40 (1985).

Brice J.C., Capper P., Jones C.L. The phase diagram of pseudo-binary system CdTe–HgTe and the segregation of CdTe, *J. Cryst. Growth*, **75**(2), 395–399 (1986).

Bublik V.T., Zaitsev A.A. Determination of the energy of mixing of HgTe–CdTe solid solutions by diffuse x-ray scattering, *Phys. Status Solidi A*, **39**(1), 345–349 (1977).

Burachek V.R., Savitski A.V., Panchuk O.E., Zaitseva T.A. Segregation of Sn in CdTe [in Russian], Deposited in UkrNIINTI, № 2603-Uk, (1985).

Busch G., Mooser E., Pearson W.B. Neue halbleitende Verbindungen mit diamantähnlicher Struktur, *Helv. Phys. Acta*, **29**(3), 192–193 (1956).

Capper P. The behaviour of selected impurities in $Cd_xHg_{1-x}Te$, *J. Cryst. Growth*, **57**(2), 280–299 (1982).

Chen A.-B., Sher A., Spices W.E. Relation between the electronic states and structural properties of $Hg_{1-x}Cd_xTe$, *J. Vac. Sci. Technol.*, **1**(3), 1674–1677 (1983).

Chen K.-T., Wiedemeier H. The temperature-composition phase diagram and the miscibility gap of $Hg_{1-x}Cd_xTe$ solid solutions by dynamic mass-loss measurements, *J. Electron. Mater.*, **24**(4), 405–411 (1995).

Cordes H., Schmid-Fetzer R. The role of interfacial reactions in Pt/CdTe contact formation, *Semicond. Sci. Technol.*, **9**(11), 2085–2096 (1994).

Cordes H., Schmid-Fetzer R. Electrical properties and contact metallurgy of elemental (Cu, Ag, Au, Ni) and compound contacts on p-$Cd_{0.95}Zn_{0.05}Te$, *Semicond. Sci. Technol.*, **10**(1), 77–86 (1995a).

Cordes H., Schmid-Fetzer R. Thermochemically stable metal contacts on CdTe: Tungsten, molybdenum and tantalum, *Z. Metallk.*, **86**(5), 304–311 (1995b).

Crocker A.J. Phase equilibria in PbTe–CdTe alloys, *J. Mater. Sci.*, **3**(5), 534–539 (1968).

Datsenko A.M., Razvazhnoy E.M., Chashchin V.A. Investigation of interaction in the Bi_2Te_3–(Cd,Pb)(S,Se,Te) systems [in Russian], *Tr. Mosk. khim. tehnol. in-t*, (120), 32–34 (1981).

Demina T.V., Derid O.P. Phase diagram of the cadmium-gallium-tellurium system [in Russian], in: *Fizika slozhn. poluprovodn. soied*. Kishinev, Moldova: Shtiintsa, pp. 118–127 (1979).

Demina T.V., Markus M.M., Nateprov A.N., Railian V.Ya. Obtaining and some properties of $CdGa_2Te_4$ single crystals [in Russian], in: *Troin. poluprovodn. i ih primienenie: Tez. dokl*. Kishinev, Moldova: Shtiintsa, p. 141 (1976).

Demina T.V., Radautsan S.I. Growth of the $CdGa_2Te_4$ single crystals and their mechanical properties [in Russian], *Izv. AN MSSR. Ser. fiz.-tekhn. i math. n.*, (2), 62–63 (1984).

Derid O.P., Demina T.V., Markus M.M. Investigation of the Cd–Ga–Te ternary system [in Russian], *Izv. AN SSSR. Neorgan. materialy*, **10**(1), 18–20 (1974).

Derid O.P., Demina T.V., Markus M.M., Radautsan S.I. The $CdTe–Ga_2Te_3$ phase diagram [in Russian], *Izv. AN MSSR. Ser. fiz.-tehn. i mat. nauk*, (2), 56–59 (1976).

Derid O.P., Kosnichan I.G. Investigation of the phase transformations in the $CdTe–Cr_2Te_3$ system [in Russian], in: *Materialy dokl. na IX nauchno-tehn. konf. Kishin. polytehn. in-ta*. Kishinev, Moldova: Shtiints, p. 104 (1973).

Dichi E., Morgant G., Legendre B. Study of the ternary system cadmium–germanium–tellurium: Phase equilibria in the ternary system, *J. Alloys Compd.*, **217**(2), 193–199 (1995).

Diehl R., Nolaeng B.I. Dry oxidation of $Hg_{1-x}Cd_xTe$: Calculation of predominance area diagrams of the oxide phases, *J. Cryst. Growth*, **66**, 91–105 (1984).

Dub Ya.F., Ivanov-Omskiy V.I., Ogorodnikov V.K., Sidorchuk P.G., Epitaxial crystallization in the Te-HgTe-CdTe system, *Inorg. Mater. (Engl. Trans.)*, **19**(1), 48–50 (1983), transl. from *Izv. Akad. Nauk SSSR, Neorg. Mater.*, **19**(1), 59–61 (1983).

Dubrovin M.N., Agranat B.A., Pelevin O.V., Sokolov A.M. Physico-chemical investigation of the HgTe–CdTe system. II. Dissolution of the solid CdTe in the HgTe melt in the ultrasonic field [in Russian], *Nauch. tr./N.-i. i proekt. in-t redkomet. prom-sti*, **65**, 164–166 (1975).

Dwarakanadha R.Y., Sreekanatha R.D., Reddy B.K., Ravikumar R.V.S.S.N., Reddy D.R. dc-Magnetic susceptibility and EPR studies of vapor phase grown $Cd_{1-z}Co_xTe$ crystals, *J. Alloys Compd.*, **470**(1–2), 12–15 (2009).

Dyntu G.M. Investigation of liquidus surface of the $CdTe–In_2Te_3–Te$ system using mathematical planning of experiment [in Russian], in: *Kristallich. i stekloobr. poluprovodn*. Kishinev, Moldova: Shtiintsa, pp. 215–221 (1977).

Dyntu G.M., Derid O.P. Phase diagram of the Cd–In–Te system and conditions of solid solution formation based on $CdIn_2Te_4$ [in Russian], in: *Troin. poluprovodn. i ih primenenie: Tez. dokl*. Kishinev, Moldova: Shtiintsa, pp. 139–141 (1976).

Dyntu G.M., Derid O.P. Obtaining of regression equation for temperatures of onset of Cd-In-Te ternary alloy crystallization [in Russian], in: *Fiz. slozhnyh poluprovodn. soied*. Kishinev, Moldova: Shtiintsa, pp. 136–141 (1979).

Dziuba E.Z. Preparation of $Cd_xHg_{1-x}Te$ crystals by vertical zone melting, *J. Electrochem. Soc.*, **116**(1), 104–106 (1969).

Farrar R.A., Gillham C.J., Bartlett B., Quelch M. The pressure-temperature phase diagrams of the HgTe and $Hg_{1-x}Cd_xTe$ systems, *J. Mater. Sci.*, **12**(4), 836–838 (1977).

Feichuk P.I., Panchuk O.E., Savitski A.V. Defect structure of CdTe doping by In [in Russian], in: *Legir. poluprovodn.* Moscow, Russia: Nauka, pp. 72–75 (1982).

Feichuk P.I., Panchuk O.E., Shcherbak L.P., Antipov I.N. The CdTe–Tl quasibinary section [in Russian], *Izv. AN SSSR. Neorgan. materialy*, **13**(1), 164–165 (1977).

Fesh R.N., Panchuk O.E., Ivanchuk R.D., Savitski A.V. Investigation of the CdTe–Fe system [in Russian], *Izv. AN SSSR. Neorgan. materialy*, **13**(1), 166–167 (1977).

Fesh R.N., Panchuk O.E., Savitski A.V., Belotski D.P. Investigation of interaction in the CdTe–Co and CdTe–Mn systems [in Russian], *Izv. AN SSSR. Neorgan. materialy*, **14**(1), 170–171 (1978).

Fochuk P.M., Panchuk O.E., Belotski D.P., Kaz'miruk I.I. Diffusion and solubility of Ga in CdTe [in Russian], Deposited in UkrNIINTI 6.09.85, N 2090-Uk (1985).

Galchenko I.E., Pelevin O.V., Sokolov A.M. Determination of the vapor pressure of mercury over melts in the Hg–Cd–Te system, *Inorg. Mater.*, **20**(7), 952–955 (1984) transl. from *Izv. Akad. Nauk SSSR, Neorg. Mater.*, **20**(7),1103–1106 (1984).

Gambino M., Vassilev V., Bros J.P. Molar heat capacities of CdTe, HgTe and CdTe–HgTe in the solid state, *J. Alloys Compd.*, **176**(1), 13–24 (1991).

Gawel W., Sztuba Z., Wojakowska A., Zaleska E. Phase studies on the quasi-binary thallium(I) telluride—cadmium telluride system, *J. Phase. Equil.*, **22**(6), 656–660 (2001a).

Gawel W., Zaleska E., Sztuba Z., Sroka A. Electrochemical studies in the subsolidus area of the Tl_2Te–CdTe solid system, *Pol. J. Chem.*, **75**(10), 1553–1559 (2001b).

Gillham C.J., Farrar R.A. Precipitation in $Cd_xHg_{1-x}Te$, *J. Mater. Sci.*, **12**(10), 1994–2000 (1977).

Glazov V.M., Nagiev V.A., Nuriev R.S. Investigation and analysis of interaction in the quasibinary systems based on zinc, cadmium and mercury tellurides with germanium telluride [in Russian], *Izv. vuzov. Ser. Tsvetnaya metallurgia*, (5), 116–120 (1972).

Glazov V.M., Nagiev V.A., Nuriev R.S. Intermolecular interaction and thermodynamics of $GeTe-A^{II}Te$ melts [in Russian], in: *Termodinamicheskiye svoistva metallicheskih splavov.* Baku, Azerbaijan: Elm, pp. 376–379 (1975).

Glazov V.M., Nagiev V.A., Zargarova M.I. Investigation of phase equilibria in the $GeTe-A^{II}Te$ systems [in Russian], *Izv. AN SSSR. Neorgan. materialy*, **6**(3), 569–571 (1970).

Glazov V.M., Pavlova L.M. Determination of the saturation boundary of the solid solutions in the cadmium–mercury–tellurium alloys of various compositions along the quasibinary section [in Russian], *Dokl. RAN*, **342**(2), 189–192 (1995).

Glazov V.M., Pavlova L.M. Equilibrium of the intrinsic point defects and component solubility in the cadmium–mercury–tellurium alloys [in Russian], *Zhurn. fiz. khimii*, **70**(3), 479–484 (1996).

Glazov V.M., Pavlova L.M. The investigation and the thermodynamic analysis of volume properties in HgTe and CdHgTe melts, *Thermochim. Acta*, **314**(1–2), 265–273 (1998).

Golovey M.I., Olekseyuk I.D., Voroshilov Yu.V. Chemical bond and solubility in the systems $A^{II}_3B^V_2-A^{II}C^{VI}$ [in Russian], in: *Khim. svyaz' v poluprovodn. i polumetallah.* Minsk, Belarus: Nauka i Tekhnika Publishing, pp. 233–239 (1972).

Golovey M.I., Rigan M.Yu. Preparation of the Cd_3As_2-2CdTe(Se, S) solid solution single crystals, in: *Khimiya i physica khalcogenidov*. K.: Naukova dumka, 38–40 (1977).

Golovey M.I., Shpyrko G.N. Enthalpy of formation and heat capacity of solid solutions in the Cd_3As_2-2CdS (Se,Te) systems [in Russian], in: *Termodinamicheskie svoistva metallicheskih splavov*. Baku, Azerbaijan: Elm, pp. 321–323 (1975).

Goncharuk L.V., Sidorko V.R. Thermodynamic properties of some solid solutions formed by $A^{II}B^{IV}$ and $A^{II}B^{V}$ semiconductor compounds, *Powder Metall. Met. Cer.*, 35(7/8), 392–396 (1996), transl. from *Poroshk. metallurgiya*, (7–8), 79–84 (1996).

Goriunova N.A., Fedorova N.N. About isomorphism of compounds with covalent bond [in Russian], *Dokl. AN SSSR*, 90(6), 1039–1041 (1953).

Goriunova N.A., Golovey M.I., Olekseyuk I.D., Rigan M.Yu., Shved V.V. Investigation of the $(Cd_3P_2)_{1-x}$–$(CdTe)_{2x}$ and $(Zn_3P_2)_{1-x}$–$(CdTe)_{2x}$ systems [in Russian], *Izv. AN SSSR. Neorgan. materialy*, 6(7), 1272–1275 (1970).

Gromakov S.D., Zoroatskaya I.V., Latypov Z.M., Chvala M.A., Eidel'man E.A., Badygina L.I., Zaripova L.G. About investigation of phase diagrams of semiconductor systems [in Russian], *Zhurn. neorgan. khimii*, 9(10), 2485–2487 (1964).

Grytsiv V.I. Investigation of copper, silver and gold behavior in cadmium telluride [in Russian], *Kand. dis., Chernovtsy*, 141 p. (1975).

Grytsiv V.I., Panchuk O.E., Belotski D.P. Phase diagram of the CdTe–Au system [in Russian], *Izv. AN SSSR. Neorgan. materialy*, 14(7), 1348–1349 (1978).

Gumiński C. On the character of the equilibrium between solid Cd–Hg–Te alloys and their saturated amalgams, *J. Less-Common Metals*, 116(2), L15–L18 (1986).

Guseinov G.D. Searching and physical investigation of new semiconductors-analogues [in Russian], *Avtoref. dis. ... doct. fiz.-mat. nauk*. Baku, Azerbaijan, 82 p. (1969).

Guseinov F.H., Babanly M.B., Kuliev A.A. Phase equilibria in the Tl_2Te–Zn(Cd)Te systems [in Russian], *Izv. AN SSSR. Neorgan. materialy*, 17(1), 31–33 (1981a).

Guseinov F.H., Babanly M.B., Kuliev A.A. Phase equilibria in the Tl_2Te_3–Zn(Cd)Te systems [in Russian], *Izv. vuzov. Khimia i khim. tehnologia*, 24(10), 1245–1248 (1981b).

Guseinov F.H., Babanly M.B., Kuliev A.A. Phase equilibria in the TlX–Cd(Zn)X (X – S, Se, Te) systems [in Russian], *Izv. AN SSSR. Neorgan. materialy*, 18(5), 759–763 (1982).

Guseinov G.D., Guseinov G.G., Ismailov M.Z., Godzhaev E.M. About structure and physical properties of new $CdTlS_2$ (Se_2, Te_2) semiconductor compounds [in Russian], *Izv. AN SSSR. Neorgan. materialy*, 5(1), 33–39 (1969).

Hahn H., Frank G., Klingler W., Störger A.D., Störger G. Über ternäre Chalkogenide des Aluminiums, Galliums und Indiums mit Zink, Cadmium und Quecksilber, *Z. anorg. und allg. Chem.*, 279(5/6), 241–270 (1955).

Hansen G.L., Schmit J.L., Casselman T.N. Energy gap versus alloy composition and temperature in $Hg_{1-x}Cd_xTe$, *J. Appl. Phys.*, 53(10), 7099–7101 (1982).

Harman T.C. Liquidus isotherms, solidus lines and LPE growth in the Te-rich corner of the Hg–Cd–Te system, *J. Electron. Mater.*, 9(6), 945–961 (1980).

Herning P.E. Experimental determination of the mercury-rich corner of the Hg–Cd–Te phase diagram, *J. Electron. Mater.*, 13(1), 1–14 (1984).

Hirai T., Kurata K. The phase diagram of pseudo-binary system PbTe–CdTe and its thermoelectric properties, *Trans. Jap. Inst. Metals*, 9(4), 301–303 (1968).

Indenbaum G.V., Dvorkin Yu.V., Stolyarov O.G. Precise determination of the mercury telluride density on the homogeneity region boundary and theoretical density of the $Cd_xHg_{1-x}Te$ solid solutions [in Russian], *Nauch. tr. Mosk. in-t stali i splavov*, (146), 118–121 (1983).

Ivanov-Omskiy V.I., Kolomiets B.T., Mal'kova A.A. Optical and photoelectrical properties of HgTe and its alloys with CdTe [in Russian], _Fizika tverdogo tela_, **6**(5), 1457–1461 (1964).

Jianrong Y. Thermodynamic study of solid-vapour equilibrium in the $(Hg_{1-x}Cd_x)_{1-y}Te_y$ system, _J. Cryst. Growth_, **126**(4), 695–700 (1993).

Jimenéz-Sandoval S., López-López S., Chao B.S., Meléndez-Lira M. On the properties of $Cu_xCd_{1-x}Te$: A novel semiconductor alloy, _Thin Solid Films_, **342**(1–2), 1–3 (1999).

Kaldis E., Krausbauer L., Widmer R. Cd_4SiS_6 and Cd_4SiSe_6, new ternary compounds, _J. Electrochem. Soc._, **114**(10), 1074–1076 (1967).

Kalisher M.H., Herning P.E., Tung T. Hg-rich liquid-phase epitaxy of $Hg_{1-x}Cd_xTe$, _Progr. Cryst. Growth Charact._, **29**, 41–83 (1994).

Karimov S.K., Gafarov S., Sultonov S. Investigation of phase equilibria in the $Cd–Tl–C^{VI}$ (C^{VI} – Se, Te) systems [in Russian], _Izv. AN SSSR. Neorgan. materialy_, **17**(8), 1346–1349 (1981)

Kikuchi R. Theoretical calculation of Hg–Cd–Te liquidus-solidus phase diagram, _J. Vac. Sci. Technol._, **21**(1), 129–132 (1982).

Korovin A.P., Vanyukov A.V., Schastlivyi V.P., Kireev P.S. Investigation of the electrophysical properties of the $Cd_xHg_{1-x}Te$ alloy [in Russian], _Izv. AN SSSR. Neorgan. materialy_, **4**(12), 2094–2100 (1968).

Kot M.V., Tyrziu V.G., Simashkevich A.V., Maronchuk Yu.E., Mshenski V.A. Dependence of energy activation in thin films from molar composition for some $A^{II}B^{VI}–A^{II}B^{VI}$ systems [in Russian], _Fizika tverdogo tela_, **4**(6), 1535–1541 (1962).

Kozyrev S.P., Vodopyanov L.K., Triboulet R. Structural analysis of the semiconductor-semimetal alloy $Cd_{1-x}Hg_xTe$ by infrared lattice-vibration spectroscopy, _Phys. Rev. B: Condens. Matter_, **58**(3), 1374–1384 (1998).

Krotov I.I., Tokmakov V.V., Vanyukov A.V., Kolobrodova N.A. Solid-vapor equilibrium in the system Te–HgTe–CdTe, _Inorg. Mater._, **15**(9), 1214–1216 (1979), transl. from _Izv. Akad. Nauk SSSR, Neorg. Mater._, **15**(9), 1542–1546 (1979).

Kutsia N.M., Stavrianidis S.A. Solubility of zinc and cadmium tellurides in germanium telluride [in Russian], _Izv. AN SSSR. Neorgan. materialy_, **19**(8), 1302–1303 (1983).

Lakiza S.N., Olekseyuk I.D. Phase interaction in the Cd–As–Te system [in Russian], _Zhurn. neorgan. khimii_, **22**(7), 1925–1931 (1977).

Laugier A. Thermodynamics and phase diagram calculations in II–VI and IV–VI ternary systems using an associated solution model, _Rev. Phys. Appl._, **8**(3), 259–270 (1973).

Lawson W.D., Nielsen S., Putley E.H., Youmg A.S. Preparation and properties of HgTe and mixed crystals of HgTe–CdTe, _J. Phys. Chem. Solids_, **9**(3–4), 325–329 (1959).

Lesyna N.V., Dremlyuzhenko S.G., Koval'chuk M.L., Ivanits'ka V.G., Rudyk N.D. Physico-chemical interaction in the Cd–CdTe–Sb system [in Ukrainian], _Nauk. visnyk Cherniv. un-tu. Fizyka. Elektronika_, (201), 72–74 (2004).

Lesyna N.V., Dremlyuzhenko S.G., Rarenko I.M. CdTe–CdSb system [in Ukrainian], _Nauk. visnyk Cherniv. un-tu. Fizyka. Elektronika_, (66), 18–19 (1999).

Lesyna N.V., Dremlyuzhenko S.G., Stets'ko Yu.P. Physico-chemical investigations of the CdTe–Te–PbTe system [in Ukrainian], _Nauk. visnyk Cherniv. un-tu. Khimiya_, (42), 6–12 (1998).

Lesyna N.V., Dremlyuzhenko S.G., Yuriychuk I.M., Ivanyts'ka V.G. Cd–Ga–CdTe system [in Ukrainian], _Voprosy khimii i khim. tekhnol._, (6), 96–99 (2007).

Leute V. Thermodynamics of solid solutions with ordering tendencies, *Calphad*, **20**(4), 407–418 (1996).

Leute V., Bolwin H. Ordering and demixing in the system $In_2Te_3/Ga_2Te_3/Cd_3Te_3$, *Solid State Ionics*, **141–142**, 279–287 (2001).

Leute V., Menge D. Thermodynamic investigations of the quasibinary systems (Cd_kSn_{1-k}) Te, $(Cd_kSn_{1-k})Se$ and $Sn(Se_lTe_{1-l})$, *Z. Phys. Chem. (Munchen)*, **176**(1), 47–64 (1992).

Litvak A.M., Charykov N.A. Thermodynamic modeling in condensed A_2B_6 phases. Model of fully associated solutions, *J. Appl. Chem. USSR (Engl. Transl.)*, **64**(8), 1488–1495 (1991), transl. from *Zh. Prikl. Khimii*, **64**(8), 1633–1640 (1991).

Liu H., Brebrick R.F. Experimental determination of tie-lines in the Hg–Cd–Te system, *J. Electron. Mater.*, **24**(10), 1377–1380 (1995).

Lusson A., Triboulet R. Liquid phase epitaxy of $Cd_xHg_{1-x}Te$ $(0,5 < x < 1)$ and phase diagram determination, *J. Cryst. Growth*, **85**(3), 503–509 (1987).

Lutsiv R.V., Pashkovskiy M.V., Svekolkina L.G., Sus' B.A., Zhukov G.A. Annealing influence on the electrical properties of the $Cd_xHg_{1-x}Te$ single crystals [in Russian], *Izv. AN SSSR. Neorgan. materialy*, **4**(5), 778–779 (1968).

Malzbender J., Jones E.D., Mullin J.B., Shaw N. Interaction between iodine and CdTe in closed tube diffusions, *J. Mater. Sci. Mater. Electron.*, **5**(6), 352–355 (1994).

Mamontov M.N. Thermodynamic properties of $Cd_{1-x}Mn_xTe$ solid solutions [in Russian], *Zhurn. fiz. khimii*, **71**(1), 9–14 (1997).

Mamontov M.N., Sevastianova L.G. Phase relations in the CdTe–MnTe–Te ternary system at 643 K [in Russian], *Neorgan. materialy*, **32**(7), 810–815 (1996).

Marbeuf A., Druilhe R., Triboulet R., Patriarche G. Thermodynamic analysis of Zn–Cd–Te, Zn–Hg–Te and Cd–Hg–Te: Phase separation in $Zn_xCd_{1-x}Te$ and $Zn_xHg_{1-x}Te$, *J. Cryst. Growth*, **117**(1–4), 10–15 (1992).

Marbeuf A., Ferah M., Janik E., Heurtel A. Consistent approach of II-VI equilibrium phase diagrams: Application to Zn–Se, Cd–Te, Hg–Te and Cd–Hg–Te systems, *J. Cryst. Growth*, **72**(1–2), 126–132 (1985).

Markovskii L.Ya., Pron' G.F. Synthesis and some properties of the zinc, cadmium and mercury tellurites [in Russian], *Zhurn. neorgan. khimii*, **13**(10), 2640–2644 (1968).

Marugin V.V., Odin I.N., Novoselova A.V., Sher A.V. Physico-chemical investigation of the Cd–Bi–Te system [in Russian], *Zhurn. neorgan. khimii*, **29**(6), 1566–1570 (1984).

Mason D.R., O'Kane D.F. Preparation and properties of some semiconducting compound, in: *Proc. Intern. Conf. Semicond. Phys.* Prague, Czech Republic, pp. 1026–1031 (1960).

Mavlonov Sh., Sherov P.N., Gulamova F. X-ray analysis of the ingots in the Sb_2Te_3–CdTe system [in Russian], *Izv. AN SSSR. Neorgan. materialy*, **26**(5), 1095–1099 (1990).

Mavlonov Sh., Sultonov S., Kurbanov H.M., Ahmedov C., Husainov B., Karimov S. Obtaining, physico-chemical and electrical properties of some $A^{II}B_2^{III}C_4^{VI}$ compounds [in Russian], in: *Troinyie poluprovodn. i ih primenienie: Tez. dokl.*, Kishinev, Moldova: Shtiinza, p. 143 (1976).

Mayer H.W., Mikhail I., Schubert K. Über einige phasen der Mischungen $ZnSb_N$ und $CdSb_N$, *J. Less-Common Metals*, **59**(1), 43–52 (1978).

Medvedev Yu.V., Berchenko N.N. Pecularities of the interfaces between solid solutions based on mercury chalcogenides and native oxides [in Russian], *Zhurn. neorgan. khimii*, **38**(12), 1940–1945 (1993).

Meschter P.J., Owens K.E., Tung T. Determination of liquidus temperatures of Hg-rich Hg–Cd–Te alloys by differential thermal analysis, *J. Electron. Mater.*, **14**(1), 33–37 (1985).

Morgant G., Legendre B., Souleau C. Étude des systèmes ternaire cadmium–tellure–éléments du groupe IV$_B$. I. Le système ternaire cadmium–plomb–tellure, *Bull. Soc. chim. Fr.*, Pt. I, (3–4), 133–136 (1980).

Morgant G., Legendre B., Souleau C. Étude des systèmes ternaires cadmium–tellure–éléments du groupe IVB. II. Le système ternaire cadmium-étain-tellure, *Bull. Soc. chim. France*, Pt.1 (1–2), 141–144 (1981).

Moskvin P., Khodakovsky V., Rashkovetskiy L., Stronski A. Polyassociative model of A^{2}B^6 semiconductor melt and p-T-x equilibria in Cd–Hg–Te system, *J. Cryst. Growth*, **310**(10), 2617–2626 (2008).

Mroczkowski J.A., Vydyanath H.R. Liquid phase epitaxial growth of $(Hg_{1-x}Cd_x)Te$ from tellurium-rich solutions using a closed tube tipping technique, *J. Electrochem. Soc.*, **128**(3), 655–661 (1981).

Nasar A., Shamsuddin M. Thermodynamic investigations of CdTe-HgTe solid solutions, *J. Less-Common Met.*, **171**(1), 83–93 (1991).

Nasirov Ya.N., Zargarova M.I., Sultanova N.R. Thermoelectric properties of the alloys of SnTe–CdTe system [in Russian], *Izv. AN SSSR. Neorgan. materialy*, **6**(9), 1711–1712 (1970).

Nikolič P.M. Solid solutions of CdSe and CdTe in PbTe and their optical properties, *Brit. J. Appl. Phys.*, **17**(3), 341–344 (1966).

Novoselova A.V., Sher A.A., Odin I.N. Heterogeneous equilibria and thermoelectrical properties of the ingots in the CdTe–Sb$_2$Te$_3$, CdTe–Bi$_2$Te$_3$ systems [in Russian], *Zhurn. neorgan. khimii*, **26**(4), 1048–1051 (1981).

Odin I.N. T-x-y diagrams for reciprocal systems PbX + CdI$_2$ = CdX + PbI$_2$ (X = S, Se, Te) [in Russian], *Zhurn. neorgan. khimii*, **46**(10), 1733–1738 (2001).

Odin I.N., Chukichev M.V., Gapanovich M.V., Kozlovskiy V.F., Nurtazin А.А., Novikov G.F. Luminescent properties of the solid solutions in the Cd–In–Te system [in Russian], *Neorgan. materially*, **45**(7), 799–805 (2009).

Odin I.N., Chukichev M.V., Gapanovich M.V., Novikov G.F. Luminescent properties of the solid solutions based on cadmium telluride in the CdTe–Ga$_2$Te$_3$, CdTe–GaTe systems [in Russian], *Neorgan. materialy*, **47**(6), 662–665 (2011).

Odin I.N., Chukichev M.V., Rubina M.E. Phase diagram and luminescent properties of the solid solutions of the CdTe–MnTe system [in Russian], *Neorgan. materialy*, **39**(4), 425–428 (2003).

Odin I.N., Grin'ko V.V., Safronov E.V., Kozlovskiy V.F. Investigation of the CdX–B$_2$X$_3$–X (X = S, Se), CdTe–B–Te systems [in Russian], *Zhurn. neorgan. khimii*, **46**(7), 1210–1214 (2001).

Odin I.N., Ivanov V.A. Liquidus surface of the Cd-Si-Te system [in Russian], *Zhurn. neorgan. khimii*, **36**(7), 1837–1841 (1991).

O'Kane D.F., Mason D.R. Semiconducting properties of peritectic compounds from the pseudo-binary system of CdTe–In$_2$Te$_3$, *J. Electrochem. Soc.*, **110**(11), 1132–1136 (1963).

Olekseyuk I.D., Golovey M.I., Rigan M.Yu., Voroshilov Yu.V., Gurzan M.I. Investigation of the Cd$_3$As$_2$–2CdTe and Cd$_3$P$_2$–2CdSe systems [in Russian], *Izv. AN SSSR. Neorgan. materialy*, **8**(4), 696–700 (1972).

Olekseyuk I.D., Hudoliy V.A., Lakiza S.N., Golovey M.I. The Cd–As–Te system [in Russian], *Zhurn. neorgan. khimii*, **21**(6), 1558–1564 (1976).

Pajaczkowska A. Physicochemical properties and crystal growth of A^{II}B^{VI}–MnB^{VI} systems, *Progr. Crystal Growth Caract.*, **1**(3), 289–326 (1978).

Panchuk O.E., Belotski D.P., Grytsiv V.I., Tomashik V.N., Antipov I.N. Investigation of phase interaction in the CdTe–Cu(Ag) systems [in Russian], *Ref. inform. o zakonch. n.-i. rabotah v vuzah USSR*, (14), 51–52 (1973a).

Panchuk O.E., Belotski D.P., Grytsiv V.I., Tomashik V.N. Investigation of the CdTe–Cu system [in Russian], *Izv. AN SSSR. Neorgan. materialy*, **10**(10), 1892–1894 (1974a).

Panchuk O.E., Feichuk P.I., Panchuk I.E. Investigation of CdTe–In system [in Russian], *Izv. AN SSSR. Neorgan. materialy*, **9**(8), 1437–1439 (1973c).

Panchuk O.E., Feichuk P.I., Shcherbak L.P. Electrical conductivity of In doping CdTe at low temperatures [in Russian], *Izv. AN SSSR. Neorgan. materialy*, **19**(3), 362–365 (1983).

Panchuk O.E., Feichuk P.I., Shcherbak L.P., Fochuk P.M., Boychuk R.F. Distribution of Ga in CdTe [in Russian], *Izv. AN SSSR. Neorgan. materialy*, **21**(7), 1118–1120 (1985).

Panchuk O.E., Feichuk P.I., Shcherbak L.P. Thallium doping of CdTe [in Russian], *Izv. AN SSSR. Neorgan. materialy*, **22**(10), 1642–1645 (1986).

Panchuk O.E., Fesh R.N., Savitski A.V., Shcherbak L.P. Investigation of Fe diffusion and solubility in CdTe [in Russian], *Izv. AN SSSR. Neorgan. materialy*, **17**(8), 1354–1356 (1981).

Panchuk O.E., Grytsiv V.I., Belotski D.P. Solubility of Ag in n-CdTe [in Russian], *Izv. AN SSSR. Neorgan. materialy*, **10**(4), 581–584 (1974c).

Panchuk O.E., Panchuk I.E., Belotski D.P., Grytsiv V.I. Investigation of the CdTe–Ag section [in Russian], *Izv. AN SSSR. Neorgan. materialy*, **10**(6), 980–982 (1974b).

Panchuk O.E., Shcherbak L.P. Solubility and diffusion of Ge in CdTe [in Russian], *Izv. AN SSSR. Neorgan. materialy*, **15**(8), 1339–1343 (1979).

Panchuk O.E., Shcherbak L.P., Feichuk P.I., Savitski A.V., Belotski D.P. Amphoteric behavior of Sn in CdTe [in Russian], *Izv. AN SSSR. Neorgan. materialy*, **14**(1), 50–53 (1978).

Panchuk O.E., Shcherbak L.P., Fesh R.N. The CdTe–Ge section of the Cd–Ge–Te ternary system [in Russian], *Izv. AN SSSR. Neorgan. materialy*, **12**(6), 1035–1038 (1976).

Panchuk O.E., Shcherbak L.P., Panchuk I.E., Antipov I.N. The cadmium telluride–tin system [in Russian], *Izv. AN SSSR. Neorgan. materialy*, **9**(4), 572–574 (1973b).

Parasyuk O., Piskach L., Morenko A., Halka V., Marchuk O. Phase equilibria in the $Cu_{2-x}Te(Se)$–Cd(Hg)Te(Se) quasibinary systems [in Ukrainian], *Nauk. visnyk Volyn. derzh. un-tu*, (4), 35–38 (1997).

Patrick R.S., Chen A.-B., Sher A., Berding M.A. Phase diagrams and microscopic structures of (Hg,Cd)Te, (Hg,Zn)Te and (Cd,Zn)Te alloys, *J. Vac. Sci. Technol.*, **6**(4), 2643–2649 (1988).

Pentin I.V., Grosheva A.A., Kozhemyakina N.V. The miscibility gap in cadmium, mercury and zinc telluride systems: Theoretical description, *Calphad*, **30**(2), 191–195 (2006).

Polistanski G.D., Ivanov Yu.M., Vaniukov A.V., Surkov E.P. Investigation of solubility and cadmium telluride growth in the salt melts [in Russian], Deposited in "Tsvetmetinformatsia", № 442Dep. M. (1978).

Pron' G.F., Markovskii L.Ya. Heats of formation of the zinc, cadmium and mercury tellurites [in Russian], *Zhurn. neorgan. khimii*, **14**(4), 880–882 (1969).

Quadri S.B., Skelton E.F., Webb A.W., Dinan J. High pressure studies of $Hg_{0.8}Cd_{0.2}Te$, *J. Vac. Sci. Technol.*, **A4**(4), 1974–1976 (1986).

Quenez P., Khodadad P. Étude du système GeTe–CdTe, *Bull. Soc. Chim. France*, (1), 3–5 (1969).

Quintero M., Guerrero E., Tovar R., Morocoima M., Grima P., Cadenas R. Temperature variation of lattice parameter and optical energy gap values of the compounds $CdIn_2Te_4$ and $MnIn_2Te_4$, *J. Phys. Chem. Solids*, **57**(3), 271–276 (1996).

Radautsan S.I., Derid O.P., Dyntu G.M. Construction of liquidus surface for Cd–In–Te ternary system using mathematical planning of experiment [in Russian], *Izv. AN LatvSSR. Ser. fiz. i tehn. n.*, (6), 15–17 (1982).

Radautsan S.I., Derid O.P., Dyntu G.M., Markus M.M. Phase interactions in the CdTe–In–InTe ternary system [in Russian], *Izv. AN MSSR. Ser. fiz.-tehn. i mat. nauk*, (3), 37–43 (1975).

Radautsan S.I., Ivanova R.A. Formation of the solid solutions based on $A^{II}B^{IV}C_3^{VI}$ compounds [in Russian], *Izv. AN MSSR*, **10**(88), 64–70 (1961).

Ray B., Spencer P.M. Phase diagram of the alloy system HgTe–CdTe, *Phys. Status Solidi*, **22**(2), 371–372 (1967).

Rhiger D.R., Kvaas R.E. Solid state quaternary phase equilibrium diagram for the Hg-Cd-Te-O system, *J. Vac. Sci. Technol.*, **A1**(3), 1712–1718 (1983).

Rigan M.Yu., Golovey M.I., Alekseenko G.F. The $(Cd_3P_2)_{1-x}-(2CdTe)_x$ phase diagram [in Russian], *Izv. vuzov. Ser. Khimia i khim. tehnologia*, **17**(12), 1865–1866 (1974).

Robertson D.S., Shaw N., Young I.M. A study of crystals in the cadmium oxide-tellurium dioxide system, *J. Mater. Sci.*, **13**(9), 1986–1990 (1978).

Rodionov Yu.I., Klokman V.R., Miakishev K.G. Solubility of $A^{II}B^{VI}$, $A^{IV}B^{VI}$ and $A^V B^{VI}$ semiconductors compounds in the melts of halogenides [in Russian], *Zhurn. neorgan. khimii*, **17**(3), 846–851 (1972).

Rogacheva E.I., Zhigareva N.K., Ivanova A.B. Solid solutions based on PbTe in the PbTe–CdTe system [in Russian], *Izv. AN SSSR. Neorgan. materialy*, **24**(10), 1629–1633 (1988).

Rosenberg A.J., Grierson R., Woolley J.C., Nikolič P. Solid solutions of CdTe and InTe in PbTe and SnTe. I. Crystal chemistry, *Trans. Metallurg. Soc. AIME*, **230**(2), 342–350 (1964).

Rubenstein M. Solubilities of some II–VI compounds in bismuth, *J. Electrochem. Soc.*, **113**(6), 623–624 (1966).

Rugg B.C., Silk N.J., Bryant A.W., Argent B.B. Calorimetric measurements of the enthalpies of formation and of mixing of II/VI and III/V compounds, *Calphad*, **19**(3), 389–398 (1995).

Rusnak N.I., Vlasiuk V.I., Beysiuk P.P. Concentration dependence of energy gap of $Cd_{1-x}Mg_xTe$ ($0 \leq x \leq 0.3$) semiconductor solid solutions [in Russian], *Elektronnaya tehnika. Ser. 6. Materialy*, **2**(247), 69–71 (1990).

Safarov M.G., Gamidov R.S., Poladov P.M., Bagirova E.M. The Sb_2Te_3–CdTe system [in Russian], *Zhurn. neorgan. khimii*, **36**(6), 1580–1584 (1991).

Sang W.-B., Wu W.-H. Thermodynamic investigation of equilibrium partial pressures over $Hg_{1-x}Cd_xTe$ melts [in Chinese], *Acta Chim. Sin.*, **54**(12), 1151–1158 (1996).

Sanz-Maudes J., Sangrador J., Rodriguez T. Numerical description of the phase diagram of the Hg–Cd–Te system in the Te-rich corner, *J. Cryst. Growth*, **102**(4), 1065–1068 (1990).

Saraie J., Kitagawa M., Ishida M., Tanaka T. Liquid phase epitaxial growth of CdTe in the CdTe–$CdCl_2$ system, *J. Cryst. Growth*, **43**(1), 13–16 (1978).

Schmit J.L. Growth, properties and applications of HgCdTe, *J. Cryst. Growth*, **65**(1–3), 249–261 (1983).

Schmit J.L. Speerschneider C.J. Phase diagram of $Hg_{1-x}Cd_xTe$, *Infrared Phys.*, **8**(3), 247–253 (1968).

Schwer H., Krämer V. Neue Überstrukturen von AB_2Z_4-Defekt-Tetraheder-Verbindungen, *Z. Kristallogr.*, **182**(1–4), 245–246 (1988).

Shamsuddin M., Nasar A. On the thermodynamic behavior of cadmium in Te-saturated HgTe–CdTe and CdSe–CdTe solid alloys, *Metall. Mater. Trans. B*, **26**(6), 569–575 (1995).

Shcherbak L., Feichouk P., Panchouk O., Lopatniouk I., Vengrenovych R. Phase equilibria in the CdTe–Ge system, *Calphad*, **21**(4), 463–468 (1997).

Shneider A.D., Gavrishchak I.V. Structure and properties of the HgTe–CdTe system [in Russian], *Fizika tverdogo tela*, **2**(9), 2079–2081 (1960).

Shuh D.W., Williams R.S. Summary abstract: Ternary solid phase equilibria in the systems (Ag, In, Au)–(Cd, Hg)–Te, *J. Vac. Sci. Technol.*, **A6**(3), Pt. II, 1564–1565 (1988).

Sidorko V.R., Goncharuk L.V. Thermodynamic properties of the solid solutions ZnTe–CdTe, ZnTe–HgTe and CdTe–HgTe, *J. Alloys Compd.*, **228**(1), 13–15 (1995).

Skauli T., Colin T. Accurate determination of the lattice constant of molecular beam epitaxial CdHgTe, *J. Cryst. Growth*, **222**(4), 719–725 (2001).

Slack G.A., Galginaitis S. Thermal conductivity and phonon scattering by magnetic impurities in CdTe, *Phys. Rev. A*, **133**(1), 253–268 (1964).

Sporken R., Sivananthan S., Reno J., Faurie J.P. X-Ray photoemission study of Hg clusters on $Hg_{1-x}Cd_xTe$ surfaces, *J. Vac. Sci. Technol. B*, **6**(4), 1204–1207 (1988).

Steininger J. Hg–Cd–Te phase diagram determination by high pressure reflux, *J. Electron. Mater.*, **5**(3), 299–320 (1976).

Su C-H. Heat capacity, enthalpy of mixing and thermal conductivity of $Hg_{1-x}Cd_xTe$ pseudobinary melts, *J. Cryst. Growth*, **78**(1), 51–57 (1986).

Sultonov C., Karimov S.K. Some electrophysical properties of $CdTl_2Se_4$ and $CdTl_2Te_4$ ternary chalcogenides [in Russian], *Zhurn. neorgan. khimii*, **15**(7), 1191–1193 (1979).

Szofran F.R., Lehoczky S.L. Liquidus temperature of Hg-rich Hg–Cd–Te alloys, *J. Electron. Mater.*, **12**(4), 713–717 (1983).

Tai H., Hori S. Equilibrium phase diagrams for the CdTe-Sb and CdTe-Bi systems [in Japanese], *J. Jap. Inst. Metals*, **34**(8), 843–846 (1970).

Tai H., Hori S. Equilibrium phase diagrams for the CdTe–Sn and CdTe–Pb systems [in Japanese], *J. Jap. Inst. Metals*, **38**(5), 451–455 (1974).

Tai H., Hori S. Equilibrium phase diagrams of the $CdTe–CdCl_2$ and $CdTe–CdBr_2$ systems [in Japanese], *J. Jap. Inst. Metals*, **40**(7), 722–725 (1976).

Tai H., Narita Y., Nakamae M., Hori S. Equilibrium phase diagram for the CdTe–In system and liquid phase epitaxial growth of CdTe from solution [in Japanese], *J. Jap. Inst. Metals*, **44**(3), 276–282 (1980).

Tairov S.M., Ormont B.F., Gartsman K.G. The SnTe–CdTe system [in Russian], *Izv. AN SSSR. Neorgan. materialy*, **5**(12), 2215–2216 (1969).

Thomassen L., Mason D.R., Rose G.D., Sarace J.C., Schmitt G.A. The phase diagram for the pseudo-binary system $CdTe–In_2Te_3$, *J. Electrochem. Soc.* **110**(11), 1127–1131 (1963).

Tomashik Z.F., Tomashik V.N. The PbTe–CdTe system [in Russian], *Izv. AN SSSR. Neorgan. materialy*, **18**(12), 1994–1996 (1982).

Tomson A.S., Men'shenina N.F., Vanyukov A.V. Equilibrium in the CdTe–HgTe–Te system [in Russian], *Izv. AN SSSR. Neorgan. materialy*, **17**(8), 1492–1493 (1981).

Triboulet R., Didier G. Growth and characterisation of $Cd_{1-x}Mn_xTe$ and MnTe crystals. Determination of the CdTe–MnTe pseudobinary phase diagram, in: *6th Mezhdunar.konf. po rosru kristallov*. M.: Rasshir. tez. V. 3. Rost iz rasplavov i vysokotemperaturnyh rastvorov. M., 229–231 (1980).

Triboulet R., Didier G. Growth and characterization of $Cd_{1-x}Mn_xTe$ and MnTe crystals; contribution to the CdTe–MnTe pseudo-binary phase diagram determination, *J. Cryst. Growth*, **52**(1), Pt. 2, 614–618 (1981).

Trishchuk L.I. Growth of the zinc and cadmium tellurides single crystals by the crystallization from the solution-melt [in Ukrainian], *Optoelektronika i poluprovodn. tekhnika*, (44), 99–106 (2009).

Trishchuk L.I., Oleinik G.S., Mizetskaya I.B. Phase equilibria in the $Cu_{2-x}Te$–ZnTe and $Cu_{2-x}Te$–CdTe systems [in Russian], *Izv. AN SSSR. Neorgan. materialy*, **20**(9), 1486–1489 (1984).

Trishchuk L.I., Oleinik G.S., Mizetskaya I.B. Thermal analysis determination of $A^{II}B^{VI}$ solid solubility in $A^I_2B^{VI}$, *Thermochim. Acta*, **92**, 611–613 (1985).

Trishchuk L.I., Oleinik G.S., Mizetskaya I.B. Physico-chemical investigation of interaction in the Ag_2Te–ZnTe and Ag_2Te–CdTe systems [in Russian], *Ukr. khim. zhurn.*, **52**(8), 799–803 (1986).

Trishchuk L.I., Oliynyk G.S., Tomashyk V.M. Phase diagrams of the $Cu_{1.95}(Ag_2)Te$–ZnTe(CdTe) quasibinary systems and liquidus surfaces of the $Cu_{1.95}(Ag_2)Te$–ZnTe–CdTe quasiternary systems, *Chem. Met. Alloys*, **1**(1), 58–61 (2008).

Tung T., Su C-H., Liao P-K., Brebrick R.E. Measurement and analysis of the phase diagram and thermodynamic properties in the Hg–Cd–Te system, *J. Vac. Sci. Technol.*, **21**(1), 117–124 (1982).

Ueda R., Ohtsuki O., Ueda I. Crystal growth of $Hg_{1-x}Cd_xTe$ using Te as a solvent, *J. Cryst. Growth*, **13/14**(5), 668–671 (1972).

Vanyukov A.V., Krotov I.I., Yermakov A.I. Investigation of CdTe and solid solutions of $Cd_xHg_{1-x}Te$ solubility in Hg, *Inorg. Mater. (Engl. Trans.)*, **13**(5), 667–671 (1977), transl. from *Izv. Akad. Nauk SSSR, Neorg. Mater.*, **13**(5), 815–819 (1977).

Vanyukov A.V., Pedos S.I., Yukhtanov E.D., Indenbaum G.V., Figel'son Yu.A. Liquidus isotherms in the mercury corner of the Hg–Cd–Te system [in Russian], in: *Polumetally i poluprovodniki s uzkimi zapreshchennymi zonami*. L'vov, Ukraine: L'vov. un-t, pp. 10–15 (1973).

Vasilyev V.P., Kholina E.N., Mamontov M.N., Morozova V.V., Bykov M.A., Khazieva R.A., Kondrakov S.V. Partial thermodynamic functions of the solid solution forming in the pseudobinary system CdTe–HgTe [in Russian], *Izv. AN SSSR. Neorgan. materialy*, **26**(8), 1632–1634 (1990b).

Vasilyev V.P., Mamontov M.N., Bykov M.A. Thermodynamic properties and stability of the solid solutions in the CdTe–HgTe–Te system [in Russian], *Vestn. MGU. Ser. 2*, **31**(3), 211–218 (1990a).

Vassilev V., Karadashka I., Parvanov S. New chalcogenide glasses in the Ag_2Te–As_2Se_3–CdTe system, *J. Phys. Chem. Solids*, **69**(7), 1835–1840 (2008).

Vengel' P.F., Tomashik V.N. Interaction of mercury telluride with cadmium [in Russian], *Izv. AN SSSR. Neorgan. materialy*, **24**(10), 1626–1628 (1988).

Vishniakov A.V., Dvoretskov G.A., Zubkovskaya V.N., Tyurin O.A., Kovtunenko P.V. Phase equilibria in the systems formed by the II-VI compounds and the elements of I and III groups of the periodic system [in Russian], *Tr. Mosk. khim.-tekhnol. in-t*, (120), 87–103 (1981).

Vishniakov A.V., Kukleva T.V., Al'tah O.L., Zubkovskaya V.N., Kovtunenko P.V. The solid state solubility of cadmium chalcogenides in copper(I) chalcogenides [in Russian], *Zhurn. neorgan. khimii*, **25**(5), 1358–1361 (1980).

Vishniakov A.V., Kukleva T.V., Kovtunenko P.V. Liquidus thermodynamic analysis in the CdCl₂–CdX (X=S, Se, Te) systems [in Russian], Deposited in VINITI, № 2289—78 Dep (1978).

Vishniakov A.V., Tiurin O.A., Kovtunenko P.V. Solubility of Ag in the cadmium chalcogenides [in Russian], *Zhurn. neorgan. khimii*, **28**(5), 1274–1280 (1983).

Vishniakov A.V., Zubkovskaya V.N., Kukleva T.V. Solubility of copper in cadmium telluride [in Russian], *Izv. AN SSSR. Neorgan. materialy*, **25**(4), 578–581 (1989).

Voronin G.F., Pentin I.V. Decomposition of the solid solutions based on cadmium, mercury and zinc tellurides [in Russian], *Zhurn, fiz. khimii*, **79**(10), 1771–1778 (2005).

Voronin G.F., Pentin I.V., Vassilev V.P. Thermodynamic stability conditions of the solid solution in the cadmium telluride—mercury telluride system [in Russian], *Zhurn. fiz. khimii*, **77**(12), 2119–2125 (2003).

Vul B.M., Ivanov V.S., Rukavishnikov V.A., Sal'man V.M., Chapnin V.A. Properties of cadmium telluride doping with iron [in Russian], *Fiz. i tehn. poluprovodn.*, **6**(7), 1264–1267 (1972).

Wei S.-H., Ferreira L.G., Zunger A. First-principles calculation of temperature-composition phase diagrams of semiconductor alloys, *Phys. Rev. B*, **41**(12), 8240–8269 (1990).

Weil M. New phases in the systems Ca-Te-O and Cd-Te-O: The calcium tellurite(IV) Ca₄Te₅O₁₄, and the cadmium compounds Cd₂Te₃O₉ and Cd₂Te₂O₇ with mixed-valent oxotellurium(IV/VI) anions, *Solid State Sci.*, **6**(1), 29–37 (2004)

Weitze D., Leute V. The phase diagrams of the quasibinary systems HgTe/In₂Te₃ and CdTe/In₂Te₃, *J. Alloys Compd.*, **236**(1–2), 229–235 (1996).

Wermke A., Boeck T., Gobel T., Jacobs K. Thermodynamic investigation on the liquid phase epitaxy of Hg₁₋ₓCdₓTe layers, *J. Cryst. Growth*, **121**(4), 571–578 (1992).

Wiedemeier H., Chen K.-T. Observation of phase separation in Hg₁₋ₓCdₓTe solid solutions by low incident angle x-ray diffraction, *J. Electron. Mater.*, **23**(9), 963–968 (1994).

Woolley J.C., Ray B. Solid solution in AᴵᴵBᵛᴵ tellurides, *J. Phys. Chem. Solids*, **13**(1/2), 151–153 (1960a).

Woolley J.C., Ray B. Effects of solid solutions of Ga₂Te₃ with AᴵᴵBᵛᴵ tellurides, *J. Phys. Chem. Solids*, **16**(1/2), 102–106 (1960b).

Woolley J.C., Ray B. Effects of solid solutions of In₂Te₃ with AᴵᴵBᵛᴵ tellurides, *J. Phys. Chem. Solids*, **15**(1/2), 27–32 (1960c).

Yang J., Silk N.J., Watson A., Bryant A.W., Chart T.G., Argent B.B. The thermodynamics and phase diagrams of the Cd–Hg and Cd–Hg–Te systems, *Calphad*, **19**(3), 415–430 (1995).

Yokozawa M., Kato H., Takayanagi S. Behaviour of In in CdTe crystals [in Japanese], *Denki kagaku*, **34**(10), 828–833 (1966).

Yu T.C., Brebrick R.F. The Hg–Cd–Zn–Te phase diagram, *J. Phase Equilib.*, **13**(5), 476–496 (1992).

Zul'fugarly Dzh.I., Agaev A.B., Azadaliev R.A. Investigation of chemical correlation in the Nd₂Te₃–CdTe system [in Azerbaijani], *Azerb. khim. zhurn.*, (3), 103–105 (1982).

Zul'fugarly Dzh.I., Agaev A.B., Azadaliev R.A., Alieva U.R. Obtaining and investigation of some electrophysical properties of solid solutions of CdTe–Gd₂Te₃ system [in Russian], *Dokl. AN AzSSR*, **40**(10), 48–50 (1984).

7

Systems Based on HgS

7.1 Mercury–Lithium–Sulfur

HgS–Li₂S: The phase diagram is not constructed. $Li_2Hg_6S_7$ could not be prepared under the synthetic flux conditions (Axtell et al. 1998).

7.2 Mercury–Sodium–Sulfur

HgS–Na₂S: The phase diagram is not constructed. $Na_2Hg_3S_4$ was obtained by the interaction of stoichiometric quantity of Na_2S and HgS at 600°C (Klepp 1992). It crystallizes in an orthorhombic structure with lattice parameters $a = 730.8 \pm 0.5$, $b = 1401.1 \pm 0.8$, and $c = 823.1 \pm 0.3$ pm.

HgS–Na₂S₄: The phase diagram is a eutectic type (Figure 7.1) (Garner and White 1970). The eutectic composition and temperature are 2 mol. % HgS and 250°C.

This system was investigated using x-ray diffraction (XRD) and optical analysis.

7.3 Mercury–Potassium–Sulfur

HgS–K₂S: The phase diagram is not constructed. $K_2Hg_3S_4$, $K_2Hg_6S_7$, and K_6HgS_4 ternary compounds are formed in this system (Sommer and Hoppe 1978, Kanarzidis and Park 1990, Axtell et al. 1998). The first of them crystallizes in an orthorhombic structure with lattice parameters $a = 1056.1 \pm 0.5$, $b = 653.4 \pm 0.3$, and $c = 1370.6 \pm 0.2$ pm and calculation density $5.68 \, g \, cm^{-3}$; the second melts congruently at 556 ± 10°C and crystallizes in a tetragonal structure

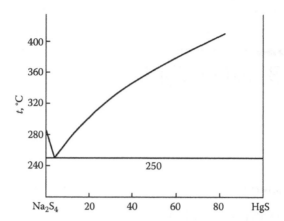

FIGURE 7.1
HgS–Na$_2$S$_4$ phase diagram. (From Garner, W.R. and White, W.B., *J. Cryst. Growth*, 7(3), 343, 1970.)

with lattice parameters $a = 1380.5 \pm 0.8$ and $c = 408.0 \pm 0.3$ pm and calculation density 6.43 g cm^{-3} and energy gap $E_g = 1.51$ eV (Kanarzidis and Park 1990, Axtell et al. 1998); and the third crystallizes in a hexagonal structure with lattice parameters $a = 998.5$ and $c = 765.2$ pm and experimental density 2.99 g cm^{-3} (Sommer and Hoppe 1978).

This system was investigated using differential thermal analysis (DTA) and XRD (Axtell et al. 1998). K$_2$Hg$_3$S$_4$ (K$_2$Hg$_6$S$_7$) was synthesized by the heating of K$_2$S + HgS + S mixtures under a vacuum at 220°C (370°C) for 99 h followed by cooling to 50°C at a rate of 2°C h^{-1} (Axtell et al. 1998) or by the interaction of K$_2$S$_4$ and HgS (K$_2$S$_3$ and Hg or HgS) in an evacuated Pyrex tube at 220° (210°C) for 4 (3) days (Kanarzidis and Park 1990). K$_2$Hg$_6$S$_7$ can also be prepared by direct interaction of K$_2$S and HgS under a vacuum and heating to 375°C for 7 days followed by cooling to 50°C at a rate of 2°C h^{-1}. K$_2$Hg$_3$S$_4$ is not stable in water and decomposes rapidly to form HgS, and K$_2$Hg$_6$S$_7$ is insoluble in water and common organic solvents (Kanarzidis and Park 1990 and Axtell et al. 1998). K$_6$HgS$_4$ was obtained by the sintering of mixtures from HgS and K$_2$S at 380°C for 7 days in argon atmosphere (Sommer and Hoppe 1978). Single crystals of K$_2$Hg$_3$S$_4$ and K$_2$Hg$_6$S$_7$ were grown from the melt of HgS and K$_2$S (Axtell et al. 1998).

7.4 Mercury–Rubidium–Sulfur

HgS–Rb$_2$S: The phase diagram is not constructed. Rb$_2$Hg$_6$S$_7$ and Rb$_6$HgS$_4$ ternary compounds are formed in this system (Sommer and Hoppe 1978, Axtell et al. 1998). The first of them melts congruently at 556 ± 10°C and crystallizes in a tetragonal structure with lattice parameters $a = 1392.21 \pm 0.08$

and $c = 412.04 \pm 0.02$ pm and calculation density 6.65 g cm^{-3} and energy gap $E_g = 1.55$ eV (Axtell et al. 1998), and the second crystallizes in a hexagonal structure with lattice parameters $a = 1034$ and $c = 794.2$ pm and experimental density 3.80 g cm^{-3} (Sommer and Hoppe 1978).

$Rb_2Hg_6S_7$ can be prepared by the reaction of Rb_2S and HgS by heating under a vacuum at 400°C for 1.5 days followed by heating at 500°C for one day, slow cooling to 415°C, and quenching to room temperature (Axtell et al. 1998). This compound is air and water stable and is insoluble in common organic solvents. Rb_6HgS_4 was obtained by the sintering of mixtures from HgS and Rb_2S at 380°C for 7 days in argon atmosphere (Sommer and Hoppe 1978). Single crystals of $Rb_2Hg_6S_7$ were grown from the melt of HgS and Rb_2S (Axtell et al. 1998).

7.5 Mercury–Cesium–Sulfur

HgS–Cs$_2$S: The phase diagram is not constructed. $Cs_2Hg_6S_7$ ternary compound is formed in this system. It melts congruently at 556 ± 10°C and crystallizes in a tetragonal structure with lattice parameters $a = 1395.8 \pm 0.4$ and $c = 415.9 \pm 0.2$ pm, calculation density 6.94 g cm^{-3}, and energy gap $E_g = 1.61$ eV (Axtell et al. 1998) [$a = 1406.3 \pm 0.3$ and $c = 418.95 \pm 0.18$ pm at 215 K and calculation density 6.789 g cm^{-3} (Bugaris and Ibers 2008)].

$Cs_2Hg_6S_7$ can be prepared by the reaction of Cs_2S and HgS by heating under a vacuum at 400°C for 1.5 days followed by heating at 500°C for one day, slow cooling to 50°C, and quenching to room temperature (Axtell et al. 1998). It can be also synthesized from a solid-state reaction of Cs_2S_3, HgS, and S (Bugaris and Ibers 2008). CsI was added to aid in the crystallization of the final product. The reaction takes place in an evacuated fused silica tube at 850°C in 24 h, with the mixture kept at this temperature for 120 h, cooled at 5°C h^{-1} to 200°C, and then cooled to room temperature in 2 h. This compound is air and water stable and is insoluble in common organic solvents. Single crystals of $Cs_2Hg_6S_7$ were grown from the melt of HgS and Cs_2S (Axtell et al. 1998).

7.6 Mercury–Copper–Sulfur

Ternary compounds were not found in the Hg–Cu–S ternary system. The peritectic decomposition of CuS at 507°C leads to different triangulations of this system at different temperatures (Figure 7.2) (Ollitrault-Fichet et al. 1984). In the HgS–Cu$_2$S–S subsystem, there are ternary peritectic with the next equilibrium, $L + Cu_{2-x}S \Leftrightarrow HgS + CuS$, which is situated near the S-rich corner at 470 ± 3°C, and ternary degenerated eutectic $L \Leftrightarrow HgS + CuS + S$.

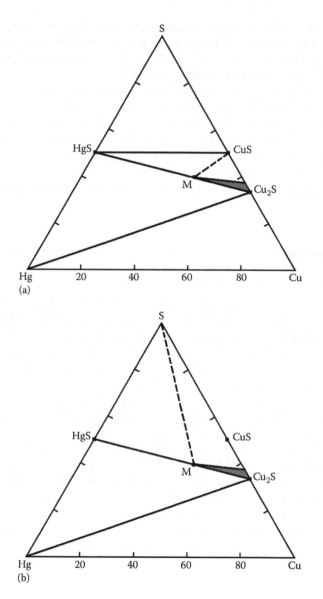

FIGURE 7.2
Triangulation of the Hg–Cu–S ternary system at 416°C–507°C (a) and 507°C–690°C (b) M—solid solution in the HgS–Cu₂S system with limit composition $Hg_{0.18}Cu_{0.42}S_{0.40}$. (From Ollitrault-Fichet, R. et al., *J. Less-Common Metals*, 96(1), 49, 1984.)

The extension of homogeneity region is variable and depends on the temperatures. At 125°C, it is situated near $Cu_{1.75}S$ and increases progressively with temperature. At 420°C, this region achieves the HgS–Cu₂S system in the eutectoid point (48 mol. % HgS), and at 435°C, it achieves Cu₂S (Ollitrault-Fichet et al. 1984).

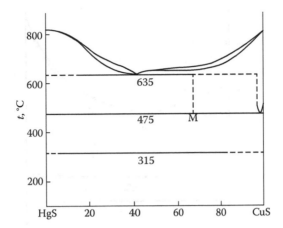

FIGURE 7.3
Phase relations in the HgS–CuS system. (From Ollitrault-Fichet, R. et al., *J. Less-Common Metals*, 96(1), 49, 1984.)

HgS–CuS: This section is a nonquasibinary section of the Hg–Cu–S ternary system (Figure 7.3) (Ollitrault-Fichet et al. 1984). Thermal effects at 315°C correspond to the polymorphous transformation of HgS, at 475°C (are not shown in Figure 7.4) to the crystallization of ternary peritectic, and at 635°C to the immiscibility, which exists within the interval of 15–100 mol. % CuS. Minimum on the liquidus line corresponds to 41 mol. % CuS.

This system was investigated using DTA, metallography, and XRD (Ollitrault-Fichet et al. 1984).

HgS–Cu₂S: The most reliable phase diagram is a eutectic type (Figure 7.4) (Ollitrault-Fichet et al. 1984). The eutectic composition and temperature are 60 mol. % HgS and 690°C, respectively. Thermal effects at 315°C correspond to the polymorphous transformation of HgS, at 420°C to the decomposition of saturated solid solutions ($Hg_{0.18}Cu_{0.42}S_{0.40}$), and at 475°C to the crystallization of ternary peritectic. Solid solutions based on HgS are not stable and cannot be quenched.

The attempts to synthesize ternary sulfides in the HgS–Cu₂S system at different Cu₂S contents had not given any results (Charbonnier 1973). According to the data from XRD, the mixtures of mercury and copper sulfides existed in the ingots, preparing by the annealing at 800°C and 1000°C for 15 and 8 days, respectively (the mixtures annealing at 1000°C were in addition heat treated at 1350°C for 6 days).

According to the data of Kulakov et al. (1985), the HgS–Cu₂S section is nonquasibinary section of Hg–Cu–S ternary system that is connected with the nonstoichiometry of copper disulfide. The ingots containing up to 20 mol. % Cu₂S crystallize as solid solutions based on high-temperature HgS modification. These solid solutions decompose at cooling. At 750°C–770°C, intermediate sulfide is formed according to the peritectic reaction with

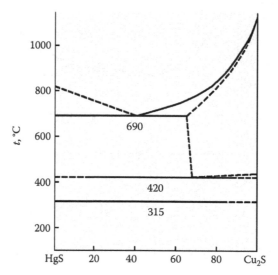

FIGURE 7.4
HgS–Cu₂S phase diagram. (From Ollitrault-Fichet, R. et al., *J. Less-Common Metals*, 96(1), 49, 1984.)

starting sulfide ratio 1:1. This structure was not determined because it has polymorphous transformation at 459°C and decomposes at 416°C. The eutectic composition and temperature are 74 mol. % HgS and 703°C. The alloys containing up to 37 mol. % HgS crystallize in a structure based on digenite. At room temperature, the solubility of Cu₂S in HgS and HgS in Cu₂S is equal to 1.1–1.5 and 0.5–1 mol. %, respectively.

This system was investigated using DTA, metallography, and XRD (Ollitrault-Fichet et al. 1984, Kulakov et al. 1985). The ingots were annealed at 350°C for 72 h (Kulakov et al. 1985).

7.7 Mercury–Silver–Sulfur

HgS–Ag₂S: The phase diagram is of the eutectic type with coordinates of the eutectic point at 629°C and 78 mol. % HgS (Figure 7.5) (Parasyuk et al. 2002). No ternary phases are formed in this system (Charbonnier 1973, Parasyuk et al. 2002). The Ag₂S solid-solution range extends up to 37 mol. % HgS at 400°C, and the HgS solid solutions remain below 5 mol. % Ag₂S at the same temperature (Parasyuk et al. 2002).

According to the data of Shao et al. (2003) and Ye et al. (2003), the ternary compound Ag₂HgS₂, which crystallizes in a monoclinic structure, was prepared at 200°C via a solvo-displacement route. This compound was

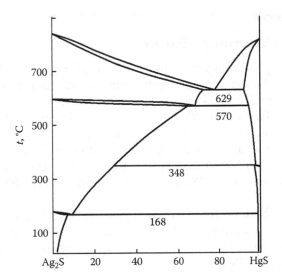

FIGURE 7.5

HgS–Ag₂S phase diagram. (From Parasyuk, O.V. et al., *J. Alloys Compd.*, 336(1–2), 213, 2002.)

characterized with XRD, x-ray photoelectron spectra, and transmission electron microscope images.

This system was investigated using DTA, metallography, and XRD (Parasyuk et al. 2002). The ingots were annealed at 400°C for 250 h.

To obtain Ag_2HgS_2, 0.002 mol HgO and Ag_2O were added to a Teflon reactor of 20 mL capacity, then 18 mL CS_2 was filled in (Shao et al. 2003, Ye et al. 2003). The reactor was maintained at 200°C for 5 days. Then the products were collected, washed with absolute ethanol, and dried in a vacuum at 40°C for 4 h.

7.8 Mercury–Barium–Sulfur

HgS–BaS: The phase diagram is not constructed. $BaHgS_2$ and Ba_2HgS_3 ternary compounds are formed in this system (Rad and Hoppe 1981a,b). $BaHgS_2$ crystallizes in an orthorhombic structure with lattice parameters $a = 421$, $b = 1438$, and $c = 733$ pm and experimental density 5.92 g cm⁻³ (Rad and Hoppe 1981b). Ba_2HgS_3 compound crystallizes in the same structure with $a = 893.1$, $b = 435.7$, and $c = 1725.7$ pm and calculation and experimental density correspondingly 5.65 and 5.59 g cm⁻³ (Rad and Hoppe 1981a).

This ternary compounds were obtained by the interaction of HgS and BaS at 700°C–750°C ($BaHgS_2$) and 850°C (Ba_2HgS_3) (Rad and Hoppe 1981a,b).

7.9 Mercury–Aluminum–Sulfur

HgS–Al$_2$S$_3$: The phase diagram is not constructed. HgAl$_2$S$_4$ is formed in this system. It crystallizes in a tetragonal structure of chalcopyrite type with lattice parameters a=550.59 and c=1019.18 pm (Schwer and Krämer 1990) [a=547.7 and c=1024 pm (Hahn et al. 1955), a=549 and c = 1026 pm (Range et al. 1968)] and calculation and experimental density 4.112 [4.11 (Range et al. 1968)] and 4.08 g cm^{-3} (Hahn et al. 1955). At high temperature and pressure (6.0 GPa or 60 kbar and 420°C), this compound can exist in a cubic structure of spinel type with lattice parameter a = 1028 pm and calculation density 4.84 g cm^{-3} (Range et al. 1968).

HgAl$_2$S$_4$ was obtained from HgS and powderlike Al and S in equimolar ratio. The mixtures were pressed and annealed at 800°C for 12–24 h (Hahn et al. 1955). Its single crystals were grown using chemical transport reactions (Schwer and Krämer 1990).

7.10 Mercury–Gallium–Sulfur

The highest temperature and sharply dropping part of Hg–Ga–S liquidus surface is projected in the crystallization field of the solid solutions based on high-temperature Ga$_2$S$_3$ modification (Figure 7.6) (Siniakova and Il'iasheva 1986). It is limited by the m_1–I line of liquid-phase decomposition and p_1–I line of A-phase formation according to the peritectic reaction (A-phase is formed in the HgS–Ga$_2$S$_3$ system). The field of HgGa$_2$S$_4$ is

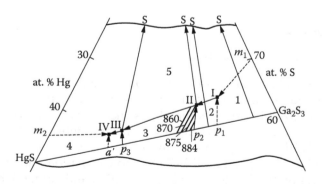

FIGURE 7.6

Projection of a part of Hg–Ga–S liquidus surface (the fields of crystallization: 1, solid solutions based on γ-Ga$_2$S$_3$; 2, A-phase; 3, HgGa$_2$S$_4$; and 4, solid solutions based on β-HgS; 5, S). (From Siniakova, E.F. and Il'iasheva, I.A., *Izv. AN SSSR. Neorgan. materialy*, 22(10), 1625, 1986.)

TABLE 7.1

Nonvariant Equilibria in the Hg–Ga–S
Ternary System

Symbol	Reaction	t, °C
m_1	$L_1 \Leftrightarrow Ga_2S_3 + L_2$	987 ± 5
m_2	$L_1 \Leftrightarrow HgS + L_2$	788 ± 3
p_1	$L_1 + Ga_2S_3 \Leftrightarrow A\text{-phase}$	962 ± 2
p_2	$L_1 + A\text{-phase} \Leftrightarrow HgGa_2S_4$	$(884–889) \pm 2$
p_3	$L_1 + HgGa_2S_4 \Leftrightarrow (\beta\text{-HgS})$	771 ± 2
a'	$L_1 \Leftrightarrow (\beta\text{-HgS})$	752 ± 2
I	$L_1 + L_2 + Ga_2S_3 \Leftrightarrow HgGa_2S_4$	955 ± 2
II	$L_1 + L_2 + A\text{-phase} \Leftrightarrow HgGa_2S_4$	871 ± 2
III	$L_1 + L_2 + HgGa_2S_4 \Leftrightarrow (\beta\text{-HgS})$	743 ± 2
IV	$L_1 \Leftrightarrow (\beta\text{-HgS}) + L_2$	—

Source: Siniakova, E.F. and Il'iasheva, I.A., *Izv. AN SSSR.
Neorgan. materialy, 22(10), 1625, 1986.

limited by the lines p_2–II and p_3–III that correspond to the peritectic reactions and II–III line of liquid-phase decomposition. The field of primary crystallization of solid solutions based on β-HgS is divided into two parts by a'–IV monovariant critical azeotropic line and III–IV–m_2 line of liquid-phase decomposition.

Siniakova and Il'iasheva (1986) determined nonvariant equilibria in the Hg–Ga–S ternary system (Table 7.1) and constructed the liquidus isotherms at 860°C, 870°C, 875°C, and 884°C. The parts of isothermal sections of the Hg–Ga–S ternary system at 880°C, 870°C, 850°C, and 780°C are shown in Figure 7.7 (Siniakova et al. 1987). The homogeneity region of $HgGa_2S_4$ is situated not only along the HgS–Ga_2S_3 quasibinary system but also on the side of the sulfur excess. It has the most width at 870°C and is contracted at the next temperature increase. The ingot containing 29.9 ± 0.3 at. % Ga and 27.7 ± 0.6 at. % S has a maximum temperature of peritectic transformation (889°C).

At 870°C, the homogeneity region of A-phase is propagated in the Hg-rich side relative to the HgS–Ga_2S_3 system.

Isotherms of $HgGa_2S_4$ homogeneity region are shown in Figure 7.8 (Siniakova et al. 1987).

This system was investigated using DTA, metallography, and XRD (Siniakova and Il'iasheva 1986, Siniakova et al. 1987). The ingots were annealed at 860°C, 870°C, 875°C, and 884°C for 1–2 months (Siniakova and Il'iasheva 1986).

HgS–Ga$_2$S$_3$: The phase diagram is shown in Figure 7.9 (Il'iasheva et al. 1985). $HgGa_2S_4$ ternary compound is formed in this system. It melts incongruently at 884°C–889°C (Il'iasheva et al. 1985) [882°C–896°C (Il'iasheva

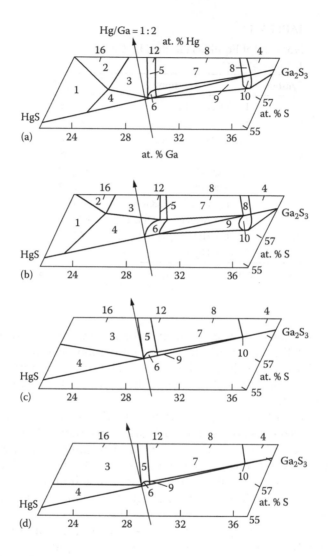

FIGURE 7.7
Parts of isothermal sections of Hg–Ga–S ternary system at (a) 880°C, (b) 870°C, (c) 850°C, and (d) 780°C: 1, L_1; 2, $L_1 + L_2$; 3, $L_1 + L_2 + HgGa_2S_4$; 4, $L_1 + HgGa_2S_4$; 5, $L_2 + HgGa_2S_4$; 6, $HgGa_2S_4$; 7, $L_2 + HgGa_2S_4 + A$-phase; 8, $L_2 + A$-phase; 9, $HgGa_2S_4 + A$-phase; and 10, A-phase. (From Siniakova, E.F. et al., *Izv. AN SSSR. Neorgan. materialy*, 23(3), 382, 1987.)

et al. 1982)] and crystallizes in a tetragonal structure of chalcopyrite type with lattice parameters that change in the homogeneity region from $a = 551.4 \pm 0.2$ and $c = 1023 \pm 0.6$ pm to $a = 549.0 \pm 0.2$ and $c = 1025.8 \pm 0.6$ pm (Il'iasheva et al. 1985) [$a = 549.6$ and $c = 1021$ pm (Hahn et al. 1955); $a = 550.0$ (550.8) ± 0.3 and $c = 1024.0$ (1024.8) ± 0.6 pm (Il'iasheva et al. 1984); $a = 552$

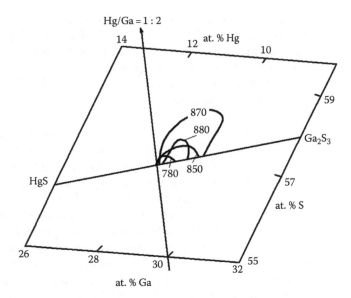

FIGURE 7.8
Isotherms of the HgGa$_2$S$_4$ homogeneity region. (From Siniakova, E.F. et al., *Izv. AN SSSR. Neorgan. materialy*, 23(3), 382, 1987.)

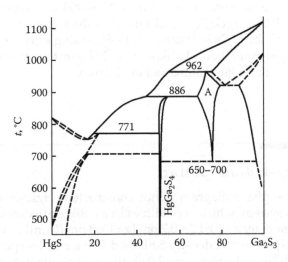

FIGURE 7.9
HgS–Ga$_2$S$_3$ phase diagram. (From Il'iasheva, N.A. et al., *Izv. AN SSSR. Neorgan. materialy*, 21(11), 1860, 1985.)

and $c = 1026$ pm (Krausbauer et al. 1965); $a = 551.06$ and $c = 1023.92$ pm (Schwer and Krämer 1990); $a = 550.6$ and $c = 1029.9$ pm (Badikov et al. 1979)] and calculation and experimental density 5.002 and 4.95 g cm^{-3}, respectively (Hahn et al. 1955), and energy gap $E_g = 2.79$–2.84 eV (Beun et al. 1961). HgGa$_2$S$_4$ has a narrow homogeneity range (Il'iasheva et al. 1985), is stable at heating in a vacuum up to 500°C, and begins to decompose at 520°C (Krausbauer et al. 1965).

A-Phase, which is formed at high temperature, has a wide homogeneity region within the interval of 65–75 mol. % Ga$_2$S$_3$ and is stable from 700°C to 962°C (Il'iasheva et al. 1985). At 650°C–700°C, A-phase decomposes forming HgGa$_2$S$_4$ and α-Ga$_2$S$_3$. It can be easily obtained by quenching, and such ingots decompose only at the exposure at 650°C for 14 days.

The solubility of Ga$_2$S$_3$ in HgS reaches 20 mol. % (Il'iasheva et al. 1985). These solid solutions interact with HgGa$_2$S$_4$ according to a peritectic reaction at 771 ± 2°C and have azeotropic point. The composition and temperature of this point are 15–20 mol. % Ga$_2$S$_3$ and 752 ± 2°C, respectively. Solid solutions based on high-temperature Ga$_2$S$_3$ contain up to 25–27 mol. % HgS, cannot be quenched, and decompose at cooling forming HgGa$_2$S$_4$ or A-phase and α-Ga$_2$S$_3$.

According to the data of Badikov et al. (1979), HgGa$_2$S$_4$ melts congruently at 880°C, and solid solutions over the entire range of concentration are formed in the HgS–HgGa$_2$S$_4$ subsystem, which decompose at cooling. The composition and temperature of azeotropic point are 30 mol. % Ga$_2$S$_3$ and 755°C, respectively.

This system was investigated using DTA, metallography, optical analysis, and XRD. The ingots were annealed at 650°C–1000°C (Badikov et al. 1979, Il'iasheva et al. 1985). HgGa$_2$S$_4$ was obtained by the annealing of HgS + Ga$_2$S$_3$ mixtures at 900°C for 12–24 h (Hahn et al. 1955). Its single crystals were grown using the Bridgman method (Badikov et al. 1979) or the chemical transport reactions (Beun et al. 1961, Krausbauer et al. 1965).

7.11 Mercury–Indium–Sulfur

HgS–In$_2$S$_3$: The phase diagram is not constructed. HgIn$_2$S$_4$ compound is formed in this system, which crystallizes in a cubic structure of spinel type with lattice parameter $a = 1081.2 \pm 0.7$ pm [$a = 1083$ pm (Donika et al. 1980)]; calculation and experimental density 5.815 and 5.79 g cm^{-3}, respectively (Hahn and Klingler 1950); and energy gap 2.0 eV (Beun et al. 1960). According to the data of (Donika et al. 1980), solid solutions are formed at the interaction of β-In$_2$S$_3$ and HgIn$_2$S$_4$ in the HgS–In$_2$S$_3$ system.

7.12 Mercury–Thallium–Sulfur

Two wide immiscibility regions exist on the liquidus surface of the Hg–Tl–S ternary system (Figure 7.10) (Asadov 1983a). The dotted line that passes through p_3 point denominates peritectic monovariant reaction of $HgTl_4S_3$ formation. Nonvariant equilibria in this ternary system are given in Table 7.2.

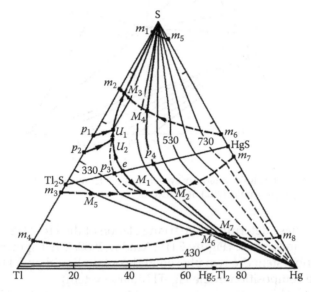

FIGURE 7.10
Liquidus surface of the Hg–Tl–S ternary system. (From Asadov, M.M., *Izv. AN SSSR. Neorgan. materialy*, 19(10), 1626, 1983a.)

TABLE 7.2

Nonvariant Equilibria in the Hg–Tl–S Ternary System

Symbol	Reaction	Tl, at. %	Hg, at. %	t, °C
e	$L \Leftrightarrow Tl_2S + \beta\text{-}Hg_3Tl_2S_4$	43	18	247
E	$L \Leftrightarrow \alpha\text{-}Hg_3Tl_2S_4 + Tl_2S_5 + S$	0.6	0.4	107
p_4	$L + \beta\text{-}HgS \Leftrightarrow \beta\text{-}Hg_3Tl_2S_4$	30	28	377
U_1	$L + Tl_2S \Leftrightarrow Tl_4S_3 + \beta\text{-}Hg_3Tl_2S_4$	40	6	237
U_2	$L + Tl_4S_3 \Leftrightarrow TlS + \beta\text{-}Hg_3Tl_2S_4$	38	6	227
p_5	$L + Tl_2S \Leftrightarrow HgTl_4S_3$	45	16	277
M_1	$L(M_1) \Leftrightarrow L(m_1) + Tl_2S + \beta\text{-}Hg_3Tl_2S_4$	40	30	227
M_2	$L(M_2) \Leftrightarrow L(m_2) + \beta\text{-}HgS + \beta\text{-}Hg_3Tl_2S_4$	28	41	362
M_3	$L(M_3) \Leftrightarrow L(m_3) + TlS + \beta\text{-}Hg_3Tl_2S_4$	26	3	157
M_4	$L(M_4) \Leftrightarrow L(m_4) + \beta\text{-}HgS + \beta\text{-}Hg_3Tl_2S_4$	18	12	337
M_5	$L_1 \Leftrightarrow L_2 + Tl_2S$	60	10	442

Source: Asadov, M.M., *Izv. AN SSSR. Neorgan. materialy*, 19(10), 1626, 1983a.

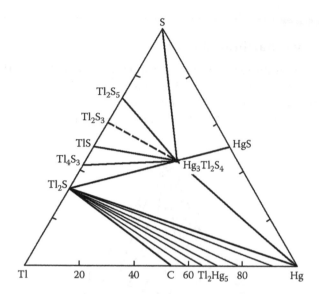

FIGURE 7.11
Isothermal section of the Hg–Tl–S ternary system at room temperature. (From Asadov, M.M., *Izv. AN SSSR. Neorgan. materialy*, 19(10), 1626, 1983a.)

α-$Hg_3Tl_2S_4$ phase determines the triangulation of the Hg–Tl–S ternary system at room temperature (Figure 7.11) (Asadov 1983a). All specimens in the Tl_2S–C–Hg subsystem are two phased and contain liquid and Tl_2S (C corresponds to the composition in the Hg–Tl binary system).

HgS–Hg_5Tl_2: This system is a nonquasibinary section of the Hg–Tl–S ternary system (Figure 7.12) (Asadov 1984). There is a limited miscibility in the liquid state along HgS–Hg_5Tl_2 section within the interval of 300°C–630°C and 25–75 mol. % 7HgS. Thallium, Tl_2S, and β-$Hg_3Tl_2S_4$ crystallize from the L_1 liquid, and β-HgS primarily segregates from the L_2 liquid. Secondary crystallization of the alloys containing up to 20 mol. % 7HgS leads to the formation of Hg_5Tl_2 + Tl mechanical mixtures. Three-phase eutectic equilibrium $L_1 \Leftrightarrow Hg_5Tl_2$ + Tl takes place at 7°C and 1 mol. % 7HgS. Thermal effects at 335°C and 220°C correspond to the polymorphous transformation of HgS and $Hg_3Tl_2S_4$, respectively. Mutual solubility of HgS and Hg_5Tl_2 is insignificant.

This system was investigated using DTA and vapor pressure measurement (Asadov 1984).

HgS–Tl: The phase diagram is not constructed. The solubility of Tl in HgS is not higher than 1 at. % (Asadov and Mustafaieva 1987).

This system was investigated using DTA, XRD, and measurement of emf of concentration chains (Asadov and Mustafaieva 1987).

HgS–TlS: This section is a nonquasibinary section of the Hg–Tl–S ternary system and is similar to the HgS–Tl_4S_3 section (Asadov 1985). It intersects the fields of Tl_2S, β-HgS, Tl_4S_3, and β-$Hg_3Tl_2S_4$ primary crystallization. Mutual solubility of HgS and TlS is not higher than 1 mol. %.

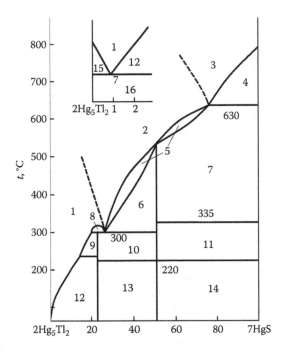

FIGURE 7.12

Phase relations in the $7HgS–2Hg_5Tl_2$ system: 1, L_1; 2, L_1+L_2; 3, L_2; 4, $L_2+\beta\text{-HgS}$; 5, $L_1+L_2+\beta\text{-HgS}$; 6, $L_1+\beta\text{-HgS}+\beta\text{-Hg}_3Tl_2S_4$; 7, $L_2+\beta\text{-HgS}+\beta\text{-Hg}_3Tl_2S_4$; 8, L_1+Tl_2S; 9, $L_1+\beta\text{-Hg}_3Tl_2S_4$; 10, $L_1+Tl_2S+\beta\text{-Hg}_3Tl_2S_4$; 11, $L_2+\alpha\text{-HgS}+\beta\text{-Hg}_3Tl_2S_4$; 12, $L_1+\beta\text{-Tl}$; 13, $L_1+Tl_2S+\alpha\text{-Hg}_3Tl_2S_4$; 14, $L_1 + \alpha\text{-HgS}+\alpha\text{-Hg}_3Tl_2S_4$; 15, $L_1+Hg_5Tl_2$; and 16, Hg_5Tl_2+Tl. (From Asadov, M.M., *Zhurn. neorgan. khimii*, 29(11), 2873, 1984.)

$HgTlS_2$ ternary compound, which according to the data (Guseinov et al., 1968, Guseinov 1969), melts at approximately 325°C and crystallizes in a tetragonal structure with lattice parameters $a = 1220 \pm 5$ and $c = 660 \pm 2$; calculation and experimental density 6.34 and 6.32 g cm^{-3}, respectively; and energy gap $E_g = 1.3$ eV, was not found in the HgS–TlS system (Asadov 1985).

This system was investigated using DTA, metallography, and XRD (Asadov 1985).

HgS–Tl$_2$S: The phase diagram is shown in Figure 7.13 (Kuliev et al. 1978). The eutectic composition and temperature are 45 mol. % HgS and 245°C, respectively. $HgTl_4S_3$ and $Hg_3Tl_2S_4$ ternary compounds are formed in this system (Kuliev et al. 1978, 1981). They melt incongruently at 375°C ($Hg_3Tl_2S_4$) and 280°C ($HgTl_4S_3$). $HgTl_4S_3$ compound is stable within the interval of 280°C–245°C.

The solubility of HgS in Tl_2S at 245°C is equal to 3 mol. % and the solubility of Tl_2S in HgS is insignificant (Kuliev et al. 1978). Thermal effects at 335°C and 220°C correspond to the polymorphous transformation of HgS and $Hg_3Tl_2S_4$, respectively.

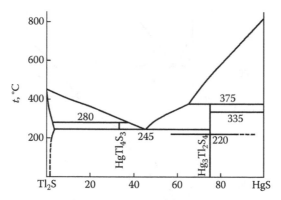

FIGURE 7.13
HgS–Tl₂S phase diagram. (From Kuliev, A.A. et al., *Zhurn. neorgan. khimii*, 23(3), 854, 1978.)

This system was investigated using DTA, metallography, XRD, and microhardness measurement. The ingots were annealed at temperatures below solidus temperatures for 400 h (Kuliev et al. 1978, 1981).

HgS–Tl₄S₃: This system is a nonquasibinary section of the Hg–Tl–S ternary system (Figure 7.14) (Asadov 1985). It intersects the fields of Tl₂S, β-HgS, and

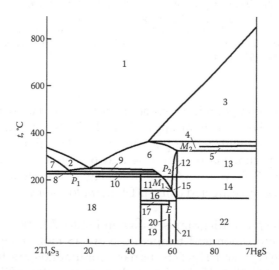

FIGURE 7.14
Phase relations in the 7HgS–2Tl₄S₃ system: 1, L; 2, L + (α-Tl₂S); 3, L + β-HgS; 4, L + β-HgS + β-Hg₃Tl₂S₄; 5, L + α-HgS + β-Hg₃Tl₂S₄; 6, L + β-Hg₃Tl₂S₄; 7, L + Tl₂S + Tl₄S₃; 8, L + TlS + Tl₄S₃; 9, L + Tl₂S + β-Hg₃Tl₂S₄; 10, TlS + Tl₄S₃ + β-Hg₃Tl₂S₄; 11, L₁ + TlS + α-Hg₃Tl₂S₄; 12, L₁ + L₂ + β-Hg₃Tl₂S₄; 13, L₁ + α-HgS + β-Hg₃Tl₂S₄; 14, L₂ + α-HgS + β-Hg₃Tl₂S₄; 15, L₁ + L₂ + α-Hg₃Tl₂S₄; 16, L₂ + TlS + α-Hg₃Tl₂S₄; 17, α-Hg₃Tl₂S₄ + TlS + Tl₂S₅; 18, Tl₄S₃ + TlS + α-Hg₃Tl₂S₄; 19, TlS + Tl₂S₃ + α-Hg₃Tl₂S₄; 20, Tl₂S₃ + Tl₂S₅ + α-Hg₃Tl₂S₄; 21, Tl₂S₃ + α-S + α-Hg₃Tl₂S₄; and 22, α-Hg₃Tl₂S₄ + α-HgS + α-S. (From Asadov, M.M., *Zhurn. neorgan. khimii*, 30(10), 2653, 1985.)

TABLE 7.3

Nonvariant Equilibria in the HgS–Tl$_4$S$_3$ System with Liquid Participation

Symbol	Reaction	7HgS, mol. %	2Tl$_4$S$_3$, mol. %	t, °C
U_1	L + Tl$_2$S ⇔ Tl$_4$S$_3$ + β-Hg$_3$Tl$_2$S$_4$	13	87	337
U_2	L + Tl$_4$S$_3$ ⇔ TlS + β-Hg$_3$Tl$_2$S$_4$	54	46	327
M_1	L$_1$ ⇔ L$_2$ + Tl + α-Hg$_3$Tl$_2$S$_4$	60	40	157
M_2	L$_1$ ⇔ L$_2$ + β-HgS + β-Hg$_3$Tl$_2$S$_4$	64	36	437
E	L ⇔ α-Hg$_3$Tl$_2$S$_4$ + Tl$_2$S$_5$ + α-S	63	37	107

Source: Asadov, M.M., *Zhurn. neorgan. khimii*, 30(10), 2653, 1985.

β-Hg$_3$Tl$_2$S$_4$ primary crystallization. Mutual solubility of HgS and TlS is not higher than 1 mol. %. Nonvariant equilibria in the HgS–Tl$_4$S$_3$ system with liquid participation are given in Table 7.3.

This system was investigated using DTA, metallography, and XRD (Asadov 1985).

7.13 Mercury–Thulium–Sulfur

HgS–Tm$_2$S$_3$: The phase diagram is not constructed. HgTm$_4$S$_7$ ternary compound was obtained in this system at the interaction of HgS and Tm$_2$S$_3$ within the temperature interval from 650°C to 800°C (Du et al. 1985). It crystallizes in a tetragonal structure with the lattice parameters $a = 1109 \pm 2$ and $c = 838 \pm 5$ pm.

7.14 Mercury–Ytterbium–Sulfur

HgS–Yb$_2$S$_3$: The phase diagram is not constructed. HgYb$_4$S$_7$ ternary compound was obtained in this system at the interaction of HgS and Yb$_2$S$_3$ within the temperature interval from 650°C to 800°C (Du et al. 1985). It crystallizes in a tetragonal structure with the lattice parameters $a = 1101 \pm 3$ and $c = 835 \pm 2$ pm.

7.15 Mercury–Lutetium–Sulfur

HgS–Lu$_2$S$_3$: The phase diagram is not constructed. HgLu$_4$S$_7$ ternary compound was obtained in this system at the interaction of HgS and Lu$_2$S$_3$ within the temperature interval from 650°C to 800°C (Du et al. 1985). It crystallizes in a tetragonal structure with the lattice parameters $a = 1103 \pm 2$ and $c = 833 \pm 2$ pm.

7.16 Mercury–Silicon–Sulfur

HgS–SiS$_2$: The phase diagram is not constructed. At 800°C–1000°C, Hg$_4$SiS$_6$ compound is formed in this system (Serment et al. 1968, Gulay et al. 2002). It crystallizes in a monoclinic structure with lattice parameters a = 1230.20 ± 0.05, b = 710.31 ± 0.04, c = 1227.91 ± 0.04 nm, and β = 109.721 ± 0.003° [a = 1229 ± 1, b = 709.6 ± 0.6, c = 1230 ± 1 pm, and β = 109°28' (Serment et al. 1968)] and experimental density 6.67 ± 0.01 g cm^{-3} (Serment et al. 1968).

The parameters of the monoclinic structure were recalculated in the parameters of rhombohedral and hexagonal structures (rhombohedral structure, a = 1230 ± 1 pm and α = 33°32' ± 0°10'; hexagonal structure, a = 709.6 ± 0.6 and c = 3480 ± 1 pm). According to the data of Serment et al. (1968), the lattice is practically rhombohedral, but x-ray spectrum of Hg$_4$SiS$_6$ compound well indexes by the parameters both of rhombohedral and hexagonal structure.

7.17 Mercury–Germanium–Sulfur

At 500°C, four ternary compounds (Hg$_2$Ge$_3$S$_5$, Hg$_2$GeS$_3$, HgGe$_2$S$_5$, and Hg$_4$GeS$_6$) exist in the Hg–Ge–S ternary system (Figure 7.15) (Motria 1991, Voroshilov et al. 1994a). They are in equilibrium with binary and elementary components and with each other.

The ingots were annealed at 500°C for 1200 h (Motria 1991, Voroshilov et al. 1994a).

HgS–GeS: The phase diagram is shown in Figure 7.16 (Motria 1991, Voroshilov et al. 1994a). The eutectic compositions and temperatures are 24 and 51 mol. % HgS and 529 ± 5°C and 510 ± 5°C, respectively. Two ternary compounds are formed in this system: Hg$_2$Ge$_3$S$_5$ melts congruently at 592 ± 5°C, and Hg$_2$GeS$_3$ decomposes peritectically at 539°C and has polymorphous transformation at 425 ± 5°C. The experimental density of Hg$_2$Ge$_3$S$_5$ and Hg$_2$GeS$_3$ is 4.51 and 6.25 g cm^{-3}, respectively. At 353 ± 5°C, there is polymorphous transformation of HgS, and polymorphous transformation of solid solutions based on GeS is observed from the GeS-rich side.

This system was investigated using DTA, metallography, XRD, and microhardness and density measurements. The ingots were annealed at 500°C for 1200 h (Motria 1991, Voroshilov et al. 1994a).

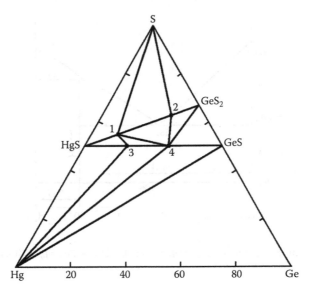

FIGURE 7.15

Isothermal section of the Hg–Ge–S ternary system at 500°C: 1, Hg_4GeS_6; 2, $HgGe_2S_3$; 3, Hg_2GeS_3; and 4, $Hg_2Ge_3S_5$. (From Motria, S.F., Mercury–germanium(tin)–sulfur(selenium) ternary systems [in Russian], in *Poluch. i svoistva slozhn. poluprovodn., Uzhgorod. gos. un-t,* Kiev, Ukraine, pp. 17–26, 1991; Voroshilov, Yu.V. et al., *Ukr. khim. zhurn.,* 60(1), 27, 1994a.)

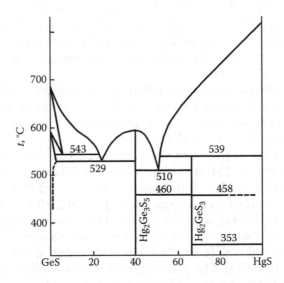

FIGURE 7.16

HgS–GeS phase diagram. (From Motria, S.F., Mercury–germanium(tin)–sulfur(selenium) ternary systems [in Russian], in *Poluch. i svoistva slozhn. poluprovodn., Uzhgorod. gos. un-t,* Kiev, Ukraine, pp. 17–26, 1991; Voroshilov, Yu.V. et al., *Ukr. khim. zhurn.,* 60(1), 27, 1994a.)

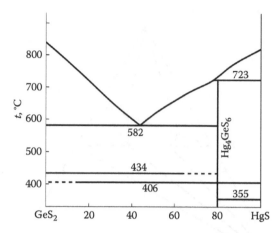

FIGURE 7.17
HgS–GeS$_2$ phase diagram. (From Olekseyuk, I.D. et al., *J. Alloys Compd.*, 417(1–2), 131, 2006.)

HgS–GeS$_2$: The most reliable phase diagram is shown in Figure 7.17 (Olekseyuk et al. 2006). Eutectic crystallizes at 582°C [580 ± 5°C (Motria 1991, Voroshilov et al. 1994a)] and contains 44 mol. % HgS. Hg$_4$GeS$_6$ is formed in this system. It melts incongruently at 723°C [720 ± 5°C (Motria 1991, Voroshilov et al. 1994a)] and has polymorphous transformation at 406°C [395 ± 5°C (Motria 1991, Voroshilov et al. 1994a)]. The peritectic point corresponds to 78 mol. % HgS. The polymorphous transitions of HgS and GeS$_2$ take place at 355°C and 434°C, respectively (Olekseyuk et al. 2006). Glassy ingots were obtained from the GeS$_2$-rich side (Motria 1991, Voroshilov et al. 1994a).

According to the data of Motria (1991) and Voroshilov et al. (1994a), once more, the ternary compound HgGe$_2$S$_5$ is formed in this system according to the solid-phase reaction at 565 ± 5°C [existence of this compound was not confirmed also by Marchuk et al. (2002)].

Low-temperature Hg$_4$GeS$_6$ crystallizes in a monoclinic structure with lattice parameters $a = 1234.51 \pm 0.08$, $b = 716.78 \pm 0.06$, $c = 1234.67 \pm 0.07$ pm, and $\beta = 109.484 \pm 0.005°$ (Olekseyuk et al. 2006) [$a = 1235 \pm 1$, $b = 717.1 \pm 0.6$, $c = 1238.2 \pm 0.8$ pm, and $\beta = 109.2 \pm 0.1°$ (Motria 1991, Voroshilov et al. 1994a); $a = 1234 \pm 1$, $b = 712.7 \pm 0.6$, $c = 1236 \pm 1$ pm, and $\beta = 109°27' \pm 0°10'$ (Serment et al. 1968)] and calculation and experimental density 6.92 and 6.90 [6.88 (Serment et al. 1968)] g cm^{-3}, respectively (Voroshilov et al. 1994a). The parameters of the monoclinic structure were recalculated in the parameters of rhombohedral and hexagonal structures (rhombohedral structure, $a = 1236 \pm 1$ pm and $\alpha = 33°30' \pm 0°10'$; hexagonal structure, $a = 712.7 \pm 0.6$ and $c = 3497 \pm 1$ pm) (Serment et al. 1968). The lattice is practically rhombohedral, whereas x-ray spectrum of Hg$_4$GeS$_6$

compound well indexes by the parameters both of rhombohedral and hexagonal structure.

According to the data of Hahn and Lorent (1958), Hg_2GeS_4 ternary compound is formed in the $HgS–GeS_2$ system. It crystallizes in a hexagonal structure with lattice parameters $a = 717$ and $c = 3490$ pm and calculation and experimental density 5.789 and 5.61 g cm^{-3}. It can be seen that the lattice parameters of this compound coincide with the lattice parameters of Hg_4GeS_6 in the hexagonal structure (Serment et al. 1968). Therefore, it is possible that composition of Hg_2GeS_4 was determined incorrectly.

This system was investigated using DTA, metallography, XRD, and microhardness and density measurements. The ingots were annealed at 400°C for 250 h (Olekseyuk et al. 2006) [at 500°C for 1200 h (Motria 1991, Voroshilov et al. 1994a)].

7.18 Mercury–Tin–Sulfur

Ternary compounds were not found in the Hg–Sn–S ternary system (Voroshilov et al. 1993). At 500°C, mercury sulfide is in equilibrium with binary phases of Sn–S binary system, and Hg is in equilibrium with SnS (Figure 7.18) (Motria 1991, Voroshilov et al. 1993).

HgS–SnS: The phase diagram is a eutectic type (Figure 7.19) (Motria et al. 1988, Motria 1991). The eutectic composition and temperature are 52 mol. % HgS and 562 ± 5°C, respectively. Thermal effects at 585 ± 5°C correspond to the polymorphous transition of SnS, and at 347 ± 5°C, there is polymorphous transformation of HgS. The solubility of HgS in SnS and SnS in HgS at the eutectic temperature is not higher than 2 and 3 mol. %, respectively. At 500°C, SnS dissolves less than 1.5 mol. % HgS and HgS dissolves not higher than 2 mol. % SnS.

This system was investigated using DTA, metallography, XRD, and microhardness measurement. The ingots were annealed at 500°C for 1200 h (Motria et al. 1988, Motria 1991).

HgS–SnS$_2$: The phase diagram is a eutectic type (Figure 7.20) (Motria et al. 1988, Motria 1991). The eutectic composition and temperature are 48 ± 2 mol. % HgS and 647 ± 5°C, respectively. At 347 ± 5°C, there is polymorphous transformation of HgS. The solubility of HgS in SnS$_2$ at the eutectic temperature is not higher than 2 mol. % and the solubility of SnS$_2$ in HgS at 500°C is less than 1 mol. %.

This system was investigated using DTA, metallography, XRD, and microhardness measurement. The ingots were annealed at 500°C for 1200 h (Motria et al. 1988, Motria 1991).

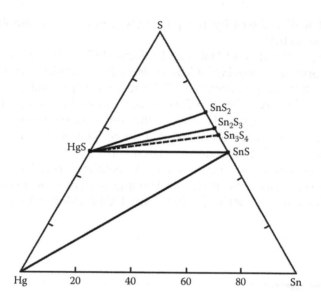

FIGURE 7.18
Isothermal section of the Hg–Sn–S ternary system at 500°C. (From Motria, S.F., Mercury-germanium(tin)–sulfur(selenium) ternary systems [in Russian], in *Poluch. i svoistva slozhn. poluprovodn., Uzhgorod. gos. un-t*, Kiev, Ukraine, pp. 17–26, 1991; Voroshilov, Yu.V. et al., *Zhurn. neorgan. khimii*, 38(6), 1061, 1993.)

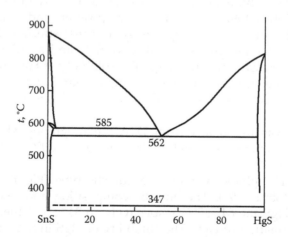

FIGURE 7.19
HgS–SnS phase diagram. (From Motria, S.F. et al., *Zhurn. neorgan. khimii*, 33(8), 2103, 1988; Motria, S.F., Mercury–germanium(tin)–sulfur(selenium) ternary systems [in Russian], in *Poluch. i svoistva slozhn. poluprovodn., Uzhgorod. gos. un-t*, Kiev, Ukraine, pp. 17–26, 1991.)

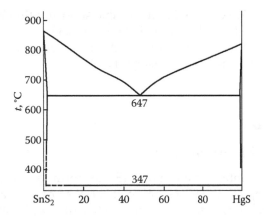

FIGURE 7.20

HgS–SnS₂ phase diagram. (From Motria, S.F. et al., *Zhurn. neorgan. khimii*, 33(8), 2103, 1988; Motria, S.F., Mercury–germanium(tin)–sulfur(selenium) ternary systems [in Russian], in *Poluch. i svoistva slozhn. poluprovodn.*, Uzhgorod. gos. un-t, Kiev, Ukraine, pp. 17–26, 1991.)

7.19 Mercury–Lead–Sulfur

HgS–PbS: The phase diagram is a eutectic type (Figure 7.21) (Kulakov and Sokolovskaya 1975). The eutectic composition and temperature are 32 mol. % PbS and 676°C, respectively. The solubility of HgS in PbS increases from 5 mol. % at 340°C to 11 mol. % at the eutectic temperature. The solubility of PbS in β-HgS is less than 0.5 mol. %, but such small solubility increases the temperature of polymorphous transformation for HgS from 344°C (Kulakov 1975) to 351 ± 2°C (Kulakov and Sokolovskaya 1975).

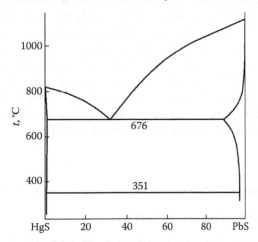

FIGURE 7.21

HgS–PbS phase diagram. (From Kulakov, M.P. and Sokolovskaya, Zh.D., *Zhurn. neorgan. khimii*, 20(8), 2290, 1975.)

Sharma et al. (1977) obtained a metastable solid-solution $Hg_xPb_{1-x}S$ in the form of thin films within the interval of $0 \leq x \leq 0.33$.

This system was investigated using DTA, metallography, XRD, and chemical analysis (Kulakov and Sokolovskaya 1975).

7.20 Mercury–Titanium–Sulfur

Mercury can be intercalated into TiS_2 to form a compound Hg_xTiS_2 ($x = 0.27–1.29$) (Ong et al. 1992, Ganal et al. 1995, Moreau et al. 1996, Sidorov et al. 1998). It crystallizes in a monoclinic structure with lattice parameters $a = 592.23$, $b(TiS_2) = 340.76$, $b(Hg) = 275.66$, $c = 886.2$ pm, and $\beta = 102.33°$ at $x = 1.24$ (Ganal et al. 1995) and $a = 592.09 \pm 0.08$, $b = 340.74 \pm 0.09$, $c = 886.62 \pm 0.12$ pm, and $\beta = 102.352 \pm 0.014°$ at $x = 1.25$ (Ong et al. 1992). Both TGA and XRD demonstrate the thermal reversibility of the intercalation process. Hg begins to slowly deintercalate at about 170°C, and substantial deintercalation occurs only above 250°C and is complete at 330°C (Ong et al. 1992).

$Hg_{1.24}TiS_2$ has unusual $(3 + 1)$-dimensional layered inadequate structure with interpenetration of 3D TiS_2 and Hg frames. An inadequacy takes place along the b-axis (Ganal et al. 1995). Differential scanning calorimetry (DSC) studies reveal the presence of a reversible thermal transition near 200°C (Moreau et al. 1996). Above the temperature of this transition, the in-plane Hg-sublattice structure and the associated intercalant Hg chains have melted to form guest layers with liquid-like disorder. The evolution of the host and Hg-sublattice cell parameters as a function of temperature exhibits the expected discontinuous behavior associated with such a first-order transition. Of particular, this compound interest is the well-defined crystalline order of both the TiS_2- and the Hg-sublattice reflections, indicating that both sublattices are well ordered at low temperature (Sidorov et al. 1998).

This compound was obtained by direct reaction of stoichiometric amounts of TiS_2 and Hg at 320°C under 1.33×10^{-7} Pa for some days followed by slow cooling to ambient temperature (Ganal et al. 1995, Sidorov et al. 1998). Samples were homogenized by annealing for 2 days at 320°C followed by slow cooling to ambient temperature (Ong et al. 1992).

7.21 Mercury–Phosphorus–Sulfur

HgS–P: The phase diagram is not constructed. Hg_3PS_3 compound is formed in this system (Olekseyuk and Golovey 1968, Golovey et al. 1969, 1977, Golovey 1975), which melts incongruently and crystallizes in a cubic structure with lattice parameter $a = 988$ pm and energy gap $E_g = 1.98$ eV (Olekseyuk and Golovey 1968, Golovey 1975, Golovey et al. 1977). There are thermal effects at 181°C for this compound that can be explicated by its polymorphous transformation (Olekseyuk and Golovey 1968). The mercury is one valence in Hg_3PS_3 compound (Olekseyuk and Golovey 1968, Golovey et al. 1969).

Single crystals of Hg_3PS_3 were grown using chemical transport reactions and sublimation (Golovey 1975, Golovey et al. 1977).

HgS–"P_2S_2": The phase diagram is not constructed. Hg_3PS_4 compound is formed in this system (Olekseyuk and Golovey 1968, Golovey et al. 1969, 1977, Golovey 1975), which melts incongruently and crystallizes in a cubic structure with lattice parameter $a = 1980$ pm and energy gap $E_g = 2.03$ eV (Olekseyuk and Golovey 1968, Golovey 1975, Golovey et al. 1977). There are thermal effects at 195°C for this compound that can be explicated by its polymorphous transformation (Olekseyuk and Golovey 1968). The mercury is one valence in Hg_3PS_4 compound (Olekseyuk and Golovey 1968, Golovey et al. 1969).

Single crystals of Hg_3PS_4 were grown using chemical transport reactions and sublimation (Golovey 1975, Golovey et al. 1977).

HgS–"P_2S_3": The phase diagram is not constructed. $Hg_4P_2S_7$ compound is formed in this system (Golovey et al. 1969, 1977, Golovey 1975). According to the data of Golovey et al. (1969), the mercury is one valence in this compound.

HgS–"P_2S_4": The phase diagram is not constructed. $Hg_2P_2S_6$ (Klinger et al. 1970, Jandali et al. 1978b) [$HgPS_3$ (Golovey et al. 1969, 1977, Golovey 1975)] compound is formed in this system, which crystallizes in a triclinic structure with lattice parameters $a = 625.2$, $b = 626.2$, $c = 712.6$ pm, $\alpha = 96.21°$, $\beta = 105.69°$, and $\gamma = 119.15°$ (Jandali et al. 1978b) [$a = 623$, $b = 624$, $c = 710$ pm, $\alpha = 96.7°$, $\beta = 105.5°$, and $\gamma = 119.2°$ (Klinger et al. 1970, 1973, Golovey 1975, Golovey et al. 1977)], calculation and experimental density 4.84 [4.90 (Klinger et al. 1973)] and 4.75 g·cm⁻³, respectively (Klinger et al. 1970, Jandali et al. 1978b), and energy gap $E_g = 2.6$ eV (Golovey 1975, Golovey et al. 1977).

Single crystals of $HgPS_3$ compound were grown by thermal decomposition of Hg_3PS_3 (Golovey 1975, Golovey et al. 1977).

HgS–P_2S_5: The phase diagram is not constructed. $Hg_2P_2S_7$ and $Hg_3P_2S_8$, or $Hg(PS_4)_2$, ternary compounds are formed in this system (Soklakov and Nechaeva 1969, Jandali et al. 1978a). $Hg_2P_2S_7$ crystallizes in a monoclinic structure with lattice parameters $a = 1088.7$, $b = 582.7$, $c = 813.2$ pm, and $\beta = 103.83°$ (Jandali et al. 1978a), and $Hg(PS_4)_2$ crystallizes in a cubic structure with the lattice parameter $a = 1356 \pm 6$ pm (Soklakov and Nechaeva 1969).

$Hg_2P_2S_7$ was obtained by heating the mixtures of HgS, red phosphorus, and sulfur in stoichiometric ratio with 5% sulfur excess for 4 weeks (Jandali et al. 1978a), and $Hg(PS_4)_2$ can be synthesized by alloying of HgS and P_2S_5 (Soklakov and Nechaeva 1969).

7.22 Mercury–Arsenic–Sulfur

HgS is in the equilibria with all phases existing in the Hg–As–S ternary system. Glass region in this system is shown in Figure 7.22 (Olekseyuk et al. 1977). The specimens for determination of this region were quenched from 400°C.

HgS–AsS: The phase diagram is not constructed. HgAsS$_2$ compound (mineral galkhaite) is formed in this system (Gruzdev et al. 1972, Kaplunnik et al. 1975). It crystallizes in a cubic structure with the lattice parameter $a = 1042.2 \pm 0.3$ pm [$a = 1041 \pm 1$ pm (Gruzdev et al. 1972)] and calculation and experimental density 5.45 and 5.4 g cm^{-3}, respectively (Gruzdev et al. 1972, Kaplunnik et al. 1975).

HgS–As$_2$S$_3$: The phase diagram is a eutectic type (Figure 7.23) (Kirilenko and Poliakov 1976). The eutectic is degenerated from the As$_2$S$_3$-rich side. The immiscibility region with monotectic temperature of 368°C exists in

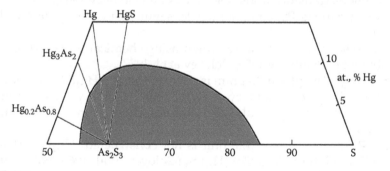

FIGURE 7.22
Glass region in the Hg–As–S ternary system. (From Olekseyuk, I.D. et al., *Kvantovaya elektronika*, (13), 93, 1977.)

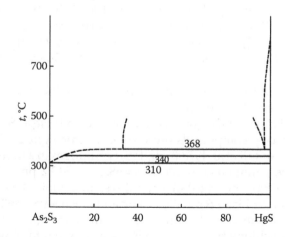

FIGURE 7.23
HgS–As$_2$S$_3$ phase diagram. (From Kirilenko, V.V. and Poliakov, Yu.A., *Izv. AN SSSR. Neorgan. materialy*, 12(2), 336, 1976.)

this system within the interval of 2.5–66.66 mol. % As_2Se_3. Thermal effects at 340°C correspond to the polymorphous transformation of HgS. Glass-forming specimens are formed within the interval of 85–100 mol. % As_2S_3.

This system was investigated using DTA and XRD. The ingots were annealed at 250°C for 720 h (Kirilenko and Poliakov 1976).

HgS–Hg₃As₂: The phase diagram is not constructed. A complex interaction takes place in this system with formation of elementary mercury (Golovey et al. 1972).

7.23 Mercury–Antimony–Sulfur

Ternary compound $HgSb_4S_8$ (mineral livingstonite) exists in the Hg–Sb–S ternary system (Richmond 1936, Gorman 1951, Craig 1970) [according to the data of Richmond (1936) and Gorman (1951), the provisional formula of this mineral is $HgSb_4S_7$]. It is stable up to $451 \pm 3°C$, melts congruently, and has experimental density 4.88 ± 0.02 g cm^{-3} (Craig 1970) [experimental and calculation density 5.00 and 4.88 g cm^{-3} (Gorman 1951)]. Livingstonite crystallizes in a triclinic structure with the lattice parameters $a = 766.5$, $b = 1084.2$, $c = 399.8$ pm, $\alpha = 99°12'$, $\beta = 102°01'$, and $\gamma = 73°48'$ (Gorman 1951) [in a monoclinic structure with the lattice parameters $a = 15.14 \pm 0.04$, $b = 3.98 \pm 0.01$, $c = 21.60 \pm 0.04$ A, and $\beta = 104°$ (Richmond 1936)].

The free energy of formation of $HgSb_4S_8$ at 300°C and 400°C is 519.2 ± 14.7 and 442.5 ± 14.7 kJ mol^{-1}, respectively (Craig 1970).

HgS–Sb₂S₃: The phase diagram is a eutectic type (Figure 7.24) (Babanly et al. 1980). The eutectic composition and temperature are 47 mol. % HgS

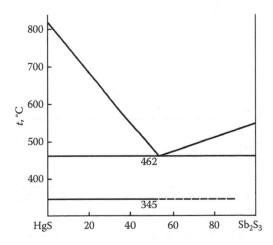

FIGURE 7.24
HgS–Sb_2S_3 phase diagram. (From Babanly, M.B. et al., *Izv. AN SSSR. Neorgan. materialy*, 16(3), 547, 1980.)

and 462°C, respectively. Thermal effects at 345°C correspond to the polymorphous transformation of HgS. Mutual solubility of HgS and Sb_2S_3 is insignificant.

This system was investigated using DTA, XRD, and microhardness measurement. The ingots were annealed at 400°C for 300 h (Babanly et al. 1980).

7.24 Mercury–Bismuth–Sulfur

HgS–Bi$_2$S$_3$: The phase diagram is a eutectic type (Figure 7.25) (Babanly et al. 1980). The eutectic composition and temperature are 62 mol. % HgS and 670°C, respectively. Thermal effects at 345°C, which are not indicated on the phase diagram, correspond to the polymorphous transformation of HgS. Mutual solubility of HgS and Bi_2S_3 is insignificant.

According to the data of Mumme and Watts (1980), Orlandi et al. (1998), Jambor and Roberts (1999), and Mandarino (1999), $HgBi_2S_4$ compound (mineral grumiplucite) is formed in this system. It crystallizes in a monoclinic structure with lattice parameters $a = 1416.4 \pm 0.5$, $b = 405.3 \pm 0.1$, $c = 1396.7 \pm 0.3$ pm, and $\beta = 118.28 \pm 0.03°$ [$a = 1417$, $b = 406$, $c = 1399$ pm, and $\beta = 118.27°$ (Mumme and Watts 1980)] and calculation density 7.02 g cm^{-3}

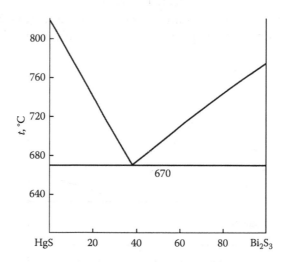

FIGURE 7.25
HgS–Bi$_2$S$_3$ phase diagram. (From Babanly, M.B. et al., *Izv. AN SSSR. Neorgan. materialy*, 16(3), 547, 1980.)

[7.00 g cm^{-3} (Mumme and Watts 1980)] (Orlandi et al. 1998, Jambor and Roberts 1999, Mandarino 1999).

This system was investigated using DTA, XRD, and microhardness measurement. The ingots were annealed at 400°C for 300 h (Babanly et al. 1980). HgBi$_2$S$_4$ compound was obtained by the interaction of binary sulfides at 700°C (Mumme and Watts 1980).

7.25 Mercury–Tantalum–Sulfur

Hg$_{1.19}$TaS$_2$ and Hg$_{1.3}$TaS$_2$ ternary compounds are formed in the Hg–Ta–S ternary system (Moreau et al. 1995). The mercury intercalation into 2H–TaS$_2$ was usually carried out at ambient temperature, whereas for 1T–TaS$_2$, an elevated reaction temperature of 150°C was necessary.

If the c-axis is perpendicular to the layers, the microdiffraction pattern is consistent with two interpenetrating c-face-centered orthorhombic sublattices and the in-plane lattice parameters of $a = 578 \pm 1$, $b = 332 \pm 1$, and $b_{Hg} = 278 \pm 1$ pm. The in-plane Hg guest layer arrangements of these two compounds bear a strong resemblance to the mercury arrangements realized in the solid-state phases of metallic Hg itself (Moreau et al. 1995).

7.26 Mercury–Oxygen–Sulfur

HgSO$_4$ and Hg$_3$O$_2$SO$_4$ (HgSO$_4 \cdot$2HgO) ternary compounds are formed in the Hg–O–S ternary system. HgSO$_4$ crystallizes in an orthorhombic structure with lattice parameters $a = 481.7$, $b = 657.7$, and $c = 478.3$ pm and calculation and experimental density 6.50 and 6.49 g cm^{-3}, respectively (Aurivillius and Malmros 1961). Hg$_3$O$_2$SO$_4$ crystallizes in a trigonal structure with lattice parameters $a = 704.9$ and $b = 1001.7$ pm [$a = 703$ and $b = 998$ pm (Nagorsen et al. 1962)] and calculation and experimental density 8.43 and 8.32 g cm^{-3} [8.52 and 8.4 g cm^{-3} (Nagorsen et al. 1962)], respectively (Aurivillius and Malmros 1961).

HgSO$_4$ was synthesized by conventional methods starting from yellow HgO and H$_2$SO$_4$, while the oxide salt Hg$_3$O$_2$SO$_4$ was obtained from Hg(C$_2$H$_3$O$_2$)$_2$ and H$_2$SO$_4$ (Aurivillius and Malmros 1961).

HgS–HgO: The phase diagram is not constructed. Hg$_3$OS$_2$ ternary compound was obtained in this system using hydrolysis of mercury halogen sulfides (Batsanov and Abaulina 1961, Batsanov et al. 1973). It crystallizes in an orthorhombic structure and its experimental density is equal to 7.5 g cm^{-3}.

Hg_3OS_2 is formed at the interaction of $Hg_3Cl_2S_2$, $Hg_3Br_2S_2$, or $Hg_3I_2S_2$ with alkalis (Batsanov and Abaulina 1961).

7.27 Mercury–Selenium–Sulfur

HgS–HgSe: The phase diagram is shown in Figure 7.26 (Asadov 1983b). Solid solutions with a sphalerite structure are easily formed in this system (Nikol'skaya and Regel' 1955). For transformation of cinnabar into the sphalerite structure 0.1 mol. % HgSe is enough.

The solubility of HgSe in HgS at 300°C, 600°C, and 800°C is equal to 16.5 [19.17 (Koreliakov 1966)], 9, and 6 mol. % respectively (Malevski and Chzhun 1965), and the solubility of HgS in HgSe at 300°C and 600°C constitutes 69.5 and 82 mol. % (Malevski and Chzhun 1965). Lattice parameters of forming solid solutions increase with HgSe contents increasing.

According to the data of Shchennikov et al. (1993) and Voronin et al. (2001), the structure of the solid solutions $HgSe_{1-x}S_x$ ($x = 0.1$–0.6) changes from a cubic of the sphalerite type to a hexagonal of the cinnabar type at the pressure increasing. The pressure transition decreases with x increasing, and a metastable two-phase region exists in the region of the phase transition (Shchennikov et al. 1993).

This system was investigated using DTA, XRD, neutron diffractometry, and measurements of microhardness and emf of concentration chains

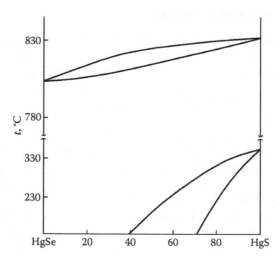

FIGURE 7.26
HgS–HgSe phase diagram. (From Asadov, M.M., *Zhurn. neorgan. khimii*, 28(7), 1812, 1983b.)

(Asadov 1983b, Voronin et al. 2001). The ingots were annealed at 200°C for 800 h (Asadov 1983b) or at 300°C, 500°C, and 600°C for 30–140 h (Malevski and Chzhun 1965).

7.28 Mercury–Tellurium–Sulfur

HgS–HgTe: The phase equilibria in this system are shown in Figure 7.27 (Malevski and Chzhun 1965). Mercury telluride and β-HgS form solid solutions in the wide region of concentrations (Nikol'skaya and Regel' 1955, Nikol'skaya 1963). At 600°C, solid solutions based on HgTe exist within the interval of 11–100 mol. % HgTe (Malevski and Chzhun 1965). These solid solutions decompose at temperatures below 600°C forming two phases with the sphalerite structures. The degradation region determined using XRD is situated at 500°C and 300°C within the interval of 16–68 and 14–88 mol. % HgTe, respectively.

According to the data of Shchennikov et al. (1995) and Kozlenko et al. (2002), the structure of the solid solutions $HgTe_{1-x}S_x$ at x up to 0.271 changes from a cubic of the sphalerite type to a hexagonal of the cinnabar type with the pressure increasing. The pressure transition decreases with

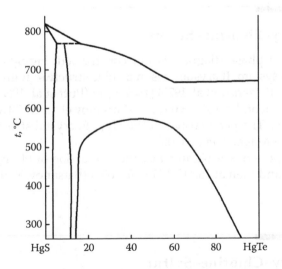

FIGURE 7.27
Phase equilibria in the HgS–HgTe system. (From Malevski, A.Yu. and Chzhun, Tszia-zhun, About isomorphous substitution of sulfur by selenium and tellurium in the mercury sulfides [in Russian], in *Eksperiment.-metodich. issled. rudnyh mineralov*, Moscow, Russia: Nauka Publish., pp. 223–226, 1965.)

x increasing (Shchennikov et al. 1995). At $x = 0.15$, solid solution begins to crystallize in the cinnabar-type structure at 1.6 GPa, and the sphalerite- and cinnabar-type structures coexist within the interval of pressure from 1.6 to 2.4 GPa (Kozlenko et al. 2002). It is possible that at pressure above 4.0 GPa, this solid solution can crystallize in the cubic structure of the NaCl type.

The solubility of HgTe in HgS decreases from 5 mol. % at 750°C to 3.5 mol. % at 300°C.

7.29 Mercury–Molybdenum–Sulfur

$Hg_xMo_6S_8$ Chevrel phases ($0 < x < 1$) are formed in the Hg–Mo–S ternary system, which crystallize in a rhombohedral structure with lattice parameters $a = 651$ pm and $\alpha = 92.53°$ (Tarascon et al. 1983).

This compound could be obtained by the interaction of the chemical elements at 1050°C–1250°C or by the oxidation of $Cu_2Mo_6S_8$ using iodine followed by Hg diffusion into Mo_6S_8 phase at 350°C for 3 days (Tarascon et al. 1983).

7.30 Mercury–Fluorine–Sulfur

HgS–HgF₂: The phase diagram is not constructed. $Hg_3S_2F_2$ compound is formed in this system. It crystallizes in a cubic structure with lattice parameter $a = 823$ pm (Batsanov et al. 1973) [$a = 814$ pm (Puff et al. 1968)] and calculation and experimental density 8.68 [8.39 (Batsanov et al. 1973)] and 8.64 g cm^{-3} (Puff et al. 1968). The peculiarities of the $Hg_3S_2F_2$ crystal structure are given in the review of Magarill et al. (2007).

$Hg_3S_2F_2$ compound was obtained by the interaction of Hg_2F_2 and sulfur at 200°C for 10 h and then at 150°C–170°C for 60 h (Batsanov et al. 1973).

7.31 Mercury–Chlorine–Sulfur

HgS–HgCl₂: The phase diagram is shown in Figure 7.28 (Pan'ko et al. 1989, Voroshilov et al. 1994b). Two eutectics are in this system. Eutectic composition and temperature for the first of them are 75 mol. % HgS and 525 ± 5°C. The second eutectic is degenerated from the HgCl₂-rich side at 280°C. The $Hg_3S_2Cl_2$ ternary compound is formed in the HgS–HgCl₂ system

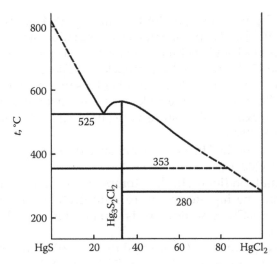

FIGURE 7.28

HgS–HgCl$_2$ phase diagram. (From Pan'ko, V.V. et al., *Zhurn. neorgan. khimii*, 34(12), 3208, 1989; Voroshilov, Yu.V. et al., *Visnyk L'viv. un-tu. Ser. khim.*, (33), 11, 1994b.)

(Batsanov and Abaulina 1961, Puff 1962, Pan'ko et al. 1989, Voroshilov et al. 1994b). It melts congruently at 548°C and has polymorphous transformation at 355 ± 5°C [340°C (Carlson 1967)].

α-Hg$_3$S$_2$Cl$_2$ crystallizes in a cubic body-centered structure with lattice parameter $a = 890.5 \pm 0.3$ pm (Voroshilov et al. 1994b) [$a = 892.5 \pm 0.3$ pm (Pan'ko et al. 1989, Voroshilov et al. 1994b), $a = 893.7$ pm (Puff and Kohlschmidt 1962, Puff and Küster 1962, Puff et al. 1966, Batsanov et al. 1973), $a = 894.9 \pm 0.2$ pm (Carlson 1967, Fruch and Gray 1968), $a = 894$ pm (Voroshilov et al. 1981), $a = 894.0 \pm 0.5$ pm (Food and Berendsen 1974), $a = 894.0 \pm 0.1$ pm (Aurivillius 1967)] and calculation and experimental density 6.827 [6.89 (Carlson 1967), 6.86 (Puff et al. 1968), 6.83 (Batsanov et al. 1973), 6.845 (Food and Berendsen 1974), 6.90 (Aurivillius 1967)] and 6.895 ± 0.086 g cm^{-3} [6.74 (Aurivillius 1967)], respectively. This compound exists in nature as the mineral corderoite (Food and Berendsen 1974, Vasilyev and Grechishchev 1979, Vasilyev and Lavrentyev 1986).

β-Hg$_3$S$_2$Cl$_2$ crystallizes in a monoclinic structure with lattice parameters $a = 1799.6 \pm 0.4$, $b = 1028.9 \pm 0.2$, $c = 928.1 \pm 0.4$ pm, and $\beta = 116.14 \pm 0.02°$ and experimental density 7.11 g cm^{-3} (Voroshilov et al. 1994b) [in a primitive cubic structure with lattice parameter $a = 1792.5 \pm 0.7$ pm (Voroshilov et al. 1996); $a = 1794.3 \pm 0.3$ pm (Pan'ko et al. 1989, Voroshilov et al. 1994b); $a = 1793.3$ pm (Puff et al. 1966)].

According to the data of Puff et al. (1966), Carlson (1967), Ďurovič (1968), Batsanov et al. (1973), and McCormack and Dickson (1998, 1999), metastable γ-Hg$_3$S$_2$Cl$_2$ modification exists in this system. It crystallizes in an orthorhombic structure with lattice parameters $a = 933.2 \pm 0.5$, $b = 1682 \pm 2$, and $c = 910.8 \pm 0.5$ pm [$a = 934.9$, $b = 1684.3$, and $c = 909.4$ pm (Puff et al. 1966, Batsanov et al. 1973)] and experimental and calculation density 6.83 ± 0.5 and 6.87 [6.76 (McCormack and Dickson 1999)] g cm^{-3}, respectively (Carlson 1967, McCormack and Dickson 1998, McCormack and Dickson 1999) [calculation density is equal to 6.82 g cm^{-3} (Puff et al. 1966, Batsanov et al. 1973)]. The structure of γ-Hg$_3$S$_2$Cl$_2$ belongs to the order–disorder structure (Ďurovič 1968). This modification exists in the nature as the mineral kenhsuite (McCormack and Dickson 1998, 1999).

The peculiarities of the Hg$_3$S$_2$Cl$_2$ crystal structure are given in the review of Magarill et al. (2007).

In hydrothermal conditions, the low-temperature modification α-Hg$_3$S$_2$Cl$_2$ is converted into the high-temperature modification β-Hg$_3$S$_2$Cl$_2$ at $300 \pm 0.5°C$, and γ-Hg$_3$S$_2$Cl$_2$ modification is metastable at all temperatures (Puff et al. 1966).

Optical energy gap of Hg$_3$S$_2$Cl$_2$ is equal to 3.2 eV (Batsanov et al. 1973).

This system was investigated using DTA and XRD. The ingots were annealed for 10 days. α-Hg$_3$S$_2$Cl$_2$ was obtained by the interaction of H$_2$S with low-concentration solution of HgCl$_2$ or by the annealing of mixtures from HgS and HgCl$_2$ at temperatures below 300°C (Pan'ko et al. 1989, Voroshilov et al. 1994b). Powder samples of this compound were synthesized by boiling freshly precipitated black HgS in an acidified solution of HgCl$_2$ for some hours (Aurivillius 1967). β-Hg$_3$S$_2$Cl$_2$ was synthesized by the annealing of mixtures from HgS and HgCl$_2$ at temperatures up to 400°C, and γ-Hg$_3$S$_2$Cl$_2$ was obtained by quenching of mixtures HgS + HgCl$_2$ from 750°C or by the interaction of diluted HgCl$_2$ alcoholic solution with CS$_2$ (Puff et al. 1966). The Hg$_3$S$_2$Cl$_2$ compound was obtained also by the interaction of Hg$_2$Cl$_2$ with S (Batsanov et al. 1973). Its single crystals were grown by the sublimation method (Takei and Hagiwara 1976, Voroshilov et al. 1981) or by the interaction of solid HgS and gaseous HCl at temperature scheme 398°C–295°C (Carlson 1967).

7.32 Mercury–Bromine–Sulfur

HgS–HgBr$_2$: The phase diagram is shown in Figure 7.29 (Pan'ko et al. 1989, Voroshilov et al. 1994b). Two eutectics are in this system that crystallize at 515°C and 240°C (eutectic from the HgBr$_2$-rich side is degenerated). The Hg$_3$S$_2$Br$_2$ ternary compound is formed in the HgS–HgBr$_2$ system (Batsanov and Abaulina 1961, Puff 1962, Puff and Kohlschmidt 1962, Puff et al. 1966, Batsanov et al. 1973, Pan'ko et al. 1989, Voroshilov et al. 1994b). It melts congruently at 532°C and has polymorphous transformation at $360 \pm 5°C$ (Pan'ko et al. 1989, Voroshilov et al. 1994b).

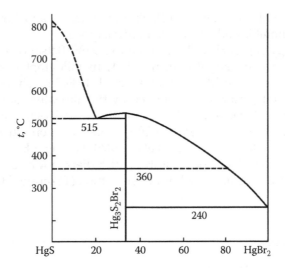

FIGURE 7.29

HgS–HgBr$_2$ phase diagram. (From Pan'ko, V.V. et al., *Zhurn. neorgan. khimii*, 34(12), 3208, 1989; Voroshilov, Yu.V. et al., *Visnyk L'viv. un-tu. Ser. khim.*, (33), 11, 1994b.)

α-Hg$_3$S$_2$Br$_2$ crystallizes in a monoclinic structure with lattice parameters $a = 1028.9 \pm 0.2$, $b = 1632.8 \pm 0.4$, $c = 928.1 \pm 0.4$ pm, and $\beta = 98.32°$ (Voroshilov et al. 1996b) [in an orthorhombic structure with lattice parameters $a = 1820 \pm 2$, $b = 919 \pm 2$, and $c = 926.2 \pm 0.8$ pm (Pan'ko et al. 1989, Voroshilov et al. 1994b); $a = 3689$, $b = 1808$, and $c = 928$ pm (Puff et al. 1966)] and calculation density 7.110 g cm^{-3} (Voroshilov et al. 1996b).

β-Hg$_3$S$_2$Br$_2$ crystallizes in a monoclinic structure with lattice parameters $a = 1727.3 \pm 0.4$, $b = 937.4 \pm 0.3$, $c = 947.3 \pm 0.3$ pm, and $\beta = 89.78°$ (Voroshilov et al. 1996b) [in a tetragonal structure with lattice parameters $a = 1322 \pm 2$ and $c = 904 \pm 4$ pm (Pan'ko et al. 1989, Voroshilov et al. 1994b); in an orthorhombic structure with lattice parameters $a = 1822$, $b = 919$, and $c = 924$ pm (Puff et al. 1966, Batsanov et al. 1973)] and calculation density 7.10 g cm^{-3} (Batsanov et al. 1973) [7.13 g cm^{-3} (Voroshilov et al. 1996b)].

γ-Hg$_3$S$_2$Br$_2$ crystallizes in a tetragonal structure with lattice parameters $a = 1314$ and $c = 889$ pm and calculation density 7.15 g cm^{-3} (Puff et al. 1966, Batsanov et al. 1973).

The peculiarities of the Hg$_3$S$_2$Br$_2$ crystal structure are given in the review of Magarill et al. (2007).

It can be seen that there are some discrepancies between the results of different authors. The γ-Hg$_3$S$_2$Br$_2$ and β-Hg$_3$S$_2$Br$_2$ modifications from Puff et al. (1966) and Batsanov et al. (1973) correspond to β-Hg$_3$S$_2$Br$_2$ and α-Hg$_3$S$_2$Br$_2$ modifications, respectively, from Pan'ko et al. (1989), Voroshilov et al. (1994b), and a and b parameters for α-Hg$_3$S$_2$Br$_2$ in Puff et al. (1966) are two times greater than in Pan'ko et al. (1989) and Voroshilov et al. (1994b). Therefore, one can conclude that this ternary compound exists only in two modifications, which are given in Pan'ko et al. (1989) and Voroshilov et al. (1994b): orthorhombic and tetragonal modifications.

Optical energy gap of $Hg_3S_2Br_2$ is equal to 2.9 eV (Batsanov et al. 1973).

This system was investigated using DTA, metallography, and XRD. The ingots were annealed for 10 days (Pan'ko et al. 1989, Voroshilov et al. 1994b). α-$Hg_3S_2Br_2$ and β-$Hg_3S_2Br_2$ were obtained by the annealing of mixtures from HgS and $HgBr_2$ at 200°C and 420°C, respectively, and γ-$Hg_3S_2Br_2$ (metastable modification) by the interaction of $HgBr_2$ water solution, enriched by KBr, with sodium thiosulfate at 80°C (Puff et al. 1966). The $Hg_3S_2Br_2$ compound was obtained also by the interaction of Hg_2Br_2 with S (Batsanov et al. 1973). Its single crystals were grown using the sublimation method (Takei and Hagiwara 1976).

7.33 Mercury–Iodine–Sulfur

HgS–HgI₂: The phase diagram is shown in Figure 7.30 (Pan'ko et al. 1989, Voroshilov et al. 1994b). Two eutectics are in this system that crystallize at 330±5 and 255±5°C (eutectic from the HgI_2-rich side is degenerated). An immiscibility region within the interval of 20–80 mol. % HgS exists in the HgS–HgI_2 system with monotectic temperature 385°C.

The $Hg_3S_2I_2$ ternary compound is formed in this system (Batsanov and Abaulina 1961, Puff 1962, Pan'ko et al. 1989, Voroshilov et al. 1994b), which

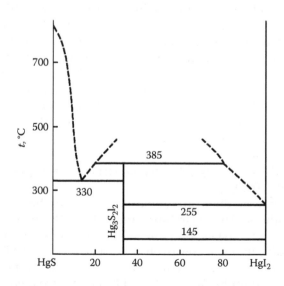

FIGURE 7.30

HgS–HgI_2 phase diagram. (From Pan'ko, V.V. et al., *Zhurn. neorgan. khimii*, 34(12), 3208, 1989; Voroshilov, Yu.V. et al., *Visnyk L'viv. un-tu. Ser. khim.*, (33), 11, 1994b.)

melts syntactically at $385 \pm 5°C$ (Pan'ko et al. 1989, Voroshilov et al. 1994b) and crystallizes in an orthorhombic structure with lattice parameters $a = 979.92 \pm 0.08$, $b = 1870.3 \pm 0.3$, and $c = 946.22 \pm 0.07$ pm (Beck and Hedderich 2000) [$a = 966.2 \pm 0.9$, $b = 1877 \pm 6$, and $c = 933 \pm 1$ pm (Pan'ko et al. 1989, Voroshilov et al. 1994b); $a = 969$, $b = 1850$, and $c = 936$ pm (Takei and Hagiwara 1976)]. Phase transitions of this compound were not determined within the temperature interval from $-160°C$ to $300°C$ (Beck and Hedderich 2000).

According to the data of Puff and Kohlschmidt (1962), Batsanov et al. (1973), and Voroshilov et al. (1981), this compound crystallizes in a cubic structure with lattice parameter $a = 9.66$ pm (Puff and Kohlschmidt 1962, Batsanov et al. 1973) [$a = 9.68$ pm (Voroshilov et al. 1981)]. Calculation density and energy gap of $Hg_3S_2I_2$ are equal to 6.77 [7.045 (Beck and Hedderich 2000)] g cm^{-3} and 2.5 eV, respectively (Batsanov et al. 1973). Thermal effects at $145°C$ correspond to the polymorphous transformation of HgI_2 (Pan'ko et al. 1989, Voroshilov et al. 1994b).

The peculiarities of the $Hg_3S_2I_2$ crystal structure are given in the review of Magarill et al. (2007).

This system was investigated using DTA, metallography, and XRD. The ingots were annealed for 10 days (Pan'ko et al. 1989, Voroshilov et al. 1994b). The $Hg_3S_2I_2$ compound was obtained by the interaction of Hg_2I_2 with S (Batsanov et al. 1973). Its single crystals were grown using the sublimation method (Takei and Hagiwara 1976, Voroshilov et al. 1981) or by the interaction of HgS and HgI_2 in evacuated glass ampoules at $260°C$ (Beck and Hedderich 2000).

7.34 Mercury–Manganese–Sulfur

HgS–MnS: The phase equilibria in this system have been discussed in Pajaczkowska (1978) based on papers that have appeared up to 1977. The phase diagram is not constructed. The solubility of MnS in HgS at $600°C$ and $300°C$ is equal to 37.5 and 10 mol. %, respectively (Pajaczkowska and Rabenau 1977). According to the data of Rodic et al. (1996), $Hg_{1-x}Mn_xS$ ($x \leq 0.33$) crystallizes in a sphalerite structure. The solubility of HgS in MnS at the same temperatures is insignificant (Pajaczkowska and Rabenau 1977).

High-pressure–high-temperature x-ray energy dispersive measurements performed for $Hg_{1-x}Mn_xS$ ($x = 0.02$ and 0.07) enabled to investigate the p-T conditions of the sphalerite–cinnabar phase transition for these solid solutions (Paszkowicz et al. 1999). It is demonstrated that at a given temperature, the transition pressure increases with increasing Mn content in the single crystal. For $Hg_{0.98}Mn_{0.02}S$, the phase transition on uploading occurs at 1.25 GPa, and for $Hg_{0.93}Mn_{0.07}S$, it occurs at 2.44 GPa. The bulk modulus value $B_0 = 38.79$ GPa and its first pressure derivative $B_0' = 2.3$. The solubility of the

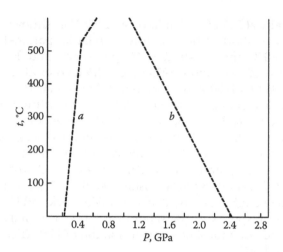

FIGURE 7.31
Tentative limits of α-Hg$_{1-x}$Mn$_x$S and β-Hg$_{1-x}$Mn$_x$S phase occurrence at the *p-T* diagram. (From Paszkowicz, W. et al., *J. Alloys Compd.*, 286(1–2), 208, 1999.)

MnS in α-HgS is at a level of 2 mol. %, or even less, that is, lower than that (10 mol. %) evaluated in Pajaczkowska and Rabenau (1977). Rodic et al. (1996) noted that a pressure of 1.8 GPa applied to Hg$_{0.95}$Mn$_{0.05}$S induces a phase separation from sphalerite to cinnabar structure type.

Tentative limits of α-Hg$_{1-x}$Mn$_x$S and β-Hg$_{1-x}$Mn$_x$S phase occurrence at the *p-T* diagram are shown in Figure 7.31 (Paszkowicz et al. 1999). β-Hg$_{1-x}$Mn$_x$S is observed to the left of line "*a*," and α-Hg$_{1-x}$Mn$_x$S is observed to the right of line "*b*." In between lines "*a*" and "*b*," either one of these phases is observed or their mixture, depending on the thermal history.

Mutual solubility was investigated using XRD (Pajaczkowska and Rabenau 1977). The crystal growth of the Hg$_{1-x}$Mn$_x$S solid solutions is given in Pajaczkowska (1978).

7.35 Mercury–Cobalt–Sulfur

HgS–CoS: The phase diagram is not constructed. Solid-solution Hg$_{0.972}$Co$_{0.028}$S with a sphalerite-type structure starts to transform at room temperature to a cinnabar-type structure at about 0.7 GPa and ends at about 2.2 GPa (Paszkowicz et al. 1998). At 650°C, the phase transition occurs at 1.7 GPa.

Hg$_{1-x}$Co$_x$S solid solutions were grown using the Bridgman method (Paszkowicz et al. 1998).

References

Asadov M.M. Phase diagram of the Hg–Tl–S system [in Russian], *Izv. AN SSSR. Neorgan. materialy*, **19**(10), 1626–1629 (1983a).

Asadov M.M. The HgS+Tl₂Se ⇔ HgSe+Tl₂S mutual system [in Russian], *Zhurn. neorgan. khimii*, **28**(7), 1812–1816 (1983b).

Asadov M.M. Mercury–thallium–sulfur system from the Hg-rich side [in Russian], *Zhurn. neorgan. khimii*, **29**(11), 2873–2876 (1984).

Asadov M.M. Interaction of liquid with solid phases in the HgS–Tl₄S₃(TlS) sections [in Russian], *Zhurn. neorgan. khimii*, **30**(10), 2653–2655 (1985).

Asadov M.M., Mustafaieva S.N. Solubility of thallium in mercury sulfide and relaxation current in the α-HgS crystals doping by the thallium [in Russian], *Izv. AN SSSR. Neorgan. materialy*, **23**(9), 1561–1563 (1987).

Aurivillius K. An x-ray single crystal study of Hg₃S₂Cl₂, *Arkiv Kemi*, **26**(6), 497–505 (1967).

Aurivillius K., Malmros B. Studies on sulphates, selenates and chromates of mercury (II), *Acta Chem. Scand.*, **15**(9), 1932–1938 (1961).

Axtell E.A. (III), Park Y., Chondroudis K., Kanatzidis M.G. Incorporation of A₂Q into HgQ and dimensional reduction to A₂Hg₃Q₄ and A₂Hg₆Q₇ (A=K, Rb, Cs; Q = S, Se), *J. Amer. Chem. Soc.*, **120**(1), 124–136 (1998).

Babanly M.B., Kurbanov A.A., Kuliev A.A. Phase equilibria and intermolecular interaction in the HgS–Sb₂S₃(Bi₂S₃) systems [in Russian], *Izv. AN SSSR. Neorgan. materialy*, **16**(3), 547–548 (1980).

Badikov V.V., Matveev I.N., Paniutin V.L., Pshenichnikov S.M., Repiahova S.M., Rychik O.V., Rozenson A.E., Trotsenko N.K., Ustinov N.D. Growth and optical properties of mercury thiogallate [in Russian], *Kvantovaya elektronika*, **6**(8), 1807–1809 (1979).

Batsanov S.S., Abaulina L.I. Interaction of mercury halogenides with chalcogens [in Russian], *Izv. SO AN SSSR*. (10), 67–73 (1961).

Batsanov S.S., Kolomiychuk V.N., Derbeneva S.S., Erenburg R.S. Synthesis and physico-chemical investigation of mercury halogenchalcogenides [in Russian], *Izv. AN SSSR. Neorgan. materialy*, **9**(7), 1098–1104 (1973).

Beck J., Hedderich S. Synthesis and crystal structure of Hg₃S₂I₂ and Hg₃Se₂I₂, new members of the Hg₃E₂X₂ family, *J. Solid State Chem.*, **151**(1), 73–76 (2000).

Beun J.A., Nitsche R., Lichtensteiger M.L. Photoconductivity in ternary sulfides, *Physica*, **26**(8), 647–649 (1960).

Beun J.A., Nitsche R., Lichensteiger M. Optical and electrical properties of ternary chalcogenides, *Physica*, **27**(5), 448–452 (1961).

Bugaris D.E., Ibers J.A. Dicaesium hexamercury heptasulfide, *Acta Crystallogr.*, **E64**(9), i55–i56 (2008).

Carlson E.H. The growth of HgS and Hg₃S₂Cl₂ single crystals by a vapor phase method, *J. Cryst. Growth*, **1**(5), 271–277 (1967).

Charbonnier M. Contribution à l'étude des sulfures simples et mixtes des métaux des groupes I B (Cu–Ag) et II B (Zn–Cd–Hg): These doct. sci. phys. Univ. Claude-Bernard.—Lyon, France, 164p. (1973).

Craig J.R. Livingstonite, HgSb₄S₈: Synthesis and stability, *Amer. Mineralogist*, **55**(5–6), 919–924 (1970).

Donika F.G., Zhitar' V.F., Radautsan S.I. *Semiconductors of the ZnS-In$_2$S$_3$ System* [in Russian], Kishinev, Shtiintsa Publish., 148 p. (1980).

Du Y., Chang H., Li Z., Tang G., Jin Z. Sulfides of rare earths with IIB and 3d elements, *New Frontiers Rare Earth Sci. and Appl. Proc. Int. Conf., Beijing*, Sept. 10–14, 1985. Vol. 1, Beijing, China, pp. 370–373 (1985).

Ďurovič S. The crystal structure of γ-Hg$_3$S$_2$Cl$_2$, *Acta Crystallogr.*, **B24**(12), 1661–1670 (1968).

Food E.E., Berendsen P. Corderoite, first mineral occurrence of α-Hg$_3$S$_2$Cl$_2$, from the Cordero Mercury Deposit, Humboldt County, Nevada, *Amer. Mineralogist*, **59**(7–8), 652–655 (1974).

Fruch A.J., Gray N. Confirmation and refinement of the structure of Hg$_3$S$_2$Cl$_2$, *Acta Crystallogr.*, *B*, **24**(1), 156–157 (1968).

Ganal P., Moreau P., Ouvrard G., Sidorov M., McKelvy M., Glaunsinger W. Structural investigation of mercury-intercalated titanium disulfide. I. The crystal structure of Hg$_{1.24}$TiS$_2$, *Chem. Mater.*, **7**(6), 1132–1139 (1995).

Garner W.R., White W.B. Growth of cinnabar (HgS) from sodium sulfide–sulfur fluxes, *J. Cryst. Growth*, **7**(3), 343–347 (1970).

Golovey M.I. Obtaining and some properties of ternary chalcogenides in the A^I(A^{II})–B^V–C^{VI} systems [in Russian], in: *XI Mendeleevski s'ezd po obshch. i prikl. khimii, Alma-Ata*, Sept. 1975: Ref. dokl. i soobshch. N 1, Moscow, Russia: Nauka Publishing, pp. 52–53 (1975).

Golovey M.I., Gurzan M.I., Olekseyuk I.D., Lada A.V., Bogdanova A.V., Pan'ko V.V., Voroshilov Yu.V., Peresh E.Yu. Obtaining, properties and application of single crystals of ternary chalcogenides based on I, II, III and IV groups of the periodical system [in Russian], in: *Fizika i khimia hal'kogenidov*. Kiev, Ukraine: Nauk. dumka, pp. 6–11 (1977).

Golovey M.I., Olekseyuk I.D., Semrad E.E., Gurzan M.I. About chemical interaction in the A^I–B^V–C^{VI} ternary systems [in Russian], in: *Khim. sviaz' v poluprovodn.*, Minsk, Belarus: Nauka i tehnika, pp. 235–243 (1969).

Golovey M.I., Olekseyuk I.D., Voroshilov Yu.V. Chemical bond and solubility in the systems A$^{II}_3$B$^V_2$–A^{II}C^{VI} [in Russian], in: *Khim. svyaz' v poluprovodn. i polumetallah*. Minsk, Belarus: Nauka i Tekhnika Publishing, pp. 233–239 (1972).

Gorman D.H. An x-ray study of the mineral livingstonite, *Amer. Mineralogist*, **36**(5–6), 480–483 (1951).

Gruzdev V.S., Stepanov V.I., Shumkova N.G., Chernitsova N.M., Yudin R.N., Bryzgalov I.A. Galhaite HgAsS$_2$—The new mineral from arsenic–antimony–mercury ores of the SSSR [in Russian], *Dokl. AN SSSR*, **205**(5), 1194–1197 (1972).

Gulay L.D., Olekseyuk I.D., Parasyuk O.V. Crystal structure of the Hg$_4$SiS$_6$ and Hg$_4$SiSe$_6$ compounds, *J. Alloys Comp.*, **347**(1–2), 115–120 (2002).

Guseinov G.D. Searching and physical investigation of new semiconductors-analogues [in Russian], Avtoref. dis. ... doct. fiz.-mat. nauk.—Baku, Azerbaijan, 82p. (1969).

Guseinov G.D., Ismailov M.Z., Talybov A.G. About structure and some properties of HgTlS$_2$ [in Russian], *Izv. AN SSSR. Neorgan. materialy*, **4**(4), 514–517 (1968).

Hahn H., Frank G., Klingler W., Störger A.D., Störger G. Über ternäre Chalkogenide des Aluminiums, Galliums und Indiums mit Zink, Cadmium und Quecksilber, *Z. anorg. und allg. Chem.*, **279**(5/6), 241–270 (1955).

Hahn H., Klingler W. Über die Kristallstruktur einiger ternärer Sulfide, die sich vom Indium (III) Sulfid ableiten, *Z. anorg. und allg. Chem.*, **263**(4), 177–190 (1950).

Hahn H., Lorent C. Untersuchungen über ternäre Chalkogenide. Über ternäre Sulfide und Selenide des Germaniums mit Zink, Cadmium und Quecksilber, *Naturwissenschaften*, **45**(24), 621–622 (1958).

Il'iasheva N.A., Nenashev B.G., Siniakov I.V. Investigations in the Hg–Ga–S system [in Russian], in: *Fiz.-khim. issled. mineraloobrazuyushch. sistem*, Novosibirsk, Russia, pp. 23–29 (1982).

Il'iasheva N.A., Nenashev B.G., Siniakov I.V., Siniakova E.F. Investigation of the HgS–Ga_2S_3 system by the oriented crystallization [in Russian], in: *Fiz.-khim. issled. sul'fid. i silikat. sistem*, Novosibirsk, Russia, pp. 15–21 (1984).

Il'iasheva N.A., Siniakova E.F., Nenashev B.G., Siniakov I.V. The HgS–Ga_2S_3 phase diagram [in Russian], *Izv. AN SSSR. Neorgan. materialy*, **21**(11), 1860–1864 (1985).

Jambor J.L., Roberts A.C. New mineral names, *Amer. Mineralogist*, **84**(9), 1464–1468 (1999).

Jandali M.Z., Eulenberger G., Hahn H. Darstellung und Kristallstruktur des Quecksilber (II)-thiodiphosphats $Hg_2P_2S_7$, *Z. anorg. und allg. Chem.*, **445**(8), 192 (1978a).

Jandali M.Z., Eulenberger G., Hahn H. Die Kristallstrukturen von $Hg_2P_2S_6$ und $Hg_2P_2Se_6$, *Z. anorg. und allg. Chem.*, **447**(10), 105–118 (1978b).

Kanarzidis M.G., Park Y. Molten salt synthesis of low-dimensional ternary chalcogenides. Novel structure types in the K/Hg/Q system (Q=S, Se), *Chem. Mater.*, **2**(2), 99–101 (1990).

Kaplunnik L.N., Pobedimskaya E.A., Belov N.V. Crystal structure of the galhaite $HgAsS_2$ [in Russian], *Dokl. AN SSSR*, **225**(3), 561–563 (1975).

Kirilenko V.V., Poliakov Yu.A. The As_2S_3–HgS system [in Russian], *Izv. AN SSSR. Neorgan. materialy*, **12**(2), 336–338 (1976).

Klepp K.O. $Na_2Hg_3S_4$: A thiomercurate with layered anions, *J. Alloys Compd.*, **182**(2), 281–288 (1992).

Klinger W., Eulenberger G., Hahn H. Über Hexachalkogeno-hypodiphosphate vom Typ $M_2P_2X_6$, *Naturwissenschaften*, **57**(2), 88 (1970).

Klinger W., Ott R., Hahn H. Über die Darstellung und Eigenschaften von Hexathio- und Hexaselenohypodiphosphaten, *Z. anorg. und allg. Chem.*, **396**(3), 271–278 (1973).

Koreliakov Yu.A., Komarov O.E., Nikol'skaya E.I. Investigation of structural properties of mercury selenide—Mercury sulfide system [in Russian], in: *Uchen. zap. Kalinin. ped. in-ta*, **40**, 94–96 (1966).

Kozlenko D.P., Shchennikov V.V., Voronin V.I., Glazkov V.P., Savenko B.N. Neutron diffraction study of a structure transition in the ternary compound $HgTe_{0.85}S_{0.15}$ at pressure [in Russian], *Fiz. tv. tela*, **44**(9), 1553–1556 (2002).

Krausbauer L., Nitsche R., Wild P. Mercury gallium sulfide, $HgGa_2S_4$, a new phosphor, *Physica*, **31**(1), 113–121 (1965).

Kulakov M.P. The melting temperature and vapor pressure of HgS [in Russian], *Izv. AN SSSR. Neorgan. materialy*, **11**(3), 553–554 (1975).

Kulakov M.P., Sokolovskaya Zh.D. Differential thermal analysis of the HgS–PbS system [in Russian], *Zhurn. neorgan. khimii*, **20**(8), 2290–2291 (1975).

Kulakov M.P., Sokolovskaya Zh.D., Sorokin V.I. The HgS–$Cu_{2-x}S$ system [in Russian], *Izv. AN SSSR. Neorgan. materialy*, **21**(4), 551–555 (1985).

Kuliev A.A., Asadov M.M., Kuliev R.A., Babanly M.B. Phase equilibria in the Tl_2S–HgS and Tl_2Se–HgS systems [in Russian], *Zhurn. neorgan. khimii*, **23**(3), 854–856 (1978).

Kuliev A.A., Babanly M.B., Asadov M.M., Kulieva N.A., Guseinov F.H., Kurbanov A.A. Phase diagrams and thermodynamic properties of the Tl–Zn(Cd,Hg,Ge)–S systems [in Russian], in: *12th Mendeleev. s'ezd po obshch. i prikl. khimii. Ref. dokl. i soobshch.* N 1, M., pp. 68–69 (1981).

Magarill S.A., Pervukhina N.V., Borisov S.V., Pal'chik N.A. Crystal chemistry and features of structure formation of mercury oxo- and chalcogenides [in Russian], *Uspekhi khimii*, **76**(2), 115–146 (2007).

Malevski A.Yu., Chzhun Tszia-zhun. About isomorphous substitution of sulfur by selenium and tellurium in the mercury sulfides [in Russian], in: *Eksperiment.-metodich. issled. rudnyh mineralov*. Moscow, Russia: Nauka Publishing, pp. 223–226 (1965).

Mandarino J.A. Grumiplucite $HgBi_2S_4$, *Mineral Rec.*, **30**(5), 401 (1999).

Marchuk O.V., Gulay L.D., Parasyuk O.V. The Cu_2S–HgS–GeS_2 system at 670 K and the crystal structure of the $Cu_6Hg_{0.92}GeS_{5.92}$ compound, *J. Alloys Compd.*, **333**(1–2), 143–146 (2002).

McCormack J.K., Dickson F.W. Kenhsuite, gamma-$Hg_3S_2Cl_2$, a new mineral species from McDermitt mercury deposit, Humboldt County, Nevada, *Can. Miner.*, **36**(1), 201–206 (1998).

McCormack J.K., Dickson F.W. Kenhsuite $Hg_3S_2Cl_2$, *Mineral Rec.*, **30**(3), 234 (1999).

Moreau P., Ganal P., Lemaux S., Ouvrard G., McKelvy M. Mercury sublattice melting transition in the misfit intercalation compound $Hg_{1.24}TiS_2$, *J. Phys. Chem. Solids*, **57**(6–8), 1129–1132 (1996).

Moreau P., Ganal P., Marie A.M., Ouvrard G. TEM study of the misfit intercalation compounds α-$Hg_{1.19}TaS_2$ and β-$Hg_{1.3}TaS_2$, *Inorg. Chem.*, **34**(22), 5496–5500 (1995).

Motria S.F. Mercury–germanium(tin)–sulfur(selenium) ternary systems [in Russian], in: *Poluch. i svoistva slozhn. poluprovodn., Uzhgorod. gos. un-t*, Kiev, Ukraine, pp. 17–26 (1991).

Motria S.F., Semrad E.E., Voroshilov Yu.V., Yatskovich I.I. Physico-chemical investigation of the HgS–SnS, HgS–SnS_2 systems [in Russian], *Zhurn. neorgan. khimii*, **33**(8), 2103–2105 (1988).

Mumme W.G., Watts J.A. $HgBi_2S_4$: Crystal structure and relationship with the pavonite homologous series, *Acta Crystallogr.*, **B36**(6), 1300–1304 (1980).

Nagorsen G., Lyng S., Weiss A., Weiss A. Zur Konstitution von $HgSO_4 \cdot 2HgO$, *Angew. Chem.*, **74**(3), 119 (1962).

Nikol'skaya E.I. Structural properties of the mercury telluride–mercury sulfide system [in Russian], *Uchen. zap. Kalinin. ped. in-ta*, **33**, 61–66 (1963).

Nikol'skaya E.I., Regel' A.R. Solid solution formation and magnetic susceptibility in the $HgTe$–$HgSe$, $HgTe$–β-HgS, $HgSe$–β-HgS systems [in Russian], *Zhurn. tekhn. fiziki*, **25**(8), 1347–1351 (1955).

Olekseyuk I.D., Golovey M.I. Obtaining and investigation of some properties of Hg_3PS_4 and Hg_3PS_3 compounds [in Russian], *Izv. AN SSSR. Neorgan. materialy*, **4**(10), 1676–1680 (1968).

Olekseyuk I.D., Mazurets I.I., Parasyuk O.V. Phase equilibria in the HgS–Ga_2S_3–GeS_2 system, *J. Alloys Compd.*, **417**(1–2), 131–137 (2006).

Olekseyuk I.D., Tsitrovskiy V.V., Turyanitsa I.D., Stoyka I.M., Chukhno T.A. Obtaining and properties of modulation and nonlinear materials based on some chalcogenides [in Russian], *Kvantovaya elektronika*, (13), 93–96 (1977).

Ollitrault-Fichet R., Rivet J., Flahaut J. Diagramme de phase du système Hg–Cu–S: étude du triangle HgS–Cu$_2$S–S, *J. Less-Common Metals*, **96**(1), 49–62 (1984).

Ong E.W., McKelvy M.J., Ouvrard G., Glaunsinger W.S. Mercury intercalates of titanium disulfide: Novel intercalation compounds, *Chem. Mater.*, **4**(1), 14–17 (1992).

Orlandi P., Dini A., Olmi F. Grumiplucite, a new mercury–bismuth sulfosalt from the Levigliani Mine, Apuan Alps, Tuscany, Italy, *Can. Miner.*, **36**(5), 1321–1326 (1998).

Pajaczkowska A. Physicochemical properties and crystal growth of A^{II}B^{VI}–MnB^{VI} systems, *Progr. Crystal Growth Caract.*, **1**(3), 289–326 (1978).

Pajaczkowska A., Rabenau A. Phase studies and hydrothermal synthesis in the system mercury sulfide–manganese sulfide, *J. Solid State Chem.*, **21**(1), 43–48 (1977).

Pan'ko V.V., Hudoliy V.A., Voroshilov Yu.V., Shelemba M.S. Phase diagrams of the HgS–HgCl$_2$(Br$_2$, I$_2$) systems [in Russian], *Zhurn. neorgan. khimii*, **34**(12), 3208–3210 (1989).

Parasyuk O.V., Gulay L.D., Piskach L.V., Galagowska O.P. The Ag$_2$S–HgS–GeS$_2$ system at 670 K and the crystal structure of the Ag$_2$HgGeS$_4$ compound, *J. Alloys Compd.*, **336**(1–2), 213–217 (2002).

Paszkowicz W., Szuszkewicz W., Dynowska E., Domagała J., Witkowska B., Marczak M., Zinn P. High pressure—High temperature study of Hg$_{1-x}$Mn$_x$S, *J. Alloys Compd.*, **286**(1–2), 208–212 (1999).

Paszkowicz W., Szuszkiewicz W., Szamota-Sadowska K., Domagala J., Witkowska B., Marczak M., Zinn P. X-ray diffraction study of sphalerite-cinnabar phase transition in Hg$_{1-x}$Co$_x$S, *Bull. Czech. Slov. Crystallogr. Assoc.*, **5**(Spec. Issue B), 180 (1998).

Puff H. Ternäre Verbindungen von Quecksilberhalogeniden mit Elementen der 5. und 6. Hauptgruppe, *Angew. Chem.*, **74**(16), 659 (1962).

Puff H., Harpain A., Hoop K.P. Polymorphie bei Quecksilberschwefel–Halogeniden, *Naturwissenschaften*, **53**(11), 274 (1966).

Puff H., Heine D., Lieck C. Quecksilberschwefelfluorid, *Naturwissenschaften*, **55**(6), 298 (1968).

Puff H., Kohlschmidt R. Quecksilberchalkogenid-halogenide, *Naturwissenschaften*, **49**(14), 299 (1962).

Puff H., Küster J. Die Kristallstruktur von Hg$_3$S$_2$Cl$_2$, *Naturwissenschaften*, **49**(13), 299 (1962).

Rad H.D., Hoppe R. Zur Kenntnis von Ba$_2$HgS$_3$, *Z. anorg. und allg. Chem.*, **483**(12), 7–17 (1981a).

Rad H.D., Hoppe R. Zur Kenntnis von Ba[HgS$_2$], *Z. anorg. und allg. Chem.*, **483**(12), 18–25 (1981b).

Range K.-J., Becker W., Weiss A. Über Hochdruckphasen des CdAl$_2$S$_4$, HgAl$_2$S$_4$, ZnAl$_2$Se$_4$, CdAl$_2$Se$_4$ und HgAl$_2$Se$_4$ mit Spinellstruktur, *Z. Naturforsch.*, **23B**(7), 1009 (1968).

Richmond W.E. Crystallography of livingstonite, *Amer. Mineralogist*, **21**(11), 719–720 (1936).

Rodic D., Spasojevic V., Bajorek A., Onnerud P. Similarity of structure properties of Hg1–xMnxS and Cd1–xMnxS (structure properties of HgMnS and CdMnS), *J. Magn. Magn. Mater.*, **152**(1–2), 159–164 (1996).

Schwer H., Krämer V. The crystal structures of $CdAl_2S_4$, $HgAl_2S_4$ and $HgGa_2S_4$, *Z. Kristallogr.*, **190**(1–2), 103–110 (1990).

Serment J., Perez G., Hagenmuller P. Les systèmes SiS_2–MS et GeS_2–MS (M = Cd, Hg) entre 800 et 1000°C, *Bull. Soc. Chim. France*, (2), 561–566 (1968).

Shao M., Kong L., Hu B., Yu G., Qian Y. Solvo-displacement route to ternary compounds Ag–M–S (M = Ga, Cu or Hg), *Inorg. Chem. Commun.*, **6**(5), 555–557 (2003).

Sharma N.C., Pandya D.K., Sehgal H.K., Chopra K.L. The structural properties of $Pb_{1-x}Hg_xS$ films of variable optical gap, *Thin Solid Films*, **42**(3), 383–391 (1977).

Shchennikov V.V., Gavaleshko N.P., Frasuniak V.M. Semiconductor–metal transitions in HgMgTe and HgSeS at superhigh pressure [in Russian], *Fiz. tv. tela*, **35**(2), 389–394 (1993).

Shchennikov V.V., Gavaleshko N.P., Frasuniak V.M. Phase transition in the HgTeS crystals at high pressure [in Russian], *Fiz. tv. tela*, **37**(11), 3532–3535 (1995).

Sidorov M.V., McKelvy M.J., Cowley J.M., Glaunsinger W.S. Novel guest-layer behavior of mercury titanium disulfide intercalates, *Chem. Mater.*, **10**(11), 3290–3293 (1998).

Siniakova E.F., Il'iasheva I.A. Investigation of the Hg–Ga–S system [in Russian], *Izv. AN SSSR. Neorgan. materialy*, **22**(10), 1625–1629 (1986).

Siniakova E.F., Il'iasheva I.A., Pavliuchenko V.S. Homogeneity region of digallium–mercury tetrasulfide and phase relations in the Hg–Ga–S system [in Russian], *Izv. AN SSSR. Neorgan. materialy*, **23**(3), 382–386 (1987).

Soklakov A.I., Nechaeva V.V. Obtaining and x-ray investigation of the Cu, Hg, W, Sr, Sb, As and Pb thiophosphates [in Russian], *Izv. AN SSSR. Neorgan. materialy*, **5**(5), 989 (1969).

Sommer H., Hoppe R. Thio- und Selenomercurate (II). $K_6[HgS_4]$, $K_6[HgSe_4]$, $Rb_6[HgS_4]$ und $Rb_6[HgSe_4]$, *Z. anorg. und allg. Chem.*, **443**(6), 201–211 (1978).

Takei K., Hagiwara H. The synthesis of single crystals of $HgX_2 \cdot 2HgS$ (X = halogen), *Bull. Chem. Soc. Jap.*, **40**(5), 1425–1426 (1976).

Tarascon J.M., Waszczak J.V., Hull G.W., Di Salvo F.J., Blitzer L.D. Synthesis and physical properties of new superconducting Chevrel phases $Hg_xMo_6S_8$, *Solid State Commun.*, **47**(12), 973–979 (1983).

Vasilyev V.I., Grechishchev O.K. The fist finding of corderoite (α-$Hg_3S_2Cl_2$) in the mercury ores of USSR [in Russian], *Dokl. AN SSSR*, **246**(4) 951–953 (1979).

Vasilyev V.I., Lavrentyev Yu.G. New finds and data on corderoite (α-$Hg_3S_2Cl_2$) composition [in Russian], *Geologiya i geofizika*, (12), 117–121 (1986).

Voronin V.I., Shchennikov V.V., Berger I.F., Glazkov V.P., Kozlenko D.P., Savenko B.N., Tikhomirov S.V. Neutron diffraction study of a structure transition in the ternary systems of the mercury chalcogenides $HgSe_{1-x}S_x$ at high pressure [in Russian], *Fiz. tv. tela*, **43**(11), 2076–2080 (2001).

Voroshilov Yu.V., Gad'mashi Z.P., Slivka V.Yu., Hudoliy V.A. Obtaining and natural optical activity of the mercury chalcogenide crystals [in Russian], *Izv. AN SSSR. Neorgan. materialy*, **17**(11), 2022–2024 (1981).

Voroshilov Yu.V., Hudoliy V.A., Pan'ko V.V. Phase equilibria in the HgS–HgTe–$HgCl_2$ system and crystal structure of β-$Hg_3S_2Cl_2$ and Hg_3TeCl_4 compounds [in Russian], *Zhurn. neorgan. khimii*, **41**(2), 287–293 (1996a).

Voroshilov Yu.V., Hudoliy V.A., Pan'ko V.V., Minets Yu.V. Phase equilibria in the HgS–HgTe–$HgBr_2$ system and crystal structure of $Hg_3S_2Br_2$ and Hg_3TeBr_4 compounds [in Russian], *Neorgan. materialy*, **32**(12), 1466–1472 (1996b).

Voroshilov Yu.V., Motria S.F., Semrad E.E. Phase equilibria in the mercury–tin–sulfur(selenium) systems [in Russian], *Zhurn. neorgan. khimii*, **38**(6), 1061–1064 (1993).

Voroshilov Yu.V., Motria S.F., Semrad E.E. Physico-chemical investigation of the Hg–Ge–S system [in Russian], *Ukr. khim. zhurn.*, **60**(1), 27–31 (1994a).

Voroshilov Yu.V., Pan'ko V.V., Pechars'kyi V.K., Hudoliy V.A. Phase equilibria in the HgS(Se, Te)–HgCl$_2$(Br$_2$, I$_2$) systems and crystal structure of compounds [in Ukraininan], *Visnyk L'viv. un-tu. Ser. khim.*, (33), 11–24 (1994b).

Ye S., Shao M., Tao X. Microwave-assisted solvothermal route to silver copper sulfide and silver mercury sulfide ternary compounds via co-displacement reaction, *Mater. Lett.*, **57**(13–14), 2056–2059 (2003).

8

Systems Based on HgSe

8.1 Mercury–Potassium–Selenium

HgSe–K$_2$Se: The phase diagram is not constructed. K$_2$Hg$_3$Se$_4$ and K$_6$HgSe$_4$ ternary compounds are formed in this system (Sommer and Hoppe 1978, Axtell et al. 1988, Kanatzidis and Park 1990). The first of them crystallizes in an orthorhombic structure with lattice parameters $a = 1082.0 \pm 0.2$, $b = 678.3 \pm 0.1$, and $c = 1404.2 \pm 0.3$ pm at 153 K and calculation density 6.42 g cm^{-3} (Axtell et al. 1988, Kanatzidis and Park 1990), and the second crystallizes in a hexagonal structure with lattice parameters $a = 1036$ and $c = 788.3$ pm and calculation and experimental density 3.404 and 3.40 g cm^{-3}, respectively (Sommer and Hoppe 1978).

K$_2$Hg$_3$Se$_4$ was synthesized by the heating of K$_2$Se + HgSe + Se mixtures under a vacuum at 250°C for 99 h followed by cooling to 50°C at a rate of 2°C/h (Axtell et al. 1988) or by the interaction of K$_2$Se$_4$ and HgSe in an evacuated Pyrex tube at 250° for 4 days (Kanatzidis and Park 1990). This compound is not stable in water and decomposes rapidly in moist air and light to form HgSe (Kanatzidis and Park 1990). Single crystals of K$_2$Hg$_3$Se$_4$ were grown from the melt of HgSe and K$_2$Se (Axtell et al. 1988). K$_6$HgSe$_4$ was obtained by the sintering of mixtures from HgSe and K$_2$Se at 360°C for 10 days in argon atmosphere (Sommer and Hoppe 1978). K$_2$Hg$_6$Se$_7$ could not be prepared in this system (Kanatzidis and Park 1990).

8.2 Mercury–Rubidium–Selenium

HgSe–Rb$_2$Se: The phase diagram is not constructed. Rb$_2$Hg$_6$Se$_7$ and Rb$_6$HgSe$_4$ ternary compounds are formed in this system (Sommer and Hoppe 1978, Axtell et al. 1988). Rb$_6$HgSe$_4$ crystallizes in a hexagonal structure with lattice parameters $a = 1072$ and $c = 719.2$ pm and calculation and experimental density 4.192 and 4.19 g cm^{-3}, respectively (Sommer and Hoppe 1978).

$Rb_2Hg_6Se_7$ melts at $556 \pm 10°C$ and displays room-temperature band gap of 1.13 eV (Axtell et al. 1988).

$Rb_2Hg_6Se_7$ can be prepared by the reaction of Rb_2Se and HgSe by heating under a vacuum at 550°C for 2 days followed by slow cooling to 150°C and quenching to room temperature (Axtell et al. 1988). This compound is air and water stable and is insoluble in common organic solvents. Rb_6HgSe_4 was obtained by the sintering of mixtures from HgSe and Rb_2Se at 360°C for 10 days in argon atmosphere (Sommer and Hoppe 1978).

8.3 Mercury–Cesium–Selenium

HgSe–Cs_2Se: The phase diagram is not constructed. $Cs_2Hg_3Se_4$ and $Cs_2Hg_6Se_7$ ternary compounds are formed in this system (Axtell et al. 1988). The first of them crystallizes in an orthorhombic structure with lattice parameters $a = 1204.7 \pm 0.4$, $b = 646.5 \pm 0.2$, and $c = 1477.1 \pm 0.6$ pm and calculation density 6.83 g cm^{-3}, and the second melts congruently at $556 \pm 10°C$ and crystallizes in a tetragonal structure with lattice parameters $a = 1450.5 \pm 0.7$ and $c = 430.8 \pm 0.2$ pm and calculation density 7.41 g cm^{-3} and energy gap $E_g = 1.17$ eV (Axtell et al. 1988).

$Cs_2Hg_3Se_4$ was obtained by the heating of $Cs_2Se + HgSe + Se$ mixtures under a vacuum at 250°C for 99 h followed by cooling to 50°C at a rate of 2°C/h (Axtell et al. 1988). This compound is relatively stable in water for a short period of time but decomposes in an hour. $Cs_2Hg_6Se_7$ can be prepared by direct synthesis from Cs_2Se and HgSe under a vacuum at 375°C for 72 h followed by cooling to 50°C at a rate of 3°C/h (Axtell et al. 1988). This compound is not soluble in water and any common organic solvents.

Single crystals of $Cs_2Hg_3Se_4$ and $Cs_2Hg_6Se_7$ were grown from the melt of HgSe and Cs_2Se (Axtell et al. 1988).

8.4 Mercury–Copper–Selenium

HgSe–$Cu_{1.8}Se$: The phase diagram is a eutectic type (Figure 8.1) (Parasyuk et al. 1997). The eutectic composition and temperature are 12 mol. % $Cu_{1.8}Se$ and 715°C, respectively. The solubility of HgSe in $Cu_{1.8}Se$ at the eutectic temperature is equal to 54 mol. %. At 100°C, a eutectoid decomposition of solid solution based on $Cu_{1.8}Se$ takes place.

This system was investigated using differential thermal analysis (DTA), metallography, and x-ray diffraction (XRD). The ingots were annealed at 400°C for 250 h (Parasyuk et al. 1997).

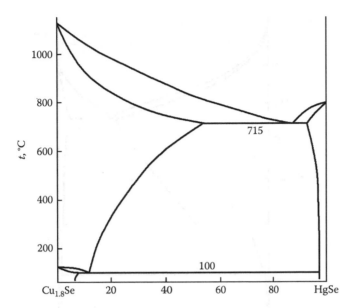

FIGURE 8.1

HgSe–Cu$_{1.8}$Se phase diagram. (From Parasyuk, O. et al., *Nauk. visnyk Volyn. derzh. un-tu*, (4), 35, 1997.)

8.5 Mercury–Silver–Selenium

HgSe–Ag$_2$Se: The phase diagram of this system belongs to the eutectic type with the eutectic point at 75 mol. % HgSe and 658°C (Figure 8.2) (Parasyuk et al. 2002). The solid solubility of HgSe in Ag$_2$Se equals 0 to ~73 mol. % at 658°C and decreases with decreasing temperature to 5 mol. % at room temperature. The HgSe solid-solution range extends to less than 5 mol. % Ag$_2$Se at 658°C and also decreases with decreasing temperature. The line at 136°C is associated with the polymorphous transformation of Ag$_2$Se.

This system was investigated using DTA, metallography, and XRD. The ingots were annealed at 400°C for 250 h (Parasyuk et al. 2002).

8.6 Mercury–Aluminum–Selenium

HgSe–Al$_2$Se$_3$: The phase diagram is not constructed. HgAl$_2$Se$_4$ ternary compound is formed in this system. It crystallizes in a hexagonal structure with lattice parameters $a = 569.6$ and $c = 1072$ pm (Hahn et al. 1955) [$a = 571.0$ and

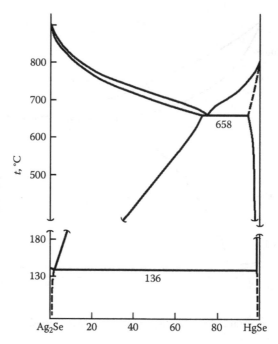

FIGURE 8.2
HgSe–Ag$_2$Se phase diagram. (From Parasyuk, O.V. et al., *J. Alloys Compd.*, 339(1–2), 140, 2002.)

$c = 1074.0$ pm (Range et al. 1968, Meenakshi et al. 2010); $a = 568.9$ and $c = 1069.5$ pm (Singh et al. 2011)] and calculation and experimental density 5.053 [5.05 (Range et al. 1968)] and 5.02 g cm^{-3} (Hahn et al. 1955). The ambient tetragonal structure is retained in HgAl$_2$Se$_4$ up to 13 GPa (Meenakshi et al. 2010). At 14.7 GPa, a structural transition takes place. The high-pressure phase could be fitted to a disordered NaCl-type structure with lattice parameter $a = 514.61$ pm, which is stable up to 22.2 GPa, the highest pressure of the measurements. The value of the bulk modulus is 66 ± 1.5 GPa [65.8 GPa and its first pressure derivative $B_0' = 4.72$ (Singh et al. 2011)] in the chalcopyrite phase of HgAl$_2$Se$_4$.

At high temperature and pressure (6.3 GPa or 63 kbar and 420°C), this compound can exist in a cubic structure of spinel type with lattice parameter $a = 1078$ pm and calculation density 6.06 g cm^{-3} (Range et al. 1968).

HgAl$_2$Se$_4$ is a direct wide band-gap semiconductor with $E_g = 2.24$ eV (Singh et al. 2011).

The analysis of the Raman data of HgAl$_2$Se$_4$ reveals a general trend observed for different defect chalcopyrite materials (Meenakshi et al. 2006). The line widths of the Raman peaks change at intermediate pressures between 4 and 6 GPa as an indication of the pressure-induced two-stage order–disorder transition observed in this material.

HgAl$_2$Se$_4$ was obtained from HgSe and powderlike Al and Se in equimolar ratio. The mixtures were pressed and annealed at 800°C for 12–24 h (Hahn et al. 1955). Its single crystals were grown using chemical transport reactions with iodine as a transport agent (Meenakshi et al. 2010). The structural properties of $HgAl_2Se_4$ were calculated using the full potential linear augmented plane-wave method based on density functional theory (Singh et al. 2011).

8.7 Mercury–Gallium–Selenium

HgSe–Ga₂Se₃: The phase diagram is shown in Figure 8.3 (Kerkhoff and Leute 2004). HgSe and Ga_2Se_3 form extended regions of solid solutions with a sphalerite structure, and at several distinct $HgSe/Ga_2Se_3$ ratios, the formation of superstructures was found. The solidus and liquidus yield a common minimum at 21 mol. % Ga_2Se_3 and 791°C [785°C (Metlinski et al. 1973)] (the system is considered as $3HgSe–Ga_2Se_3$). This phase diagram can be modeled by a Gibbs energy function for a subregular system including ordering terms.

Lattice parameters of the solid solutions based on HgSe decrease linearly with the increase of Ga_2Se_3 contents from 608.6 pm for HgSe to 591.0 and 557.2 pm at 25 and 50 mol. % Ga_2Se_3, respectively (Metlinski et al. 1973).

According to the data of Metlinski et al. (1973), ordered solid solutions are formed within the interval of 29–40 mol. % Ga_2Se_3. Maximum of ordering is situated at 37.5 mol. % Ga_2Se_3, where $Hg_5Ga_2Se_8$ ternary compound is formed. This compound crystallizes in a cubic structure with lattice parameter $a = 1168.76 \pm 0.02$ pm and calculation density 7.3806 ± 0.0003 g cm^{-3}

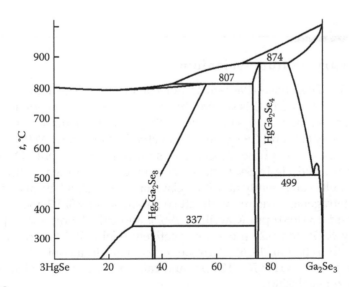

FIGURE 8.3
$3HgSe–Ga_2Se_3$ phase diagram. (From Kerkhoff, M. and Leute, V., *J. Alloys Compd.*, 385(1–2), 148, 2004.)

(Kozer et al. 2010) [according to the data of Kerkhoff and Leute (2004), it crystallizes in a tetragonal structure with lattice parameters $a = 828.0$ and $c = 1170.9$ pm, and its tetragonal lattice is undistorted].

The $HgGa_2Se_4$ ternary compounds melt incongruently at 877°C [870°C (Metlinski et al. 1973)] and 75 mol. % Ga_2Se_3 and have a narrow homogeneity region (Kerkhoff and Leute 2004). $HgGa_2Se_4$ crystallizes also in a tetragonal structure with lattice parameters $a = 571.3$, 571.0, 571.0, 571.6, and 571.7 and $c = 1083.1$, 1082.7, 1082.2, 1080.9, and 1080.4 pm at 1000°C, 900°C, 800°C, 700°C, and 600°C, respectively Kerkhoff and Leute (2004) [$a = 569.3$ and $c = 1082.6$ pm (Gastaldi and Pardo 1984, Gastaldi et al. 1985); $a = 570.3$ and $c = 1076$ pm (Hahn et al. 1955)], and calculation and experimental density 6.185 and 6.10 g cm⁻³, respectively (Hahn et al. 1955).

According to the data of Metlinski et al. (1973) at 640°C–660°C, the ordering of $HgGa_2Se_4$ takes place and this transformation is accompanied by the considerable enlargement of ingots. The ordered phase crystallizes in a tetragonal structure with changeable lattice parameters from $a = 572.0$ and $c = 1076.2$ pm at 75 mol. % Ga_2Se_3 to $a = 570.1$ and $c = 1082.2$ pm at 77 mol. % Ga_2Se_3.

Solid solutions based on Ga_2Se_3 crystallize in a defective low-ordered structure of the sphalerite type with changeable lattice parameter from $a = 548.4$ pm for 90 mol. % Ga_2Se_3 to 542.2 pm for Ga_2Se_3 (Metlinski et al. 1973).

This system was investigated using DTA, metallography, and XRD (Metlinski et al. 1973, Kerkhoff and Leute 2004). The single crystals of $HgGa_2Se_4$ were grown using chemical transport reactions (Gastaldi and Pardo 1984, Gastaldi et al. 1985).

8.8 Mercury–Indium–Selenium

HgSe–InSe: The phase diagram is not constructed. The $HgInSe_2$ ternary compound is formed in this system that crystallizes in a tetragonal structure with lattice parameters $a = 1161$ and $c = 588$ pm and calculation and experimental density 6.97 and 6.96 g cm⁻³, respectively (Guseinov 1969).

HgSe–In₂Se₃: The phase diagram is shown in Figure 8.4 (Derid 1978). The $HgIn_2Se_4$ and $Hg_5In_2Se_8$ ternary compounds are formed in this system. $HgIn_2Se_4$ melts congruently at 880°C (Gavrilitsa 1967, Derid 1978) and crystallizes in a tetragonal structure with lattice parameters $a = 575.2$ and $c = 1178$ pm [$a = 577.0$ and $c = 1166.0$ pm (Gavrilitsa 1967)] and calculation and experimental density 6.331 and 6.26 g cm⁻³, respectively (Hahn 1955), and energy gap $E_g = 0.6$ eV (Busch et al. 1956). $Hg_5In_2Se_8$ crystallizes in a cubic structure with lattice parameter $a = 1188.76 \pm 0.02$ pm and calculation density 7.39176 ± 0.00004 g cm⁻³ (Kozer et al. 2010).

Alloys of the $(3HgSe)_x–(In_2Se_3)_{1-x}$ system from the HgSe-rich side ($x \geq 0.5$) crystallize in very narrow interval of temperatures and have the

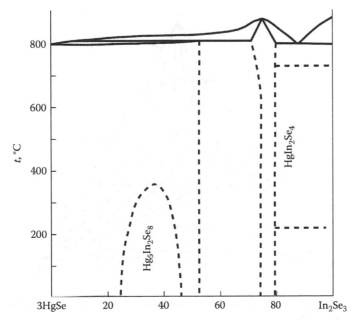

FIGURE 8.4
3HgSe–In$_2$Se$_3$ phase diagram. (From Derid, O.P., Investigation of the phase diagrams of the A^{II}–B^{III}–C^{VI} ternary system [in Russian], in *Teoret. i eksp. issled. slozhnyh poluprovodn. soed.,* Kishinev, Shtiintsa Publish., pp. 44–64, 1978.)

sphalerite structure (Radautsan and Gavrilitsa 1961, Gavrilitsa 1967, Derid 1978). Appreciable quantities of mercury were remaining in the ampoules after synthesis of the alloys containing great contents of HgSe.

This system was investigated using DTA, metallography, XRD, and micro-hardness measurement. The ingots were annealed at 500°C for 100–150 h (Radautsan and Gavrilitsa 1961, Gavrilitsa 1967, Derid 1978).

8.9 Mercury–Thallium–Selenium

Liquidus surface of the Hg–Tl–Se ternary system includes the fields of primary crystallization of all starting and intermediate phases (Figure 8.5) (Asadov et al. 1982). The Hg$_3$Tl$_2$Se$_4$ ternary compound is formed in this system (Asadov et al. 1982, Asadov 1983). The fields of Tl and Hg crystallization are degenerated. Two immiscibility regions exist in this system: one of them is situated in the Se-rich corner and the second in the region of Tl- and Hg-rich side. They form a continuous band of immiscibility between the Hg–Se and the Tl–Se binary systems. Nonvariant equilibria in this ternary system are given in Table 8.1 (Asadov et al. 1982).

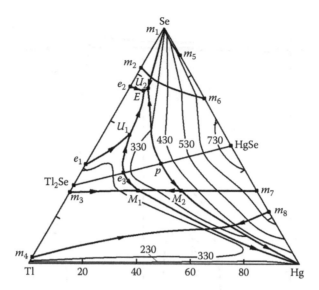

FIGURE 8.5
Liquidus surface of the Hg–Tl–Se ternary system. (From Asadov, M.M. et al., *Zhurn. neorgan. khimii*, 27(12), 3173, 1982.)

TABLE 8.1

Nonvariant Equilibria in the Hg–Tl–Se Ternary System

Symbol	Reaction	Se, at. %	Hg, at. %	t, °C
e	$L \Leftrightarrow Tl_2Se + \beta\text{-}Hg_3Tl_2Se_4$	38	16	320
p	$L + HgSe \Leftrightarrow \beta\text{-}Hg_3Tl_2Se_4$	42	28	417
E	$L + \Leftrightarrow TlSe + \alpha\text{-}Hg_3Tl_2Se_4 + Se$	73	6	170
U_1	$L + Tl_2Se \Leftrightarrow TlSe + \alpha\text{-}Hg_3Tl_2Se_4$	55	10	282
U_2	$L + HgSe \Leftrightarrow \alpha\text{-}Hg_3Tl_2Se_4 + Se$	74	7	180
M_1	$L_1 \Leftrightarrow Tl_2Se + \beta\text{-}Hg_3Tl_2Se_4 + L_2$	31	25	297
M_2	$L_1 + HgSe \Leftrightarrow \beta\text{-}Hg_3Tl_2Se_4 + L_2$	31	41	392

Source: Asadov, M.M. et al., in *Zhurn. neorgan. khimii*, 27(12), 3173, 1982.

Isothermal section of the Hg–Tl–Se ternary system at room temperature is shown in Figure 8.6 (Asadov et al. 1982, Asadov 1983).

HgSe–TlSe: This section is a nonquasibinary section of the Hg–Tl–Se ternary system (Figure 8.7) (Asadov et al. 1982). The HgTlSe$_2$ ternary compound [a tetragonal structure; $a = 786$ and $c = 684$ pm; calculation and experimental density 7.89 and 7.88 g cm^{-3}, respectively; energy gap $E_g = 0.30$ eV (Guseinov 1969)] was not found in this system (Kuliev et al. 1971, Asadov et al. 1982).

According to the data of Kuliev et al. (1971), the phase diagram of the HgSe–TlSe system is a eutectic type and the solubility of HgSe in TlSe reaches 70 mol. %, but these data are not reliable.

This system was investigated using DTA, metallography, XRD, and measurements of microhardness and emf of concentration chains (Kuliev 1971,

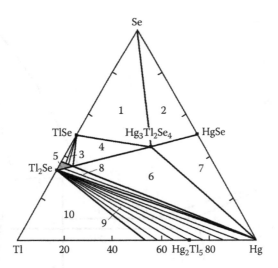

FIGURE 8.6
Isothermal section of the Hg–Tl–Se ternary system at room temperature: 1, TlSe + Se + α-Hg$_3$Tl$_2$Se$_4$; 2, HgSe + Se + α-Hg$_3$Tl$_2$Se$_4$; 3, TlSe + (Tl$_2$Se); 4, TlSe + (Tl$_2$Se) + α-Hg$_3$Tl$_2$Se$_4$; 5, (Tl$_2$Se); 6, (Tl$_2$Se) + α-Hg$_3$Tl$_2$Se$_4$ + Hg; 7, α-Hg$_3$Tl$_2$Se$_4$ + HgSe + Hg; 8, (Tl$_2$Se) + Hg; 9, L + (Tl$_2$Se); and 10, L + Tl + (Tl$_2$Se). (From Asadov, M.M. et al., *Zhurn. neorgan. khimii*, 27(12), 3173, 1982; Asadov, M.M., *Zhurn. fiz. khimii*, 57(7), 1795, 1983.)

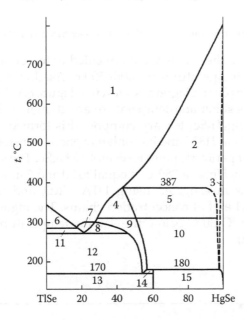

FIGURE 8.7
Phase relations in the HgSe–TlSe system: 1, L; 2, L + (HgSe); 3, (HgSe); 4, L + β-Hg$_3$Tl$_2$Se$_4$; 5, L + (HgSe) + β-Hg$_3$Tl$_2$Se$_4$; 6, L + TlSe; 7, L + (Tl$_2$Se); 8, L + (Tl$_2$Se) + α-Hg$_3$Tl$_2$Se$_4$; 9, L + α-Hg$_3$Tl$_2$Se$_4$; 10, L + (HgSe) + α-Hg$_3$Tl$_2$Se$_4$; 11, L + (Tl$_2$Se) + TlSe; 12, L + TlSe + α-Hg$_3$Tl$_2$Se$_4$; 13, TlSe + α-Hg$_3$Tl$_2$Se$_4$ + Se; 14, L + α-Hg$_3$Tl$_2$Se$_4$ + Se; and 15, (HgSe) + α-Hg$_3$Tl$_2$Se$_4$ + Se. (From Asadov, M.M. et al., *Zhurn. neorgan. khimii*, 27(12), 3173, 1982.)

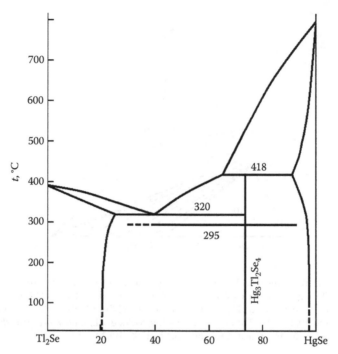

FIGURE 8.8

HgSe–Tl₂Se phase diagram. (From Asadov, M.M. et al., *Zhurn. neorgan. khimii*, 27(12), 3173, 1982.)

Asadov et al. 1982). The ingots were annealed at temperatures 20°C–30°C below the solidus temperatures for 300–500 h (Asadov et al. 1982).

HgSe–Tl₂Se: The phase diagram is shown in Figure 8.8 (Asadov et al. 1982). The eutectic composition and temperature are 40 mol. % HgSe and 320°C, respectively. The $Hg_3Tl_2Se_4$ ternary compound is formed in this system. It melts incongruently at 418°C and has polymorphous transformation at 295°C. The composition of peritectic point is 65 mol. % HgSe. The solubility of HgSe in Tl₂Se and Tl₂Se in HgSe at 280°C is equal to 20 and 5 mol. %, respectively.

This system was investigated using DTA, XRD, and measurements of microhardness and emf of concentration chains. The ingots were annealed at temperatures 20°C–30°C below the solidus temperatures for 300–500 h (Asadov et al. 1982).

8.10 Mercury–Silicon–Selenium

HgSe–SiSe₂: The phase diagram of this system was constructed in the range 0–60 mol. % SiSe₂ (Figure 8.9) (Parasyuk et al. 2003). It was established that Hg_4SiSe_6 compound is formed in this system. It melts congruently at

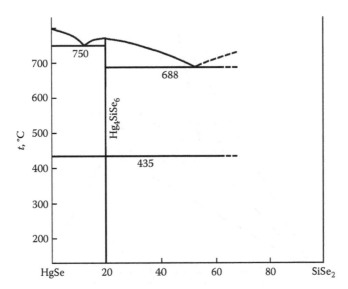

FIGURE 8.9
HgSe–SiSe$_2$ phase diagram. (From Parasyuk, O.V. et al., *J. Alloys Compd.*, 348(1–2), 157, 2003.)

771°C, forms eutectics at ~12 (750°C) and ~52 mol. % SiSe$_2$ (688°C), has a polymorphous transformation at 435°C (Parasyuk et al. 2003), and crystallizes in a monoclinic structure with lattice parameters $a = 1281.10 \pm 0.04$, $b = 740.34 \pm 0.04$, $c = 1274.71 \pm 0.01$ nm, and $\beta = 109.605 \pm 0.003°$ (Gulay et al. 2002).

This system was investigated using DTA, metallography, and XRD (Gulay et al. 2002, Parasyuk et al. 2003).

8.11 Mercury–Germanium–Selenium

The Hg$_2$GeSe$_3$ and Hg$_2$GeSe$_4$ ternary compounds are formed in the Hg–Ge–Se ternary system (Figure 8.10) (Motria et al. 1986, Motria 1991). The regions of solid solutions were not found.

Two glass regions exist in this system (Figure 8.11) (Feltz et al. 1976, 1980). The vitreous region from the Se-rich side increases up to 15 at. % Hg as the GeSe$_2$ concentration increases. The second glass region begins in the GeSe–GeSe$_2$ system and extends in the ternary system up to 25 at. % Hg.

HgSe–GeSe: This section is a nonquasibinary section of Hg–Ge–Se ternary system (Figure 8.12) (Motria 1991). The Hg$_2$GeSe$_3$ compound is formed, which melts incongruently at $568 \pm 5°C$, crystallizes in a cubic structure with lattice parameter $a = 1103.4 \pm 3$ pm, and has polymorphous transformation at

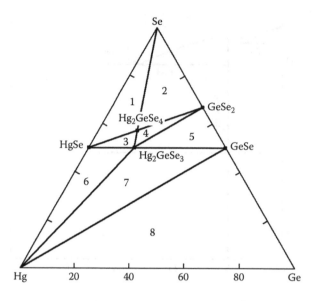

FIGURE 8.10
Isothermal section of the Hg–Ge–Se ternary system at 380°C. (From Motria, S.F. Mercury–germanium(tin)–sulfur(selenium) ternary systems [in Russian], in *Poluch. i svoistva slozhn. poluprovodn.*, Uzhgorod. gos. un-t, Kiev, 17–26, 1991.)

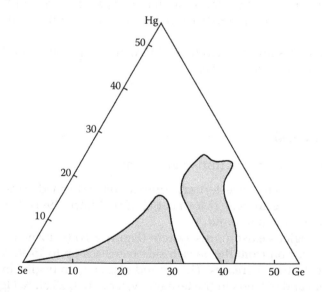

FIGURE 8.11
Glass regions in the Hg–Ge–Se ternary system. (From Feltz, A. et al., New vitreous semiconductors, *Tr. 6th Mezhdunar. konf. po amorf. i zhidkim poluprovodn: Struktura i sv-va nekristal. poluprovodn.* Leningrad, November 18–24, 1975, Leningrad, Nauka Publish., pp. 24–31, 1976.)

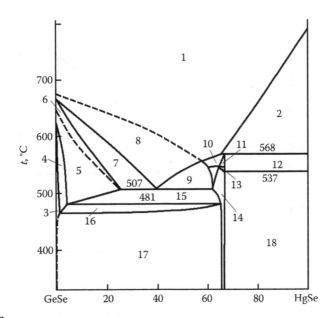

FIGURE 8.12

Phase relations in the HgSe–GeSe system: 1, L; 2, L+HgSe; 3, α-GeSe; 4, α-GeSe+β-GeSe; 5, β-GeSe; 6, L+β-GeSe; 7, L+GeSe₂+β-GeSe; 8, L+GeSe₂; 9, L+GeSe₂+α-Hg₂GeSe₃; 10, L+β-Hg₂GeSe₃; 11, L+α-Hg₂GeSe₃; 12, HgSe+β-Hg₂GeSe₃; 13, β-Hg₂GeSe₃; 14, α-Hg₂GeSe₃; 15, β-GeSe+α-Hg₂GeSe₃; 16, α-GeSe+β-GeSe+α-Hg₂GeSe₃; 17, α-GeSe+α-Hg₂GeSe₃; and 18, HgSe+α-Hg₂GeSe₃. (From Motria, S.F., Mercury–germanium(tin)–sulfur(selenium) ternary systems [in Russian], in *Poluch. i svoistva slozhn. poluprovodn.*, Uzhgorod. gos. un-t, Kiev, pp. 17–26, 1991.)

537±5°C. Homogeneity region of Hg₂GeSe₃ at the temperature of a ternary eutectic (507°C) is within the interval of 63.0–66.7 mol. % HgS.

The solubility of HgSe in α- and β-GeSe reaches 1 and 28 mol. %, respectively, and the solubility of GeSe in HgSe is insignificant (Motria 1991).

This system was investigated using DTA, metallography, XRD, and micro-hardness measurement. The ingots were annealed at 380°C–430°C for 1200 h (Motria 1991).

HgSe–GeSe₂: The phase diagram is shown in Figure 8.13 (Motria 1991, Motria et al. 1996). The eutectic composition and temperature are 42±3 mol. % HgSe and 600±5°C, respectively. Hg₂GeSe₄ ternary compound is formed in this system, which melts incongruently at 620±5°C (the peritectic composition is 60±3 mol. % HgSe) and has polymorphous transformation at 557±5°C. Low-temperature Hg₂GeSe₄ modification crystallizes in a tetragonal structure with lattice parameters $a=567.9$ and $c=1126.9$ pm (Motria 1991) [$a=567±2$ and $c=1132±2$ pm (Motria et al. 1986), $a=569.1$ and $c=1128$ pm (Hahn and Lorent 1958)] and calculation and experimental density 7.20 and 6.97 g cm⁻³, respectively (Motria et al. 1986) [7.179 and 7.09 g cm⁻³ (Hahn and Lorent 1958)].

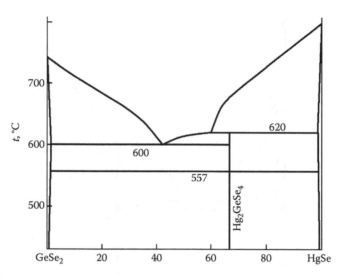

FIGURE 8.13

HgSe–GeSe$_2$ phase diagram. (From Motria, S.F. et al., *Ukr. khim. zhurn.*, 52(8), 807, 1986; Motria, S.F. Mercury–germanium(tin)–sulfur(selenium) ternary systems [in Russian], in *Poluch. i svoistva slozhn. poluprovodn.*, Uzhgorod. gos. un-t, Kiev, pp. 17–26, 1991.)

Glass-forming region in the HgSe–GeSe$_2$ system within the interval from GeSe$_2$ compound to 56 mol. % HgSe was determined by Olekseyuk et al. (1999). The temperatures of softening and crystallization of these glasses decrease at the HgSe addition.

Mutual solubility of HgSe and GeSe$_2$ is insignificant (Motria et al. 1986).

This system was investigated using DTA, metallography, XRD, and microhardness and density measurements. The ingots were annealed at 500°C for 1200 h (Motria et al. 1986, Motria 1991, Olekseyuk et al. 1999).

8.12 Mercury–Tin–Selenium

The isothermal section of the Hg–Sn–Se ternary system at 400°C is shown in Figure 8.14 (Motria 1991, Voroshilov et al. 1993). Hg$_2$SnSe$_4$ ternary compound exists at this temperature and it is in equilibria with HgSe, Se, SnSe$_2$, and SnSe. Homogeneity region of this compound is insignificant.

This system was investigated using DTA, metallography, XRD, and microhardness and density measurements. The ingots were annealed at 400°C for 1200 h (Motria 1991, Voroshilov et al. 1993).

HgSe–SnSe: The phase diagram is a eutectic type (Figure 8.15) (Motria 1991, Voroshilov et al. 1993). The eutectic composition and temperature are 39 mol. % HgSe and 607 ± 5°C, respectively. The solubility of SnSe in

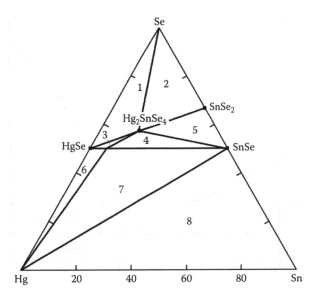

FIGURE 8.14

Isothermal section of the Hg–Sn–Se ternary system at 400°C: 1, $HgSe + Se + Hg_2SnSe_4$; 2, $SnSe_2 + Se + Hg_2SnSe_4$; 3, $Hg_{1-x}Sn_xSe + Hg_2SnSe_4$; 4, $Hg_{1-x}Sn_xSe + Hg_2SnSe_4 + SnSe$; 5, $SnSe + SnSe_2 + Hg_2SnSe_4$; 6, $Hg_{1-x}Sn_xSe + Hg$; 7, $SnSe + Hg_{1-x}Sn_xSe + Hg$; and 8, $SnSe + Hg + Sn$. (From Motria, S.F. Mercury–germanium(tin)–sulfur(selenium) ternary systems [in Russian], in *Poluch. i svoistva slozhn. poluprovodn.*, Uzhgorod. gos. un-t, Kiev, pp. 17–26, 1991; Voroshilov, Yu.V. et al., *Zhurn. neorgan. khimii*, 38(6), 1061, 1993.)

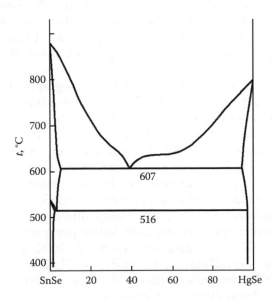

FIGURE 8.15

HgSe–SnSe phase diagram. (From Motria, S.F. Mercury–germanium(tin)–sulfur(selenium) ternary systems [in Russian], in *Poluch. i svoistva slozhn. poluprovodn.*, Uzhgorod. gos. un-t, Kiev, pp. 17–26, 1991; Voroshilov, Yu.V. et al., *Zhurn. neorgan. khimii*, 38(6), 1061, 1993.)

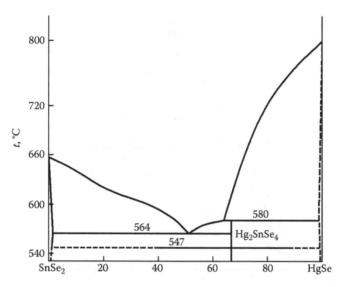

FIGURE 8.16

HgSe–SnSe$_2$ phase diagram. (From Motria S.F., et al., *Ukr. khim. zhurn.*, 52(8), 807, 1986; Motria, S.F., Mercury–germanium(tin)–sulfur(selenium) ternary systems [in Russian], in *Poluch. i svoistva slozhn. poluprovodn.*, Uzhgorod. gos. un-t, Kiev, pp. 17–26, 1991; Voroshilov, Yu.V. et al., *Zhurn. neorgan. khimii*, 38(6), 1061, 1993.)

HgSe at eutectic temperature reaches 7 mol. % and decreases to 5 mol. % at $497 \pm 5°C$. The solubility of HgSe in SnSe at the same temperatures is equal to 3 and not higher than 2 mol. %, respectively. In the solid-solution region, the temperature of SnSe polymorphous transformation decreases to $516 \pm 5°C$.

This system was investigated using DTA, metallography, XRD, and microhardness and density measurements. The ingots were annealed at 400°C and 500°C for 1200 h (Motria 1991, Voroshilov et al. 1993).

HgSe–SnSe$_2$: The phase diagram is shown in Figure 8.16 (Motria et al. 1991, Voroshilov et al. 1993). The eutectic composition and temperature are 49 mol. % HgSe and $564 \pm 5°C$, respectively. Hg$_2$SnSe$_4$ ternary compound is formed in this system, which melts incongruently at $580 \pm 5°C$ (the composition of peritectic point is 64 ± 2 mol. % HgSe) and has polymorphous transformation at $547 \pm 5°C$. This compound crystallizes in a thiogallate structure (defect chalcopyrite) with lattice parameters $a = 577.70 \pm 0.02$ and $c = 1155.70 \pm 0.07$ pm (Gulay and Parasyuk 2002) [$a = 577.9 \pm 0.2$ and $c = 1155.8 \pm 0.3$ pm (Voroshilov et al. 1993), $a = 578 \pm 2$ and $c = 1155 \pm 2$ pm (Motria et al. 1986)] and calculation and experimental density 7.13 and 7.03 g cm^{-3}, respectively (Voroshilov et al. 1993).

The solubility of Hg$_2$SnSe$_4$ in HgSe and SnSe$_2$ is not higher than 2 mol. %.

This system was investigated using DTA, metallography, XRD, and microhardness and density measurements. The ingots were annealed at

400°C and 500°C for 1200 h (Motria 1991, Voroshilov et al. 1993). Single crystals of Hg_2SnSe_4 were grown by crystallization from the solution in the melt (Voroshilov et al. 1993).

8.13 Mercury–Lead–Selenium

HgSe–PbSe: The phase diagram is a eutectic type (Figure 8.17) (Vaniarho et al. 1967, 1968, 1969, Leute and Köller 1986). The eutectic composition and temperature are 70 mol. % HgSe and $690 \pm 10°C$ (Vaniarho et al. 1967, 1968, 1969) [696°C (Leute and Köller 1986)]. At the eutectic temperature, the solubility of HgSe in PbSe reaches 6.7 mol. % and decreases to 5, 3, 2.5, and 1.5 mol. % at 630°C, 530°C, 480°C, and 430°C, respectively, and the solubility of PbSe in HgSe at the same temperatures is not higher than 0.5 mol. % (Leute and Köller 1986). According to the data of Vaniarho et al. (1967, 1968, 1969), solid solutions based on PbSe contain 9 mol. % at 820°C, 12 mol. % at 690°C, and 7 mol. % at 650°C, and the solubility of PbSe in HgSe at the eutectic temperature is less than 5 mol. %.

This system was investigated using DTA, metallography, XRD, and microhardness measurement (Vaniarho et al. 1967, 1968, 1969, Leute and Köller 1986). The ingots were annealed at 820°C (Vaniarho et al. 1967, 1968, 1969).

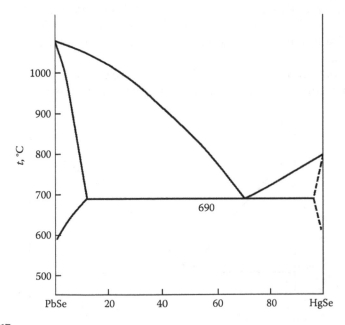

FIGURE 8.17
HgSe–PbSe phase diagram. (From Vaniarho, V.G. et al., *Izv. AN SSSR. Neorgan. materialy*, 3(7), 1276–1277 (1967); Vaniarho, et al., *Vestn. Mosk. un-ta. Ser. Khimia*, (6), 108, (1968); Vaniarho, V.G. et al., *Izv. AN SSSR. Neorgan. materialy*, 5(11), 2025, 1969.)

8.14 Mercury–Phosphorus–Selenium

HgSe–"P₂Se₄": The phase diagram is not constructed. $Hg_2P_2Se_6$ is formed in this system (Klinger et al. 1970, 1973, Jandali et al. 1978), which crystallizes in a monoclinic structure with lattice parameters $a = 654.5$, $b = 1137.7$, $c = 1316.0$ pm, and $\beta = 98.47°$ (Jandali et al. 1978) [$a = 652$, $b = 1152$, $c = 1364$ pm, and $\beta = 99.1°$ (Klinger et al. 1970, 1973)] and calculation and experimental density 6.20 [6.15 (Klinger et al. 1973)] and 6.05 g cm⁻³, respectively (Jandali et al. 1978).

8.15 Mercury–Arsenic–Selenium

HgSe–As: The phase diagram is a eutectic type (Figure 8.18) (Olekseyuk et al. 1987). The eutectic composition is 40 at. % As, and it crystallizes at 592°C. The liquid solutions of this system are close to regular solutions.

This system was investigated using DTA, metallography, and XRD (Olekseyuk et al. 1987).

HgSe–As₂Se₃: The phase diagram is a eutectic type (Figure 8.19) (Kurbanov et al. 1985, Olekseyuk et al. 1987). The eutectic composition and temperature are 10 mol. % HgSe and 375°C (Olekseyuk et al. 1987) [4 mol. % HgSe and 349°C (Babanly et al. 1983, Kurbanov et al. 1985)], respectively. The liquid solutions of this system are close to regular solutions (Olekseyuk et al. 1987).

Alloys containing more than 40 mol. % As₂Se₃ solidify in glassy state. The temperatures of softening and crystallization of these glasses decrease

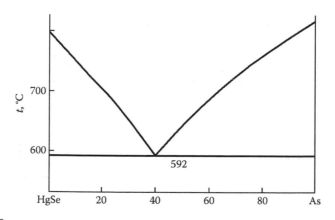

FIGURE 8.18

HgSe–As phase diagram. (Olekseyuk, I.D., et al., *Izv. AN SSSR. Neorgan. materialy*, 23(8), 1282, 1987.)

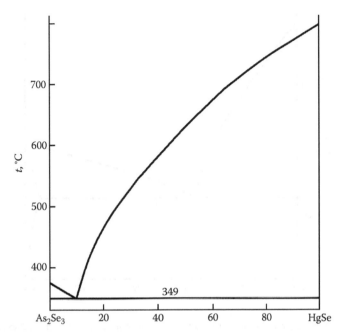

FIGURE 8.19

HgSe–As_2Se_3 phase diagram. (From Kurbanov, A.A. et al., *Azerb. khim. zhurn.*, (2), 99, 1985; Olekseyuk, I.D., et al., *Izv. AN SSSR. Neorgan. materialy*, 23(8), 1282, 1987.)

at the HgSe addition. The glasses of this system crystallize at the constant temperature (277°C) (Babanly et al. 1983, Kurbanov et al. 1985).

This system was investigated using DTA, metallography, XRD, and microhardness measurement (Babanly et al. 1983, Kurbanov et al. 1985, Olekseyuk et al. 1987). The ingots were annealed at 320°C for 200 h (Kurbanov et al. 1985).

HgSe–Hg_3As_2: The phase diagram is not constructed. A complex interaction takes place in this system with formation of elementary mercury (Golovey et al. 1972).

8.16 Mercury–Antimony–Selenium

HgSe–Sb_2Se_3: The phase diagram is a eutectic type (Figure 8.20) (Kurbanov et al. 1981). The eutectic composition and temperature are 62 mol. % HgSe and 526°C, respectively. At 462°C within the interval of 5–90 mol. % HgSe, there are small thermal effects, the nature of which was not defined.

This system was investigated using DTA, XRD, and microhardness measurement. The ingots were annealed at 500°C for 450 h (Kurbanov et al. 1981).

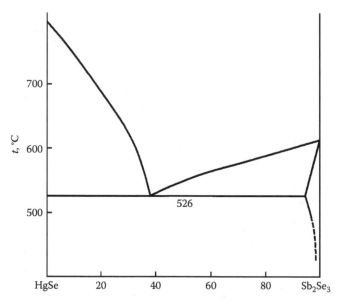

FIGURE 8.20

HgSe–Sb$_2$Se$_3$ phase diagram. (From Kurbanov, A.A. et al., *Zhurn. neorgan. khimii*, 26(5), 1438, 1981.)

8.17 Mercury–Bismuth–Selenium

HgSe–Bi$_2$Se$_3$: The phase diagram is a eutectic type (Figure 8.21) (Kurbanov et al. 1981). The eutectic composition and temperature are 68 mol. % HgSe and 620°C, respectively.

This system was investigated using DTA, XRD, and microhardness measurement. The ingots were annealed at 500°C for 450 h (Kurbanov et al. 1981).

8.18 Mercury–Oxygen–Selenium

The phase diagram is not constructed. Isothermal section of the Hg–O–Se ternary system at room temperature, constructing according to thermodynamic calculations, is shown in Figure 8.22 (Medvedev Yu and Berchenko 1993).

HgSeO$_4$ and Hg$_3$O$_2$SeO$_4$ (HgSeO$_4$ 2HgO) ternary compounds are formed in the Hg–O–Se ternary system. HgSeO$_4$ crystallizes in an orthorhombic structure with lattice parameters $a = 497.9$, $b = 672.1$, and $c = 492.8$ pm and calculation and experimental density 6.92 and 6.88 g cm^{-3}, respectively (Aurivillius and Malmros 1961). Hg$_3$O$_2$SeO$_4$ crystallizes in a trigonal structure with lattice

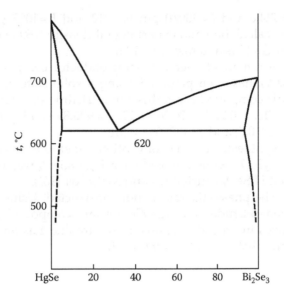

FIGURE 8.21
HgSe–Bi$_2$Se$_3$ phase diagram. (From Kurbanov, A.A. et al., *Zhurn. neorgan. khimii*, 26(5), 1438, 1981.)

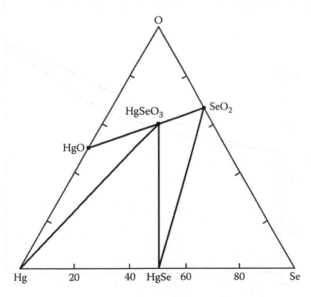

FIGURE 8.22
Isothermal section of the Hg–O–Se ternary system at room temperature. (From Medvedev, Yu.V. and Berchenko, N.N., *Zhurn. neorgan. khimii*, 38(12), 1940, 1993.)

parameters $a = 714.6$ and $b = 1007.0$ pm [$a = 712$ and $b = 1005$ pm (Nagorsen et al. 1962)] and calculation and experimental density 8.69 and 8.63 g cm^{-3}, respectively (Aurivillius and Malmros 1961).

A new mixed-framework mercury semiconductor compound $Hg_3O_5Se_4$, or $(Hg_3Se_2)(Se_2O_5)$, has been prepared using a solid-state reaction in Zou et al. (2007). This compound crystallizes in a triclinic structure with lattice parameters $a = 748.7 \pm 0.2$, $b = 776.26 \pm 0.12$, $c = 906.89 \pm 0.18$ pm, $\alpha = 95.528 \pm 0.016°$, $\beta = 108.55 \pm 0.02°$, and $\gamma = 107.759 \pm 0.016°$.

$HgSeO_4$ was synthesized using conventional methods starting from yellow HgO and H_2SeO_4, while the oxide salt $Hg_3O_2SeO_4$ was obtained from $Hg(C_2H_3O_2)_2$ and H_2SeO_4 (Aurivillius and Malmros 1961).

HgSe–HgO: The phase diagram is not constructed. Using hydrolysis of mercury halogen–selenides, the Hg_3OSe_2 ternary compound was obtained, which crystallizes in an orthorhombic structure and has an experimental density 5.2 g cm^{-3} (Batsanov et al. 1961, 1973).

8.19 Mercury–Tellurium–Selenium

HgSe–HgTe: The phase diagram is shown in Figure 8.23 (Strauss et al. 1960, Leute and Plate 1989). Solid solutions with sphalerite structure over the entire range of concentrations are formed in this system (Nikol'skaya and Regel' 1955,

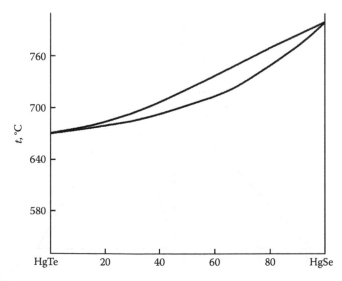

FIGURE 8.23
HgSe–HgTe phase diagram. (From Strauss, A.J. et al., HgTe–HgSe system, *Quart Progr. Rept. Solid State Res. Lincoln Lab. Mass. Inst. Technol.*, Massachusetts, Apr., 24–25, 1960; Leute, V. and Plate, H., *Ber. Bunsenges. Phys. Chem.*, 93(7), 757, 1989.)

Elpatyevskaya 1956, Strauss 1960, Malevski et al. 1965, Pajaczkowska and Dziuba 1971, Leute and Plate 1989). The liquidus and solidus lines are situated too close to one other: within the interval of 25–75 mol. % HgSe, the distance between them is about 3 mol. %. The lattice parameter changes linearly with composition (Malevski et al. 1965, Leute and Köller 1986, Leute and Plate 1989).

The calculated critical values for spinodal demixing are 33 mol. % HgSe and 160 K (Leute and Plate 1989).

This system was investigated using DTA, metallography, XRD, and chemical analysis (Strauss 1960, Leute and Köller 1986, Leute and Plate 1989). The phase diagram was calculated using the model of regular associated solutions. The solid/liquid equilibria were calculated with the assumption of a melt ideality (Leute and Plate 1989).

8.20 Mercury–Chromium–Selenium

HgSe–Cr$_2$Se$_3$: The phase diagram is shown in Figure 8.24 (Kovaleva et al. 1979, 1980). The eutectic is degenerated from the HgSe-rich side and crystallizes at 785°C. HgCr$_2$Se$_4$ ternary compound is formed in this system.

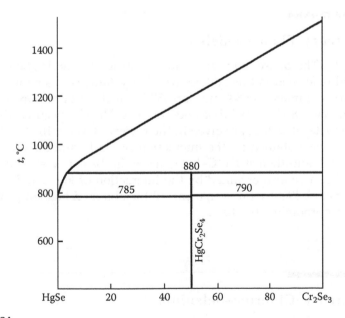

FIGURE 8.24
HgSe–Cr$_2$Se$_3$ phase diagram. (From Kovaleva, I.S. and Kuznetsova, I.Ya., Investigation of interaction in the HgSe–Cr$_2$Se$_3$–Se system [in Russian], in *Troinyie poluprovodn. i ih primenienie: Tez. dokl.*, Kishinev, Shtiintsa, pp. 70–71, 1979; Kovaleva, I.S. et al., *Zhurn. neorgan. khimii*, 25(4), 1156, 1980.)

It melts incongruently at 885±5°C (Kovaleva et al. 1979, 1980), crystallizes in a cubic structure with lattice parameter $a = 1075.3$ pm (Abdinov 1976), and decomposes by heating in the air (Zhukov 1997). At temperatures less than 410°C, $HgCr_2Se_{4-y}$ loses selenium according to the next equation, $y = (0.51 \pm 0.01)$ T—(0.160 ± 0.005), and at the higher temperatures, this compound oxidizes forming Cr_2O_3.

$HgCr_2Se_4$ has a narrow homogeneity range with Hg and Se deficiency (Chebotaev et al. 1985, Kovaleva et al. 1996): for $Hg_{1-x}Cr_2Se_4$ and $HgCr_2Se_{4-y}$, the composition is within the interval of $0 < x < 0.10$ and $0 < y \leq 0.03$, respectively (Kovaleva et al. 1996).

Thermal effects at 790°C correspond to the polymorphous transformation of Cr_2Se_3 (Kovaleva et al. 1979, 1980).

This system was investigated using DTA, metallography, XRD, and magnetic susceptibility measurement (Kovaleva et al. 1979, 1980, 1996, Chebotaev et al. 1985). Thermal stability of $HgCr_2Se_4$ was investigated using thermogravimetry and DTA (Zhukov et al. 1997). The ingots were annealed at 550°C–650°C for 5–8 days (Chebotaev et al. 1985). $Hg_{1-x}Cr_2Se_4$ was annealed at 650°C for 3 days and $HgCr_2Se_{4-y}$ at 500°C and 700°C for 2 days (Kovaleva et al. 1996).

8.21 Mercury–Fluorine–Selenium

$HgSe–HgF_2$: The phase diagram is not constructed. The $Hg_3Se_2F_2$ ternary compound is formed in this system, which crystallizes in a cubic structure with lattice parameter $a = 889$ pm [$a = 838.7$ pm (Puff et al. 1968)] and calculation density 8.47 g cm^{-3} (Batsanov et al. 1973). The peculiarities of the $Hg_3Se_2F_2$ crystal structure are given in the review of Magarill et al. (2007).

$Hg_3Se_2F_2$ was obtained by the interaction of Hg_2F_2 and Se (annealing at 220°C for 10 h and then at 170°C–190°C for 60 h) (Batsanov et al. 1973). This compound can be also obtained by the interaction of HgSe and (C_5H_5NH) HgF_4 (Puff et al. 1968); moreover, a part of the obtained $Hg_3Se_2F_2$ has apparently an orthorhombic structure.

8.22 Mercury–Chlorine–Selenium

$HgSe–HgCl_2$: The phase diagram is shown in Figure 8.25 (Pan'ko 1989 Voroshilov 1994). The eutectic from the HgSe-rich side contains 86 mol. % HgSe and crystallizes at 580°C, and the eutectic from $HgCl_2$-rich side is degenerated and crystallizes at 275°C. The $Hg_3Se_2Cl_2$ ternary compound is

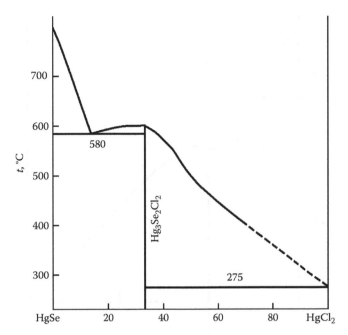

FIGURE 8.25

HgSe–HgCl$_2$ phase diagram. (From Pan'ko, V.V. et al., *Zhurn. neorgan. khimii*, 34(9), 2331, 1989; Voroshilov, Yu.V., et al., *Visnyk L'viv. un-tu. Ser. khim.*, (33), 11, 1994.)

formed in this system, which melts congruently at 600°C (Puff 1962, Pan'ko et al. 1989, Voroshilov 1994) and crystallizes in a cubic structure with lattice parameter $a = 906 \pm 1$ pm (Batsanov and Abaulina 1961, Puff and Kohlschmidt 1962, Puff and Küster 1962, Batsanov et al. 1973, Voroshilov et al. 1981, 1994) and calculation density 7.65 g cm^{-3} and energy gap $E_g = 3.2$ eV (Batsanov et al. 1973). The peculiarities of the Hg$_3$Se$_2$Cl$_2$ crystal structure are given in the review of Magarill et al. (2007).

This system was investigated using DTA, metallography, and XRD. The ingots were annealed for 650–700 h (Pan'ko et al. 1989, Voroshilov et al. 1994). Hg$_3$Se$_2$Cl$_2$ was obtained by the interaction of Hg$_2$Cl$_2$ and Se (annealing at 210°C for 10 h and then at 160°C–180°C for 60 h) (Batsanov et al. 1973). Its single crystals were grown by sublimation method (Voroshilov et al. 1981).

8.23 Mercury–Bromine–Selenium

HgSe–HgBr$_2$: The phase diagram is shown in Figure 8.26 (Pan'ko et al. 1989, Voroshilov et al. 1994). The eutectic from the HgSe-rich side contains 75 mol. % HgSe and crystallizes at 550°C, and the eutectic from HgBr$_2$-rich

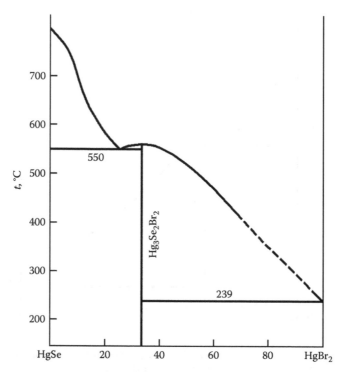

FIGURE 8.26
HgSe–HgBr$_2$ phase diagram. (From Pan'ko, V.V. et al., *Zhurn. neorgan. khimii*, 34(9), 2331, 1989.)

side is degenerated and crystallizes at 239°C. The Hg$_3$Se$_2$Br$_2$ ternary compound is formed in this system, which melts congruently at 560°C (Batsanov
and Abaulina 1961, Puff 1962, Pan'ko et al. 1989) and crystallizes in an orthorhombic structure with lattice parameters $a = 1770.1 \pm 0.6$, $b = 942.4 \pm 0.3$, and
$c = 1002.9 \pm 0.3$ pm (Pan'ko et al. 1989, Hudoliy et al. 1990, Voroshilov et al.
1994) [$a = 942$, $b = 974$, and $c = 878$ pm (Puff and Kohlschmidt 1962); $a = 945$,
$b = 975$, and $c = 880$ pm (Batsanov et al. 1973)]; calculation density 7.58 g cm^{-3};
and energy gap $E_g = 2.7$ eV (Batsanov et al. 1973). According to the data
of Minets et al. (2004), Hg$_3$Se$_2$Br$_2$ crystallizes in a monoclinic structure
with lattice parameters $a = 1752.9 \pm 0.6$, $b = 940.8 \pm 0.4$, $c = 977.5 \pm 0.4$ pm, and
$\beta = 89.51 \pm 0.03°$ and calculation density 7.577 ± 0.009 g cm^{-3}. The peculiarities of the Hg$_3$Se$_2$Br$_2$ crystal structure are given in the review of Magarill
et al. (2007).

This system was investigated using DTA, metallography, and XRD.
The ingots were annealed for 650–700 h (Pan'ko et al. 1989, Voroshilov
et al. 1994). Hg$_3$Se$_2$Br$_2$ was obtained by the interaction of Hg$_2$Br$_2$ and Se
(annealing at 200°C for 10 h and then at 150°C–170°C for 60 h) (Batsanov
et al. 1973).

8.24 Mercury–Iodine–Selenium

HgSe–HgI₂: The phase diagram is shown in Figure 8.27 (Pan'ko et al. 1989, Voroshilov et al. 1994). The eutectic is degenerated from the HgI_2-rich side and crystallizes at 240°C. $Hg_3Se_2I_2$ is formed in this system (Batsanov and Abaulina 1961, Puff 1962, Puff and Kohlschmidt 1962, Batsanov et al. 1973), which melts incongruently at 420°C (the peritectic composition is 54 mol. % HgSe) (Pan'ko et al. 1989, Voroshilov et al. 1994) and crystallizes in a tetragonal structure with lattice parameters $a = 1023.1 \pm 0.4$ and $c = 1431 \pm 2$ pm and experimental density 6.74 g cm⁻³ (Voroshilov et al. 1994). According to the data of Beck and Hedderich (2000), this compound crystallizes in an orthorhombic structure with the lattice parameters $a = 976.60$ and $b = 1938.1$ pm and calculation density 7.384 g cm⁻³. Phase transitions of this compound were not determined within the temperature interval from –160°C to 300°C (Beck and Hedderich 2000). Minets et al. (2004) indicated that $Hg_3Se_2I_2$ crystallizes in a monoclinic structure with lattice parameters $a = 1939.2 \pm 0.7$, $b = 965.8 \pm 0.7$, $c = 1091.8 \pm 0.3$ pm, and $\beta = 116.64 \pm 0.07°$ and calculation density 7.37 ± 0.01 g cm⁻³. The peculiarities of the $Hg_3Se_2I_2$ crystal structure are given in the review of Magarill et al. (2007).

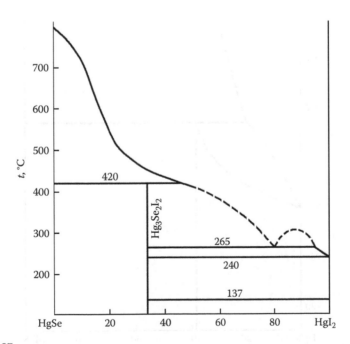

FIGURE 8.27
HgSe–HgI₂ phase diagram. (From Pan'ko, V.V. et al., *Zhurn. neorgan. khimii*, 34(9), 2331, 1989; Voroshilov, Yu.V., et al., (33), 11, 1994.)

The immiscibility region with monotectic temperature 265°C exists in the HgSe–HgI$_2$ system within the interval of 5–20 mol. % HgSe. Thermal effects at 137°C correspond to a polymorphous transformation of HgI$_2$.

Optical energy gap of Hg$_3$Se$_2$I$_2$ is equal to 2.3 eV (Batsanov et al. 1973).

This system was investigated using DTA, metallography, and XRD. The ingots were annealed for 650–700 h (Pan'ko et al. 1989, Voroshilov et al. 1994). Hg$_3$Se$_2$I$_2$ was obtained by the interaction of Hg$_2$I$_2$ and Se (annealing at 125°C for 10 h and then at 75°C–95°C for 60 h) (Batsanov et al. 1973). Its single crystals were grown by the interaction of HgS and HgI$_2$ in evacuated glass ampoules at 260°C (Beck and Hedderich 2000).

8.25 Mercury–Manganese–Selenium

HgSe–MnSe: The phase equilibria in this system have been discussed in Pajaczkowska (1978) based on papers that have appeared up to 1977. The phase diagram is shown in Figure 8.28 (Pajaczkowska and Rabenau 1977). The peritectic temperature is 868°C. Solid solutions based on HgSe at the peritectic

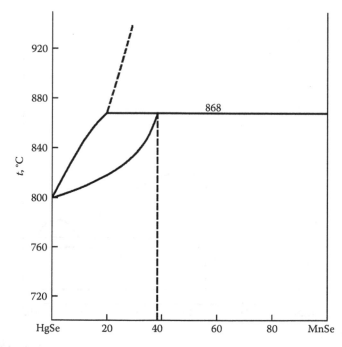

FIGURE 8.28
HgSe–MnSe phase diagram. (From Pajaczkowska, A. and Rabenau, A., *Mater. Res. Bull.*, 12(2), 183, 1977.)

temperature contain 38.5 mol. % MnSe and crystallize in a cubic structure of the sphalerite type. The solubility of HgSe in MnSe is insignificant.

This system was investigated using DTA and XRD (Pajaczkowska and Rabenau 1977). The crystal growth of the $Hg_{1-x}Mn_xSe$ solid solutions is given in Pajaczkowska (1978).

8.26 Mercury–Iron–Selenium

HgSe–FeSe: The phase diagram is not constructed. $Hg_{1-x}Fe_xSe$ solid solutions within the interval of $0 \leq x \leq 0.1$ are formed in this system (Paranchich et al. 1989, Szuszkiewicz et al. 1992). Infrared reflectivity was investigated for these solid solutions. The theoretical analysis of the experimental curves suggests opening of the energy gap for the composition $x = 0.087$ at a temperature close to 80 K.

This system was investigated using the XRD (Paranchich et al. 1989). The single crystals of the solid solutions were grown using the Bridgman method (Szuszkiewicz et al. 1992).

References

Abdinov D.Yu., Mehtieva S.I., Guseinov A.F., Dzhalilov N.S. Thermal conductivity of $HgCr_2Se_4$, $CuCr_2Se_4$ compounds [in Russian], in: *Troinyie poluprovodn. i ih primenienie: Tez. dokl.* Kishinev: Shtiintsa, 113 (1976).

Asadov M.M. Thermodynamic properties of Hg–Tl–Se ternary system [in Russian], *Zhurn. fiz. khimii*, **57**(7), 1795–1796 (1983).

Asadov M.M., Babanly M.B., Kuliev A.A. Phase equilibria and thermodynamic properties of the Hg–Tl–Se system [in Russian], *Zhurn. neorgan. khimii*, **27**(12), 3173–3178 (1982).

Aurivillius K., Malmros B. Studies on sulphates, selenates and chromates of mercury (II), *Acta Chem. Scand.*, **15**(9), 1932–1938 (1961).

Axtell E.A. (III), Park Y., Chondroudis K., Kanatzidis M.G. Incorporation of A_2Q into HgQ and dimensional reduction to $A_2Hg_3Q_4$ and $A_2Hg_6Q_7$ (A=K, Rb, Cs; Q=S, Se), *J. Amer. Chem. Soc.*, **120**(1), 124–136 (1998).

Babanly M.B., Kulieva N.A., Guseynov F.H., Kurbanov A.A. Phase equilibria and glass formation in the Tl–Cd(Ge)–S and Hg–As–Se ternary system [in Russian], in: *Neorgan. soyedin.—sintez i svoystva*, Baku, 114–118 (1983).

Batsanov S.S., Abaulina L.I. Interaction of mercury halogenides with chalcogens [in Russian], *Izv. SO AN SSSR.* (10), 67–73 (1961).

Batsanov S.S., Kolomiychuk V.N., Derbeneva S.S., Erenburg R.S. Synthesis and physico-chemical investigation of mercury halogenchalcogenides [in Russian], *Izv. AN SSSR. Neorgan. materialy*, **9**(7), 1098–1104 (1973).

Beck J., Hedderich S. Synthesis and crystal structure of $Hg_3S_2I_2$ and $Hg_3Se_2I_2$, new members of the $Hg_3E_2X_2$ family. *J. Solid State Chem.*, **151**(1), 73–76 (2000).

Busch G., Mooser E., Pearson W.B. Neue halbleitende Verbindungen mit diamantähnlicher Struktur, *Helv. phys. acta*, **29**(3), 192–193 (1956).

Chebotaev N.M., Simonova M.I., Arbuzova T.I., Gizhevski B.A., Samohvalov A.A. Deviation from stoichiometry and some physical properties of $HgCr_2Se_4$ magnetic semiconductor [in Russian], *Izv. AN SSSR. Neorgan. materialy*, **21**(9), 1468–1470 (1985).

Derid O.P. Investigation of the phase diagrams of the A^{II}–B^{III}–C^{VI} ternary system [in Russian], in: *Teoret. i eksp. issled. slozhnyh poluprovodn. soed.*, Kishinev: Shtiintsa Publish., 44–64 (1978).

Elpatyevskaya O.D., Konikova R.A., Regel' A.R., Yavorskiy I.V. About stability of the crystal structure of the HgSe–HgTe solid solutions [in Russian], *Zhurn. tekhn. fiz.*, **26**(10), 2154–2156 (1956).

Feltz A., Burckhardt W. Über Glasbildung und Eigenschaften von Chalkogenidsystemen. XIX. Zur Glasbildung von HgSe mit Germaniumseleniden, *Z. anorg. und allg. Chem.*, **461**(2), 35–47 (1980).

Feltz A., Burckhardt W., Senf L. New vitreous semiconductors, *Tr. 6th Mezhdunar. konf. po amorf. i zhidkim poluprovodn.: Struktura i sv-va nekristal. poluprovodn.* Leningrad, Nov. 18–24, 1975, Leningrad: Nauka Publish., 24–31 (1976).

Gastaldi L., Pardo M.-P. Croissance cristalline et affinement de la structure $HgGa_2Se_4$, *C. r. Acad. sci. Sèr. 2*, **298**(2), 37–38 (1984).

Gastaldi L., Simeone L.G., Viticoli S. Cation ordering and crystal structures in AGa_2X_4 compounds ($CoGa_2S_4$, $CdGa_2S_4$, $CdGa_2Se_4$, $HgGa_2Se_4$, $HgGa_2Te_4$), *Solid State Commun.*, **55**(7), 605–607 (1985).

Gavrilitsa E.I. Some physico-chemical properties of alloys from the $(HgSe)_{3x}$–$(In_2Se_3)_{1-x}$ section [in Russian], in: *Materialy dokl. 3rd nauchno-tehn. konf. Kishin. polytehn. in-ta*, Kishinev: Kishin. polutehn. in-t, 107–109 (1967).

Golovey M.I., Olekseyuk I.D., Voroshilov Yu.V. Chemical bond and solubility in the systems $A^{II}_3B^V_2$–$A^{II}C^{VI}$ [in Russian], in: *Khim. svyaz' v poluprovodn. i polumetallah.* Minsk, Nauka i Tekhnika Publish., 233–239 (1972).

Gulay L.D., Olekseyuk I.D., Parasyuk O.V. Crystal structure of the Hg_4SiS_6 and Hg_4SiSe_6 compounds, *J. Alloys Compd*, **347**(1–2), 115–120 (2002).

Gulay L.D., Parasyuk O.V. Crystal structure of the $Ag_{2.66}Hg_2Sn_{1.34}Se_6$ and Hg_2SnSe_4 compounds, *J. Alloys and Compd*, **337**(1–2), 94–98 (2002).

Guseinov G.D. Searching and physical investigation of new semiconductors-analogues [in Russian], *Avtoref. dis. ... doct. fiz.-mat. nauk.—Baku*, 82 p. (1969).

Hahn H., Frank G., Klingler W., Störger A.D., Störger G. Über ternäre Chalkogenide des Aluminiums, Galliums und Indiums mit Zink, Cadmium und Quecksilber, *Z. anorg. und allg. Chem.*, **279**(5/6), 241–270 (1955).

Hahn H., Lorent C. Untersuchungen über ternäre Chalkogenide. Über ternäre Sulfide und Selenide des Germaniums mit Zink, Cadmium und Quecksilber, *Naturwissenschaften*, **45**(24), 621–622 (1958).

Hudoliy V.A., Pan'ko V.V., Voroshilov Yu.V., Shelemba M.S. Phase equilibria in the HgTe–HgSe–$HgBr_2$ system [in Russian], *Izv. AN SSSR. Neorgan. materialy*, **26**(11), 2425–2428 (1990).

Jandali M. Z., Eulenberger G. Hahn H. Die Kristallstrukturen von $Hg_2P_2S_6$ und $Hg_2P_2Se_6$, *Z. anorg. und allg. Chem.*, **447**(10), 105–118 (1978).

Kanatzidis M.G., Park Y. Molten salt synthesis of low-dimensional ternary chalco-genides. Novel structure types in the K/Hg/Q system (Q=S, Se), *Chem. Mater.*, **2**(2), 99–101 (1990).

Kerkhoff M., Leute V. The phase diagram of the quasibinary system Hg_3Se_3/Ga_2Se_3, *J. Alloys Compd.*, **385**(1–2), 148–155 (2004).

Klinger W., Eulenberger G., Hahn H. Über Hexachalkogeno-hypodiphosphate vom Typ $M_2P_2X_6$, *Naturwissenschaften*, **57**(2), 88 (1970).

Klinger W., Ott R., Hahn H. Über die Darstellung und Eigenschaften von Hexathio- und Hexaselenohypodiphosphaten, *Z. anorg. und allg. Chem.*, **396**(3), 271–278 (1973).

Kovaleva I.S., Kuznetsova I.Ya. Investigation of interaction in the $HgSe–Cr_2Se_3–Se$ system [in Russian], in: *Troinyie poluprovodn. i ih primenienie: Tez. dokl.* Kishinev: Shtiintsa, 70–71 (1979).

Kovaleva I.S., Kuznetsova I.Ya., Kalinnikov V.T. Interaction of HgSe with Cr_2Se_3 [in Russian], *Zhurn. neorgan. khimii*, **25**(4), 1156–1158 (1980).

Kovaleva I.S., Kuznetsova I.Ya., Novotortsev V.M. Deviation from stoichiometry of chalkochromite $HgCr_2Se_4$ [in Russian], *Zhurn. neorgan. khimii*, **41**(1), 27–28 (1996).

Kozer V.R., Fedorchuk A.O., Olekseyuk I.D., Parasyuk O.V. Crystal structure of the phases $Hg_5C^{III}_2X_8$ (C^{III}=Ga, In; X=Se, Te), *J. Alloys Compd.*, **503**(1), 40–43 (2010).

Kuliev A.A., Kagramanian Z.G., Suleimanov D.M. Investigation of phase diagram of TlSe–HgSe system [in Russian], *Uchen. zap. Azerb. un-t. Ser. khim. nauk.* (4), 66–68 (1971).

Kurbanov A.A., Babanly M.B., Kuliev A.A. Interaction of mercury selenide with anti-mony and bismuth selenides [in Russian], *Zhurn. neorgan. khimii*, **26**(5), 1438–1439 (1981).

Kurbanov A.A., Babanly M.B., Kuliev A.A. Phase equilibria in the HgSe(Te)–$As_2Se_3(Te_3)$ systems [in Russian], *Azerb. khim. zhurn.*, (2), 99–101 (1985).

Leute V., Köller H.-J. Thermodynamic properties of the quasiternary system $Hg_kPb_{(1-k)}Se_lTe_{(1-l)}$, *Z. phys. Chem.* (BRD), **150**(2), 227–243 (1986).

Leute V., Plate H. The phase diagram of the semiconductor alloy $Zn_kHg_{(1-k)}Se_lTe_{(1-l)}$, *Ber. Bunsenges. Phys. Chem.*, **93**(7), 757–763 (1989).

Magarill S.A., Pervukhina N.V., Borisov S.V., Pal'chik N.A. Crystal chemistry and fea-tures of structure formation of mercury oxo- and chalcogenides [in Russian], *Uspekhi khimii*, **76**(2), 115–146 (2007).

Malevski A.Yu., Chzhun Tszia-zhun. About isomorphous substitution of sulfur by selenium and tellurium in the mercury sulfides [in Russian], in: *Eksperiment.-metodich. issled. rudnyh mineralov*, Moscow: Nauka Publish., 223–226 (1965).

Medvedev Yu.V., Berchenko N.N. Peculiarities of the interface between solid solu-tions based on mercury chalcogenides and native oxides [in Russian], *Zhurn. neorgan. khimii*, **38**(12), 1940–1945 (1993).

Meenakshi S., Vijayakumar V., Eifler A., Hochheimer H.D. Pressure-induced phase transition in defect chalcopyrites $HgAl_2Se_4$ and $CdAl_2S_4$, *J. Phys. Chem. Solids*, **71**(5), 832–835 (2010).

Meenakshi S., Vijayakumar V., Godwal B.K., Eifler A., Orgzall I., Tkachev S., Hochheimer H.D. High pressure X-ray diffraction study of $CdAl_2Se_4$ and Raman study of AAl_2Se_4 (A=Hg, Zn) and $CdAl_2X_4$ (X=Se, S), *J. Phys. Chem. Solids*, **67**(8), 1660–1667 (2006)

Metlinski N.N., Tyrziu V.G., Markus M.M., Derid O.P. Phase diagram of the HgSe–Ga_2Se_3 system [in Russian], *Monokristally i tehnika, Khar'kov*, [1(8)], 52–56 (1973).

Minets Yu.V., Voroshilov Yu.V., Pan'ko V.V., Khudolii V.A. Phase equilibria in the HgSe–HgBr$_2$–HgI$_2$ system and crystal structure of Hg$_3$Se$_2$Br$_2$ and Hg$_3$Se$_2$I$_2$, *J. Alloys Compd.*, **365**(1–2), 121–125 (2004).

Motria S.F. Mercury–germanium(tin)–sulfur(selenium) ternary systems [in Russian], in: *Poluch. i svoistva slozhn. poluprovodn.*, Uzhgorod. gos. un-t, Kiev, 17–26 (1991).

Motria S.F., Voroshilov Yu.V., Potoriy M.V., Semrad E.E. Phase equilibria in the Ge(Sn)Se$_2$–HgSe systems [in Russian], *Ukr. khim. zhurn.*, **52**(8), 807–809 (1986).

Nagorsen G., Lyng S., Weiss Alarich, Weiss Armin. Zur Konstitution von HgSO$_4$ 2HgO, *Angew. Chem.*, **74**(3), 119 (1962).

Nikol'skaya E.I., Regel' A.R. Solid solution formation and magnetic susceptibility in the HgTe–HgSe, HgTe–β-HgS, HgSe–β-HgS systems [in Russian], *Zhurn. tekhn. fiziki*, **25**(8), 1347–1351 (1955).

Olekseyuk I.D., Bozhko V.V., Parasyuk O.V., Galyan V.V., Petrus' I.I. Physico-chemical and physical properties of glasses of the HgSe–GeSe$_2$ system, *Funct. Mater.*, **6**(3), 550–553 (1999).

Olekseyuk I.D., Sopko T.V., Mel'nichenko T.N. Phase equilibria in the HgSe(HgTe)–As, HgSe(HgTe)–As$_2$Se$_3$(As$_2$Te$_3$) systems [in Russian], *Izv. AN SSSR. Neorgan. materialy*, **23**(8), 1282–1285 (1987).

Pajaczkowska A. Physicochemical properties and crystal growth of A^{II}B^{VI}–MnB^{VI} systems, *Progr. Crystal Growth Caract.*, **1**(3), 289–326 (1978).

Pajaczkowska A., Dziuba E. Z. The solubility of HgS, HgSe and HgTe in Hg, *J. Cryst. Growth.*, **11**(1), 21–24 (1971).

Pajaczkowska A., Rabenau A. Phase studies in the system mercury selenide–manganese selenide, *Mater. Res. Bull.*, **12**(2), 183–188 (1977).

Pan'ko V.V., Hudoliy V.A., Voroshilov Yu.V. The HgSe(Te)–HgHal$_2$ systems [in Russian], *Zhurn. neorgan. khimii*, **34**(9), 2331–2335 (1989).

Paranchich S.Yu., Paranchich L.D., Makogonenko V.N., Lototski V.B. Structural, electrical and thermal properties of Fe$_x$Hg$_{1-x}$Se [in Russian], *Izv. AN SSSR. Neorgan. materialy*, **25**(2), 233–236 (1989).

Parasyuk O.V., Gulay L.D., Piskach L.V., Kumanska Yu.O. The Ag$_2$Se–HgSe–SnSe$_2$ system and the crystal structure of the Ag$_2$HgSnSe$_4$ compound, *J. Alloys Compd.*, **339**(1–2), 140–143 (2002).

Parasyuk O.V., Gulay L.D., Romanyuk Ya.E., Olekseyuk I.D. The Ag$_2$Se–HgSe–SiSe$_2$ system in the 0–60 mol. % SiSe$_2$ region, *J. Alloys Compd.*, **348**(1–2), 157–166 (2003).

Parasyuk O., Piskach L., Morenko A., Halka V., Marchuk O. Phase equilibria in the Cu$_{2-x}$Te(Se)–Cd(Hg)Te(Se) quasibinary systems [in Ukrainian], *Nauk. visnyk Volyn. derzh. un-tu*, (4), 35–38 (1997).

Puff H. Ternäre Verbindungen von Quecksilberhalogeniden mit Elementen der 5. und 6. Hauptgruppe, *Angew. Chem.*, **74**(16), 659 (1962).

Puff H., Heine D., Lieck C. Quecksilberschwefelfluorid, *Naturwissenschaften*, **55**(6), 298 (1968).

Puff H., Kohlschmidt R. Quecksilberchalkogenid-halogenide, *Naturwissenschaften*, **49**(14), 299 (1962).

Puff H., Küster J. Die Kristallstruktur der kubischen Triquecksilber-dichalkonium-dihalogenide, *Naturwissenschaften*, **49**(20), 464–465 (1962).

Radautsan S.I., Gavrilitsa E.I. Solid solutions in the HgSe–In$_2$Se$_3$ system [in Russian], *Izv. AN MSSR*, [10(88)], 95–97 (1961).

Range K.-J., Becker W., Weiss A. Über Hochdruckphasen des $CdAl_2S_4$, $HgAl_2S_4$, $ZnAl_2Se_4$, $CdAl_2Se_4$ und $HgAl_2Se_4$ mit Spinellstruktur, *Z. Naturforsch.*, **23B**(7), 1009 (1968).

Singh P., Verma U.P., Jensen P. Electronic and optical properties of defect chalcopyrite $HgAl_2Se_4$, *J. Phys. Chem. Solids*, **72**(12), 1414–1418 (2011).

Sommer H., Hoppe R. Thio- und Selenomercurate (II). $K_6[HgS_4]$, $K_6[HgSe_4]$, $Rb_6[HgS_4]$ und $Rb_6[HgSe_4]$, *Z. anorg. und allg. Chem.*, **443**(6), 201–211 (1978).

Strauss A. J., Farrel L. B., Harman T. C. HgTe–HgSe system, *Quart Progr. Rept. Solid State Res. Lincoln Lab. Mass. Inst. Technol, Massachusetts*, Apr., 24–25 (1960).

Szuszkiewicz W., Dybko K., Witkowska B., Julien C., Balkanski M. Composition dependence of the energy gap in HgFeSe—optical verification, *Acta Phys. Pol., A*, **82**(5), 757–760 (1992).

Vaniarho V.G., Zlomanov V.P., Novoselova A.V. On the question about solid solution formation in the PbSe–MnSe and PbSe–HgSe systems [in Russian], *Izv. AN SSSR. Neorgan. materialy*, **3**(7), 1276–1277 (1967).

Vaniarho V.G., Zlomanov V.P., Novoselova A.V. T-x phase diagrams of the lead selenide–mercury selenide and lead telluride–mercury telluride systems [in Russian], *Vestn. Mosk. un-ta. Ser. Khimia*, (6), 108–109 (1968).

Vaniarho V.G., Zlomanov V.P., Novoselova A.V. Investigation of the PbSe–HgSe system [in Russian], *Izv. AN SSSR. Neorgan. materialy*, **5**(11), 2025–2026 (1969).

Voroshilov Yu.V., Gad'mashi Z.P., Slivka V.Yu., Hudoliy V.A. Obtaining and natural optical activity of the mercury chalcogenides crystals [in Russian], *Izv. AN SSSR. Neorgan. materialy*, **17**(11), 2022–2024 (1981).

Voroshilov Yu.V., Motria S.F., Semrad E.E. Phase equilibria in the mercury–tin–sulfur(selenium) systems [in Russian], *Zhurn. neorgan. khimii*, **38**(6), 1061–1064 (1993).

Voroshilov Yu.V., Pan'ko V.V., Pechars'kyi V.K., Hudoliy V.A. Phase equilibria in the $HgS(Se, Te)–HgCl_2(Br_2, I_2)$ systems and crystal structure of compounds [in Ukrainian], *Visnyk L'viv. un-tu. Ser. khim.*, (33), 11–24 (1994).

Zhukov E.G., Poluliak E.S., Varnakova E.S., Fedorov V.A. Investigation of chalcogenide spinel stability at the heating in the air [in Russian], *Neorgan. materialy*, **33**(8), 939–941 (1997).

Zou J.-P., Guo G.-C., Guo S.-P., Lu Y.-B., Wu K.-J., Wang M.-S., Huang J.-S. Synthesis, crystal and band structures and optical properties of a new mixed-framework mercury selenide-diselenite $(Hg_3Se_2)(Se_2O_5)$, *Dalton Trans.*, (42), 4854–4858 (2007).

9

Systems Based on HgTe

9.1 Mercury–Rubidium–Tellurium

HgTe–Rb₂Te: The phase diagram is not constructed. $Rb_2Hg_3Te_4$ ternary compound is formed in this system, which crystallizes in an orthorhombic structure with lattice parameters $a = 1217.7 \pm 0.2$, $b = 724.5 \pm 0.2$, and $c = 1454.5 \pm 0.2$ pm (Li et al. 1997).

Single crystals of $Rb_2Hg_3Te_4$ were grown from solvothermal reaction using ethylenediamine as a solvent (Li et al. 1997).

9.2 Mercury–Copper–Tellurium

HgTe–Cu: This section is a nonquasibinary section of the Hg–Cu–Te ternary system. Copper telluride and Hg are formed at the interaction of HgTe with Cu (Tomashik et al. 1987).

HgTe–Cu₁.₈Te: The phase diagram belongs to the eutectic type (Figure 9.1) (Parasyuk et al. 1997). The eutectic crystallizes at 542°C and contains 63 mol. % HgTe. Solid solutions based on HgTe reach 5 mol. % $Cu_{1.8}Te$, and high-temperature $Cu_{1.8}Te$ modification dissolves 42 mol. % HgTe at 542°C. Four eutectoid reactions at 414°C, 310°C, 285°C, and 225°C and one peritectoid reaction at 208°C take place in this system.

This system was investigated using differential thermal analysis (DTA), metallography, and x-ray diffraction (XRD). The ingots were annealed at 400°C for 250 h (Parasyuk et al. 1997).

9.3 Mercury–Silver–Tellurium

HgTe–Ag: This section is a nonquasibinary section of the Hg–Ag–Te ternary system. Silver telluride and Hg are formed at the interaction of HgTe with Ag (Tomashik et al. 1987, Shuh and Williams 1988).

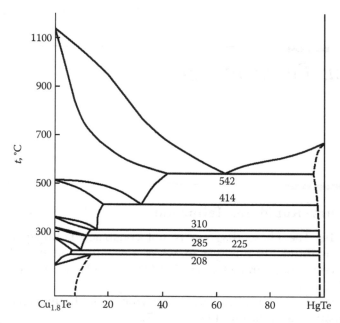

FIGURE 9.1
The HgTe–Cu$_{1.8}$Te phase diagram. (From Parasyuk, O. et al., *Nauk. visnyk Volyn. derzh. un-tu,* (4), 35, 1997.)

9.4 Mercury–Gold–Tellurium

HgTe–Au: According to the data of thermochemical calculations and XRD data, this section is a quasibinary section of the Hg–Au–Te ternary system (Shuh and Williams 1988).

9.5 Mercury–Magnesium–Tellurium

HgTe–MgTe: The phase diagram is shown in Figure 9.2 (Gavaleshko et al. 1982, 1983). This system was investigated within the interval of 0–20 mol. % MgTe. Solid solutions with sphalerite structure are formed in this concentration region. Lattice parameters decrease with the increase of MgTe contents (Frasuniak et al. 1982).

With the increase in pressure, two phase transitions take place for the Hg$_{1-x}$Mg$_x$Te ($x < 0.1$) solid solutions (Shchennikov et al. 1993). At 1.3 ± 0.1 GPa, a sphalerite-type structure changes in a cinnabar-type structure, which transforms into a NaCl-type structure at 9 ± 1 GPa. Substitution of Hg by Mg

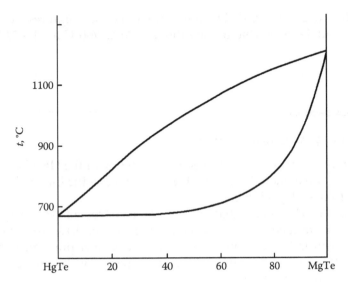

FIGURE 9.2
The HgTe–MgTe phase diagram. (From Gavaleshko, N.P. et al., *Fizika kristalliz.*, *Kalinin*, 1–57, 1982; Gavaleshko, N.P. et al., *Izv. AN SSSR. Neorgan. materialy*, 10(2), 327, 1983.)

in the sphalerite structure does not influence the pressure of the "sphalerite–cinnabar" phase transition.

The phase diagram was calculated using the model of regular associated solutions (Gavaleshko et al. 1982, 1983).

9.6 Mercury–Zinc–Tellurium

HgTe–Zn: This section is a nonquasibinary section of the Hg–Zn–Te ternary system. Solid solutions $Zn_xHg_{1-x}Te$ and Hg are formed at the interaction of HgTe with Zn (Tomashik and Vengel 1990).

9.7 Mercury–Aluminum–Tellurium

HgTe–Al₂Te₃: The phase diagram is not constructed. $HgAl_2Te_4$ ternary compound is formed in this system, which crystallizes in a tetragonal structure of chalcopyrite type with lattice parameters $a = 599.2$ and $c = 1209$ pm ($c = 604.5$ pm) [$a = 850.8$ and $c = 4866.5$ pm (Schwer and Kramer 1988)] and calculation and experimental density 5.816 and 5.79 g cm^{-3}, respectively (Hahn et al. 1955).

HgAl$_2$Te$_4$ was obtained by the annealing of mixtures, containing HgTe, Al, and Te in stoichiometric ratio, at 600°C–700°C for 12–24 h (Hahn et al. 1955).

9.8 Mercury–Gallium–Tellurium

HgTe–Ga: This section is a nonquasibinary section of the Hg–Ga–Te ternary system. Gallium monotelluride and Hg are formed at the interaction of HgTe with Ga (Tomashik et al. 1987).

HgTe–GaTe: The phase diagram is not constructed. HgGaTe$_2$ ternary compound is formed in this system, which crystallizes in a tetragonal structure with lattice parameters $a = 1185$ and $c = 684$ pm, calculation and experimental density 7.31 and 7.30 g cm^{-3}, respectively, and energy gap $E_g = 2.4$ eV (Guseinov 1969).

HgTe–Ga$_2$Te$_3$: The calculated phase diagram of the 3HgTe–Ga$_2$Te$_3$ is shown in Figure 9.3 (Leute et al. 1999). The only remarkable difference between calculated and experimental data occurs at the Ga$_2$Te$_3$-rich boundary of the miscibility gap between the chalcopyrite phase and the disordered cubic region. This calculated boundary is somewhat beyond the region where tetragonal reflections could be detected. Nevertheless, the calculated phase diagram yields a satisfactory description of the experimental results.

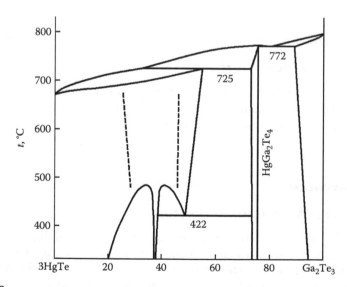

FIGURE 9.3
The calculated phase diagram for the quasibinary system 3HgTe–Ga$_2$Te$_3$. (From Leute, V. et al., *J. Alloys Compd.*, 289(1–2), 233, 1999.)

Earlier, the phase diagram of this system was investigated experimentally by Woolley and Ray (1960b), Ray et al. (1970), and Wensierski von et al. (1997), and the obtained results do not contradict to the calculated phase diagram in Leute et al. (1999).

$Hg_5Ga_2Te_8$ crystallizes in a cubic structure of sphalerite type with lattice parameter $a = 1247.38 \pm 0.02$ pm and calculation density 7.33047 ± 0.0004 g cm^{-3} (Kozer et al. 2010) or in the orthorhombic structure with a: b: $c = 1:2:\sqrt{2}$ (Ray et al. 1970). $HgGa_2Te_4$ has two polymorphous modifications: one of them crystallizes in a defect tetragonal structure of chalcopyrite type with lattice parameters $a = 600.5$ and $c = 1201$ pm [$a = 602.5$ and $c = 1203.7$ pm (Agostinelli et al. 1985, Gastaldi et al. 1985] and calculation and experimental density 6.481 and 6.42 g cm^{-3}, respectively (Hahn et al. 1955). The second modification crystallizes in a cubic structure with lattice parameter $a = 600.5$ pm (Hahn et al. 1955) [$a = 600.2$ pm (Agostinelli et al. 1985, Gastaldi et al. 1985)].

Hailing et al. (1983) investigated phase transformation of $Hg_5Ga_2Te_8$ and $Hg_3Ga_2Te_6$ compounds under hydrostatic pressure. Phase transitions for $Hg_5Ga_2Te_8$ and $Hg_3Ga_2Te_6$ with formation of metallic phases take place at 5.2 ± 0.2 and 6.0 ± 0.2 GPa ($52 + 2$ and 60 ± 2 kbar), respectively. At the pressure decreasing, the inverse transition for these compounds occurs at 4.3 ± 0.2 and 5.3 ± 0.2 GPa (43 ± 2 and 53 ± 2 kbar), respectively.

This system was investigated using DTA, metallography, XRD, and electron probe microanalyzer (EPMA) (Ray et al. 1970, Wensierski et al. 1997). The ingots were annealed at 650°C for 100 h (Ray et al. 1970). Single crystals $(Ga_2Te_3)_x(Hg_3Te_3)_{1-x}$ ($0 \leq x \leq 0.2$) were grown using the Bridgman method (Lototski et al. 1986). The cubic modification of $HgGa_2Te_4$ was obtained from the solution in the KBr melt, and tetragonal modification was grown using chemical transport reactions (Agostinelli et al. 1985, Gastaldi et al. 1985).

9.9 Mercury–Indium–Tellurium

HgTe–In: This section is a nonquasibinary section of the Hg–In–Te ternary system. Indium monotelluride and indium amalgam are formed at the interaction of HgTe with In (Vengel' and Tomashik 1987, Shuh and Williams 1988).

HgTe–InTe: The phase diagram is shown in Figure 9.4 (Vengel' and Tomashik 1989). The eutectic and azeotropic compositions and temperatures are 83 mol. % HgTe and 630°C and 66.7 mol. % HgTe and 713°C, respectively. At the eutectic temperature, solid solutions based on HgTe contain up to 78 mol. %. The solubility of HgTe in InTe at the same temperature is equal to 7 mol. % and 3 mol. % at room temperature. α-Solid solutions tend to disorder and to decompose within the interval of 40–50 mol. % HgTe, forming two solid solutions that crystallize in a cubic structure of sphalerite type.

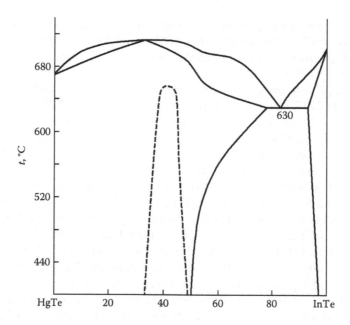

FIGURE 9.4
The HgTe–InTe phase diagram. (From Vengel', P.F. and Tomashik, V.N., *Izv. AN SSSR. Neorgan. materialy*, 25(3), 388, 1989.)

This system was investigated using DTA, metallography, XRD, and microhardness measurement. The ingots were annealed at 600°C for 700 h (Vengel' and Tomashik 1989).

HgTe–In₂Te₃: The calculated phase diagram of the 3HgTe–In₂Te₃ is shown in Figure 9.5 (Leute et al. 1999). The experimentally determined liquidus and solidus data can be described fairly well. Earlier, the phase diagram of this system was investigated experimentally by Woolley and Ray (1960a), Spencer (1964), Ray and Spencer (1965), Spencer (1968), Anan'ina et al. (1976), and Leute and Schmidtke (1988a,b), and the obtained results do not contradict to the calculated phase diagram in Leute et al. (1999).

The disordered modification of In₂Te₃ forms solid solutions by alloying with HgTe (Weitze and Leute 1996). The ordered modifications of In₂Te₃ below 620°C show no measurable solubility for HgTe. The extent of the miscibility gap between the orthorhombic form of In₂Te₃ and its cubic solution with HgTe increases linearly with temperature decreasing.

HgIn₂Te₄ ternary compound melts incongruently at 714°C (Leute et al. 1999) and crystallizes in a tetragonal structure of the chalcopyrite type with lattice parameters $a = 620.6$ and $c = 1244.7$ pm (Leute and Schmidtke 1988a) [$a = 617.4$ and $c = 1235$ pm (Hahn et al. 1955); $a = 621.0 \pm 0.5$ pm (Grushka et al. 1982a)] and calculation and experimental density 6.595 and 6.34 g cm⁻³, respectively (Hahn et al. 1955).

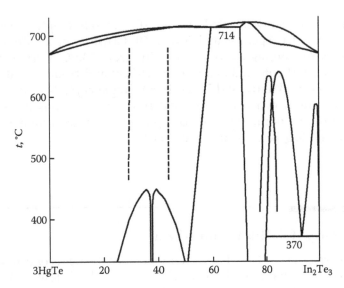

FIGURE 9.5
The calculated phase diagram for the quasibinary system $3HgTe-In_2Te_3$. (From Leute, V. et al., *J. Alloys Compd.*, 289(1–2), 233, 1999.)

Hailing et al. (1983) investigated phase transformation of the $HgIn_2Te_4$ compound under hydrostatic pressure and showed that phase transition for $HgIn_2Te_4$ with formation of metallic phase takes place at 4.8 ± 0.2 GPa (48 ± 2 kbar). With the decrease in pressure, the inverse transition for this compound occurs at 4.3 ± 0.2 GPa (43 ± 2 kbar).

$Hg_5In_2Te_8$ has phase transition at 447°C (Leute and Schmidtke 1988a) and crystallizes in a cubic structure of the sphalerite type with lattice parameter $a=1267.23\pm0.02$ pm and calculation density 7.4069 ± 0.0004 g cm^{-3} (Kozer et al. 2010) [a = 1266.6 pm (Leute and Schmidtke 1988a), $a=633\pm1$ pm (Grushka et al. 1982a)].

$Hg_3In_2Te_6$ melts congruently at 711°C (Spencer 1964, Ray and Spencer 1965) and crystallizes in a cubic structure of the sphalerite type with lattice parameter $a=628.0\pm0.2$ pm (Grushka et al. 1982a) [in a tetragonal structure with lattice parameters $a=1024$ and $c=2950$ pm (Mayet and Roubin 1980)] and energy gap $E_g=0.78$ eV at 100 K (Bakumenko et al. 1983). According to the data of Maynell et al. (1970), this compound has the order–disorder transition at 312 ± 5°C and high-temperature and low-temperature modifications crystallize in a cubic structure of the sphalerite type with lattice parameter $a=628.9\pm0.2$ and 1887.0 ± 0.3 pm, respectively. Further investigations showed (Grushka et al. 1982a,b, Leute and Schmidtke 1988a) that thermal treatment of the $Hg_5In_2Te_8$ crystals leads to the decomposition of this compound into two phases ($Hg_5In_2Te_8$ and $HgIn_2Te_4$) but not to its ordering.

At the small In_2Te_3 concentrations (up to 2 mol. %), anomalous behaviors of microhardness, electrical conductivity, and lattice parameter take place in the $3HgTe-In_2Te_3$ system (Grushka et al. 1983).

This system was investigated using DTA, metallography, EPMA, and XRD (Spencer 1964, Spencer and Ray 1968, Grushka et al. 1982a, Weitze and Leute 1996). The ingots were annealing from 330°C to 680°C between 6 weeks and 1 h (Leute and Schmidtke 1988a,b) and from 21 h at 863°C to 280 days at 330°C (Weitze and Leute 1996). Single crystals of the $HgIn_2Te_4$ compound were grown using chemical transport reactions (Gilevich and Zhukova 1974) and of the $Hg_3In_2Te_6$ compound using the Bridgman method (Maynell et al. 1970, Bakumenko et al. 1983).

9.10 Mercury–Thallium–Tellurium

Liquidus surface of the Hg–Tl–Te system includes the fields of primary crystallization of next phases: HgTe, β-$Hg_3Tl_2Te_4$, Tl_2Te, Tl_5Te_3, TlTe, Tl_2Te_3, Hg, Tl, and Te (Figure 9.6) (Babanly et al. 1983). Nonvariant equilibria in this system are given in Table 9.1. At room temperature, γ-solid solutions based on Tl_5Te_3 have noticeable extension and reach the HgTe–Tl_2Te section (Figure 9.7) (Asadov 1983, Babanly et al. 1983). Below 180°C, solid solutions based on HgSe are practically absent.

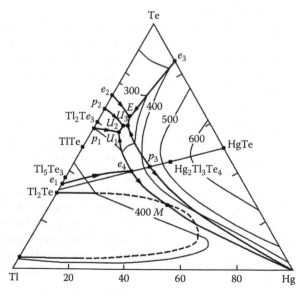

FIGURE 9.6

Liquidus surface of the Hg–Tl–Te ternary system. (From Babanly, M.B. et al., *Izv. AN SSSR. Neorgan. materialy*, 19(4), 583, 1983.)

TABLE 9.1

Nonvariant Equilibria in the Hg–Tl–Te Ternary System

Symbol	Reaction	Hg, at. %	Te, at. %	t, °C
e_4	$L \Leftrightarrow (Tl_5Te_3) + Hg_3Tl_2Te_4$	22	40	350
E	$L \Leftrightarrow Tl_2Te_3 + Te + HgTe$	11	61	200
p_3	$L + HgTe \Leftrightarrow Hg_3Tl_2Te_4$	28	42	460
U_1	$L + (Tl_5Te_3) \Leftrightarrow TlTe + Hg_3Tl_2Te_4$	9	60	250
U_2	$L + TlTe \Leftrightarrow Tl_2Te_3 + Hg_3Tl_2Te_4$	10	58	235
U_3	$L + Hg_3Tl_2Te_4 \Leftrightarrow HgTe + Tl_2Te_3$	10.5	59	230
M	$L_1 \Leftrightarrow L_2 + Hg_3Tl_2Te_4 + (Tl_5Te_3)$	41	27	410

Source: Babanly, M.B. et al., *Izv. AN SSSR. Neorgan. materialy*, 19(4), 583, 1983.

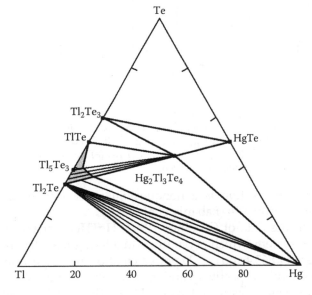

FIGURE 9.7
Isothermal section of the Hg–Tl–Te ternary system at room temperature. (From Asadov, M.M., *Zhurn. neorgan. khimii*, 28(2), 539, 1983; Babanly, M.B. et al., *Izv. AN SSSR. Neorgan. materialy*, 19(4), 583, 1983.)

This system was investigated using DTA, XRD, and measurements of microhardness and emf of concentrated chains (Asadov 1983, Babanly et al. 1983).

HgTe–Tl: This section is a nonquasibinary section of the Hg–Tl–Te ternary system (Figure 9.8) (Asadov 1988). The maximum solubility of Tl in HgTe is approximately 1.2 at. % at 470°C.

The section was investigated using DTA, XRD, and microhardness measurement (Asadov 1988).

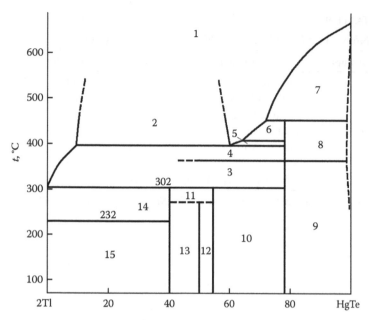

FIGURE 9.8

Phase relations in the HgTe–2Tl system: 1, L; 2, L_1+L_2; 3, $L_1+Tl_2Te+\alpha\text{-}Hg_3Tl_2Te_4$; 4, $L_1+Tl_2Te+\beta\text{-}Hg_3Tl_2Te_4$; 5, $L_2+\beta\text{-}Hg_3Tl_2Te_4+Tl_2Te$; 6, $L_2+\beta\text{-}Hg_3Tl_2Te_4$; 7, $L_2+(HgTe)$; 8, $L_2+\beta\text{-}Hg_3Tl_2Te_4+(HgTe)$; 9, $L_1+\alpha\text{-}Hg_3Tl_2Te_4+(HgTe)$; 10, $L_1+\alpha\text{-}Hg_3Tl_2Te_4+(Tl_5Te_3)$; 11, $L_1+Tl_2Te+(Tl_5Te_3)$; 12, $L_1+(Tl_5Te_3)$; 13, L_1+Tl_2Te; 14, $L_1+Tl_2Te+\beta\text{-}Tl$; and 15, $L_1+Tl_2Te+\alpha\text{-}Tl$. (From Asadov, M.M., *Azerb. khim. zhurn.*, (2), 134, 1988.)

HgTe–TlTe: This section is a nonquasibinary section of the Hg–Tl–Te ternary system (Figure 9.9) (Babanly et al. 1983).

According to the data of Guseinov (1969), $HgTlTe_2$ ternary compound is formed in this system, which crystallizes in a tetragonal structure with lattice parameters $a = 863.4$ and $c = 729$ pm, calculation and experimental density 8.37 and 8.35 g cm^{-3}, respectively, and energy gap $E_g = 0.20$ eV.

This system was investigated using DTA, XRD, and measurements of microhardness and emf of concentrated chains. The ingots were annealed at temperatures 20°C–30°C below the solidus temperatures for 300–400 h (Babanly et al. 1983).

HgTe–Tl₂Te: Three ternary compounds are formed in this system (Figure 9.10) (Sztuba et al. 2000, 2001). $HgTl_{10}Te_6$ melts congruently at 423 ± 1°C. $Hg_3Tl_2Te_4$ and $Hg_9Tl_2Te_{10}$ melt incongruently at respective temperatures of 380 ± 1°C and 440 ± 1°C and respective peritectic points of 51 ± 1 and 68 ± 1 mol. % HgTe. From obtained data, it may be supposed that $Hg_3Tl_2Te_4$ has a polymorphic transformation at ~153°C. $HgTl_{10}Te_6$ shows semiconductive properties. The eutectic on the Tl_2Te side is degenerated and the other eutectic was found to be at 46 ± 0.5 mol. % and at 348 ± 1°C [46 mol. % HgTe and 347°C (Asadov et al. 1978)]. No evidence has been obtained for solid solutions in this system.

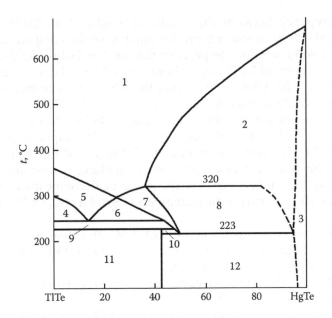

FIGURE 9.9
Phase relations in the HgTe–TlTe system: 1, L; 2, L + (HgTe); 3, (HgTe); 4, L + TlTe + (Tl$_5$Te$_3$); 5, L + (Tl$_5$Te$_3$); 6, L + (Tl$_5$Te$_3$) + α-Hg$_3$Tl$_2$Te$_4$; 7, L + α-Hg$_3$Tl$_2$Te$_4$; 8, L + (HgTe) + α-Hg$_3$Tl$_2$Te$_4$; 9, L + TlTe + α-Hg$_3$Tl$_2$Te$_4$; 10, L + Tl$_2$Te$_3$ + α-Hg$_3$Tl$_2$Te$_4$; 11, TlTe + Tl$_2$Te$_3$ + α-Hg$_3$Tl$_2$Te$_4$; and 12, Tl$_2$Te$_3$ + (HgTe) + α-Hg$_3$Tl$_2$Te$_4$. (From Babanly, M.B. et al., *Izv. AN SSSR. Neorgan. materialy*, 19(4), 583, 1983.)

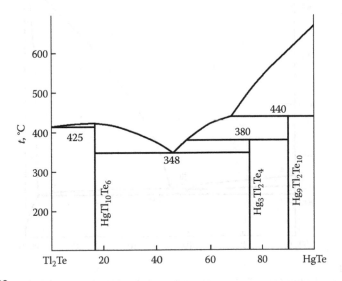

FIGURE 9.10
The HgTe–Tl$_2$Te phase diagram. (From Sztuba, Z. et al., *J. Phase Equil.*, 21(5), 447, 2000; Sztuba, Z. et al., *Pol. J. Chem.*, 75(1), 135, 2001.)

The discrepancy between the results of Sztuba et al. (2000, 2001) and Asadov et al. (1978) concerns both the number of the compounds and the existence of solid solutions (especially that on the Tl_2Te side: 0–22 mol. % HgTe). The compound $HgTl_{10}Te_6$ is characterized by a flat maximum on the liquidus curve, which is only 8°C higher than the melting point of pure Tl_2Te and therefore could be not observed by Asadov et al. (1978).

The formation of the compound $Hg_3Tl_2Te_4$ found by Asadov et al. (1978) was confirmed by Sztuba et al. (2000, 2001), but this compound melts incongruently at 380°C, while Asadov et al. (1978) regarded this temperature as a polymorphic transition point. Instead, the melting temperature of 458°C (Asadov et al. 1978) is believed to be the peritectic decomposition of the compound $Hg_9Tl_2Te_{10}$, which the measurements of Sztuba et al. (2000, 2001) place at 440°C.

This system was investigated using DTA, DSC, and measurement of emf of galvanic cells (Sztuba et al. 2000, 2001) and using DTA, XRD, and microhardness measurement (Asadov et al. 1978).

$HgTe–Tl_2Te_3$: This section is a nonquasibinary section of the Hg–Tl–Te ternary system (Figure 9.11) (Babanly et al. 1983).

The system was investigated using DTA, XRD, and measurements of microhardness and emf of concentrated chains. The ingots were annealed at temperatures 20°C–30°C below the solidus temperatures for 300–400 h (Babanly et al. 1983).

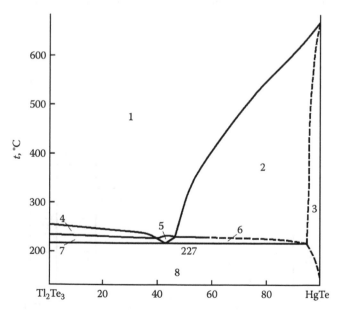

FIGURE 9.11

Phase relations in the $HgTe–Tl_2Te_3$ system: 1, L; 2, L + (HgTe); 3, (HgTe); 4, L + TlTe; 5, L + α-$Hg_3Tl_2Te_4$; 6, L + (HgTe) + α-$Hg_3Tl_2Te_4$; 7, L + TlTe + Tl_2Te_3; and 8, Tl_2Te_3 + (HgTe). (From Babanly, M.B. et al., *Izv. AN SSSR. Neorgan. materialy*, 19(4), 583, 1983.)

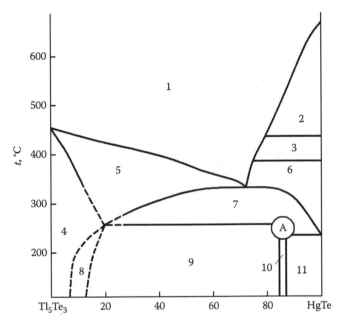

FIGURE 9.12
Phase relations in the HgTe–Tl$_5$Te$_3$ system: 1, L; 2, L+(HgTe); 3, L+β-Hg$_3$Tl$_2$Te$_4$; 4, (Tl$_5$Te$_3$); 5, L+(Tl$_5$Te$_3$); 6, L+α-Hg$_3$Tl$_2$Te$_4$; 7, L+(Tl$_5$Te$_3$)+α-Hg$_3$Tl$_2$Te$_4$; 8, (Tl$_5$Te$_3$)+TlTe; 9, (Tl$_5$Te$_3$)+ TlTe+α-Hg$_3$Tl$_2$Te$_4$; 10, (Tl$_5$Te$_3$)+Tl$_2$Te$_3$+α-Hg$_3$Tl$_2$Te$_4$; and 11, Tl$_2$Te$_3$+(HgTe)+α-Hg$_3$Tl$_2$Te$_4$. (From Babanly, M.B. et al., *Izv. AN SSSR. Neorgan. materialy*, 19(4), 583, 1983.)

HgTe–Tl$_5$Te$_3$: This section is a nonquasibinary section of the Hg–Tl–Te ternary system (Figure 9.12) (Babanly et al. 1983). The phase boundaries in the A region have not been determined.

The system was investigated using DTA, XRD and measurements of microhardness and emf of concentrated chains. The ingots were annealed at temperatures 20°C–30°C below the solidus temperatures for 300–400 h (Babanly et al. 1983).

9.11 Mercury–Europium–Tellurium

HgTe–EuTe: The phase diagram is shown in Figure 9.13 (Gavaleshko et al. 1985). The eutectic temperature is 655°C. The solubility of EuTe in HgTe at the eutectic temperature is equal to 12.5 mol. % and does not change with decreasing temperature. A new phase appears in the HgTe–EuTe system at the contents of EuTe more than 12.5 mol. %.

This system was investigated using DTA and XRD (Gavaleshko et al. 1985).

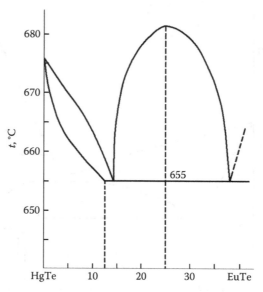

FIGURE 9.13
Part of the HgTe–EuTe phase diagram. (From Gavaleshko, N.P. et al., *Izv. AN SSSR. Neorgan. materialy*, 21(11), 1965, 1985.)

9.12 Mercury–Germanium–Tellurium

HgTe–GeTe: The phase diagram is a eutectic type (Figure 9.14) (Glazov et al. 1970, 1972, 1975). The eutectic temperature is 560°C. At 340°C, there is a eutectoid transformation in the HgTe–GeTe system. The mutual solubility of HgTe and GeTe is not higher than 4 mol. % (Glazov et al. 1972).

This system was investigated using DTA, metallography, and XRD (Glazov et al. 1970, 1972, 1975).

9.13 Mercury–Tin–Tellurium

HgTe–Sn: This section is a nonquasibinary section of the Hg–Sn–Te ternary system. Tin telluride and Hg are formed at the interaction of HgTe with Sn (Vengel' et al. 1986a).

HgTe–SnTe: The phase diagram is a eutectic type (Figure 9.15) (Vengel' et al. 1983). The eutectic composition and temperature are 45 mol. % SnTe and 584 ± 3°C, respectively. The solubility of HgTe and SnTe at the eutectic temperature is equal to 7 mol. %, and the solubility of SnTe in HgTe is not higher than 1 mol. %.

This system was investigated using DTA, metallography, XRD, and microhardness measurement (Vengel' et al. 1983).

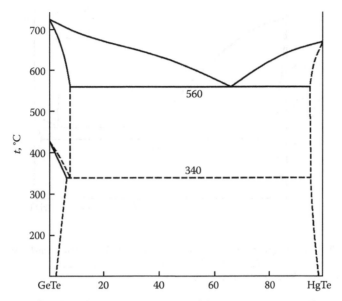

FIGURE 9.14
The HgTe–GeTe phase diagram. (From Glazov, V.M. et al., *Izv. AN SSSR. Neorgan. materialy*, 6(3), 569, 1970; Glazov, V.M. et al., *Izv. vuzov. Ser. Tsvetnaya metallurgia*, (5), 116, 1972; Glazov, V.M. et al., *Termodinamicheskiye svoistva metallicheskih splavov*, 376, 1975.)

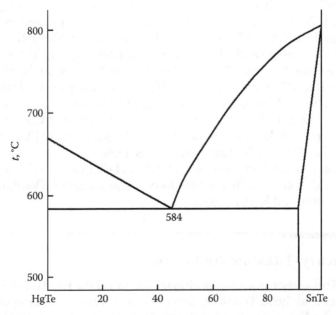

FIGURE 9.15
The HgTe–SnTe phase diagram. (From Vengel', P.F. et al., *Ukr. khim. zhurn.*, 49(12), 1247, 1983.)

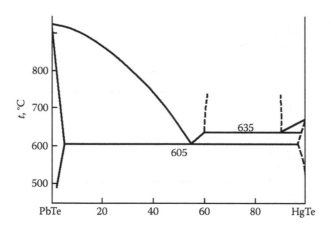

FIGURE 9.16
The HgTe–PbTe phase diagram. (From Vaniarho, V.G. et al., *Vestn. Mosk. un-ta. Ser. Khimia*, (6), 108, 1968; Vaniarho, V.G. et al., *Izv. AN SSSR. Neorgan. materialy*, 6(1), 133, 1970; Leute V. and Köller H.-J., *Z. phys. Chem. (BRD)*, 150(2), 227, 1986.)

9.14 Mercury–Lead–Tellurium

HgTe–Pb: This section is a nonquasibinary section of the Hg–Pb–Te ternary system. Lead telluride and Hg are formed at the interaction of HgTe with Pb (Tomashik et al. 1986a).

HgTe–PbTe: The phase diagram is a eutectic type (Figure 9.16) (Vaniarho et al. 1968, 1970, Leute and Köller 1986). The eutectic composition and temperature are 55 mol. % HgTe and 605 ± 5°C, respectively (Vaniarho et al. 1968, 1970) [76 mol. % HgTe and 610°C (Leute and Köller 1986)]. The immiscibility region within the interval of 60–90 mol. % HgTe exists in the HgTe–PbTe system at 635°C (Vaniarho et al. 1968, 1970) [according to the data of Leute and Köller (1986), there is not any immiscibility region in this system]. The solubility of HgTe in PbTe at 430°C, 530°C, 610°C, and 630°C is equal correspondingly to 1.5, 3, 4.6, and 4.0 mol. %, and the solubility of PbTe in HgTe is not higher than 0.5 mol. % (Leute and Köller 1986).

This system was investigated using DTA, metallography, XRD, and microhardness and thermoelectromotive force measurements (Vaniarho et al. 1968, 1970, Leute and Köller 1986).

9.15 Mercury–Titanium–Tellurium

HgTe–Ti: This section is a nonquasibinary section of the Hg–Ti–Te ternary system (Mizera et al. 1980). Titanium ditelluride was formed in this system when the crystals had been grown using the Bridgman method from the HgTe+Te+Ti starting material at the Ti contents higher than its solubility limit in HgTe.

9.16 Mercury–Arsenic–Tellurium

HgTe–As: The phase diagram is a eutectic type (Figure 9.17) (Olekseyuk et al. 1987). The eutectic composition and temperature are 31 at. % As and 606°C, respectively (the data are taken from Figure 9.17). The liquid solutions of this system are close to regular solutions.

This system was investigated using DTA, metallography, and XRD (Olekseyuk et al. 1987).

HgTe–As$_2$Te$_3$: The phase diagram is a eutectic type (Figure 9.18) (Kurbanov et al. 1985, Olekseyuk et al. 1987). The eutectic composition and

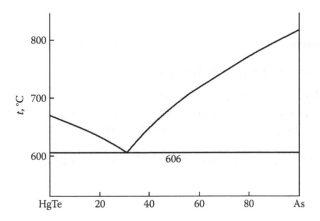

FIGURE 9.17
The HgTe–As phase diagram. (From Olekseyuk, I.D. et al., *Izv. AN SSSR. Neorgan. materialy*, 23(8), 1282, 1987.)

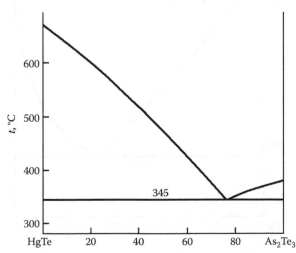

FIGURE 9.18
The HgTe–As$_2$Te$_3$ phase diagram. (From Kurbanov, A.A., *Azerb. khim. zhurn.*, (2), 99, 1985.)

temperature are approximately 24 mol. % HgTe [15 mol. % HgTe (Olekseyuk et al. 1987) and 345°C, respectively (Kurbanov et al. 1985)]. The liquid solutions of this system are close to regular solutions Olekseyuk et al. 1987).

This system was investigated using DTA, metallography, XRD, and micro-hardness measurement (Kurbanov et al. 1985, Olekseyuk et al. 1987). The ingots were annealed at 320°C for 200 h (Kurbanov et al. 1985).

HgTe–Hg$_3$As$_2$: The phase diagram is not constructed. A complex inter-action takes place in this system with formation of elementary mercury (Golovey et al. 1972).

9.17 Mercury–Antimony–Tellurium

HgTe–Sb: The phase diagram is a eutectic type (Figure 9.19) (Tomashik et al. 1986b). The eutectic composition and temperature are 67 at. % Sb and 467°C, respectively. The mutual solubility of HgTe and Sb is not higher than 0.5 mol. %.

This system was investigated using DTA and metallography (Tomashik et al. 1986b).

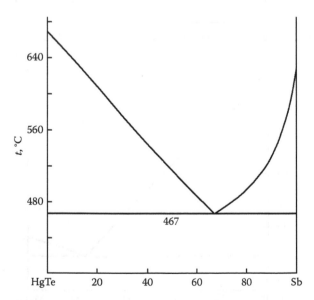

FIGURE 9.19
The HgTe–Sb phase diagram. (From Tomashik, V.N. et al., *Ukr. khim. zhurn.*, 52(7), 708, 1986a.)

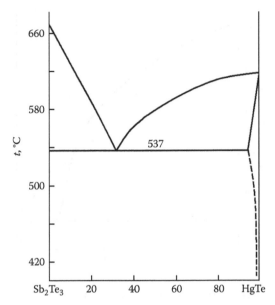

FIGURE 9.20
The HgTe–Sb$_2$Te$_3$ phase diagram. (From Babanly, M.B. et al., *Zhurn. neorgan. khimii*, 24(8), 2293, 1979.)

HgTe–Sb$_2$Te$_3$: The phase diagram is a eutectic type (Figure 9.20) (Babanly et al. 1979). The eutectic composition and temperature are 68 mol. % HgTe and 537°C, respectively. The mutual solubility of HgTe and Sb$_2$Te$_3$ is insignificant.

This system was investigated using DTA and microhardness measurement. The ingots were annealed at 450°C for 400 h (Babanly et al. 1979).

9.18 Mercury–Bismuth–Tellurium

HgTe–Bi: The phase diagram is a eutectic type (Figure 9.21) (Vengel' et al. 1986). The eutectic composition and temperature are 81 at. % Bi and 173°C, respectively. The solubility of HgTe in Bi and Bi in HgTe is not higher than 3 and 1 mol. % correspondingly.

This system was investigated using DTA and metallography (Vengel' et al. 1986b).

HgTe–Bi$_2$Te$_3$: The phase diagram is a eutectic type (Figure 9.22) (Babanly et al. 1979). The eutectic composition and temperature are 63 mol. % HgTe and 538°C, respectively. The mutual solubility of HgTe and Bi$_2$Te$_3$ is insignificant.

This system was investigated using DTA and microhardness measurement. The ingots were annealed at 450°C for 400 h (Babanly et al. 1979).

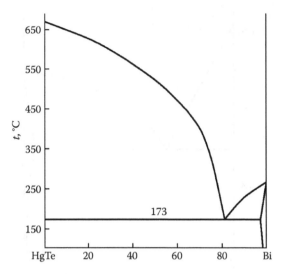

FIGURE 9.21
The HgTe–Bi phase diagram. (From Vengel', P.F. and Tomashik V.N., *Izv. AN SSSR Neorgan. materialy*, 22(7), 1212, 1986b.)

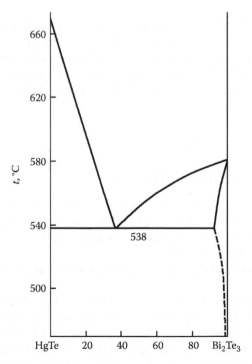

FIGURE 9.22
The HgTe–Bi$_2$Te$_3$ phase diagram. (From Babanly, M.B. et al., *Zhurn. neorgan. khimii*, 24(8), 2293, 1979.)

9.19 Mercury–Vanadium–Tellurium

HgTe–V: This section is a nonquasibinary section of the Hg–V–Te ternary system (Mizera et al. 1980). Vanadium ditelluride was formed in this system when the crystals had been grown using the Bridgman method from the HgTe+Te+V starting material at the V contents higher than its solubility limit in HgTe.

9.20 Mercury–Oxygen–Tellurium

The mixed-valent mercury(II) tellurite(IV) tellurate(VI), $Hg_2Te_2O_7$, which is formed in the Hg–O–Te ternary system, is dimorphous and exists in two modifications with slightly different density (Weil 2003). α-$Hg_2Te_2O_7$ crystallizes in a monoclinic structure with lattice parameters $a = 1291.0 \pm 0.4$, $b = 740.7 \pm 0.2$, $c = 1325.6 \pm 0.4$ pm, and $\beta = 112.044(5)°$, and β-$Hg_2Te_2O_7$ crystallizes in an orthorhombic structure with lattice parameters $a = 744.05 \pm 0.12$, $b = 2371.3 \pm 0.4$, and $c = 1352.2 \pm 0.2$ Å.

The predominance area diagram of the phase relationships in the system Hg–O–Te at 330°C is shown in Figure 9.23 (Diehl and Nolaeng 1984).

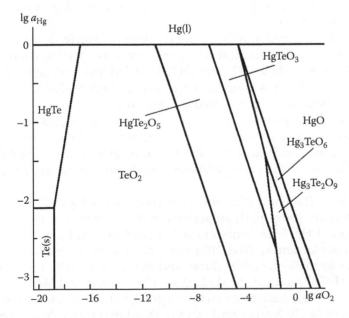

FIGURE 9.23
Predominance area of the condensed phases in equilibrium with the gas phase in the system Hg–O–Te at 330°C. (From Diehl, R. and Nolaeng, B.I., *J. Cryst. Growth*, 66(1), 91, 1984.)

For activities of oxygen of 10^{-20} and of Hg in the range of 10^{-2} to 1, HgTe is the predominant condensed species. With decreasing a_{Hg}, the tellurium activity increases and becomes unity at a_{Hg} of about 0.8 10^{-2}. At $a_{Te}=1$, the lower boundary of the HgTe predominance area is reached, and along the boundary line, HgTe coexists with the solid Te and the gas phase. At $a_{Hg}=1$, the predominance area of HgTe is confined by a second boundary line along which HgTe coexists with liquid Hg and the gas phase. For values of a_{Hg} in excess of unity, there is a forbidden area.

With the increase of oxygen activity, HgTe starts to react with O_2 to form oxide compounds, and the first oxide, which exists in equilibrium with HgTe, is TeO_2. In the lower left corner of Figure 9.23, the TeO_2 area is confined by the Te/TeO_2 boundary line. TeO_2 is stable in equilibrium with gas phase over a large range of Hg and O_2 activity values, and for an oxygen activity lower than 10^{-10}, it coexists with liquid Hg. With the increase of the oxygen activity, TeO_2 reacts with the gas phase to form $HgTe_2O_5$, which itself reacts further with the gas phase to form $HgTeO_3$. Further increase of $a(O_2)$ leads to the formation of $Hg_3Te_2O_9$, and then the reaction proceeds via Hg_3TeO_6 to the formation of HgO, which is stable in coexistence with the gas phase for high values of oxygen activity (Diehl and Nolaeng 1984).

According to the data of Grice (1989) and Pervukhina et al. (1999), Hg_2TeO_3 ternary compound (mineral magnolite) is formed in the Hg–O–Te ternary system, which crystallizes in an orthorhombic structure with lattice parameters $a = 595.8 \pm 0.1$, $b = 1057.6 \pm 0.2$, and $c = 374.9 \pm 0.1$ pm and calculation density 8.108 g cm^{-3}. This mercury oxytelluride is the product of Hg oxidation at low oxygen partial pressure.

Colorless single crystals of both modifications of $Hg_2Te_2O_7$ were prepared simultaneously using chemical transport reactions, starting from stoichiometric mixtures of HgO, TeO_2, and TeO_3 in sealed and evacuated silica glass ampoules (Weil 2003). A temperature gradient of 600°C–550°C was applied and small amounts of $HgCl_2$ served as transport agent.

HgTe–HgO: The phase diagram is not constructed. Hg_3Te_2O ternary compound was obtained in this system using hydrolysis of mercury halogentellurides (Batsanov and Abaulina 1961, Batsanov et al. 1973). It crystallizes in a cubic structure with lattice parameter $a = 648$ pm and has an experimental density 8.7 g cm^{-3}.

HgTe–TeO_2: The phase diagram is not constructed. $HgTeO_3$, $HgTe_2O_5$, and Hg_2TeO_4 ternary compounds are formed in this system, and the existence of the compound $Hg_2Te_3O_8$ is questionable (Diehl and Nolaeng 1984). $HgTeO_3$ has experimental density 7.66 ± 0.01 g cm^{-3}, melts at 510°C, and has a polymorphous transformation at 395°C (Pron' and Markovskii 1969). The heat of formation of $HgTeO_3$ is equal to 441.62 ± 0.08 kJ/mol (Pron' and Markovskii 1969). This compound has been obtained by the precipitation from water solutions at the mixing of Na_2TeO_3 and $Hg(CH_3COO)_2$ (Markovskii and Pron' 1968).

HgTe–TeO_3: The phase diagram is not constructed. Hg_3TeO_6 and $Hg_3Te_2O_9$ ternary compounds are formed in this system (Diehl and Nolaeng 1984).

9.21 Mercury–Chromium–Tellurium

HgTe–Cr: This section is a nonquasibinary section of the Hg–Cr–Te ternary system (Mizera et al. 1980). Chromium ditelluride was formed in this system when the crystals had been grown using the Bridgman method from the HgTe+Te+Cr starting material at the Cr contents higher than its solubility limit in HgTe.

9.22 Mercury–Chlorine–Tellurium

HgTe–HgCl$_2$: The phase diagram is shown in Figure 9.24 (Pan'ko et al. 1989, Voroshilov et al. 1994). The eutectic compositions and temperatures are 85 mol. % HgTe and 565°C and 10 mol. % HgTe and 261°C, respectively. Hg$_3$Te$_2$Cl$_2$ and Hg$_3$TeCl$_4$ ternary compounds are formed in this system (Batsanov and Abaulina 1961, Puff and Kohlschmidt 1962, Puff 1962, Puff and Küster 1962, Batsanov et al. 1973, Voroshilov et al. 1981, Pan'ko et al. 1989,

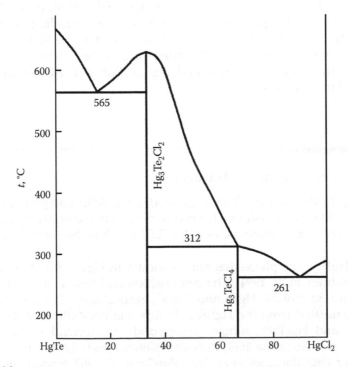

FIGURE 9.24
The HgTe–HgCl$_2$ phase diagram. (Pan'ko, V.V. et al., *Zhurn. neorgan. khimii*, 34(9), 2331, 1989; Voroshilov, Yu.V. et al., *Visnyk L'viv. un-tu. Ser. khim.*, (33), 11, 1994.)

Voroshilov et al. 1994). The first of them melts congruently at 630°C (Pan'ko et al. 1989) [626°C (Voroshilov et al. 1981)] and crystallizes in a cubic structure with lattice parameter $a = 932.6 \pm 0.3$ pm (Voroshilov et al. 1981, 1994) [$a = 933$ pm (Puff 1962, Puff and Kohlschmidt 1962, Batsanov et al. 1973)], calculation density 7.79 g cm^{-3}, and energy gap $E_g = 3.0$ eV (Batsanov et al. 1973).

Hg_3TeCl_4 melts incongruently at 312°C and crystallizes in an orthorhombic structure with lattice parameters $a = 1152.2 \pm 0.4$, $b = 1214.0 \pm 0.4$, and $c = 1268.3 \pm 0.2$ pm (Voroshilov et al. 1994, 1996a) [$a = 644.4 \pm 0.6$, $b = 1191 \pm 1$, and $c = 1247.0 \pm 0.9$ pm (Pan'ko et al. 1989)] and experimental density 6.52 g cm^{-3} (Voroshilov et al. 1994).

According to the data of Puff and Kohlschmidt (1962) and Aravamudan et al. (1979), two other ternary compounds (Hg_5TeCl_8 and $Hg_5Te_2Cl_6$) are formed in the $HgTe–HgCl_2$ system. Hg_5TeCl_8 crystallizes in a tetragonal structure with lattice parameters $a = 894.4$ and $c = 1084.6$ pm and sublimates at 250°C–350°C without decomposition (Aravamudan et al. 1979). Pan'ko et al. (1989) and Voroshilov et al. (1994) did not confirm the formation of these two ternary compounds.

The peculiarities of the crystal structure of the ternary compounds forming in this system are given in the review of Magarill et al. (2007).

This system was investigated using DTA and XRD. The ingots were annealed for 650–700 h (Pan'ko et al. 1989, Voroshilov et al. 1994). $Hg_3Te_2Cl_2$ was obtained by the interaction of Hg_2Cl_2 and Te [annealing at 300°C for 10 h and then at 250°C–270°C for 60 h (Batsanov et al. 1973)]. Hg_5TeCl_8 and $Hg_5Te_2Cl_6$ were synthesized by the annealing of the mixtures containing corresponding quantity of HgTe and $HgCl_2$ (Puff and Kohlschmidt 1962, Aravamudan et al. 1979).

9.23 Mercury–Bromine–Tellurium

According to the data of Chen et al. (2007), Hg_2TeBr_3 compound exists in the Hg–Br–Te ternary system. This compound is characterized by a 2D layer structure and is a semiconductor ($\Delta E_g = 2.03$ eV). It is thermally stable up to 173°C.

HgTe–HgBr$_2$: The phase diagram is shown in Figure 9.25 (Pan'ko et al. 1989, Voroshilov et al. 1994). The composition and temperature of the first eutectic are 80 mol. % HgTe and 552°C, respectively. The second eutectic is degenerated from the $HgBr_2$-rich side and crystallizes at 239°C. The $Hg_3Te_2Br_2$ and Hg_3TeBr_4 ternary compounds are formed in this system (Batsanov and Abaulina 1961, Puff and Kohlschmidt 1962, Puff 1962, Puff and Küster 1962, Batsanov et al. 1973, Pan'ko et al. 1989, Hudoliy et al. 1990, Voroshilov et al. 1994). The first of them melts congruently at 580°C (Pan'ko et al. 1989, Hudoliy et al. 1990, Voroshilov et al. 1994) and crystallizes in

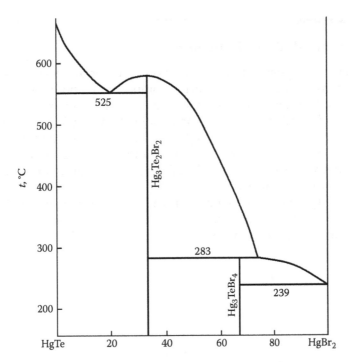

FIGURE 9.25
The HgTe–HgBr₂ phase diagram. (From Pan'ko, V.V. et al., *Zhurn. neorgan. khimii*, 34(9), 2331, 1989.)

a cubic structure with lattice parameter $a = 958.9$ pm (Hudoliy et al. 1990) [$a = 953 \pm 1$ pm (Voroshilov et al. 1981, 1994), $a = 962$ pm (Batsanov et al. 1973), $a = 954$ pm (Puff 1962, Puff and Kohlschmidt 1962)], calculation density 7.59 g cm⁻³, and energy gap $E_g = 2.7$ eV (Batsanov et al. 1973).

Hg₃TeBr₄ melts incongruently at 283°C (Pan'ko et al. 1989, Hudoliy et al. 1990, Voroshilov et al. 1994) (the peritectic composition is 25 mol. % HgTe) and crystallizes in an orthorhombic structure with lattice parameters $a = 1236.0 \pm 0.5$, $b = 1252.3 \pm 0.4$, and $c = 1286.8 \pm 0.5$ pm [$a = 672.3 \pm 0.5$, $b = 1202.7 \pm 0.8$, and $c = 1262 \pm 1$ pm (Pan'ko et al. 1989)] and calculation density 7.00 g cm⁻³ (Voroshilov et al. 1994, 1996b).

The Hg₅Te₂Br₆ ternary compound (Puff and Kohlschmidt 1962) does not exist in the HgTe–HgBr₂ systems (Pan'ko et al. 1989). The peculiarities of the crystal structure of the ternary compounds forming in this system are given in the review of Magarill et al. (2007).

This system was investigated using DTA and XRD. The ingots were annealed for 650–700 h (Pan'ko et al. 1989). Hg₃Te₂Br₂ was obtained by the interaction of Hg₂Br₂ and Te (annealing at 290°C for 10 h and then at 240°C–260°C for 60 h) (Batsanov et al. 1973). Hg₅Te₂Br₆ was synthesized by the annealing of the mixtures containing corresponding quantity of HgTe and HgBr₂ (Puff and Kohlschmidt 1962). Hg₂TeBr₃ was obtained by solid-state reactions (Chen et al. 2007).

9.24 Mercury–Iodine–Tellurium

HgTe–HgI$_2$: The phase diagram is shown in Figure 9.26 (Wiedemeier and Hutchins 1996a,b). The eutectic composition and temperature are 82 ± 3 mol. % HgI$_2$ and $250 \pm 3°C$, respectively. The α to β transition of HgI$_2$ at 133°C is independent of sample composition between 33.3 and 100 mol. % HgI$_2$. Solid solutions of HgTe and HgI$_2$ with a cubic, sphalerite-type structure exist above 300°C, having a maximum solubility of 11.7 ± 0.8 mol. % HgI$_2$ in HgTe at $501 \pm 5°C$. Numerical results of the solid solubility of HgI$_2$ in HgTe are summarized in Table 9.2.

The Hg$_3$Te$_2$I$_2$ ternary compound is formed by a peritectic reaction upon cooling at $501 \pm 5°C$ with the peritectic point at approximately 37 ± 4 mol. % HgI$_2$ (Batsanov and Abaulina 1961, Puff 1962, Puff and Kohlschmidt 1962,

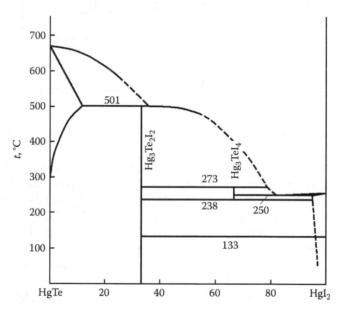

FIGURE 9.26
The HgTe–HgI$_2$ phase diagram. (From Wiedemeier, H. and Hutchins, M.A., *Z. anorg. und allg. Chem.*, 622(1), 157, 1996a; Wiedemeier, H. and Hutchins, M.A., *Z. anorg. und allg. Chem.*, 622(7), 1150, 1996b.)

TABLE 9.2

Solubility Limits of HgI$_2$ in HgTe

t, °C	344	393	449	492	501	542	592	623
mol. % HgTe	1.5	3.1	7.1	10.0	11.7	8.9	5.1	3.2

Pan'ko et al. 1989, Voroshilov et al. 1994, Wiedemeier and Hutchins 1996a,b, Voroshilov et al. 2000). It crystallizes in a monoclinic structure with lattice parameters a = 1427.6 ± 0.3, b = 972.2 ± 0.3, c = 1438.2 ± 0.3 pm, and β = 100.11 ± 0.02° (Voroshilov et al. 2000) [a = 977.9 ± 0.8, b = 1408.2 ± 0.9, c = 1417 ± 2 pm, and β = 96.2° (Pan'ko et al. 1989); a = 970 ± 3, b = 1434 ± 2, c = 1422 ± 4 pm, and β = 79.9 ± 0.2° (Liahovitskaya et al. 1989); a = 978, b = 1409, c = 1414 pm, and β = 96° (Puff and Kohlschmidt 1962, Batsanov et al. 1973)], calculation density 7.64 [7.508 ± 0.006 (Voroshilov et al. 2000)] g cm^{-3}, and energy gap 2.1 eV (Batsanov et al. 1973). This compound has a very narrow homogeneity range (Wiedemeier and Hutchins 1996a).

The Hg$_3$TeI$_4$ ternary compound is formed by a eutectoid reaction at 238 ± 3°C and is stable up to 273 ± 3°C, where it melts by a peritectic reaction with the peritectic point at approximately 79 ± 3 mol. % HgI$_2$. It crystallizes in a cubic structure with lattice parameter a = 624.4 ± 0.1 pm [a = 624.0 ± 0.3 pm (Wiedemeier and Hutchins 1996b)] and calculation and experimental density 8.45 and 7.03 g cm^{-3}, respectively (Wiedemeier et al. 1997). The difference between calculation and experimental density is explained by the fact that the unit cell of Hg$_3$TeI$_4$ compound is Hg$_{2.48}$Te$_{0.8}$I$_{3.20}$.

The peculiarities of the crystal structure of the ternary compounds forming in this system are given in the review of Magarill et al. (2007).

The HgTe–HgI$_2$ phase diagram was also constructed in Pan'ko et al. (1989) and Voroshilov et al. (1994). The results in Figure 9.26 differ considerably from those reported by Pan'ko et al. (1989) and Voroshilov et al. (1994). The main differences are the reported insignificant solubility of HgI$_2$ in HgTe, the absence of Hg$_3$TeI$_4$ compound, and the interpretation of the physical phenomena occurring at 238°C and 273°C transition isotherms. Their assignment to eutectic and monotectic transitions at 240°C and 277°C, respectively, is not in agreement with the results of Wiedemeier and Hutchins (1996a,b).

The known condensed phases in the Hg–I–Te ternary system at room temperature and likely regions of three-phase equilibria within the ternary system are indicated in Figure 9.27 (Wiedemeier and Hutchins 1996b).

This system was investigated using DTA and XRD (Pan'ko et al. 1989, Voroshilov et al. 1994, Wiedemeier and Hutchins 1996a,b). The starting mixtures for DTA were first annealed for two days at 700°C, cooled to 450°C over a period of 3 days, and then annealed at 450°C for 10–40 days. The samples were annealed at selected temperature for 4–28 days: 344°C—28 days, 393°C—15 days, 449°C—11 days, 492°C—10 days, 542°C—7 days, and 592°C and 623°C—4 days (Wiedemeier and Hutchins et al. 1996a) [for 650–700 h (Pan'ko et al. 1989, Voroshilov et al. 1994)]. Hg$_3$Te$_2$I$_2$ was obtained by the interaction of Hg$_2$I$_2$ and Te (annealing at 180°C for 10 h and then at 130°C–150°C for 60 h) (Batsanov et al. 1973). Its single crystals were grown by sublimation (Liahovitskaya et al. 1989).

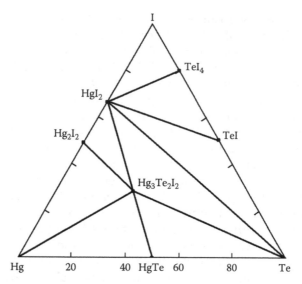

FIGURE 9.27
Isothermal section of the Hg–I–Te ternary system at room temperature. (From Wiedemeier, H. and Hutchins M.A., *Z. anorg. und allg. Chem.*, 622(7), 1150, 1996b.)

9.25 Mercury–Manganese–Tellurium

Figure 9.28 presents the experimental results, which characterized equilibrium of the melt with $Hg_{1-x}Mn_xTe$ solid solutions only (Danilov et al. 1992). The calculation liquidus surface along the sections with constant Te concentration is shown also. All calculations were accomplished using a model of fully associated solutions. An excess of Te in the Hg–Mn–Te system is accumulated in the liquid phase and forms $MnTe_2$, resulting in a reduction of the $Hg_{1-x}Mn_xTe$ region with the sphalerite-type structure (Becla et al. 1985).

The calculated liquidus curves indicating the melt equilibrium with $MnTe_2$ are also given in Figure 9.28 (Danilov et al. 1992). It is shown that this region is metastable at Te content in the melt less than 75 at. % and liquidus surface corresponds to the equilibria with $Hg_{1-x}Mn_xTe$ solid solutions. Superposition of the calculated liquidus surfaces gives the possibility to obtain a general dependence of the liquidus temperature in the HgTe–MnTe–Te subsystem (Figure 9.29).

The results of the calculations of the liquidus and solidus surfaces at 60–86 at. % Te and 0–35 at. % Mn are given in Figures 9.30 and 9.31 (Becla et al. 1985, Mironov et al. 1986, Zhovnir et al. 1989). For the practical convenience, the Mn isoconcentration lines along the Te isoconcentration lines in the liquid phase are shown in Figure 9.30. The field on the left side of the line corresponding to 35 at. % Mn is the region of the $Hg_{1-x}Mn_xTe$ solid solutions' primary crystallization.

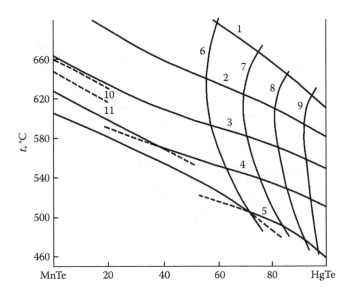

FIGURE 9.28
Isoconcentration lines of two-phase equilibria in the HgTe–MnTe–Te system at (1) 60, (2) 65, (3) 70, (4) 75, and (5) 80 at. % Te and isosolidus lines at (6) 70, (7) 80, (8) 83, and (9) 90 mol. % HgTe (dashed lines—calculations of the metastable equilibria; 10 and 11—metastable liquidus lines concerning MnTe$_2$ crystallization at 65 and 70 at. %, respectively). (From Danilov, M.A. et al. *Neorgan. materialy*, 28(9), 1860, 1992.)

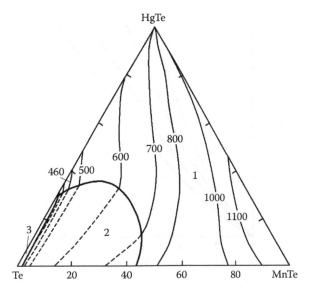

FIGURE 9.29
Liquidus surface of the HgTe–MnTe–Te subsystem: primary crystallization of the (1) Hg$_{1-x}$Mn$_x$Te solid solutions, (2) MnTe$_2$, and (3) Te. (From Danilov, M.A. et al., *Neorgan. materialy*, 28(9), 1860, 1992.)

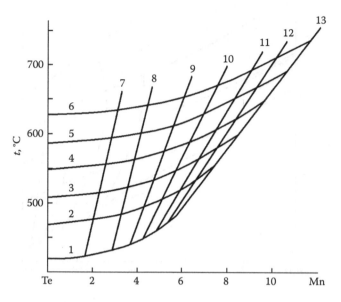

FIGURE 9.30
Polythermal sections of the Te-rich part of the Hg–Mn–Te system at (1) 86, (2) 80, (3) 75, (4) 70, (5) 65, and (6) 60 at. % Te and (7) 5, (8) 10, (9) 15, (10) 20, (11) 25, (12) 30, and (13) 35 at. % Mn. (From Becla, P. et al., *J. Vac. Sci. Technol.*, A3(1), 116, 1985; Mironov, K.E. et al., *P. I, L'vov*, 148, 1986; Zhovnir, G.I. et al., *Izv. AN SSSR. Neorgan. materialy*, 25(7), 1216, 1989.)

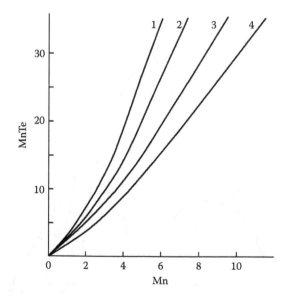

FIGURE 9.31
Dependences of the MnTe content in a solid phase from the Mn concentration in the melts at (1) 86, (2) 80, (3) 70, and (4) 60 at. %Te. (From Becla, P. et al., *J. Vac. Sci. Technol.*, A3(1), 116, 1985; Mironov, K.E. et al., *P. I, L'vov*, 148, 1986; Zhovnir, G.I. et al., *Izv. AN SSSR. Neorgan. materialy*, 25(7), 1216, 1989.)

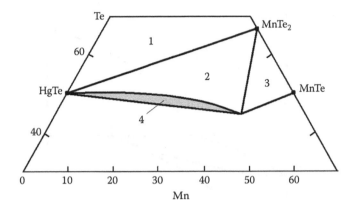

FIGURE 9.32
Region of the $Hg_{1-x}Mn_xTe$ solid solutions in the Hg–Mn–Te ternary system: 1, $HgTe + MnTe_2 + Te$; 2, $Hg_{1-x}Mn_xTe + MnTe_2$; 3, $Hg_{1-x}Mn_xTe + MnTe_2 + MnTe$; and 4, $Hg_{1-x}Mn_xTe$. (From Bykov, M.A. et al., *Zhurn. neorgan. khimii*, 43(3), 504, 1998.)

The part of the Hg–Mn–Te phase diagram in the region of the $Hg_{1-x}Mn_xTe$ solid solutions is shown in Figure 9.32 (Bykov et al. 1998). The obtained results are in good agreement with the results of Delves and Lewis (1963).

The liquidus temperatures were determined using DTA, XRD, and visual–thermal analysis and verified by the growth experiments (Becla et al. 1985, Zhovnir et al. 1989, Danilov et al. 1992, Bykov et al. 1998). The ingots were annealed at 450°C during 20 days and then were homogenized at 400°C during 60 days (Bykov et al. 1998). Phase equilibria in the Hg–Mn–Te system were calculated according to the model of regular associated solutions (Zhovnir et al. 1989).

HgTe–MnTe: The phase equilibria in this system have been discussed in Pajaczkowska (1978) on the base of papers, which have appeared up to 1977. The first experimental results concerning the phase equilibria in this system were obtained by Delves and Lewis (1963). $MnTe_2$ was determined in the alloys, containing more than 35 mol. % MnTe, and this fact leads to the conclusion about nonquasibinarity of this system (Delves and Lewis 1963, Bykov et al. 1998).

The phase diagram of this system according to the data of Koneshova et al. (1994) is shown in Figure 9.33. The solubility of MnTe in HgTe at room temperature reaches 10 mol. % and increases with the increasing temperature. The peritectic temperature was estimated as 800°C. The data about continuous series of the solid solutions in this system (Danilov et al. 1992) and about existing of an immiscibility region (Lopatinskiy 1976) were not confirmed (Koneshova et al. 1994).

The concentration lattice dependence of the $Hg_{1-x}Mn_xTe$ solid solutions within the interval up to 80 mol. % MnTe obeys Vegard's law (Delves and Lewis 1963).

This system was investigated using DTA, XRD, and metallography (Delves and Lewis 1963, Koneshova et al. 1994). The ingots were annealed at 600°C during 390, 790, and 1140 h (Koneshova et al. 1994). $Hg_{1-x}Mn_xTe$ single crystals have been obtained using the Bridgman method (up to $x = 0.2$) (Kaniewski et al. 1982),

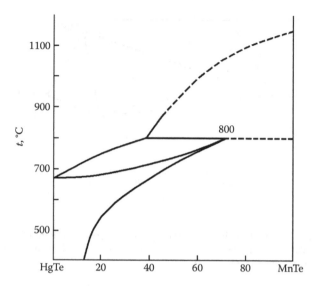

FIGURE 9.33
The HgTe–MnTe phase diagram. (From Koneshova, T.I. and Kholina, E.N., *Neorgan. materialy*, 30(8), 1101, 1994.)

or by the solid-state crystallization at 250°C–280°C ($x = 0.05$ and 0.12) (Paranchich et al. 2005), or using simple zone melting ($0.04 \leq x \leq 0.2$) (Bodnaruk et al. 1992) and zone melting through the Te melt (Garbuz et al. 1990). The crystal growth of the $Hg_{1-x}Mn_xTe$ solid solutions is also given in Pajaczkowska (1978).

9.26 Mercury–Iron–Tellurium

HgTe–FeTe: The phase diagram is not constructed. In the sintered materials HgTe+Fe+Te (in the proportions according to the HgTe–FeTe system), the products of reaction are HgTe and $FeTe_2$ phases (Mizers et al. 1980). When powdered HgTe and FeTe in proportion according to the stoichiometric formula $Hg_{1-x}Fe_xTe$ ($0.03 \leq x \leq 0.95$) were annealed at 600°C for 14–19 days, HgTe, FeTe, and $FeTe_2$ phases were observed. The content of the $FeTe_2$ phase was significant. The solubility of FeTe in HgTe is less than 0.5 mol. %.

9.27 Mercury–Cobalt–Tellurium

HgTe–Co: This section is a nonquasibinary section of the Hg–Co–Te ternary system (Mizera et al. 1980). Cobalt ditelluride was formed in this system when the crystals had been grown using the Bridgman method from the HgTe+Te+Co starting material at the Co contents higher than its solubility limit in HgTe.

9.28 Mercury–Nickel–Tellurium

HgTe–Ni: This section is a nonquasibinary section of the Hg–Ni–Te ternary system (Mizera et al. 1980). Nickel ditelluride was formed in this system when the crystals had been grown using the Bridgman method from the HgTe+Te+Ni starting material at the Ni contents higher than its solubility limit in HgTe.

References

Agostinelli E., Gastaldi L., Viticoli S. Crystal growth and x-ray structural investigation of two forms of $HgGa_2Te_4$, *Mater Chem. Phys.*, **12**(4), 303–312 (1985).

Anan'ina D.B., Bakumenko V.L., Bonakov A.K., Grushka G.G., Kurbatov L.N. Investigation of the $(In_2Te_3)_x$–$(Hg_3Te_3)_{1-x}$ system within the interval of $0.15 \le x \le 0.37$ [in Russian], *Izv. AN SSSR. Neorgan. materialy*, **12**(4), 763–764 (1976).

Aravamudan G., Venkatasubramanian P.N., Sethuraman P.R., Ramadass N. Preparation and characterization of a novel mercury (II) telluride chloride Hg_5TeCl_8, *Z. anorg. und allg. Chem.*, **457**(10), 238–240 (1979).

Asadov M.M. About physico-chemical peculiarities of thallium dissolution in mercury telluride [in Russian], *Azerb. khim. zhurn.*, (2), 134–137 (1988).

Asadov M.M. The Hg–Tl–Te system [in Russian], *Zhurn. neorgan. khimii*, **28**(2), 539–541 (1983).

Asadov M.M., Babanly M.B., Kuliev A.A. Phase equilibria in the Tl_2Te–HgTe (HgSe) systems [in Russian], *Izv. AN SSSR. Neorgan. materialy*, **14**(5), 960–961(1978).

Babanly M.B., Asadov M.M., Kuliev A.A. Phase equilibria and thermodynamic properties of the Hg–Tl–Te system [in Russian], *Izv. AN SSSR. Neorgan. materialy*, **19**(4), 583–587 (1983).

Babanly M.B., Kurbanov A.A., Kuliev A.A. The $HgTe$–Sb_2Te_3 and $HgTe$–Bi_2Te_3 systems [in Russian], *Zhurn. neorgan. khimii*, **24**(8), 2293–2294 (1979).

Bakumenko V.L., Bonakov V.L., Grushka G.G., Rudnevski V.S. Recombination radiation of $Hg_3In_2Te_6$ [in Russian], *Izv. AN SSSR. Neorgan. materialy*, **19**(7), 1210–1211 (1983).

Batsanov S.S., Abaulina L.I. Interaction of mercury halogenides with chalcogens [in Russian], *Izv. SO AN SSSR.* (10), 67–73 (1961).

Batsanov S.S., Kolomiychuk V.N., Derbeneva S.S., Erenburg R.S. Synthesis and physico-chemical investigation of mercury halogenchalcogenides [in Russian], *Izv. AN SSSR. Neorgan. materialy*, **9**(7), 1098–1104 (1973).

Becla P., Wolff P.A., Aggarwal P.L., Yuen S.Y. LPE growth conditions for $Cd_{1-x}Mn_xTe$ and $Hg_{1-x}Mn_xTe$ epitaxial layers, *J. Vac. Sci. Technol.*, **A3**(1), 116–118 (1985).

Bodnaruk O.A., Gorbatyuk I.N., Kalenik V.I., Pustyl'nik O.D., Rarenko I.M., Shafranyuk V.P. Crystal structure and electrophysical parameters of the $Mn_xHg_{1-x}Te$ crystals, *Neorgan. materialy*, **28**(2), 335–339 (1992).

Bykov M.A., Shevel'kov A.V., Shpanchenko R.V. Structure features of the solid solutions based on mercury telluride in the Hg–Mn–Te system [in Russian], *Zhurn. neorgan. khimii*, **43**(3), 504–507 (1998).

Chen W.-T., Li X.-F., Luo Q.-Y., Xu Y.-P., Zhou G.-P. Synthesis, structure and properties of a novel metal tellutobromide—Hg_2TeBr_3, *Inorg. Chem. Commun.*, **10**(4), 427–431 (2007).

Danilov M.A., Litvak A.M., Mironov K.E. Phase diagram of the MnTe–HgTe–Te system [in Russian], *Neorgan. materialy*, **28**(9), 1860–1870 (1992).

Delves R.T., Lewis B. Zinc blende type HgTe–MnTe solid solutions, *J. Phys. Chem. Solids*, **24**(4), 549–556 (1963).

Diehl R., Nolaeng B.I. Dry oxidation of $Hg_{1-x}Cd_xTe$: Calculation of predominance area diagrams of the oxide phases, *J. Cryst. Growth*, **66**(1), 91–105 (1984).

Frasuniak V.M., Gavaleshko N.P., Pareniuk I.A. Physico-chemical properties of alloys in the HgTe-MgTe system [in Russian], *Izv. AN SSSR. Neorgan. materialy*, **18**(6), 1045–1047 (1982).

Garbuz N.G., Kondrakov S.V., Popov S.A., Susov E.V., Filatov A.V., Khazieva R.A., Kholina E.N. Microstructure and galvanomagnetic properties of the HgTe–MnTe and $Hg_{1-x}Cd_xTe–Hg_{1-y}Mn_yTe$ solid solutions [in Russian], *Izv. AN SSSR. Neorgan. materialy*, **26**(3), 536–539 (1990).

Gastaldi L., Simeone L.G., Viticoli S. Cation ordering and crystal structures in AGa_2X_4 compounds ($CoGa_2S_4$, $CdGa_2S_4$, $CdGa_2Se_4$, $HgGa_2Se_4$, $HgGa_2Te_4$), *Solid State Commun.*, **55**(7), 605–607 (1985).

Gavaleshko N.P., Frasuniak V.M., Gorichok M.N. Obtaining and some properties of $Mg_xHg_{1-x}Te$ crystals [in Russian], in: *Fizika kristalliz.*, Kalinin, 53–57 (1982).

Gavaleshko N.P., Gorley P.N., Paranchich S.Yu., Frasuniak V.M., Homiak V.V. Phase diagrams of CdSe–HgSe, ZnSe–HgSe, MgTe–HgTe quasibinary systems [in Russian], *Izv. AN SSSR. Neorgan. materialy*, **10**(2), 327–329 (1983).

Gavaleshko N.P., Kav'yuk P.V., Lototski V.B., Solonchuk L.S., Homiak V.V. Investigation of interaction in the HgTe–EuTe system [in Russian], *Izv. AN SSSR. Neorgan. materialy*, **21**(11), 1965–1966 (1985).

Gilevich M.P., Zhukova L.S. Investigation of chemical transport and doping of the $HgIn_2Te_4$ crystals [in Russian], *Vestn. Belorus. un-ta, Ser. 2*, (1), 11–15 (1974).

Glazov V.M., Nagiev V.A., Nuriev R.S. Intermolecular interaction and thermodynamics of GeTe–A^{II}Te melts [in Russian], in: *Termodinamicheskiye svoistva metallicheskih splavov*, Baku, Azerbaijan: Elm, pp. 376–379 (1975).

Glazov V.M., Nagiev V.A., Nuriev R.S. Investigation and analysis of interaction in the quasibinary systems based on zinc, cadmium and mercury tellurides with germanium telluride [in Russian], *Izv. vuzov. Ser. Tsvetnaya metallurgia*, (5), 116–120 (1972).

Glazov V.M., Nagiev V.A., Zargarova M.I. Investigation of phase equilibria in the GeTe–A^{II}Te systems [in Russian], *Izv. AN SSSR. Neorgan. materialy*, **6**(3), 569–571 (1970).

Golovey M.I., Olekseyuk I.D., Voroshilov Yu.V. Chemical bond and solubility in the systems $A^{II}_3B^V_2$–$A^{II}C^{VI}$ [in Russian], in: *Khim. svyaz' v poluprovodn. i polumetallah*. Minsk, Belarus: Nauka i Tekhnika Publishing, pp. 233–239 (1972).

Grice J.D. The crystal structure of magnolite, $Hg_2^{1+}Te^{4+}O_3$, *Can. Miner.*, **27**(1), 133–136 (1989).

Grushka G.G., Gerasimenko V.S., Grushka Z.M. Investigation of physico-chemical interaction in the HgTe–In_2Te_3 system [in Russian], *Zhurn. neorgan. khimii*, **28**(7), 1878–1880 (1983).

Grushka G.G., Gerasimenko V.S., Grushka Z.M., Skulish E.D. Properties of the alloys in the In_2Te_3–HgTe system [in Russian], *Izv. AN SSSR. Neorgan. materialy*, **18**(12), 1989–1993 (1982a).

Grushka G.G., Skulish E.D., Grushka Z.M. Phase transition in $Hg_3In_2Te_6$ [in Russian], *Izv. AN SSSR. Neorgan. materialy*, **18**(8), 1388–1390 (1982b).

Guseinov G.D. Searching and physical investigation of new semiconductors-analogues [in Russian], *Avtoref. dis. ... doct. fiz.-mat. nauk.—Baku*, 82p. (1969).

Hahn H., Frank G., Klingler W., Störger A.D., Störger G. Über ternäre Chalkogenide des Aluminiums, Galliums und Indiums mit Zink, Cadmium und Quecksilber, *Z. anorg. und allg. Chem.*, **279**(5/6), 241–270 (1955).

Hailing T., Saunders G.A., Penfold J.W. Pressure-induced phase transitions in the vacancy compounds $Hg_5Ga_2\square Te_8$, $Hg_3Ga_2\square Te_6$ and $Hg_2In_2\square Te_4$, High Temp.—High Pressures, **15**(5), 533–538 (1983).

Hudoliy V.A., Pan'ko V.V., Voroshilov Yu.V., Shelemba M.S. Phase equilibria in the $HgTe–HgSe–HgBr_2$ system [in Russian], *Izv. AN SSSR. Neorgan. materialy*, **26**(11), 2425–2428 (1990).

Kaniewski J., Witkowska B., Giriat W. Preparation of $Mn_xHg_{1-x}Te$ mixed crystals, $x <$ 0.2, *J. Cryst. Growth*, **60**(1), 179–181 (1982).

Koneshova T.I., Kholina E.N. Investigation of the MnTe–HgTe polythermic quasibinary section of the Mn–Hg–Te ternary system, *Neorgan. materialy*, **30**(8), 1101–1102 (1994).

Kozer V.R., Fedorchuk A.O., Olekseyuk I.D., Parasyuk O.V. Crystal structure of the phases $Hg_5C^{III}_2X_8$ (C^{III} = Ga, In; X = Se, Te), *J. Alloys Compd.*, **503**(1), 40–43 (2010).

Kurbanov A.A., Babanly M.B., Kuliev A.A. Phase equilibria in the $HgSe(Te)–As_2Se_3(Te_3)$ systems [in Russian], *Azerb. khim. zhurn.*, (2), 99–101 (1985).

Leute V., Köller H.-J. Thermodynamic properties of the quasiternary system $Hg_kPb_{(1-k)}Se_lTe_{(1-l)}$, *Z. Phys. Chem. (BRD)*, **150**(2), 227–243 (1986).

Leute V., Schmidtke H.M. Thermodynamics and kinetics of the quasibinary system $Hg_{(3-3k)}In_{2k}Te_3$. I. Investigations by x-ray diffraction and differential thermoanalysis, *J. Phys. Chem. Solids*, **49**(4), 409–420 (1988a).

Leute V., Schmidtke H.M. Thermodynamics and kinetics of the quasibinary system $Hg_{(3-3k)}In_{2k}Te_3$. II. Investigations by electron microprobe measurements, *J. Phys. Chem. Solids*, **49**(11), 1317–1327 (1988b).

Leute V., Weitze D., Zeppenfeld A. Phase diagrams of II-VI/III-VI solid solutions with ordering tendency, *J. Alloys Compd.*, **289**(1–2), 233–243 (1999).

Li J., Chen Z., Lam K.-C., Mulley S., Proserpio D.M. $Rb_2Hg_3Te_4$: A new layered compound synthesized from solvothermal reactions, *Inorg. Chem.*, **36**(4), 684–687 (1997).

Liahovitskaya V.A., Sorokina N.I., Safonov A.A., Verin I.A., Andrianov V.I. Growth and structure of the $Hg_3Te_2I_2$ crystals [in Russian], *Kristallographia*, **34**(4), 835–838 (1989).

Lopatinskiy I.E. Crystallization peculiarities of the two-phase melts in the HgTe–MnTe system [in Russian], *Izv. AN SSSR. Neorgan. materialy*, **12**(2), 344–346 (1976).

Lototski V.B., Radevich E.I., Gavaleshko N.P. Physico-chemical properties of the $(Ga_2Te_3)_x(Hg_3Te_3)_{1-x}$ solid solutions [in Russian], *Izv. AN SSSR. Neorgan. materialy*, **22**(11), 1916–1918 (1986).

Magarill S.A., Pervukhina N.V., Borisov S.V., Pal'chik N.A. Crystal chemistry and features of structure formation of mercury oxo- and chalcogenides [in Russian], *Uspekhi khimii*, **76**(2), 115–146 (2007).

Markovskii L.Ya., Pron' G.F. Synthesis and some properties of the zinc, cadmium and mercury tellurites [in Russian], *Zhurn. neorgan. khimii*, **13**(10), 2640–2644 (1968).

524 Ternary Alloys Based on II-VI Semiconductor Compound

Mayet F., Roubin M. Contribution a l'étude des système ternaire $Ag_2Te-In_2Te_3-HgTe$. Mise en evidence de deux phases nouvelles, *C. r. Acad. sci. C*, **291**(13), 291–294 (1980).

Maynell C.A., Saunders G.A., Seddon T. An order-disorder transformation in $Hg_3In_2Te_6$, *Phys. Lett.*, **A31**(6), 338–339 (1970).

Mironov K.E., Untila P.G., Zelenova O.V. $Mn_xCd_yHg_{1-x-y}Te$ epitaxial layers [in Russian], in: *Materialy VII Vsesoyuz. simpoz. Poluprovodniki s uzkoy zapreshchennoy zonoy i polumetally*, P. I, L'vov, pp. 148–150 (1986).

Mizera A., Klimkiewicz M., Pajączkowska A., Godwod K. Structural investigations of Hg–Te–Fe alloys, *Phys. Status Solidi (a)*, **58**(2), 361–366 (1980).

Olekseyuk I.D., Sopko T.V., Mel'nichenko T.N. Phase equilibria in the HgSe(HgTe)–As, $HgSe(HgTe)-As_2Se_3(As_2Te_3)$ systems [in Russian], *Izv. AN SSSR. Neorgan. materialy*, **23**(8), 1282–1285 (1987).

Pajaczkowska A. Physicochemical properties and crystal growth of $A^{II}B^{VI}-MnB^{VI}$ systems, *Progr. Crystal Growth Caract.*, **1**(3), 289–326 (1978).

Pan'ko V.V., Hudoliy V.A., Voroshilov Yu.V. The $HgSe(Te)-HgHal_2$ systems [in Russian], *Zhurn. neorgan. khimii*, **34**(9), 2331–2335 (1989).

Paranchich S.Yu., Paranchich L.D., Makogonenko V.N., Andriychuk M.D., Romanyuk V.R., Ivonyak Yu.I., Sinilo S.V. Obtaining and properties of $Mn_xHg_{1-x}Te$ [in Russian], *Neorgan. materialy*, **41**(12), 1452–1455 (2005).

Parasyuk O., Piskach L., Morenko A., Halka V., Marchuk O. Phase equilibria in the $Cu_{2-x}Te(Se)-Cd(Hg)Te(Se)$ quasibinary systems [in Ukrainian], *Nauk. visnyk Volyn. derzh. un-tu*, (4), 35–38 (1997).

Pervukhina N.V., Romanenko G.V., Borisov S.V., Magarill S.A., Pal'chik N.A. Crystal chemistry of mercury (I)- and mercury (I, II)-containing minerals [in Russian], *Zhurn. strukt. khimii*, **40**(3), 561–581 (1999).

Pron' G.F., Markovskii L.Ya. Heats of formation of the zinc, cadmium and mercury tellurites [in Russian], *Zhurn. neorgan. khimii*, **14**(4), 880–882 (1969).

Puff H. Ternäre Verbindungen von Quecksilberhalogeniden mit Elementen der 5. und 6. Hauptgruppe, *Angew. Chem.*, **74**(16), 659 (1962).

Puff H., Kohlschmidt R. Quecksilberchalkogenid-halogenide, *Naturwissenschaften*, **49**(14), 299 (1962).

Puff H., Küster J. Die Kristallstruktur der kubischen Triquecksilber-dichalkonium-dihalogenide, *Naturwissenschaften*, **49**(20), 464–465 (1962).

Ray B., Spencer P.M. Phase diagram of the pseudo-binary system $In_2Te_3-Hg_3Te_3$, *Solid State Commun.*, **3**(12), 389–391 (1965).

Ray B., Spencer P.M., Younger P.A. Temperature-composition section of the equilibrium diagram of the pseudo-binary alloy system $Hg_3Te_3-Ga_2Te_3$, *J. Phys.*, **D3**(1), 37–44 (1970).

Schwer H., Krämer V. Neue Überstrukturen von AB_2Z_4-Defekt-Tetraheder-Verbindungen, *Z. Kristallogr.*, **182**(1–4), 245–246 (1988).

Shchennikov V.V., Gavaleshko N.P., Frasuniak V.M. Semiconductor–metal transitions in HgMgTe and HgSeS at superhigh pressure [in Russian], *Fiz. tv. tela*, **35**(2), 389–394 (1993).

Shuh D.W., Williams R.S. Summary abstract: Ternary solid phase equilibria in the systems (Ag, In, Au)–(Cd, Hg)–Te, *J. Vac. Sci. Technol.*, **A6**(3), Pt. II, 1564–1565 (1988).

Spencer P.M. The semiconducting properties of $HgTe-In_2Te_3$ alloys, *Brit. J. Appl. Phys.*, **15**(6), 625–632 (1964).

Spencer P.M., Ray B. Phase diagram of the alloy system $Hg_3Te_3–In_2Te_3$, *J. Phys. (Brit. J. Appl. Phys) D*, **1**(3), 299–301 (1968).

Sztuba Z., Gawel W., Zaleska E., Mädge H., Matyjasik S. Thermal and electrical studies of the system thallium (I) telluride–mercury (II) telluride, *J. Phase Equil.*, **21**(5), 447–450 (2000).

Sztuba Z., Gawel W., Zaleska E., Sroka A. Electrochemical studies of the thallium (I) telluride—mercury (II) telluride solid system, *Pol. J. Chem.*, **75**(1), 135–140 (2001).

Tomashik V.N., Vengel' P.F. Interaction of mercury telluride with zinc [in Russian], *Izv. AN SSSR. Neorgan. materialy*, **26**(1), 212–214 (1990).

Tomashik V.N., Vengel' P.F., Mizetskaya I.B. About interaction of HgTe with Cu and Ag [in Russian], *Izv. AN SSSR. Neorgan. materialy*, **23**(10), 1646–1648 (1987).

Tomashik V.N., Vengel' P.F., Mizetskaya I.B. The HgTe–Sb system [in Russian], *Ukr. khim. zhurn.*, **52**(7), 708–710 (1986a).

Tomashik V.N., Vengel' P.F., Mizetskaya I.B., Kurbanov K.R. Interaction of mercury telluride with lead [in Russian], *Izv. AN SSSR. Neorgan. materialy*, **22**(2), 224–227 (1986b).

Tomashik V.N., Vengel' P.F., Mizetskaya I.B., Nizkova A.I. Interaction of mercury telluride with gallium [in Russian], *Izv. AN SSSR. Neorgan. materialy*, **23**(2), 218–221 (1987).

Vaniarho V.G., Zlomanov V.P., Novoselova A.V. Investigation of the PbTe–HgTe system [in Russian], *Izv. AN SSSR. Neorgan. materialy*, **6**(1), 133–134 (1970).

Vaniarho V.G., Zlomanov V.P., Novoselova A.V. T-x phase diagrams of the lead selenide–mercury selenide and lead telluride–mercury telluride systems [in Russian], *Vestn. Mosk. un-ta. Ser. Khimia*, (6), 108–109 (1968).

Vengel' P.F., Tomashik V.N. Interaction in the HgTe–In system [in Russian], *Izv. AN SSSR. Neorgan. materialy*, **23**(2), 214–217 (1987).

Vengel' P.F., Tomashik V.N. The HgTe–Bi phase diagram [in Russian], *Izv. AN SSSR. Neorgan. materialy*, **22**(7), 1212–1214 (1986).

Vengel' P.F., Tomashik V.N. The HgTe–InTe system [in Russian], *Izv. AN SSSR. Neorgan. materialy*, **25**(3), 388–392 (1989).

Vengel' P.F., Tomashik V.N., Mizetskaya I.B. Physico-chemical interaction in the HgTe–SnTe system [in Russian], *Ukr. khim. zhurn.*, **49**(12), 1247–1250 (1983).

Vengel' P.F., Tomashik V.N., Mizetskaya I.B., Kurbanov K.R. The exchange interaction of mercury telluride with tin [in Russian], *Ukr. khim. zhurn.*, **52**(4), 359–362 (1986).

Voroshilov Yu.V., Gad'mashi Z.P., Slivka V.Yu., Hudoliy V.A. Obtaining and natural optical activity of the mercury chalcogenides crystals [in Russian], *Izv. AN SSSR. Neorgan. materialy*, **17**(11), 2022–2024 (1981).

Voroshilov Yu.V., Hudoliy V.A., Pan'ko V.V. Phase equilibria in the HgS–HgTe–$HgCl_2$ system and crystal structure of $β-Hg_3S_2Cl_2$ and Hg_3TeCl_4 compounds [in Russian], *Zhurn. neorgan. khimii*, **41**(2), 287–293 (1996a).

Voroshilov Yu.V., Hudoliy V.A., Pan'ko V.V., Minets Yu.V. Phase equilibria in the HgS–HgTe–$HgBr_2$ system and crystal structure of $Hg_3S_2Br_2$ and Hg_3TeBr_4 compounds [in Russian], *Neorgan. materialy*, **32**(12), 1466–1472 (1996b).

Voroshilov Yu.V., Pan'ko V.V., Minets' Yu.V. The crystal structure of $Hg_3Te_2I_2$ [in Ukrainian], *Nauk. visnyk Uzhgorod. un-tu. Ser. Khimiya*, (5), 6–8 (2000).

Voroshilov Yu.V., Pan'ko V.V., Pechars'kyi V.K., Hudoliy V.A. Phase equilibria in the HgS(Se, Te)–$HgCl_2(Br_2, I_2)$ systems and crystal structure of compounds [in Ukrainian], *Visnyk L'viv. un-tu. Ser. khim.*, (33), 11–24 (1994).

Weil M. Dimorphism in mercury(II) tellurite(IV) tellurate(VI): Preparation and crystal structures of α- and β-Hg$_2$Te$_2$O$_7$, *Z. Kristallogr.*, **218**(10), 691–698 (2003).

Weitze D., Leute V. The phase diagrams of the quasibinary systems HgTe/In$_2$Te$_3$ and CdTe/In$_2$Te$_3$, *J. Alloys Compd.*, **236**(1–2), 229–235 (1996).

Wensierski von H., Bolwin H., Zeppenfeld A., Leute V. Ordering phenomena and demixing in the quasiternary system Ga$_2$Te$_3$/Hg$_3$Te$_3$/In$_2$Te$_3$, *J. Alloys Compd.*, **255**, 169–177 (1997).

Wiedemeier H., Hutchins M.A. The temperature-composition phase diagram of the HgTe–HgI$_2$ pseudobinary system on the HgTe rich side, *Z. anorg. und allg. Chem.*, **622**(1), 157–163 (1996a).

Wiedemeier H., Hutchins M.A. The temperature-composition phase equilibria in HgTe–HgI$_2$ pseudobinary system, *Z. anorg. und allg. Chem.*, **622**(7), 1150–1160 (1996b).

Wiedemeier H., Hutchins M.A., Grin Y., Feldmann C., von Schnering H.G. The synthesis and crystal structure of Hg$_3$TeI$_4$, *Z. anorgan. und allg. Chem.*, **623**(11), 1843–1846 (1997).

Woolley J.C., Ray B. Effects of solid solutions of In$_2$Te$_3$ with A^{II}B^{VI} tellurides, *J. Phys. Chem. Solids*, **15**(1/2), 27–32 (1960a).

Woolley J.C., Ray B. Effects of solid solutions of Ga$_2$Te$_3$ with A^{II}B^{VI} tellurides, *J. Phys. Chem. Solids*, **16**(1/2), 102–106 (1960b).

Zhovnir G.I., Kletskiy S.V., Sochinskiy N.V., Frasunyak V.M. Phase equilibria in the Mn–Hg–Te system [in Russian], *Izv. AN SSSR. Neorgan. materialy*, **25**(7), 1216–1218 (1989).

Index